Lecture Notes in Computer Science 13327

More information about this series at https://link.springer.com/bookseries/558

Fiona Fui-Hoon Nah · Keng Siau (Eds.)

HCI in Business, Government and Organizations

9th International Conference, HCIBGO 2022
Held as Part of the 24th HCI International Conference, HCII 2022
Virtual Event, June 26 – July 1, 2022
Proceedings

Editors
Fiona Fui-Hoon Nah
City University of Hong Kong
Hong Kong, Hong Kong

Keng Siau
City University of Hong Kong
Hong Kong, Hong Kong

ISSN 0302-9743 ISSN 1611-3349 (electronic)
Lecture Notes in Computer Science
ISBN 978-3-031-05543-0 ISBN 978-3-031-05544-7 (eBook)
https://doi.org/10.1007/978-3-031-05544-7

This Springer imprint is published by the registered company Springer Nature Switzerland AG
The registered company address is: Gewerbestrasse 11, 6330 Cham, Switzerland

Foreword

Human-computer interaction (HCI) is acquiring an ever-increasing scientific and industrial importance, as well as having more impact on people's everyday life, as an ever-growing number of human activities are progressively moving from the physical to the digital world. This process, which has been ongoing for some time now, has been dramatically accelerated by the COVID-19 pandemic. The HCI International (HCII) conference series, held yearly, aims to respond to the compelling need to advance the exchange of knowledge and research and development efforts on the human aspects of design and use of computing systems.

The 24th International Conference on Human-Computer Interaction, HCI International 2022 (HCII 2022), was planned to be held at the Gothia Towers Hotel and Swedish Exhibition & Congress Centre, Göteborg, Sweden, during June 26 to July 1, 2022. Due to the COVID-19 pandemic and with everyone's health and safety in mind, HCII 2022 was organized and run as a virtual conference. It incorporated the 21 thematic areas and affiliated conferences listed on the following page.

A total of 5583 individuals from academia, research institutes, industry, and governmental agencies from 88 countries submitted contributions, and 1276 papers and 275 posters were included in the proceedings to appear just before the start of the conference. The contributions thoroughly cover the entire field of human-computer interaction, addressing major advances in knowledge and effective use of computers in a variety of application areas. These papers provide academics, researchers, engineers, scientists, practitioners, and students with state-of-the-art information on the most recent advances in HCI. The volumes constituting the set of proceedings to appear before the start of the conference are listed in the following pages.

The HCI International (HCII) conference also offers the option of 'Late Breaking Work' which applies both for papers and posters, and the corresponding volume(s) of the proceedings will appear after the conference. Full papers will be included in the 'HCII 2022 - Late Breaking Papers' volumes of the proceedings to be published in the Springer LNCS series, while 'Poster Extended Abstracts' will be included as short research papers in the 'HCII 2022 - Late Breaking Posters' volumes to be published in the Springer CCIS series.

I would like to thank the Program Board Chairs and the members of the Program Boards of all thematic areas and affiliated conferences for their contribution and support towards the highest scientific quality and overall success of the HCI International 2022 conference; they have helped in so many ways, including session organization, paper reviewing (single-blind review process, with a minimum of two reviews per submission) and, more generally, acting as goodwill ambassadors for the HCII conference.

This conference would not have been possible without the continuous and unwavering support and advice of Gavriel Salvendy, founder, General Chair Emeritus, and Scientific Advisor. For his outstanding efforts, I would like to express my appreciation to Abbas Moallem, Communications Chair and Editor of HCI International News.

June 2022 Constantine Stephanidis

HCI International 2022 Thematic Areas and Affiliated Conferences

Thematic Areas

- HCI: Human-Computer Interaction
- HIMI: Human Interface and the Management of Information

Affiliated Conferences

- EPCE: 19th International Conference on Engineering Psychology and Cognitive Ergonomics
- AC: 16th International Conference on Augmented Cognition
- UAHCI: 16th International Conference on Universal Access in Human-Computer Interaction
- CCD: 14th International Conference on Cross-Cultural Design
- SCSM: 14th International Conference on Social Computing and Social Media
- VAMR: 14th International Conference on Virtual, Augmented and Mixed Reality
- DHM: 13th International Conference on Digital Human Modeling and Applications in Health, Safety, Ergonomics and Risk Management
- DUXU: 11th International Conference on Design, User Experience and Usability
- C&C: 10th International Conference on Culture and Computing
- DAPI: 10th International Conference on Distributed, Ambient and Pervasive Interactions
- HCIBGO: 9th International Conference on HCI in Business, Government and Organizations
- LCT: 9th International Conference on Learning and Collaboration Technologies
- ITAP: 8th International Conference on Human Aspects of IT for the Aged Population
- AIS: 4th International Conference on Adaptive Instructional Systems
- HCI-CPT: 4th International Conference on HCI for Cybersecurity, Privacy and Trust
- HCI-Games: 4th International Conference on HCI in Games
- MobiTAS: 4th International Conference on HCI in Mobility, Transport and Automotive Systems
- AI-HCI: 3rd International Conference on Artificial Intelligence in HCI
- MOBILE: 3rd International Conference on Design, Operation and Evaluation of Mobile Communications

List of Conference Proceedings Volumes Appearing Before the Conference

1. LNCS 13302, Human-Computer Interaction: Theoretical Approaches and Design Methods (Part I), edited by Masaaki Kurosu
2. LNCS 13303, Human-Computer Interaction: Technological Innovation (Part II), edited by Masaaki Kurosu
3. LNCS 13304, Human-Computer Interaction: User Experience and Behavior (Part III), edited by Masaaki Kurosu
4. LNCS 13305, Human Interface and the Management of Information: Visual and Information Design (Part I), edited by Sakae Yamamoto and Hirohiko Mori
5. LNCS 13306, Human Interface and the Management of Information: Applications in Complex Technological Environments (Part II), edited by Sakae Yamamoto and Hirohiko Mori
6. LNAI 13307, Engineering Psychology and Cognitive Ergonomics, edited by Don Harris and Wen-Chin Li
7. LNCS 13308, Universal Access in Human-Computer Interaction: Novel Design Approaches and Technologies (Part I), edited by Margherita Antona and Constantine Stephanidis
8. LNCS 13309, Universal Access in Human-Computer Interaction: User and Context Diversity (Part II), edited by Margherita Antona and Constantine Stephanidis
9. LNAI 13310, Augmented Cognition, edited by Dylan D. Schmorrow and Cali M. Fidopiastis
10. LNCS 13311, Cross-Cultural Design: Interaction Design Across Cultures (Part I), edited by Pei-Luen Patrick Rau
11. LNCS 13312, Cross-Cultural Design: Applications in Learning, Arts, Cultural Heritage, Creative Industries, and Virtual Reality (Part II), edited by Pei-Luen Patrick Rau
12. LNCS 13313, Cross-Cultural Design: Applications in Business, Communication, Health, Well-being, and Inclusiveness (Part III), edited by Pei-Luen Patrick Rau
13. LNCS 13314, Cross-Cultural Design: Product and Service Design, Mobility and Automotive Design, Cities, Urban Areas, and Intelligent Environments Design (Part IV), edited by Pei-Luen Patrick Rau
14. LNCS 13315, Social Computing and Social Media: Design, User Experience and Impact (Part I), edited by Gabriele Meiselwitz
15. LNCS 13316, Social Computing and Social Media: Applications in Education and Commerce (Part II), edited by Gabriele Meiselwitz
16. LNCS 13317, Virtual, Augmented and Mixed Reality: Design and Development (Part I), edited by Jessie Y. C. Chen and Gino Fragomeni
17. LNCS 13318, Virtual, Augmented and Mixed Reality: Applications in Education, Aviation and Industry (Part II), edited by Jessie Y. C. Chen and Gino Fragomeni

18. LNCS 13319, Digital Human Modeling and Applications in Health, Safety, Ergonomics and Risk Management: Anthropometry, Human Behavior, and Communication (Part I), edited by Vincent G. Duffy
19. LNCS 13320, Digital Human Modeling and Applications in Health, Safety, Ergonomics and Risk Management: Health, Operations Management, and Design (Part II), edited by Vincent G. Duffy
20. LNCS 13321, Design, User Experience, and Usability: UX Research, Design, and Assessment (Part I), edited by Marcelo M. Soares, Elizabeth Rosenzweig and Aaron Marcus
21. LNCS 13322, Design, User Experience, and Usability: Design for Emotion, Well-being and Health, Learning, and Culture (Part II), edited by Marcelo M. Soares, Elizabeth Rosenzweig and Aaron Marcus
22. LNCS 13323, Design, User Experience, and Usability: Design Thinking and Practice in Contemporary and Emerging Technologies (Part III), edited by Marcelo M. Soares, Elizabeth Rosenzweig and Aaron Marcus
23. LNCS 13324, Culture and Computing, edited by Matthias Rauterberg
24. LNCS 13325, Distributed, Ambient and Pervasive Interactions: Smart Environments, Ecosystems, and Cities (Part I), edited by Norbert A. Streitz and Shin'ichi Konomi
25. LNCS 13326, Distributed, Ambient and Pervasive Interactions: Smart Living, Learning, Well-being and Health, Art and Creativity (Part II), edited by Norbert A. Streitz and Shin'ichi Konomi
26. LNCS 13327, HCI in Business, Government and Organizations, edited by Fiona Fui-Hoon Nah and Keng Siau
27. LNCS 13328, Learning and Collaboration Technologies: Designing the Learner and Teacher Experience (Part I), edited by Panayiotis Zaphiris and Andri Ioannou
28. LNCS 13329, Learning and Collaboration Technologies: Novel Technological Environments (Part II), edited by Panayiotis Zaphiris and Andri Ioannou
29. LNCS 13330, Human Aspects of IT for the Aged Population: Design, Interaction and Technology Acceptance (Part I), edited by Qin Gao and Jia Zhou
30. LNCS 13331, Human Aspects of IT for the Aged Population: Technology in Everyday Living (Part II), edited by Qin Gao and Jia Zhou
31. LNCS 13332, Adaptive Instructional Systems, edited by Robert A. Sottilare and Jessica Schwarz
32. LNCS 13333, HCI for Cybersecurity, Privacy and Trust, edited by Abbas Moallem
33. LNCS 13334, HCI in Games, edited by Xiaowen Fang
34. LNCS 13335, HCI in Mobility, Transport and Automotive Systems, edited by Heidi Krömker
35. LNAI 13336, Artificial Intelligence in HCI, edited by Helmut Degen and Stavroula Ntoa
36. LNCS 13337, Design, Operation and Evaluation of Mobile Communications, edited by Gavriel Salvendy and June Wei
37. CCIS 1580, HCI International 2022 Posters - Part I, edited by Constantine Stephanidis, Margherita Antona and Stavroula Ntoa
38. CCIS 1581, HCI International 2022 Posters - Part II, edited by Constantine Stephanidis, Margherita Antona and Stavroula Ntoa

39. CCIS 1582, HCI International 2022 Posters - Part III, edited by Constantine Stephanidis, Margherita Antona and Stavroula Ntoa
40. CCIS 1583, HCI International 2022 Posters - Part IV, edited by Constantine Stephanidis, Margherita Antona and Stavroula Ntoa

Preface

The use and role of technology in the business and organizational context have always been at the heart of human-computer interaction (HCI) since the start of management information systems. In general, HCI research in such a context is concerned with the ways humans interact with information, technologies, and tasks in the business, managerial, and organizational contexts. Hence, the focus lies in understanding the relationships and interactions between people (e.g., management, users, implementers, designers, developers, senior executives, and vendors), tasks, contexts, information, and technology. Today, with the explosion of the metaverse, social media, big data, and the Internet of Things, new pathways are opening in this direction, which need to be investigated and exploited.

The 9th International Conference on HCI in Business, Government and Organizations (HCIBGO 2022), an affiliated conference of the HCI International (HCII) conference, promoted and supported multidisciplinary dialogue, cross-fertilization of ideas, and greater synergies between research, academia, and stakeholders in the business, managerial, and organizational domain.

HCI in business, government, and organizations ranges across a broad spectrum of topics from digital transformation to customer engagement. The HCIBGO conference facilitates the advancement of HCI research and practice for individuals, groups, enterprises, and the society at large. The topics covered include emerging areas such as artificial intelligence and machine learning, blockchain, service design, live streaming in electronic commerce, visualization, and workplace design.

One volume of the HCII 2022 proceedings is dedicated to this year's edition of the HCIBGO conference and it focuses on topics related to digital transformation in business, government, and organizations; intelligent data analysis and business analytics; user experience and innovation design; HCI in the workplace; and retail, commerce, and customer engagement.

Papers of this volume are included for publication after a minimum of two single-blind reviews from the members of the HCIBGO Program Board or, in some cases, from members of the Program Boards of other affiliated conferences. We would like to thank all of them for their invaluable contribution, support, and efforts.

June 2022

Fiona Fui-Hoon Nah
Keng Siau

9th International Conference on HCI in Business, Government and Organizations (HCIBGO 2022)

Program Board Chairs: **Fiona Fui-Hoon Nah** and **Keng Siau,** City University of Hong Kong, Kowloon Tong, Kowloon, Hong Kong

- Kaveh Abhari, San Diego State University, USA
- Andreas Auinger, University of Applied Sciences Upper Austria, Austria
- Michel Avital, Copenhagen Business School, Denmark
- Denise Baker, Missouri University of Science and Technology, USA
- Valerie Bartelt, University of Denver, USA
- Kaveh Bazargan, Allameh Tabataba'i University, Iran
- Langtao Chen, Missouri University of Science and Technology, USA
- Constantinos K. Coursaris, HEC Montréal, Canada
- Brenda Eschenbrenner, University of Nebraska at Kearney, USA
- JM Goh, Simon Fraser University, Canada
- Netta Iivari, University of Oulu, Finland
- Qiqi Jiang, Copenhagen Business School, Denmark
- Yi-Cheng Ku, Fu Jen Catholic University, Taiwan
- Murad Moqbel, University of Texas Rio Grande Valley, USA
- Natalie Pang Lee San, National University of Singapore, Singapore
- Norman Shaw, Ryerson University, Canada
- Jia Shen, Rider University, USA
- Martin Stabauer, Johannes Kepler University Linz, Austria
- Werner Wetzlinger, University of Applied Sciences Upper Austria, Austria
- I-Chin Wu, National Taiwan Normal University, Taiwan
- Dezhi Wu, University of South Carolina, USA
- Jie Yu, University of Nottingham Ningbo China, China

The full list with the Program Board Chairs and the members of the Program Boards of all thematic areas and affiliated conferences is available online at

http://www.hci.international/board-members-2022.php

HCI International 2023

The 25th International Conference on Human-Computer Interaction, HCI International 2023, will be held jointly with the affiliated conferences at the AC Bella Sky Hotel and Bella Center, Copenhagen, Denmark, 23–28 July 2023. It will cover a broad spectrum of themes related to human-computer interaction, including theoretical issues, methods, tools, processes, and case studies in HCI design, as well as novel interaction techniques, interfaces, and applications. The proceedings will be published by Springer. More information will be available on the conference website: http://2023.hci.international/.

General Chair
Constantine Stephanidis
University of Crete and ICS-FORTH
Heraklion, Crete, Greece
Email: general_chair@hcii2023.org

http://2023.hci.international/

Contents

Intelligent Data Analysis and Business Analytics

User Experience and Innovation Design

Digital Transformation in Business, Government, and Organizations

Explore the Influence of Smart Contract on Online Lending

Cheng-Hsin Chiang[1], Vipin Saini[2], Yu-Chen Yang[2(✉)], and Tsai-Wen Shih[2]

[1] Feng Chia University, Taichung, Taiwan
[2] National Sun Yat-Sen University, Kaohsiung, Taiwan
ycyang@mis.nsysu.edu.tw

Abstract. The investment behavior of the lender becomes critical when a smart contract in combination with blockchain technology comes into play the role of intermediary to self-execute the transaction. Compare to general online lending platform where the lender have a high risk of bad debt, information asymmetry building of smart contract platform provide the solution to track historical transactions and immutable records. This study explores the impact of smart contracts on lender investment intention in online P2P lending platforms by combining the theoretical constructs of TPB and TAM. An experiment was conducted which collected 70 samples from two sub-groups of general online P2P lending platform and smart contract P2P lending platform. The preliminary results of experiment suggest perceived ease of use, and perceived usefulness have a significant impact in intention to use and investment by using smart contract embedded P2P lending platform. Both theoretical and practical implications of these findings are discussed.

Keywords: Fintech · Blockchain · Smart contract · Online lending · Technology adoption

1 Introduction

The demand for peer-to-peer (P2P) lending or micro loans is increasing in financial markets. More and more individuals tend to obtain loans directly from others individuals via an online platform, cutting out the bank or the financial institution as the middleman [1]. It is more efficient than the traditional loans. However, for individuals, it may increase the lending risk due to the design of the platform or bad debts. For example, lending fraud occurs quite often in China, and this forces individuals to take high risk [2]. Plus, there also exists some concerns about information security without the supervision of third parties.

In order to solve the above problem of information security, "smart contract" is a possible solution. It is a computer protocol based on blockchains and designed to safely and transparently facilitate the credible transactions without third parties. These transactions are trackable and immutable. Individuals are allowed to track historical transactions via smart contracts and reduce the lending risk. It seems to provide a secure

F. Fui-Hoon Nah and K. Siau (Eds.): HCII 2022, LNCS 13327, pp. 3–10, 2022.
https://doi.org/10.1007/978-3-031-05544-7_1

solution for online lending. Therefore, we aim to design a lending platform based on smart contract and to analyze whether the platform is possible and is likely to work.

We employ the theory of planned behavior and the technology acceptance model to examine the effect of smart contracts on individuals' investing behaviors. A platform is built based on Ethereum to compare general online P2P lending and smart-contract-based online P2P lending. Each subject is assigned to manipulate these two lending systems in a random sequence and required to complete the survey after the manipulation. For each subject, his/her investment portfolios are stored on the platform for further comparing the difference between these two systems.

The preliminary results show that smart contracts have a significant effect on individuals' investment behaviors. It influences not only the way individuals lend and borrow money, but also their decisions on investment portfolios. Given a similar portfolio, individuals tend to make transactions on smart contract lending platform rather than general online lending platform. We also find that even though most subjects are not familiar with smart contracts in the beginning, they end up with putting more trusts on it after the experiment. When smart contracts are automatically executed, subjects perceive a great improvement in the efficiency of lending and borrowing money. Most subjects also perceive that it is available to obtain more information for risk reduction when choosing a portfolio via smart contracts. All the above outcomes indicate that the adoption of smart contract in online P2P lending helps individuals lower the risk of bad debts, increase the trust of transactions, and improve blockchain-related financial services.

The remainder of the paper is organized as follows. Section 2 conducts a comprehensive review of the related work. Section 3 outlines the stylized research models of smart contracts and online lending. Section 4 concludes the work by summarizing the preliminary results and discussion of our study.

2 Related Work

2.1 Smart Contracts and P2P Lending

Smart contract is an agreement that can be established without a third-party organization to conduct trusted transactions, and the transaction has the characteristics of traceable transactions and irreversible transaction processes. Smart contract is based on the framework of blockchain technology. When a new block is added to the system, the old block data will be introduced, and the integrity of the entire chain structure will be verified by the block [3]. The entire transaction process cannot be reversed, and new transactions can only be restarted. Therefore, smart contracts must be set according to the transaction contract rules. If the contract content needs to be modified, a new contract must be written into the blockchain again. Smart contracts will automatically execute the loan and repayment at the written contract, which provides convenient and reduce transaction costs to smart contracts users.

The smart contracts can be applied to peer-to-peer (P2P) lending. P2P lending is a financial model that matches the demand for loans. According to the investment plan proposed by the borrower based on the terms of principal, repayment term, interest, etc., investors willing to provide funds are found through the P2P lending platform, and loans are traded on the Internet platform, forming a peer-to-peer financial lending model.

This lending mode enables investors to make full use of idle funds and use the lending platform as a medium to meet the demand of the lending market and earn extra interest income. P2P lending provides a new way for individual to raise funds, but the person who borrows the money will repay the funds on time remains unknown. Smart contracts provide an automatic, safer and easier way to manage the lending and repaying process.

The biggest problem with P2P lending is information asymmetry, which is especially bad for investors and is prone to default risks. Under the state of information asymmetry, the convenience of holding more information will lead to adverse selection and moral hazard [4]. Borrowers with higher risk may cause adverse selection and lead to higher default risk [5]. By reducing information asymmetry, it can help less attractive borrowers to improve their credit profile [6]. In the P2P lending market, in order to reduce investment risks, investors have followed others to choose the same investment solution, so the herd effect in the lending market, many investors judge the probability of investment success through observation [7]. Limited borrower information is not enough to help investors reduce the investment risk. Investors have to evaluate the risk when they decide to invest in P2P lending platform. P2P lending platforms provide risk grading methods and remind whether high-risk or high-reward investment solutions are screened for high-risk borrower. Using smart contract in P2P lending, investors can do a risk assessment before they invest, and the P2P lending platform will automatically execute the smart contract to reduce the burden of the investors to manage the repayment process and tracking efforts.

2.2 Theory of Planned Behavior (TPB)

Theory of planned behavior based on theory of reasoned action and added the psychological factors into the theory. TPB increase the individual's perceived behavioral control constructs, and later adds behavioral beliefs, normative beliefs, and control beliefs to explain individual behavior patterns by adding psychological factors [8]. The generated behavior depends on the personal intention and behavior control, and personal intention comes from the positive or negative feelings of the individual's attitude toward the behavior [9].

In TPB model, people need to consider the impacts of attitude, subjective norm, perceived behavior control, intentions and decide their behaviors. Attitude mainly explores the individual's feelings and views on things, and mainly focuses on self-thought, and explores the personal subjective feelings of the behavior. The subjective norm is to explore what individuals think of the people around them about their behaviors, and the objects they identify must be those who have influence on themselves, which helps to understand the degree of influence of others on personal behaviors. Perceived behavior control refers to whether the individual feels that it is easy or difficult to perform the function or behavior, which is the personal operation feeling. The intention is generated after the first three conditions of attitude, subjective norm, and perceived behavioral control are met. Individuals evaluate whether there is sufficient motivation to achieve the implementation of the thing or behavioral process, and eventually produce actual behavior. Behavioral constructs are factors that measure willingness and continued use. When the motivation is quite sufficient and clear, individuals will be willing to continue to perform the behavior. The users with more experience in operating information systems, the more subjective norms and perceived behavioral controls will gradually change [10]. For users to continuously operate the information system, personal attitude is the most important factor. After continuous use of the information system, accumulated experience will affect subjective norm and perceived behavior control. Since smart contracts are not yet popular, if users are to continue to use the system, attitudes can be used to evaluate whether they have an impact on user operation investment, and then affect personal use intentions.

2.3 Technology Acceptance Model (TAM)

TAM explores the behavior of users of information systems. TAM explains or predicts why users adopt or does not accept information systems [11]. TAM model uses perceived usefulness and perceived ease of use to explain the behavioral intention of the information system. Perceived usefulness is defined as the use of information systems by individuals to increase their behavioral efficiency. Perceived ease of use is defined as the difficulty of personal use of information systems, and perceived ease of use affects the performance of perceived usefulness [12]. When operating the system, it is necessary to consider the actual feelings of the user and whether it can bring benefits to the user. Perceived ease of use and perceived usefulness in TAM have a direct impact on user's attitudes towards adopting new technologies [13]. Perceived ease of use and perceived usefulness influence each other. If the information system is easier to use, it will be more useful to users [14]. Since the usefulness and ease of use of the system will affect the attitude of adopting technology, this study will explore whether intelligent contracts can reduce the complexity of system operation through transaction automation, which is conducive to the improvement of transaction efficiency of P2P lending.

3 Research Model

The study explores how consumer's investment behavior are influenced by new information technology and smart contracts. We propose the research model as Fig. 1. The proposed model includes five constructs: perceived ease of use, perceived usefulness, perceived risk, attitude and intention to use.

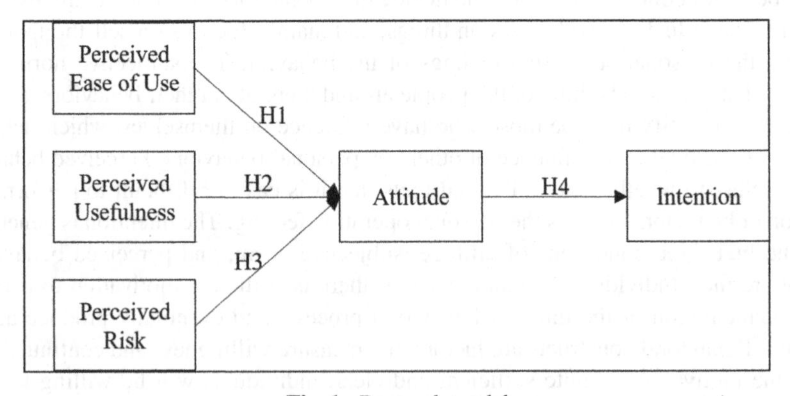

Fig. 1. Research model

Perceived Ease of Use. According to the study of [12], perceived ease of use can be interpreted as individuals' capabilities to operate a information system without too much effort. The study aims to investigate whether the use of smart contracts will simplify the overall operation processes of online lending. Therefore, we have the following hypothesis:

H1: Perceived ease of use has a significantly positive impact on attitude towards the use of a smart-contract-based system.

Perceived Usefulness. Perceived usefulness can be regarded as the degree to which individuals improve their performance via the use of a information system [15]. If smart contracts are effective to improve the collection and operation efficiency in the online lending process, it will have a positive impact on attitude.

H2: Perceived usefulness has a significantly positive impact on attitude towards the use of a smart-contract-based system.

Perceived Risk. Perceived risk can be considered as investors' fear of default and fraud [16]. It is caused by the design of the platform or the platform's trading mechanism. We believe smart contracts bring a stable and secure trading mechanism, and thus users' attitude towards the use of smart contracts becomes positive.

H3: Perceived risk has a significantly positive impact on attitude towards the use of a smart-contract-based system.

Attitude. Attitude can be considered as the positive and negative emotions that individuals react when using a system [17]. How users experience with smart contracts will influence their willingness of adopting a smart-contract-based lending system. We thus derive the following hypothesis:

H4: Attitude has a significant effect on intention to use a smart-contract-based system.

4 Preliminary Analytical Results

4.1 Sample Descriptions

Within three months of launching the lab experiment, this study collected a total of 70 valid samples from equally divided subjects of two groups' general lending platform and smart contract lending platform. Each group contains 35 samples, we applied A/B testing to compare the differences of investment behavior between two groups. The respondents were 51.4% male and 48.5% female, their age was between 21 and 40 years old, and were more engaged in the information technology industry. Among them, 43% of the subjects have investment experience, and 58% of the subjects know P2P lending, but did not borrow or invest through online P2P lending platforms. Therefore, in this experiment, it is the first contact with P2P lending, and there may be differences in the behavior of analyzing whether to use smart contracts to assist P2P lending. According to the personal investment risk appetite survey used by banks to test individual's risk attitude, most of the respondents were robust investors. Regarding the preference of choosing investment objectives, the subjects mostly choose stable returns or earn investment spreads, which means that they are willing to bear investment risks and expect to get higher returns. In this study, we will analyze whether all participants choose to increase or decrease their investment risk.

According to the background information survey in the research questionnaire, a total of five subjects have used the investment experience of online P2P lending, two subjects have used online P2P loans to initiate loans, and around 90% of the subjects did not use P2P loans. 80% of the subjects know the blockchain, and 50% of the subjects know what the smart contracts are. Many subjects are not clear about the actual application of smart contracts. Because the smart contract mechanism is not popular, all subjects have never used smart contracts for financial transactions or digital asset transfers. In this study, the test platform will be used to generate operating behaviors, and the subjects will be divided into two groups: One group use general online lending platform first, then the use smart contract assisted lending platform; the other group use smart contract assisted lending first, then they use general lending platform. Through the method of A/B Test, we compare whether the two groups have an impact on the behavioral construct, and whether the smart contract has an effect on the assistance of P2P lending, and find out the key factors influencing the smart contract on P2P lending.

4.2 Questionnaire Results

We use SmartPLS 3.2.4 to analyze the questionnaire. All the five constructs' Cronbach's alpha are larger than 0.8, and the composite reliability are all larger than 0.88, which meets the criteria of 0.6. The discriminant validity of five constructs are measured by the value of average variance extracted (AVE) should be greater than other values [18]; in this study, all the AVE of the diagonal values are greater than other construct, which meet the acceptable criteria. For the factor analysis, the factor loadings indicate that all the questions can be categorized into the construct we intended to assign.

After doing these indices analyses, the PLS results of our research framework is shown as the following Fig. 2.

The $R^2 = 0.554$, explains that attitude was influenced by perceived ease of use, perceived usefulness, and perceived risk. The R^2 of attitude explains intention is 0.553. The results indicate our research framework in P2P lending platform with smart contracts applied.

The path analysis of our research framework is shown in Table 1. The path coefficient of perceived risk to attitude was not supported. Perceived risk did not reduce the individual's attitude toward their P2P lending with smart contract technology.

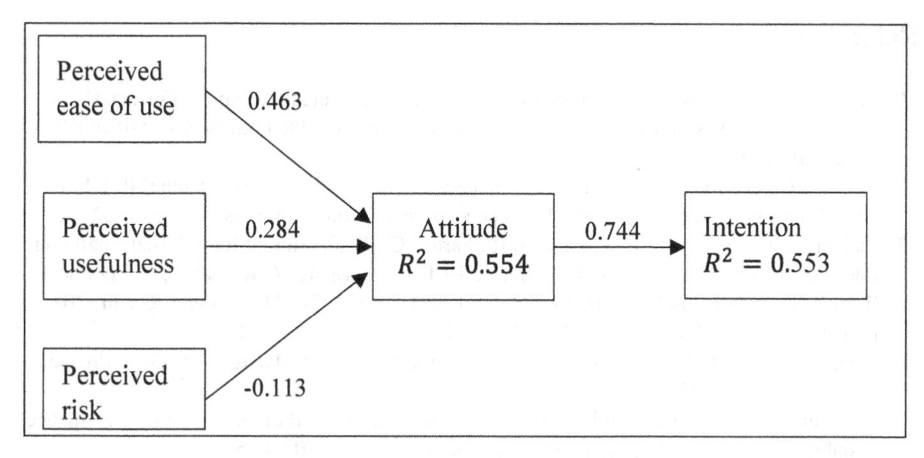

Fig. 2. The PLS results of research framework

Table 1. Path coefficient analysis of research framework

Path relationship	Average	Standard deviation	t-value
PEU → A	0.457	0.125	3.703
PU → A	0.289	0.126	2.244
PR → A	−0.122	0.079	1.432
A → IN	0.745	0.044	16.724

Note: PEU = perceived ease of use A = attitude PU = perceived usefulness PR = perceived risk IN = intention

5 Discussion and Conclusion

This study construct a research framework explains the lending behavior, align with perceived ease of use, perceived usefulness, investors will form their attitude and intention to use P2P lending platform which has smart contracts supported. In this study, we applied experiment to test if the smart contract embedded in P2P lending platform and pure P2P lending platform would affect the attitude and the intention of investors' decision. The research framework we proposed indicate that perceived ease of use and perceived usefulness indeed form the attitude of investors to use P2P lending platform, as well as their intention to invest. However, perceived risk, as a critical factor of investment decision, does not have significant effect on the investors' attitude to adopt P2P lending platform with smart contract applied. The possible explanation would be smart contract automatically execute the contract and recorded all the past transactions, so the investors feel safer when they use P2P lending platform, so the perceived risk was not the influential factors to form the investors' attitude to invest.

References

1. Feng, E.: Chinese government faces peer-to-peer lending scandals dilemma. Financial Times, 12 November 2018. https://www.ft.com/content/c71eea4a-c198-11e8-84cd-9e601db069b8. Accessed 6 Feb 2022
2. Zinn, D., Cetera, M.: Your one stop guide to peer-to-peer lending. Forbes, 1 June 2021. https://www.forbes.com/advisor/personal-loans/peer-to-peer-lending/. Accessed 6 Feb 2022
3. Idelberger, F., Governatori, G., Riveret, R., Sartor, G.: Evaluation of logic-based smart contracts for blockchain systems. In: Alferes, J.J.J., Bertossi, L., Governatori, G., Fodor, P., Roman, D. (eds.) RuleML 2016. LNCS, vol. 9718, pp. 167–183. Springer, Cham (2016). https://doi.org/10.1007/978-3-319-42019-6_11
4. Holmstrom, B.: The provision of services in a market economy. Manag. Serv. Econ. Prospect. Probl. **7**, 183–213 (1985)
5. Emekter, R., Tu, Y., Jirasakuldech, B., Lu, M.: Evaluating credit risk and loan performance in online peer-to-peer (P2P) lending. Appl. Econ. **47**(1), 54–70 (2015)
6. Berger, S.C., Gleisner, F.: Emergence of Financial Intermediaries in Electronic Markets: The Case of Online P2P Lending. BuR - Bus. Res. **2**(1), 39–65 (2009)
7. Lee, E., Lee, B.: Herding behavior in online P2P lending: an empirical investigation. Electron. Commer. Res. Appl. **11**(5), 495–503 (2012)
8. Ajzen, I.: The theory of planned behavior. Organ. Behav. Hum. Decis. Process. **50**(2), 179–211 (1991)
9. Taylor, S., Todd, P.A.: Understanding information technology usage: a test of competing models. Inf. Syst. Res. **6**(2), 144–176 (1995)
10. Hsu, M.-H., Yen, C.-H., Chiu, C.-M., Chang, C.-M.: A longitudinal investigation of continued online shopping behavior: an extension of the theory of planned behavior. Int. J. Hum Comput Stud. **64**(9), 889–904 (2006)
11. Davis, F.D., Bagozzi, R.P., Warshaw, P.R.: User acceptance of computer technology: a comparison of two theoretical models. Manage. Sci. **35**(8), 982–1003 (1989)
12. Davis, F.D.: Perceived usefulness, perceived ease of use, and user acceptance of information technology. MIS Q. **13**(3), 319–340 (1989)
13. Wanga, Y.-S., Wub, S.-C., Linc, H.-H., Wangd, Y.-M., Hee, T.-R.: Determinants of user adoption of web ATM: an integrated model of TCT and IDT. Serv. Ind. J. **32**(9), 1505–1525 (2012)
14. Venkatesh, V., Davis, F.D.: A theoretical extension of the technology acceptance model: four longitudinal field studies. Manage. Sci. **46**(2), 186–204 (2000)
15. Salisbury, W.D., Pearson, R.A., Pearson, A.W., Miller, D.W.: Perceived security and World Wide Web purchase intention. Ind. Manage. Data Syst. **101**, 165–177 (2001)
16. Yang, M., Li, H., Shao, Z., Shang, W.: Influencing lenders' repeat investment intention in P2P lending platforms in China through signaling. In: Pacific Asia Conference on Information Systems (PACIS). Association for Information Systems (2017)
17. Cheng, T.E., Lam, D.Y., Yeung, A.C.: Adoption of internet banking: an empirical study in Hong Kong. Decis. Support Syst. **42**(3), 1558–1572 (2006)
18. Fornell, C., Larcker, D.F.: Evaluating structural equation models with unobservable variables and measurement error. J. Mark. Res. **18**, 39–50 (1981)

Better Decision-Making Through Collaborative Development of Proposals

Björn Ebbinghaus(✉) and Martin Mauve

Heinrich-Heine-Universität, Universitätsstraße 1, 40225 Düsseldorf, Germany
{ebbinghaus,mauve}@hhu.de
https://cn.hhu.de

Abstract. In a traditional decision-making process, proposals are made and usually commented on by the participants, and finally a vote is taken. We have found in past public decision-making processes that there are also discussions about negative sides of proposals together with possible improvements of them. We have taken this as an opportunity to model the improvement of proposals within the decision-making process.

Our new model for this is similar to version control systems like git work. Proposals in a decision process can have none, one or several predecessors. This structure allows different constructs of the real world to be modelled. Where previously only proposals could be made without reference and structure among each other, the system presented in this work allows modelling of further developments of proposals or even coalitions and compromises, in a real-time collaborative decision-making process.

To validate our ideas, we tested them in two controlled experiments and concluded that our modifications are useful.

Keywords: Online-participation · Decision-making · Applications

1 Introduction

Our goal is to understand how a potentially large group of participants can be supported to conduct collaborative online decision-making processes with a minimum amount of external governing factors. Since there exists a multitude of systems that are regularly used for tasks such as online participation, much of the required functionality is well known. This includes mechanisms for proposing an item, discussing it and voting on alternatives. However, there is one aspect of collaborative online decision-making that is not yet well understood: how can a large group of participants work *together* to come up with and decide on proposals? In real-world politics, there is a vast array of instruments to refine proposals and form coalitions in order to reach a final decision.

Our idea is to take two of the key underlying elements of these instruments and translate them to an online setting. Namely, improving an existing proposal to gather more support for it and forming a coalition to join the support of two

F. Fui-Hoon Nah and K. Siau (Eds.): HCII 2022, LNCS 13327, pp. 11–23, 2022.
https://doi.org/10.1007/978-3-031-05544-7_2

or more proposals. We believe that these elements can be translated to mechanisms that are similar to collaborative software development through distributed version control systems.

Based on this idea, we have developed *decide*, an application for collaborative online decision-making that supports the participants to work together to come up with the best possible solution for a given issue. This does not require the participants to have homogenous interests and agree on a solution. It gives them all the tools they need to come up with good proposals, improve them, form coalitions and decide on the result.

We have conducted several experiments using *decide*. The results of those experiments are encouraging (see Sect. 4). Participants understand and use the new mechanisms provided for improving and merging proposals. Our experiments also suggest that the specific voting system employed may have a significant impact on how the participants use the option to improve and merge proposals.

The overall idea is inspired by observations of shortcomings in a recent participatory-budgeting process of ours [2,3] where we asked participants to make proposals on how to improve the study course of computer science at Heinrich-Heine University (HHU). For that process, we used D-BAS [4], a dialog-based argumentation system, to collect proposals and let participants argue about them. They were able to present pro- and contra-arguments to each other and to help inform newcomers to the process. The proposals made were used in a final vote. In this process, we noticed two behaviours:

During the process, participants made proposals that were very similar. Some proposals were slight amendments of other proposals, or they overlapped in their intentions. This poses a problem in a later vote as it can lead to vote splitting—i.e. votes are split between two candidates where there would have been a majority with only one candidate—or, on the contrary, if both overlapping proposals win at the same time, they cannot be implemented as they represent mutually exclusive issues.

Secondly, we found that when discussing proposals, participants often defended (or attacked) the proposal in question by providing possible implementation details or alternative solutions. They lacked the option of modifying or deleting an existing proposal, so their only option to add modifying details would be to make a completely new and separate proposal, which means more friction in the participation process.

1.1 Related Work

The field of collaborative decision-making has a long history. Most of it, however, deals with formal processes and (semi-)automatic decisions, often in the context of business.

This paper deals with a mainly social domain. It does not present a new decision-making framework, but an extension of traditional participatory processes in which proposals are made by participants, accompanied by a software prototype.

With the ever-increasing availability and use of the internet, online participation processes have become increasingly popular, sometimes involving large segments of the population. From informal opinion polls to legally binding participatory budgets.

In a traditional participation process, where proposals can be made by the population, these proposals are submitted by the participants and, often after a review by moderators, are presented to the public. Afterwards, there is the possibility to comment on these proposals and possibly already give one's approval. This can be possible for everyone who visits the respective website in order to keep the hurdle to participation low, or it is secured via postal invitation codes or e-passports; this usually only happens when legally binding decisions are involved.

A similar real world process is called *consensus decision-making* or *proposal-based decision-making*, which is similar to our system in that it focuses on the development of proposals, but since it is offline, it is synchronous, meaning all participants collaborate on the same step in the decision-making process at once. This is simply not feasibly for large, online processes.

1.2 Structure

In this paper we will first explain the core idea of this paper and how it has been integrated into a complete decision-making process.

We will then briefly look at voting methods, as these are relevant to the way participants interact with the system and are then used to make the final decision. Among other things, we have learned how much influence the voting method has on the motivation to cooperate and the perception of fairness.

We then present how we have experimentally tested our idea for feasibility. In doing so, we consider our idea as part of the complete system. We also explain and discuss the observations, conclusions and results.

Finally, we will draw a conclusion and show what further steps can be taken with this work.

2 The *decide* Decision-Making Platform

As a proof of concept for our idea, we developed decide. A web application, tailored for groups of non-expert participants to develop proposals together, and eventually make a decision.[1]

A decision-making process starts with someone putting forward a topic for decision and defining certain parameters, such as the desired time frame, restrictions on certain participants and visibility. Care should be taken here to describe the topic as precisely as possible and to do justice to the issue to be addressed.

[1] The *decide* application is open-source, and the code can be found here: https://link.cs.hhu.de/decide3-repository. We also have an in-development live instance here: https://link.cs.hhu.de/decide3-repository/decide3.

A decision-making process with a handful of friends will have different characteristics than a decision-making process of governmental and administrative institutions with an indefinite, large number of participants. Although our idea itself is independent of the number of participants, it is intended for use in larger groups of individual participants. Even though we use terms like *coalition* and *compromise*, it does not mean a coalition between two fixed, defined groups such as parties or other interest groups. We aim at grassroots democratic and self-regulated processes.

After a process has started, participants interact with the system by making proposals to solve the issue, arguing about them and at the same time giving their agreement to the proposals made. These approvals serve both as indications during the process of which proposals are popular and where it is worth working together (see Sect. 3), and as votes for the final decision, if it is desired that the process has a fixed end at all.

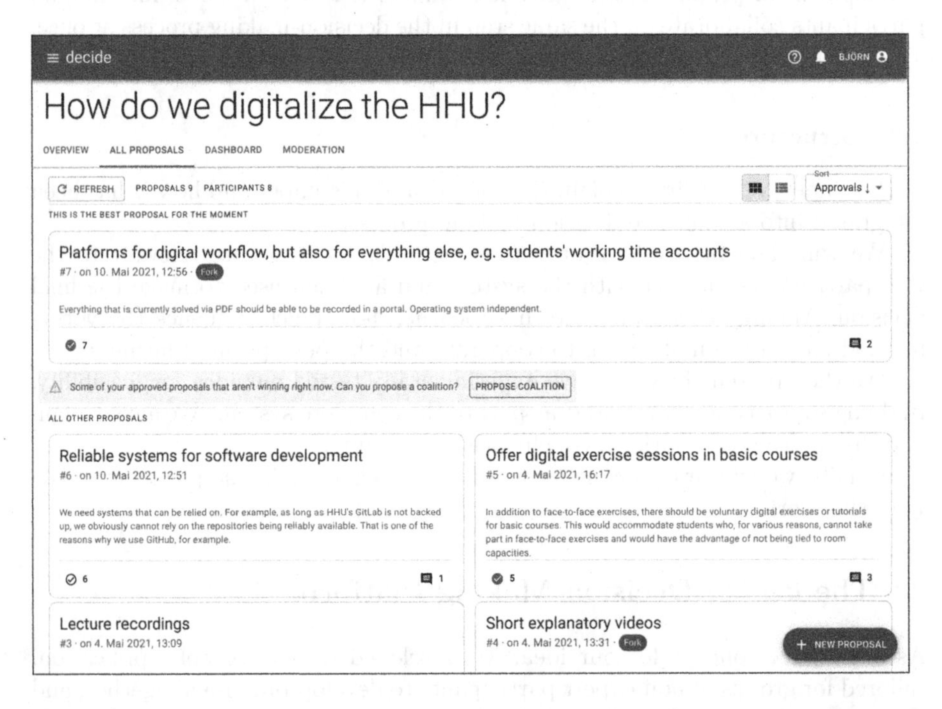

Fig. 1. The main view presented to participants in a process. The current best proposal is displayed on top. Proposals are sorted by approval by default.

2.1 The Inner Workings

We base the system of the notion of a *proposal*. A *proposal* is a text entry made by a participant. Every participant can make a *proposal* at any time during a process.

Important to note is, that proposals are immutable, no one, not even the authors can modify or even remove them. In this sense the author of a proposal hosts no special power over this proposal, nor is there a notion of stewardship. We even hid the author of a proposal, after observations of internal tests in our group, as not to associate a proposal with a specific person. This way, we foster a sense of belonging to the idea of a proposal rather than to the author. Having immutable proposals is therefore not just a convenience, but a requirement.

Another reason why proposals cannot change is our live voting system. In order to develop proposals, we need a way to communicate interest and approval of proposals between the participants. Therefore, we decided that having a traditional vote at the end of a participation process is not sufficient. Until the end of such a process it would not be visible how much interest and approval there is in a proposal. Approvals in our application are therefore possible and publicly visible to every other participant immediately after the creation of a proposal. This allows participants to assess the current climate of a process and make better decisions about which proposals are worth to develop further. We explain our voting system in more detail in Sect. 3.

2.2 Processes

Every *proposal* belongs to exactly one (decision-)*process*. A *process* is a topic that frames a decision. It has a title and a description. Optionally there can be a start and/or an end time in which a process is active and participants can participate. Usually there is at least an end time for when a decision process has to come to an end, but nothing speaks against a process that continues forever with ever evolving proposals.

Processes can be public or private and only be visible for invited participants. They can also modify rules about the process, like how much of the approvals are visible. Either the number of approvals is not visible, shown as the sum (this is the default) or with total transparency where everyone can see exactly who approves what. The last setting was popular for more informal decision-processes in small groups.

2.3 Development of Proposals

As we explained, we want to support decision-making by developing a set of proposals to a greater, better set of proposals. For this we allow and encourage multiple ways of adding new proposals.

First, every participant can submit a new proposal with an informative and unique title and a detailed description. In the best case a proposal is solely identifiable by the title, with the description only serving as implementation details or clarification.

In addition to this there are multiple ways of creating new proposals from existing proposals. They are all based on the idea that we want to model the relationship of a new proposal and the proposals that have led to it. Figure 2 shows one way, how this is presented to participants.

We call proposals with a single parent *fork* and proposals with multiple parents *merge*. These are terms from version control software, where the state of a project is tracked by a chain of annotated changes. This allows to track every development that ever occurred in the said project. Figure 3 portrays these relationships.

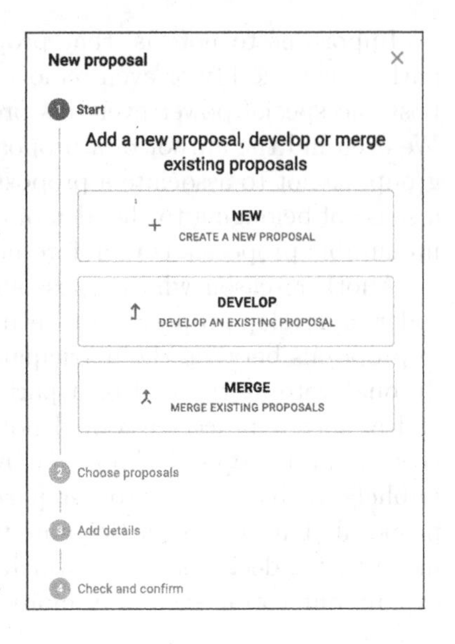

Fig. 2. Step by step interface to add a new proposal

Semantically, a *fork* can represent a variety of relationships. It would represent an enhancement of a proposal, an alternative or really any related semantic modification of a proposal. This is the intended way of modifying a proposal. Modifying it in place would not be possible as every proposal is a candidate in an ongoing vote. For the moment, we leave the exact meaning to the author of the new proposal. Of course, a more specific annotation would be conceivable, but for now we want to focus on different groups, according to the underlying system.

Merges, proposals with two or more parents, are the construct with the most potential in our system for collaboration. By enabling participants to link multiple proposals into a new one, real-world constructs are made possible. These can be, for example, compromises and coalitions. It should be noted that the content of a new *merge* proposal is not only a technical linkage of existing proposals. In terms of content, an independent proposal is created here, with its own title, content and arguments. This is a necessary freedom, as participants should not be restricted by the system as to how their *merges* should be. The application should only play an advisory role and not restrict the participants.

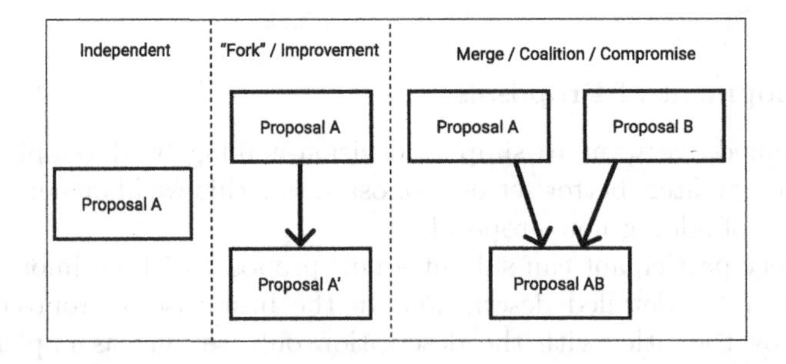

Fig. 3. Different kinds of relationships between proposals. Note, that *merges* are not limited to only two parents.

3 Voting

To come to a final decision, we use votes. So far, we already have the advantage that our system is designed to put up more candidates for a voting, which allows participants to better reflect their opinion. However, we have some special requirements in our system. In order for the participants to be able to work effectively with each other, it is necessary for them to be able to communicate amongst themselves in one way or another. We offer three different ways to do this in our system:

1. The creation of proposals itself (portrayed in Sect. 2)
2. Giving approval to one (or more) proposals
3. Nested comments on each proposal

This section deals exclusively with Point 2. It is necessary for the participants to recognize during the ongoing process which proposals are liked and which are disliked. Only then will participants be able to suggest useful enhancements and coalitions.

We have looked at different voting methods and weighed up which is suitable for the processes. For this, we have established the following requirements.

1. It must be easy to cast a vote.
2. It must be possible to see how popular a proposal is compared to another proposal.
3. The current ranking of the proposals (and thus also the current winner) must be transparent.

We first used approval voting [8]. It is a good match regarding our requirements. By today most people are familiar with a notion of a *like*-Button. Just a single click is necessary to cast a vote, fulfilling the first requirement. It is easy to aggregate and display the votes by summing them and thus communicate the current popularity of a proposal. Fulfilling the second requirement, it allows sorting the proposal by this sum, which fulfils the third requirement. This leaves the possibility of a draw, especially in processes with fewer participants. We have decided not to consider this problem further, as draws are an anomaly that can—and often should—be solved in the context of the situation. For our experiments we reliably broke the tie with the creation time of the proposals, letting more recent proposals win over older proposals. We decided on the basis of a gut feeling that newer proposals had less time to convince and woo away potential voters.

We also considered more complex voting methods. In particular, a preference-based voting method, such as Instant-runoff Voting (IRV) [1]. In such a method participants have to enter more precise votes. They have to rank proposals against each other. The winner is determined by repeatedly eliminating the worst candidate and recounting the remaining votes, until only the winner remains. While this may result in a vote that better matches their real intentions, it also requires more effort from the participants, especially if the process involves many proposals. Furthermore, with preference voting, it is no longer possible to display a simple aggregate of votes of the other participants. This makes it difficult to compare two proposals and thus the overall state of the process is no longer easily tractable.

The fact, that voting is iterative [6] in our system creates unique situations. While the effect of one's own vote in a traditional election is only visible after the counting, the participants in our system see the effects immediately.

This means that, for example, strategic voting can be experienced first-hand by participants. Participants can experiment with how they cast their votes and immediately see the impact this has on the current interim result of the process. In a process that uses plurality voting, i.e. a process where only one proposal can be approved, participants immediately see that giving their vote to a proposal has a direct negative effect on their previous favourite. We will see in Sect. 4 that this quickly leads participants to perceive the voting method as unfair, a perception that is in line with reality but usually only noticeable to people who have studied the voting method more closely.

Although in this case the weaknesses of the voting method are made more tangible by the immediate feedback, there are also possible advantages. Participants experience the direct influence of their vote, which may motivate a more detailed, further engagement with the candidates.

4 Experiments

To evaluate our platform, we tested our idea in two experiments with untrained students. After the first run, we made modifications based on our observations and feedback from the participants and then reviewed them in the second run. This led to a significant improvement in some parts of the surveys following the experiments.

4.1 Environment

We conducted the two decision-making processes with a duration of five days each, with a subsequent survey via a questionnaire. Both processes had the same theme namely: *How can we improve digitalization at the university?*. This is a current topic, which affects students from different study programmes and even staff. It lends itself well as a topic for our experiments, as there is still much need for development here. Large-scale digitalization of various aspects of university life is still uncharted territory and therefore lends itself well to input and ideas from students and staff.

In addition to payment for participation, as a special incentive, it was promised that we would forward the most approved proposal to the university's Pro-Rector for Digitalization, who would then comment on the proposal and the arguments relating to it. We then recruited the participants through advertising in lectures, the student council and personal letters. The majority of participants came from the Faculty of Natural Sciences at our university.

4.2 Setup and Execution

Both groups had the same topic and the same length for their run. Before starting the experiments, the participants were informed about what features the application provides and what the special features are. However, the task was deliberately kept simple. The participants were asked to visit the application daily, learn about new developments and participate in the process. No guidelines were given on how or when to participate. In particular, it was not required that the novel features of the application had to be used.

The processes took place completely remotely, partly due to the current pandemic. Although it is more difficult to observe the process and the participants, the process is closer to reality.

The processes started on Monday at noon. On Wednesday, there was another appeal and a brief overview of the current progress. On Friday at noon, the process ended and we released the questionnaire. At least 15 participants took part in each of the processes, although in both cases only 10 participants were actually active during the entire week like planned.

During the processes we fixed software and minor quality-of-life issues, like improved loading times and better contrast for people with impaired eyesight, based on participant feedback and observations, but the rules and content of the processes were not touched.

In the first experiment, participants had only one approval they could give to a proposal. If they agreed to another proposal, the approval to the former proposal was removed. We decided to do this because we noticed in internal, technical tests that participants tended to just agree to everything. The following working hypothesis was formulated for this:

Hypothesis 1. *Some pressure exerted on participants by the decision-making mechanism motivates collaboration with other participants.*

Participants of the second experiment could approve as many proposals as they liked. In both cases the winner with the most votes won.

4.3 Observations in the First Experiment

This made for a more active process, as participants had to think more about which proposal they wanted to vote for and which proposals they did not. They could not just approve to everything, a behaviour we noticed in earlier, internal tests.

It is possible for us to track the changes in votes in order to understand the voting behaviour of the participants. This allows us to observe and answer questions such as: *Do participants stay in one strand of proposals or do they switch back and forth?*, *Are participants more likely to shift their approvals to newer, developed proposals?*.

Based on the final survey and feedback during the process, we noted that some participants have found the voting procedure unfair or dissatisfactory, as they have had to take a risk with their vote to support a new proposal they like better. Hurting their current favourite proposal by giving their only approval to the new proposal with fewer votes.

Although vote splitting is not a new problem, in our public, live voting system it is directly experienced by the participants and thus poses a bigger problem. An open, live voting system makes strategic voting more visible and thus also the advantages and the disadvantages it entails.

We have thus made the following hypothesis for the next iteration of our system:

Hypothesis 2. *The voting mechanism is highly influential to the satisfaction with the decision result in our system.*

Based on this knowledge we ran the second experiment with simple approval voting. Each participant can give as many or as few approvals as they want. Approval voting, which is a cardinal voting method, is immune against vote splitting [7]. This removes the risk for participants to vote for a new proposal that has not yet received much attention.

4.4 Observed Differences in the Use of Our New Features

Figure 4 shows the proposals and relations between parents and children that emerged during each experiment. It can be seen that in the first experiment much more use of our tools to develop proposals was made, with the same number of participants and proposals.

We therefore assume that our original Hypothesis 1, that more collaboration takes place with a little pressure, is pointing in the right direction. Although this behaviour can be attributed to chance with the low numbers of participants, we believe that the difference in behaviours is significant enough to support our hypothesis.

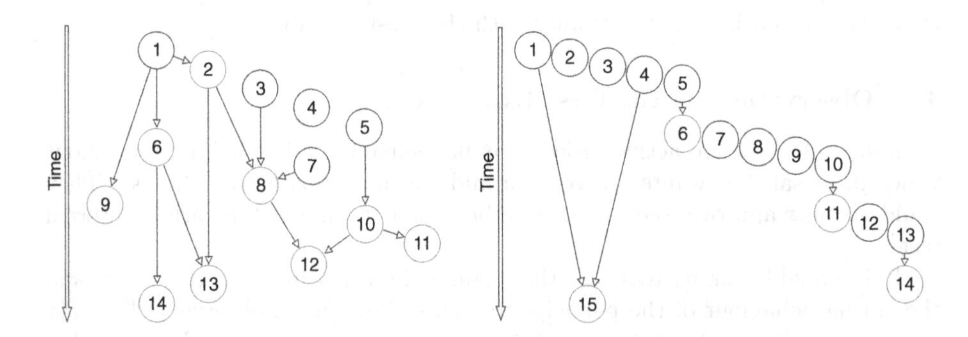

Fig. 4. Proposals and their relations made by participants in the two experiments. **Left**: First experiment with plurality voting. (3 *merges*, 6 *forks*, 5 independent) **Right**: Second experiment with approval voting. (1 *merge*, 3 *forks*, 11 independent)

4.5 Survey Results

The final survey that both groups completed after their process, consists of several sections with different statements. Participants should indicate how much they agree with these statements. We used a Likert scale [5] from 1 (strongly disagree) to 5 (strongly agree).[2]

The topics of the sections are:

S1. First encounter with the application.
S2. Feelings regarding the outcome.
S3. Statements about the process in general.
S4. Statements about argumentation.
S5. Statements about *forks*.
S6. Statements about *merges*.

Both groups do not differ significantly in their answers except for statements in one section. Overall the reception to the experiment, the application and the features it comes with was positive. The biggest criticism of the application by the participants was the confusing nature given a larger number of proposals or arguments. There was a wish for clearer indications of which proposals came from which other proposals.

Questionnaire-Section S2, about the participants feelings regarding the outcome, had significantly more positive grades with the second experiment, where participants had unlimited approvals. We compared the statements of both groups with a two-sided t-test for different variances [9], due to our low n of 10 and 12 respectively. The values are presented in Table 1.

Table 1. Comparison the statements in question-section S2 of the first experiment with a single approval and second experiment with multiple approvals. 1 = strongly disagree, 5 = strongly agree

Statement	Plurality			Approval			
	n	Avg.	Var.	n	Avg.	Var.	p(TT)
I am satisfied with the outcome of the decision	12	3,75	0,75	10	4,70	0,23	0,005
I think my opinion contributed to the final decision	12	3,50	0,64	10	4,60	0,49	0,003
I find my opinion reflected in the decision	12	3,67	1,15	10	4,70	0,23	0,009
I think my voice was heard	12	4,00	0,73	10	4,80	0,18	0,011
I felt the process was fair	12	4,00	1,64	10	4,80	0,18	0,061
I think I was correctly informed about the process	12	4,50	0,45	10	4,70	0,90	0,584
I think the decision is a joint product of everyone	12	3,75	1,48	10	4,80	0,18	0,014

The clear difference in the results of the two experimental groups reinforces our belief that the perceived fairness of a voting system used is a key determinant

[2] The full, translated data, can be found here: https://link.cs.hhu.de/data-ebbinghaus22_collabproposal.

of satisfaction with the process and outcome of a process in our application. We accept Hypothesis 2. This should therefore be considered and investigated in further research.

5 Conclusion and Future Work

In this paper, we have shown a system with features that provide participants in a decision-making process with tools for developing proposals together.

We have shown that these features are understood and used by participants. However, we have also learned through the second experiment that such a system, in its current state, needs to encourage participants to collaborate.

In our first experiment, this happened involuntarily due to constraints in the voting method. Participants were motivated by the characteristics of the voting method not to deviate further from the currently strong proposals and therefore tend to stick with the existing ones. By being limited to one approval, exploring new proposals became a risk for participants. Although effective, we believe this is not good for future processes because it goes against our overarching goal of providing the tools to make more and better proposals.

A suitable methodology to encourage participants to cooperate and at the same time provide a voting process that is perceived as fair is thus indispensable for further research on this system.

We believe that the system itself could provide more guidance and advice to participants in a process. Currently, the system itself is not *smart*, it does not actively support participants, e.g. by recommending suitable proposals or worthwhile coalitions. We have already made first attempts in this direction. By using a public, iterative voting method, the system already has information about which voter groups support which proposals during a decision-making process. This can be used for collaborative filtering, among other things, in the future.

Acknowledgements. Björn Ebbinghaus is a member of the PhD-programme *Online Participation*, supported by the North Rhine-Westphalian funding scheme *Forschungskollegs*.

References

1. Brandt, F., Conitzer, V., Endriss, U., Lang, J., Procaccia, A.D.: Handbook of Computational Social Choice. Cambridge University Press, New York (2016)
2. Ebbinghaus, B.: Decision making with argumentation graphs. Master's thesis, Department of Computer Science, Heinrich-Heine-University Düsseldorf, May 2019. https://doi.org/10.13140/RG.2.2.12515.09760/1
3. Ebbinghaus, B., Mauve, M.: Decide: supporting participatory budgeting with online argumentation. In: Prakken, H., Bistarelli, S., Francesco, S., Taticchi, C. (eds.) Computational Models of Argument. Proceedings of COMMA 2020. Frontiers in Artificial Intelligence and Applications, vol. 326, pp. 463–464. IOS Press, September 2020. https://doi.org/10.3233/FAIA200535

4. Krauthoff, T., Meter, C., Betz, G., Baurmann, M., Mauve, M.: D-BAS - a dialog-based online argumentation system. In: Computational Models of Argument. pp. 325–336, September 2018. https://doi.org/10.3233/978-1-61499-906-5-325
5. Likert, R.: A technique for the measurement of attitudes. Arch. Psychol. **22**, 55 (1932)
6. Meir, R.: Iterative voting. In: Endriss, U. (ed.) Trends in Computational Social Choice, vol. 4, pp. 69–86. AI Access (2017)
7. Poundstone, W.: Gaming the Vote: Why Elections Aren't Fair (and What We Can Do About It). Macmillan (2009)
8. Weber, R.J.: Approval voting. J. Econ. Perspect. **9**(1), 39–49 (1995)
9. Welch, B.L.: The generalization of 'student's' problem when several different population varlances are involved. Biometrika **34**(1–2), 28–35 (1947)

Design and Implementation of a Collaborative Idea Evaluation System

Andreas Hermann(✉)

European Research Center for Information Systems (ERCIS), University of Münster, Münster, Germany
andreas.hermann@ercis.uni-muenster.de

Abstract. Digital transformation is a complex endeavor with unforeseen pitfalls. However, small and medium-sized enterprises (SMEs) seem to struggle with digital transformation. To address this problem, governments have started initiatives, i.e., publicly-funded support units, with the purpose to support digital transformation through facilitating idea management in SMEs. Studies suggest that the application of (IT-based) idea evaluation systems could impact the digital transformation success of SMEs. However, contributions addressing the interplay of support units and SMEs in the context of idea evaluation are scarce. Therefore, this article addresses the following problems (1) SMEs require external support in the context of digital transformation, and (2) idea evaluation is often improperly executed. Bringing together these areas leads to the objective to design and implement a collaborative idea evaluation system. The system development follows the methodological pathway for design science research. Observations from seven small-scale projects with SMEs complemented with selected literature inform the system design. Two focus group discussions serve as an evaluation of the system. The results are a set of design features and an interaction concept. Finally, a prototypical web application puts the conceptual design into action to tackle the identified practical problem.

Keywords: Idea management · Software design · Digital transformation · Idea evaluation · Small and medium-sized enterprises · Publicly-funded support units

1 Introduction

Rapid technological advancements drive the digital transformation of organizations and industries [49]. This technology-enabled process brings along fundamental improvements of organizations' strategies, processes, and business models [39]. At the same time, digital transformation on an organizational level is a complex endeavor with unforeseen pitfalls [4]. Indeed, this high complexity leads to organizations failing with their digital transformation ventures [48].

In particular, SMEs seem overwhelmed by this complexity and struggle with progressing in their digital transformation journey [25,58]. This assumption has

F. Fui-Hoon Nah and K. Siau (Eds.): HCII 2022, LNCS 13327, pp. 24–40, 2022.
https://doi.org/10.1007/978-3-031-05544-7_3

been verified in several studies, for instance, in a survey conducted by Berghaus & Back [6]. Based on a sample of 417 organizations, their study revealed that SMEs score low to medium in terms of digital maturity [6]. Further emphasizing the struggles of SMEs is the prominent narrative of SMEs' resource constraints and knowledge gaps [6,12,25,58].

Consequently, SMEs are reliant on successful projects and cannot afford to waste their precious resources on dead-end endeavors [25]. This raises the question of how SMEs can be supported in their digital transformation efforts, despite their limitations. In an attempt to address this problem, governments have started initiatives, i.e., publicly-funded support units, with the purpose to support the digital transformation of SMEs, in particular, idea management [46]. The collaborations between SMEs and such support units mainly focus on managing digital innovation ideas [3]. Indeed, one determining factor for digital transformation success is a properly executed idea management [24,53,61]. Idea management is concerned with the generation, evaluation, and selection of ideas [11]. In particular, careful idea evaluation and selection play a critical role as the implementation phase directly follows. This criticality lies in the potential hazard of accepting a bad or rejecting a good idea for implementation [24]. Put differently, organizations can minimize losses through a properly executed idea evaluation and selection [24].

What can be observed, however, is that organizations tend to "carry out the selection of ideas ad hoc or intuitively" [53]. This leads to another important question as to how SMEs can approach the evaluation of digital innovation ideas to inform the subsequent idea selection and avoid intuitive decisions.

To provide assistance, researchers have studied idea evaluation from various perspectives (cf. [24]). Among others, authors focus on processes and methods [22,50], conceptual models [11,56], evaluation criteria [24,52] and (IT-supported) procedures for idea evaluation [22,50,52]. The findings from those studies suggest that the application of adequate (IT-based) idea evaluation systems is beneficial [8,11,22,24,50,52,53].

So far, two problem areas have emerged, i.e. (1) SMEs require external support for managing digital innovation ideas, and (2) idea evaluation is often improperly executed. Bringing together these areas leads to the following research question: *How should idea evaluation systems be conceptualized and implemented to inform the selection of digital innovation ideas in the context of the collaboration of publicly-funded support units and SMEs?* Accordingly, the research objective is to conceptualize and implement a collaborative idea evaluation system involving support units and SMEs. To achieve this objective, a set of design features, an interaction concept, and a prototypical web application are proposed as central outcomes of this article.

The remainder of this paper is structured as follows. First, the research background on digital transformation, idea management, and externally supported idea management in SMEs is discussed in Sect. 2. Subsequently, the underlying research method is described in Sect. 3. Section 4 of this article presents the set of design features, the interaction concept, and the system implementation.

Following, Sect. 5 outlines how the system concept and implementation contribute to practice and research. Finally, a conclusion is provided in Sect. 6.

2 Research Background

Designing and implementing an idea evaluation system involving SMEs and external support units necessitates a discourse on two fundamental themes. First, the understanding of idea management in the context of digital transformation needs to be introduced. Second, idea management and its particular phase idea evaluation, with an explicit focus on the domain of SMEs and external support, needs to be addressed.

2.1 Digital Transformation and Idea Management in SMEs

Digital transformation can be considered an ongoing change process that is concerned with the implementation of digital innovations in organizations [3,25,39]. This process is triggered through contextual factors, such as market developments, and enabled through the continuous development of new technology [32,39]. Digital transformation is proactive and requires an intention to search and implement digital innovations [62]. The continuous implementation of innovations can result in various outcomes, i.e., process, product, or service innovations [37,59,63] and, thereby, drive the digital transformation of SMEs [3]. What precedes these implementations are ideas, the "basic building blocks for innovation" [24]. In other words, a digital innovation can be referred to as "the successful implementation of a (creative) idea" [59]. Following this logic, the concepts digital transformation, digital innovation, and innovation idea are tightly intertwined with one another. The understanding that innovation ideas are at the root of digital transformation makes the management of ideas an important moderator for digital transformation success [3,24,61]. The management of ideas is concerned with the generation, evaluation, and selection of innovation ideas [11,24]. In particular, the evaluation of ideas is a critical stage as the selection, and subsequent implementation directly follow. Because of the often proclaimed resource constraints of SMEs, accepting a bad idea for implementation can go along with detrimental consequences [6,12,24,25,58]. Therefore, a properly executed idea management, idea evaluation in particular, could minimize potential losses for SMEs [24].

However, it appears that SMEs struggle with the complexities of idea management and, as a consequence thereof, digital transformation [25,53,58]. This assumption has been verified in several studies. For instance, Berghaus & Back [6] surveyed 417 organizations. Their results show that, in particular, SMEs score low to medium in terms of digital maturity [6]. To mitigate this situation, idea management is deemed a promising approach to leverage digital maturity and impact digital transformation success for SMEs [3,53,61]. However, many organizations tend to "carry out the selection of ideas ad hoc or intuitively" [53] rather than following clear, IT-based evaluation processes.

The importance to support idea management in organizations has been widely acknowledged in academic literature. For instance, literature of the field proposes supportive processes and methods [22,50], conceptual models [11,56], or IT solutions for idea management [22,30,52]. Idea evaluation as a sub-process of idea management has as well received considerable attention. Authors have investigated and proposed, for example, success factors for idea evaluation [1,24], idea evaluation criteria [24,52], or idea evaluation methods [7,8,36,52,53]. In this context, literature suggests that idea management and, thereby, digital transformation success can be improved through the application of IT-based idea evaluation systems [8,11,22,24,42,50,52,53].

2.2 Externally Supported Idea Management in SMEs

Despite researchers' efforts to support idea management and evaluation in SMEs, it appears that SMEs require further assistance. In fact, disregarding external support seems to be a relevant predictor for business failures in early stages of new ventures [15]. In turn, SMEs can greatly benefit from external support in the course of idea management [58,61].

As a response, governments have started publicly-funded initiatives [20] to support idea management in SMEs [3,46]. Those initiatives manifest in designated external support units, which are often a collaboration of academic and non-academic institutions. Among others, these units are responsible for supporting idea management tasks such as idea generation, idea evaluation, and idea selection in the context of SMEs' digital transformation. Moreover, they act as facilitators or counselors rather than consultants [15]. Along these lines, publicly-funded support units offer guidance and mentoring in the context of idea management and take over some of an SME's duties themselves [15].

To complete the discourse on support units facilitate SMEs' idea management, the question as to how this is achieved has yet to be addressed. The daily business of support units is often characterized by intensive interactions with SMEs in the context of workshops, small-scale digital innovation project, mentoring dialogues, or networking events[16,27,30,46,54]. To support their daily activities, support units develop and employ tools and concepts, such as procedure models, ideation tools, technology scouting, business model development tools, or digital maturity assessments [3,6,23,30,46,54,55]. Considering the interplay of support units and SMEs, the tools and concept of this list mainly address the collaboration from either the perspective of supported SMEs or the involved support unit. Tools that acknowledge both parties and their information needs in a balanced way appear to be rare [26]. So far, research at the intersection of SMEs and support units focuses on business model innovation [23], collaboration drivers and effects [13,14,38,61], or collaboration matters [46,51]. However, from reviewing the literature, it appears that little attention has been paid to improving this collaboration with designated IT-solutions. This observation holds true for idea evaluation systems addressing the interplay of support units and SMEs.

3 Research Method

The development of the system follows the methodological pathway for design science research [33,34]. Observations from seven small-scale case projects with local SMEs together with the problem definition mentioned above motivate and inform this research (cf. Table 1). All projects have been implemented in the period from October 2019 to March 2021 by a support unit (cf. [5]), which has been publicly-funded through the German *Mittelstand Digital* initiative. The general aim of those projects was to generate, improve, evaluate, prioritize, select, and conceptualize (cf. [5,24,37]) possible digital innovation ideas for the involved SMEs. In each case project several meetings and workshops were organized and conducted. From these interventions, meeting protocols, workshop material, and interview notes have been analyzed and informed the system conceptualization. Eventually, core, condensed observations pointing at a specific idea evaluation themes have been derived from each project.

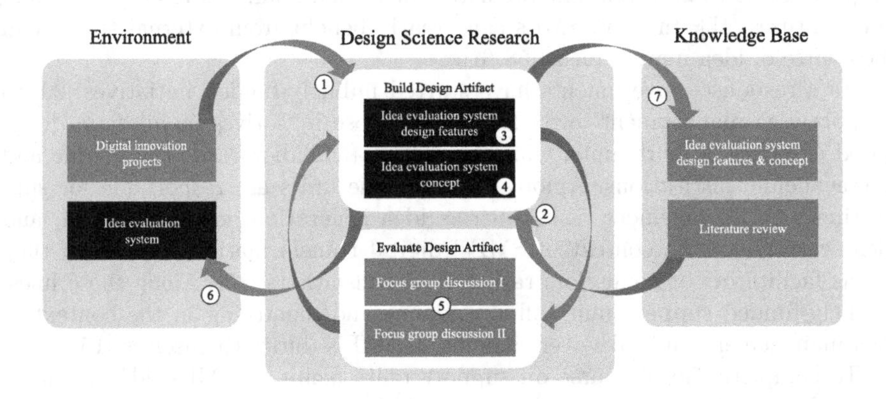

Fig. 1. Research design [34]

A synthesis of insights from relevant literature complement these practical observations and has led to the initial concept of the idea evaluation system. The literature search has been conducted in a hermeneutic fashion [9]. In this way, relevant publications from research areas such as 'digital transformation', 'idea management', 'idea evaluation', 'innovation management', or 'digital innovation' have been identified.

Finally, as specified in the system concept, the evaluation system was implemented as a web application to be applied in real-world collaborations between support units and SMEs. Throughout and after the implementation phase of the system, two focus group discussions have been conducted. A focus group discussion is based on guided interactions among the group's members. The qualitative method allows to evaluate and refine artifacts [57], such as the proposed idea evaluation system. For the conduct of the focus group discussions, the guidelines as proposed by Tremblay et al. [57] have been followed. The first focus group discussion (June 2021) involved seven academics from related subject areas. The second discussion (July 2021) has been conducted with five practitioners and three academics. The practitioners have advanced experiences in digital innovation projects and are employed at publicly-funded support units from the *Mittelstand Digital* initiative. The focus group discussions were recorded and documented to inform the refinement of the idea evaluation system.

Table 1. Overview of the case pProjects.

#	Business	Employees	Project outcomes	Observations
P1	Stationary retailer for electronics	45+	Digitalization roadmap; digitalization status assessment; idea implementation suggestions	Idea evaluation requires management involvement; idea evaluation should enable idea prioritization; collaborative idea evaluation workshops involving SME and support unit are valuable
P2	Manufacturer for truck trailers	200+	Digitalization roadmap; digital container chassis concept; idea implementation suggestions	Evaluation criteria should distinguish between discontinuous and continuous innovations; continuous employee involvement is important along the entire idea management process
P3	Manufacturer for road bikes	10+	Digitalization roadmap; marketing concept	Idea evaluation requires both management and employee involvement; idea evaluation should be continuous and repetitive; idea evaluation performance improves with experience; idea evaluation should be criteria-based
P4	Bakery business	470+	Digitalization roadmap; customer app concept	Idea evaluation should be continuous and repetitive; external expert evaluator improve idea management; idea evaluation summaries are vital
P5	Carpentry business for interior design	25+	Digital sales desk concept for jewelry stores	Idea evaluation requires substantial domain knowledge; employee involvement improves idea management success; idea evaluation should be a continuous process
P6	Stationary retailer for fashion	130+	Digitalization roadmap; digital last mile delivery concept	Collaborative idea management requires trust among the parties; idea evaluation should be performed early and often
P7	Studio for wellness services	1+	Digitalization roadmap, digital wellness services concept	External triggers and support can facilitate idea management; idea management is promising regardless of company size

4 A Collaborative Idea Evaluation System

The results can be broken down into two components: (1) a set of design features (cf. Sect. 4.1) (2) a conceptual design of the intended system, including a prototypical web application (cf. Sect. 4.2). (1) The a set of design features prescribes the system's logic (cf. [35]) and informs the implementation of a collaborative idea evaluation system (cf. Table 2). (2) The conceptual design captures the collaboration between support units and SMEs (cf. Fig. 2). The prototypical web application makes the collaborative idea evaluation system accessible for real-world collaborations between support units and SMEs (cf. Fig. 3).

4.1 Design Features

Based on the observations from the case projects, a first set of design features has been derived. This initial set has been refined throughout the two focus group discussions. The final, refined design features are presented in this section (cf. Table 2). Moreover, selected scientific sources are referenced to provide further evidence to the empirically derived design features. The feature collection is structured into three clusters, **General features**, **Idea evaluation features**, and **Collaboration features**. Idea evaluation features capture subject-matter specific features that directly prescribe the idea evaluation logic of the system. Features of the cluster collaboration features describe the essence of the intended interaction logic of the system. Last but not least, general features are generic system features without a particular focus on the idea evaluation domain.

Idea Evaluation Features: The case projects have shown that idea evaluation should always be based on reasonable criteria (DFI1) [17,21,24,52,53,61]. More specifically, a "formal [idea] evaluation process with clear and transparent criteria" [61] appears to improve idea management in general.

A formal idea evaluation process should also enable a repeated, continuous evaluation (DFI2) [11,18,21,31]. Often, highly relevant ideas (e.g., low-hanging fruits [3]) get selected over other ideas, which are added to an idea pool or backlog [50]. The continuous implementation of innovation ideas alters the current status quo in terms of digital maturity or available technologies. As a consequence thereof, the re-evaluation of ideas becomes necessary.

The need for continuous evaluations leads up to the actual evaluation act. The system should be implemented in a way that multiple users are allowed to concurrently or synchronously evaluate the very same idea (DFI3) [24,27]. This observation stems from the intended setting for which the evaluation system is designed. For particular idea management tasks, the collaboration of publicly-funded support units and SMEs is typically performed on-site [27]. In such settings, multiple devices may be used by the workshop participants. That, in turn, requires the system to support concurrency. However, even virtual settings may benefit from a synchronous idea evaluation, for instance, when ideas are being evaluated by a larger number of people.

Having idea evaluated by selected reviewers constitutes another design feature, i.e., invite-based idea evaluation (DFI4). The system should integrate a mechanism to invite employees, or external reviewers, to perform the task of idea evaluation for one or more selected ideas [3,19,24,47,56]. Often, employees or external stakeholder are specialized in certain areas so that their ratings equal those of experts [24]. In fact, in the front-end of innovation, i.e., in the context of idea management, "high external collaboration is [...] related to high front-end success for SMEs" [61].

From both focus group discussions, it turned out that anonymity when evaluating ideas is preferred over an open approach (DFI5) [24,43]. This observation is confirmed on a general level for idea management activities, for idea generation in particular, where anonymity appears to be beneficial (cf. [24]). Moreover, there

Table 2. Design features derived from the case projects.

ID	Short description	Empirical Validation	Theoretical Validation
Idea evaluation design features			
DFI1	Criteria-based idea evaluation	P2, P3	[17,21,24,52,53,61]
DFI2	Continuous idea evaluation	P3, P4, P5, P6	[11,18,21,31]
DFI3	Synchronous idea evaluation	P1	[24,27]
DFI4	Invite-based idea evaluation	P2, P3, P4, P5	[3,19,24,47,56]
DFI5	Anonymous idea evaluation	P6	[24,43]
DFI6	Evaluation feedback	P1, P4	[2,24]
Collaboration design features			
DFC1	Role concept	P1, P2, P3, P4, P5	[10,27,30,47]
DFC2	Support requests	P4, P6	[27,46,54]
DFC3	Virtual assistance	P3, P4	[16,27,42,46]
DFC4	Guided mode	P1, P6	[10,18,21,31,47]
DFC5	Exploration mode	P4	[2,3,27,30]
General design features			
DFG1	User management	P1, P2, P3, P4, P5	[10,27,30]
DFG2	Web-based idea evaluation system	–	[27]
DFG3	Customizable user interfaces	P1, P4, P6	[27]

seems to be a significant link between anonymity in idea evaluation and evaluation scores [43]. In other words, anonymity may reduce a potential response bias and lead to more authentic evaluations from the evaluator's perspective. That seems to make sense in the context of SMEs, where exposing ones identity in the course of idea evaluation may not be favoured for both the idea generator and evaluator.

While the system should not expose who performed the evaluation, feedback should be provided and evaluation results presented (DFI6) [2,24]. Both the idea generator and evaluator must have access to the evaluation results. That can be achieved by, for instance, visualizing evaluation averages on a dashboard. Thereby, idea selection can be supported and, ultimately, facilitated [24].

Collaboration Features: One major abstract requirement for the intended system is to support collaboration between SMEs and publicly-funded support units (DFC1). The case projects and focus group discussions revealed that both parties have different needs in terms of system features. For that, a fundamental prerequisite is for the system to feature a corresponding role concept [10,27,30,47] or, at least, acknowledge different roles [11,24,52,61].

The main role of involved support units is, in abstract terms, to provide support for SMEs [27,46,54]. Usually, support is provided in the course of on-site meetings. However, SMEs may require additional assistance outside of such settings [46]. Therefore, the system should integrate a mechanism that enables SMEs to request support units on demand [10,47] (DFC2). Available support units would receive that request and provide idea evaluation assistance, e.g., by scheduling an on-site meeting.

Requests could either be fulfilled through actual evaluations or some sort of counseling provided by a support unit. In the course of idea evaluation, a support unit can act as a counselor, whose "primary function [...] is to facilitate the performance of a task" [15]. Correspondingly, the system should allow support units to provide virtual counseling assistance, e.g., through a virtual messaging service (DFC3) [16,27,42,46].

The case projects have also highlighted the importance of guidance in idea evaluation (DFC4). In fact, having well-defined processes in place improves idea management in general and, thereby, idea evaluation performance [41,53,61]. Guidance can be achieved through the implementation of a mode that leads through the evaluation process step by step [18,21,31] and involves the possibility for interventions by support units [10,47].

In addition to such guided mode, the system should allow users to freely manage and evaluate ideas (DFC5). For instance, the system could feature some sort of dashboard that presents available ideas for evaluation at a glance [2,3,27,30].

General Features: An essential feature to facilitate the intended collaboration of support units and SMEs is a properly integrated user management (DFG1) [10,27,30]. According to the role concept, the system should allow both employees of SMEs and support units to register a user account. The case projects revealed that it makes sense to predetermine permissions depending on the user's role (cf. [27,30]). In other words, the role concept specified before integrates into the user management and prescribes default user rights.

To achieve this user management functionality, among others, the system should be implemented based on state-of-the-art technologies (DFG2) [27]. Moreover, the intended application scenario requires the system to be readily accessible without installation effort. Therefore, the system should be implemented as a web-based application that can be accessed anywhere at anytime. Indeed, it has been shown recently that modern web-based idea management systems are likely to go along with a multitude of benefits, such as overall idea management performance or creation of an innovation culture [42].

Another feature contributing to overall idea management success is the customizability of the system's user interfaces (DFG3). The rationale for this feature is twofold. First, the role concept demands for a differentiation of user interfaces, e.g., by hiding particular components [27]. Second, the collaboration between SMEs and external supporters may suffer from a lack of trust toward supporting units [44]. This lack, in turn, is often the main reason to not seek out external support [44]. In this sense, it is important that the supported SME keeps some sense of ownership. That, again, could be achieved by a customization of the system's interfaces and layouts (e.g., color schemes and logos matching to the SME's corporate identity).

4.2 System Concept and Implementation

The conceptual design of the idea evaluation system fulfils two purposes. First, it captures reusable, abstract design knowledge through the explicit mapping of design features to the system concept (cf. [28,35]). Second, it informs the system implementation at hand.

The system is conceptualized according to the design features listed in the previous section. Design features are indicated as dark gray circles and their respective ID (cf. Table 2). The system concept depicts the interaction between support unit and SME in the course of the idea evaluation phase of idea management. The idea evaluation phase is situated after the idea generation phase and before the idea selection [24]. In fact, the evaluation phase acts as a filtering stage, which informs the selection of ideas for subsequent implementation. The conceptualized evaluation system intends to create a continuous collaborative environment for both support unit and SME. The SME may request idea evaluation support and receives corresponding assistance by the unit. Moreover, a support unit assists in the evaluation process by providing expert knowledge and continuous feedback mediated via the depicted idea evaluation system. At the same time, the support unit evaluates ideas. This can happen, for instance, synchronously, anonymously, and continuously. As specified in the feature set, idea evaluations are based on clear criteria. SMEs access this very system and involve selected staff in the idea evaluation process. The evaluation may be performed stepwise, or in an explorative fashion considering multiple ideas at once. Eventually, based on comprehensive evaluation feedback presented by the system, promising innovation ideas are selected for realization.

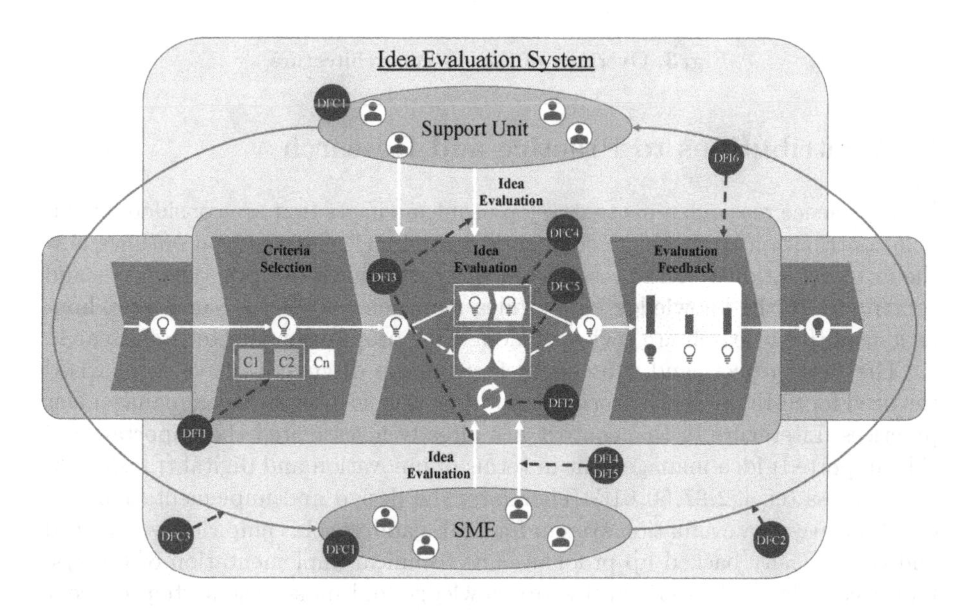

Fig. 2. Collaborative idea evaluation system concept.

Building upon the evaluation system concept and the corresponding design features, the system has been implemented as follows. The idea evaluation system is implemented based on the Python framework *Django*. *Django* is a popular web framework that allows developing state-of-the-art web applications. The framework enforces a modular system architecture, is thoroughly documented, and has a large community. The underlying web server is *Nginx*. By utilizing the *Django Rest Framework*, backend and frontend are connected via designated APIs. The frontend is implemented using the open-source Javascript library *React*. The database for the evaluation system is a *PostgreSQL* database. To ease software delivery and increase modularity, the single applications are containerized using *Docker*. The overall architecture is visualized in Fig. 3.

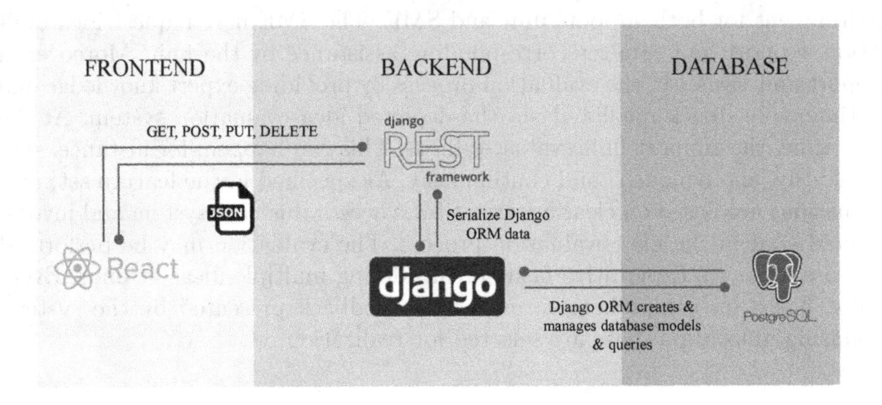

Fig. 3. Overview of the system architecture.

5 Contributions to Practice and Research

Design science research aims to develop usable artifacts that are considered valuable contributions to both practice and research [34]. This section outlines how the idea evaluation system presented in this paper addresses practical needs and contributes to the knowledge base of idea management. At the same time, limitations of this research are reviewed along with propositions for future research.

The case projects and subsequent focus group discussions have emphasised the need for SMEs to seek external support as well as integrate idea management practices. Literature in this context has already investigated the importance of (IT-supported) idea management in terms of innovation and digital transformation success (cf. [22,27,50,61]). Therefore, the design and implementation of a collaborative idea evaluation system for SMEs address this empirically observed and theoretically backed up problem. The technical implementation of the system encapsulates the derived design knowledge and makes the system concept applicable in practice. The design features can be reused by implementers to come up with alternative instantiations, ultimately, to address the identified problem. In addition to leveraging idea management and digital transformation in SMEs, the developed system addresses selected needs of involved support units. The integrated role concept and clearly defined intervention points allow support units to work more efficiently in their role as supporters for SMEs. The system can, for instance, be applied in on-site or virtual workshop settings and improve the performance of idea evaluation support provided by support units.

Additions to the knowledge base include new meta-artifacts and experiences gained during the design and evaluation an artifact [34]. In this sense, the set of design features that has been derived in the course of seven small-scale digital innovation projects is proposed as an addition to the knowledge base. More specifically, the design features capture empirical observations in the form of condensed, abstract design knowledge, which can be reused by researchers and act upon by implementers and enactors [35,45].

Moreover, the concept of the idea evaluation system extends the knowledge base as follows. The system concept integrates the identified design features and proposes an abstract interaction mechanism involving both support units and SMEs. While the interaction between external supporters and organizations, SMEs in particular, has been investigated thoroughly (e.g. [5,15,29,38,44]), less attention is paid to externally supported idea management and evaluation in SMEs. Studies that address the interaction of SMEs and external supporters focus more on topics such as business model innovation [23], collaboration drivers and effects [13,14,38,61], or collaboration matters [46,51]. The proposed concept for an idea evaluation system, hence, extends this research field by integrating the area of idea management and evaluation into the collaboration of SMEs and support units.

Experiences gained throughout the artifact design process are also important additions to the knowledge base [34]. Here, two focal points should be stressed. First, from the focus group discussions it turned out that it is crucial to regard a particular organizational context in idea evaluation. That implies that the process should allow a certain degree of freedom and configurability. Studies suggest to pay particular attention to the context-sensitive selection and weighing of idea evaluation criteria [18,21,24,40,50]. Second, from the case projects and focus group discussions alike, one central theme that occurred repeatedly is knowledge boundaries. In fact, practitioners mentioned the difficulty to build a common understanding of company goals, digital transformation ambitions, technology competencies and the like. Such knowledge boundaries are of particular relevance in the context of external units supporting SMEs [26,44]. The existence of knowledge and information boundaries may be a precursor to trust issues between the parties and prevent SMEs from seeking external support [44].

Besides these contributions to practice and research, the following three matters must be regarded as limitations. First, the initial design feature set has been empirically derived based on observation from seven small-scale digital innovation projects. To improve the robustness of these observations, the design features have been validated through selected scientific sources. To confidentially draw general conclusions, additional empirical data might be necessary. Therefore, future research could review and revise the proposed set of features through the continuous implementation of innovation projects in SMEs. Moreover, a structured literature review aiming at the identification of additional design knowledge could be performed. Second, the focus group discussions only involved participants from one support unit. Employees from target organizations have not been involved in the discussions. To improve the system evaluation, research could focus on conducting summative, naturalistic evaluations, i.e., evaluating the proposed system in real-world environments with various SMEs [60]. Third, this article analyzed the collaboration of support units and SMEs in the context of idea evaluation. However, external support for SMEs seems promising along the entire idea management process [27,52,61]. Consequently, future efforts could extend the list of design features to cover further idea management phases.

6 Conclusion

Digital transformation is undeniably one major topic for organizations regardless of size or industry. The continuous implementation of digital innovations in this transformational process is complex and cost intensive. Forced to embark on the digital transformation, SMEs implement projects to become more digital, despite their general lack of resources. Those who disregard the digital shift may become extinct.

Against this backdrop, both idea management and external support promise remedy for SMEs. However, observation from practice have revealed that idea management and evaluation in SMEs is yet in an early stage. In addition, external support units often rely on basic, non-digital tools to facilitate idea management. Literature has suggested how both concepts are effective in leveraging digital transformation success. Yet, contributions addressing the interplay of external support units and SMEs in the context of idea management, idea evaluation in particular, are scarce.

To address this practical problem and perceived research gap, this article proposes a collaborative idea management system involving both publicly-funded, external support units and SMEs. The system implementation is informed through 14 empirically derived and theoretically validated design features. An additional interaction concept encapsulates the newly generated design knowledge for future research and applications.

Through the application of the proposed idea evaluation system, two central matters are addressed. First, SMEs are equipped with a tool to overcome existing deficiencies in the context of managing and evaluating ideas. Second, support units may benefit from increased efficiency in their role as supporters. Often, such support units are publicly funded. Therefore, the funding initiatives may indirectly benefit from disseminating the proposed evaluation system. The newly created design knowledge captured as design features can be reviewed, revised, and reused in future research endeavors.

References

1. Aschehoug, S.H., Ringen, G.: Open innovation and idea generation in SMEs. In: Proceedings of the International Conference on Engineering Design, ICED, vol. 1, pp. 169–178, August 2013
2. Baez, M., Convertino, G.: Innovation cockpit: a dashboard for facilitators in idea management. In: Proceedings of the ACM Conference on Computer Supported Cooperative Work, CSCW, pp. 47–48 (2012)
3. Barann, B., Hermann, A., Cordes, A.K., Chasin, F., Becker, J.: Supporting digital transformation in small and medium-sized enterprises: a procedure model involving publicly funded support units. In: Proceedings of the 52nd Hawaii International Conference on System Sciences, HICSS, Wailea, Hawaii, pp. 4977–4986 (2019)
4. Barthel, P.: What is meant by digital transformation success? investigating the notion in IS literature. In: 16th International Conference on Wirtschaftsinformatik (2021)
5. Barthel, P., Fuchs, C., Birner, B., Hess, T.: Embedding digital innovations in organizations: a typology for digital innovation units. In: Proceedings of the 15th International Conference on Wirtschaftsinformatik, Potsdam, Germany, pp. 780–795, March 2020
6. Berghaus, S., Back, A.: Stages in digital business transformation: results of an empirical maturity study. In: Proceedings of the 10th Mediterranean Conference on Information Systems, MCIS 2016, Paphos, Cyprus (2016)
7. Binz, H., Reichle, M.: Evaluation method to determine the success potential and the degree of innovation of technical product ideas and products. In: 15th International Conference on Engineering Design, Melbourne, Australia (2005)
8. Boeddrich, H.J.: Ideas in the workplace: a new approach towards organizing the fuzzy front end of the innovation process. Creativity Innov. Manage. **13**(4), 274–285 (2004)
9. Boell, S.K., Cecez-Kecmanovic, D.: A hermeneutic approach for conducting literature reviews and literature searches. Commun. Assoc. Inf. Syst. **34**(1), 257–286 (2014)
10. Bothos, E., Apostolou, D., Mentzas, G.: Idea selection and information aggregation markets. In: 2008 IEEE International Engineering Management Conference, pp. 1–5. IEEE, Estoril, June 2008
11. Brem, A., Voigt, K.I.: Innovation management in emerging technology ventures - the concept of an integrated idea management. Int. J. Technol. Policy Manage. **7**(3), 304–321 (2007)
12. Brunswicker, S., Ehrenmann, F.: Managing open innovation in SMEs: a good practice example of a German software firm. Int. J. Ind. Eng. Manage. **4**(1), 33–41 (2013)
13. Chen, C.L., Lin, Y.C., Chen, W.H., Chao, C.F., Pandia, H.: Role of government to enhance digital transformation in small service business. Sustainability (Switzerland) **13**(3), 1–26 (2021)
14. Chen, J., Zhu, Z., Zhang, Y.: A study of factors influencing disruptive innovation in Chinese SMEs. Asian J. Technol. Innov. **25**(1), 140–157 (2017)
15. Chrisman, J.J., McMullan, W.E.: Outsider assistance as a knowledge resource for new venture survival. J. Small Bus. Manage. **42**(3), 229–244 (2004)
16. Crupi, A., et al.: The digital transformation of SMEs - a new knowledge broker called the digital innovation hub. J. Knowl. Manage. **24**(6), 1263–1288 (2020)

17. Dean, D.L., Hender, J.M., Rodgers, T.L., Santanen, E.L.: Identifying quality, novel, and creative ideas: constructs and scales for idea evaluation. J. Assoc. Inf. Syst. **7**(10), 646–699 (2006)
18. Dunstheimer, M.: Idea Management in technology development evaluation criteria for value proposition, technology and strategy. Ph.D. thesis, Halmstad University (2019)
19. Ebel, P., Bretschneider, U., Leimeister, J.M.: Leveraging virtual business model innovation: a framework for designing business model development tools. Inf. Syst. J. **26**(5), 519–550 (2016)
20. European commission: digital transformation scoreboard 2018 - EU businesses go digital: opportunities, outcomes and uptake. Tech. rep., Executive Agency for Small and Medium-sized Enterprises (2019)
21. Ferioli, M., Dekoninck, E., Culley, S., Roussel, B., Renaud, J.: Understanding the rapid evaluation of innovative ideas in the early stages of design. Int. J. Prod. Dev. **12**(1), 67–83 (2010)
22. Flynn, M., Dooley, L., O'Sullivan, D., Cormican, K.: Idea management for organisational innovation. Int. J. Innov. Manage. **07**(04), 417–442 (2003)
23. Gebauer, M., Tangour, C., Zeitschel, D.: Digital business model innovation in SMEs - case studies with DIH support from brandenburg (Germany). In: Bach Tobji, M.A., et al. (eds.) ICDEc 2020. LNBIP, vol. 395, pp. 155–165. Springer, Cham (2020). https://doi.org/10.1007/978-3-030-64642-4_13
24. Gerlach, S., Brem, A.: Idea management revisited: a review of the literature and guide for implementation. Int. J. Innov. Stud. **1**(2), 144–161 (2017)
25. Goerzig, D., Bauernhansl, T.: Enterprise architectures for the digital transformation in small and medium-sized enterprises. In: Proceedings of the 11th CIRP - Conference on Intelligent Computation in Manufacturing Engineering, Gulf of Naples, Italy, pp. 540–545 (2018)
26. Gollhardt, T.: Towards business model tools for SMES - knowledge boundaries in business model innovation projects. In: Proceedings of the 17th International Conference on Wirtschaftsinformatik, WI 2022, Nürnberg, Germany [virtual conference] (2022)
27. Gollhardt, T., Hermann, A., Cordes, A.K., Barann, B., Kruse, P.: Design of a software tool supporting orientation in the context of digital transformation. In: Proceedings of the 55th Hawaii International Conference on System Sciences, HICSS 2022, Maui, HI [virtual conference], pp. 4859–4868 (2022)
28. Gregor, S., Chandra Kruse, L., Seidel, S.: Research perspectives: the anatomy of a design principle. J. Assoc. Inf. Syst. **21**(6), 1622–1652 (2020)
29. Guimarães, L.G.d.A., Blanchet, P., Cimon, Y.: Collaboration Among Small and Medium-Sized Enterprises as Part of Internationalization: A Systematic Review, December 2021
30. Hermann, A., Gollhardt, T., Cordes, A.K., Kruse, P.: PlanDigital: a software tool supporting the digital transformation. In: Chandra Kruse, L., Seidel, S., Hausvik, G.I. (eds.) Proceedings of the 16th International Conference on Design Science Research in Information Systems and Technology, DESRIST 2021, Kristiansand, Norway [hybrid conference], pp. 356–361 (2021)
31. Herrmann, T., Binz, H., Roth, D.: Necessary extension of conventional idea processes by means of a method for the identification of radical product ideas. In: Proceedings of the International Conference on Engineering Design, ICED, vol. 8 (2017)
32. Hess, T., Benlian, A., Matt, C., Wiesböck, F.: Options for formulating a digital transformation strategy. MIS Q. Exec. **15**(2), 123–139 (2016)

33. Hevner, A., March, S.T., Park, J., Ram, S.: Design science in information systems research. MIS Q. **28**(1), 75 (2004)
34. Hevner, A.R.: A three cycle view of design science research. Scand. J. Inf. Syst. **19**(2), 87–92 (2007)
35. Hönigsberg, S., Dias, M., Dinter, B.: Design principles for digital transformation in traditional SMEs - an antipodean comparison. In: Chandra Kruse, L., Seidel, S., Hausvik, G.I. (eds.) DESRIST 2021. LNCS, vol. 12807, pp. 375–386. Springer, Cham (2021). https://doi.org/10.1007/978-3-030-82405-1_36
36. Khastehdel, M., Mansour, S.: Developing a dynamic model for idea selection during fuzzy front end of innovation. In: 7th International Conference on Industrial Technology and Management (ICITM), pp. 78–82. IEEE, Oxford (2018)
37. Kohli, R., Melville, N.P.: Digital innovation: a review and synthesis. Inf. Syst. J. **29**(1), 200–223 (2019)
38. Lu, C., Yu, B.: The effect of formal and informal external collaboration on innovation performance of SMEs: evidence from China. Sustainability (Switzerland) **12**(22), 1–21 (2020)
39. Matt, C., Hess, T., Benlian, A.: Digital transformation strategies. Bus. Inf. Syst. Eng. **57**(5), 339–343 (2015)
40. Messerle, M., Binz, H., Roth, D.: Elaboration and assessment of a set of criteria for the evaluation of product ideas. In: International Conference on Engineering Design, Seoul, Korea (2013)
41. Mikelsone, E., Liela, E.: Literature review of idea management: focuses and gaps. J. Bus. Manage. **9**, 107–121 (2015)
42. Mikelsone, E., Spilbergs, A., Segers, J.P.: Benefits of web-based idea management system application. Euro. J. Manage. Issues **29**(3), 151–161 (2021)
43. Milder, R.: The effect of anonymity in idea evaluation. Ph.D. thesis, Rotterdam School of Management (2019)
44. Mole, K., North, D., Baldock, R.: Which SMEs seek external support? business characteristics, management behaviour and external influences in a contingency approach. Eviron. Plann. C. Gov. Policy **35**(3), 476–499 (2017)
45. Möller, F., Guggenberger, T.M., Otto, B.: Towards a method for design principle development in information systems. In: Hofmann, S., Müller, O., Rossi, M. (eds.) DESRIST 2020. LNCS, vol. 12388, pp. 208–220. Springer, Cham (2020). https://doi.org/10.1007/978-3-030-64823-7_20
46. Müller, E., Hopf, H.: Competence center for the digital transformation in small and medium-sized enterprises. In: FAIM 2017–27th International Conference on Flexible Automation and Intelligent Manufacturing, Modena, Italy, vol. 11, pp. 1495–1500 (2017)
47. Murah, M.Z., Abdullah, Z., Hassan, R., Bakar, M.A., Mohamed, I., Amin, H.M.: Kacang cerdik: a conceptual design of an idea management system. Int. Educ. Stud. **6**(6), 178–184 (2013)
48. Orji, C.I.: Digital business transformation: towards an integrated capability framework for digitization and business value generation. J. Glob. Bus. Technol. **15**(1), 47–57 (2019)
49. Reis, J., Amorim, M., Melão, N., Matos, P.: Digital transformation: a literature review and guidelines for future research. In: Rocha, Á., Adeli, H., Reis, L.P., Costanzo, S. (eds.) WorldCIST'18 2018. AISC, vol. 745, pp. 411–421. Springer, Cham (2018). https://doi.org/10.1007/978-3-319-77703-0_41
50. Sandström, C., Björk, J.: Idea management systems for a changing innovation landscape. Int. J. Prod. Dev. **11**(3–4), 310–324 (2010)

51. Sassanelli, C., Terzi, S., Panetto, H., Doumeingts, G.: Digital innovation hubs supporting smes digital transformation. In: 2021 IEEE International Conference on Engineering, Technology and Innovation, ICE/ITMC 2021 - Proceedings, pp. 1–8 (2021)
52. Stevanović, M., Marjanović, D., Štorga, M.: Decision support system for idea selection. In: International Design Conference, Dubrovnik, Croatia, pp. 1951–1960 (2012)
53. Stevanovic, M., Marjanovic, D., Storga, M.: A model of idea evaluation and selection for product innovation. In: Proceedings of the International Conference on Engineering Design, ICED, Milano, Italy (2015)
54. Stich, V., Zeller, V., Hicking, J., Kraut, A.: Measures for a successful digital transformation of SMEs. In: Procedia CIRP, vol. 93, pp. 286–291. Elsevier B.V. (2020)
55. Szopinski, D., Schoormann, T., John, T., Knackstedt, R., Kundisch, D.: How software can support innovating business models: a taxonomy of functions of business model development tools. In: AMCIS 2017 - America's Conference on Information Systems: A Tradition of Innovation 2017-Augus (Recker 2012) (2017)
56. Thom, N.: Idea management in Switzerland and Germany: past, present and future. Die Unternehmung **69**(3), 238–254 (2015)
57. Tremblay, M.C., Hevner, A.R., Berndt, D.J.: Focus groups for artifact refinement and evaluation in design research. Commun. Assoc. Inf. Syst. **26**, 27 (2010)
58. Van Goolen, R., Evers, H., Lammens, C.: International innovation labs: an innovation meeting ground between SMEs and business schools. Proc. Econ. Finan. **12**, 184–190 (2014)
59. Vandenbosch, B., Saatcioglu, A., Fay, S.: Idea management: a systemic view. J. Manage. Stud. **43**(2), 259–288 (2006)
60. Venable, J., Pries-Heje, J., Baskerville, R.: FEDS: a framework for evaluation in design science research. Eur. J. Inf. Syst. **25**(1), 77–89 (2016)
61. Wagner, S., Bican, P.M., Brem, A.: Critical success factors in the front end of innovation: results from an empirical study. Int. J. Innov. Manage. **25**(04), 2150046 (2021)
62. Wessel, L., Baiyere, A., Ologeanu-Taddei, R., Cha, J., Blegind Jensen, T.: Unpacking the difference between digital transformation and it-enabled organizational transformation. J. Assoc. for Inf. Syst. **22**(1), 102–129 (2021)
63. Wiesböck, F., Hess, T.: Digital innovations: embedding in organizations. Electron. Mark. **30**(1), 75–86 (2020)

COVID-19 AI Inspector

Carlos Alexander Jarquin[1], Ryan Collin De Leon[2], and Yung-Hao Wong[1(✉)]

[1] Minghsin University of Science and Technology, 30401 Xinfeng, Hsinchu County, Taiwan
yvonwong@must.edu.tw
[2] Adamson University, 900 Ermita, Manila, Philippines

Abstract. In this paper, we aimed to aid the control measure that is implemented during the COVID-19 in Taiwan. As the virus spreads rapidly throughout the world, the Taiwanese government imposed three restrictions that help Taiwan to control the spread immediately. One of the restrictions that they imposed is to always wear a face mask. To avoid economic breakdown and still consider the general health of the public, Taiwan limits mass gatherings like in the food industry, entertainment, public transport, religious activities, etc. To be able to increase health security during a mass gathering, we developed an AI software to be able to detect people who are properly wearing a face mask, improperly wearing, and not wearing at all. The data that we used is from Kaggle to be able to use and process the data during image recognition, we use a raspberry pi board and camera. With the algorithm we used; we came up with an outstanding system where we could present excellent results due to the detection accuracy.

Keywords: Face mask · Detection · Raspberry Pi · Image recognition

1 Introduction

When World Health Organization (WHO) declared COVID-19 as a pandemic in December 2019 which started in Wuhan, China, and rapidly spread across other countries and continents, every nation has no choice but to limit the movement of people entering their countries due to the rate of spreading, the vast amount of infected people, and a huge sum of deaths in the world. Whereas many countries were preventing the spread of the disease by declaring lockdown, closing borders, etc. Taiwan decided to execute its prevention such us: (1) restrict massive gatherings since this would spread the virus with a mind-boggling speed, (2) encourage people to keep social distance everywhere to avoid any contact with infected people, and (3) wear face masks always.

In the middle of 2020, the Taiwan government lifts restrictions on massive gatherings to prevent economic breakdown because many companies were harming their businesses. However, to keep the country safe and at a minimum risk, everyone needs to always wear face masks especially in the event of massive gatherings such as in movie theaters, train and bus stations, conference areas, colleges, parks, shopping malls, etc. To have utmost health security during mass gatherings, we develop AI-Inspector to help find people who violate the third rule as stated above. An AI algorithm which its main objective is to take pictures as evidence of those violators that were not following the rules accordingly. With

F. Fui-Hoon Nah and K. Siau (Eds.): HCII 2022, LNCS 13327, pp. 41–55, 2022.
https://doi.org/10.1007/978-3-031-05544-7_4

this technology, we can be able to secure the health of everyone during mass gatherings with the ease of continuously monitoring the event.

2 Convolutional Neural Network

Artificial Intelligence is a wide-running tool that empowers every person to reorganize how we incorporate the data, analyze it, and utilize the output insights to enhance decision making and now is changing all social status [1].

A CNN is basically an algorithm which takes any type of images as input, these images could be related to any field such us, animals, people, object, plants, etc. The images will start going throughout the process in which it detects the differences between each image. This algorithm could some time to do its task because it will have to take the input image and divided in multiple pieces to start analyzing every part of it, so it throws the right output. This Deep Learning algorithm with enough training and input images could understand attributes as shown in Fig. 1 [1].

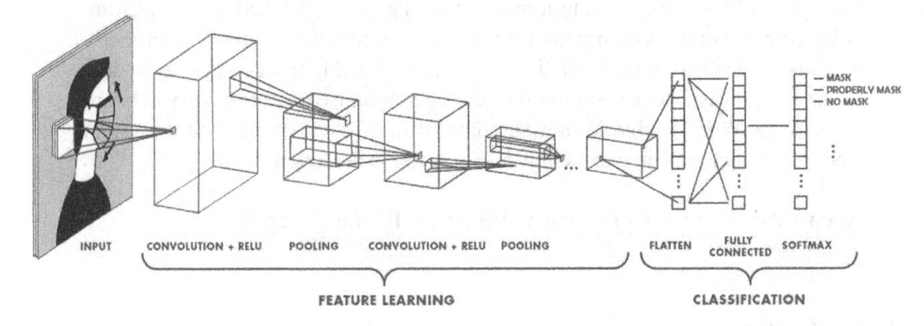

Fig. 1. Convolutional Neural Network

In Fig. 2 we see just a matrix of pixels; what we can do in this case is basically convert the 3 × 3 matrix to a 9 × 1 spatial vector and feedforward its linear algorithm that can be applied to binary classification tasks. However, this technique does not meet the main purpose since it could pop up a precision estimation score, on the other hand we

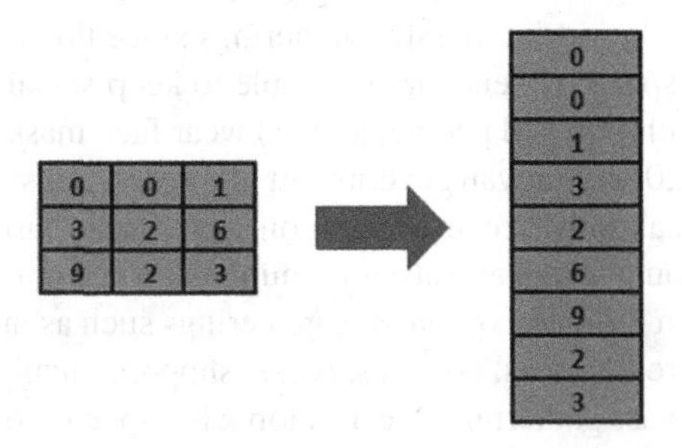

Fig. 2. Flatten matrix to vector

have an acceptable performance with low precision for complex images. This algorithm could perfectly catch up every pixel of the input image through the relevant filters [2].

2.1 Convolutional Layer

This is the most important layer of the Deep Learning algorithm because here is where the real computation is taking place. It order for this to happen it needs 3 main elements which are the input image, a linear filter, and the activation map [2].

Now, we will see an image divided by its three colors which are Blue, Green and Red as shown in Fig. 3. There are several images with different color scales such us RGB, HSV, RAL, CMYK, etc. You might an idea of how the computational process works deeply here and even more when there's an image with a huge dimension like 8000 pixels. Here the main task of the CNN is to lessen the input data into an image of less dimension like 300 × 300 pixels without losing the quality since this is important when we need to predict new images [2].

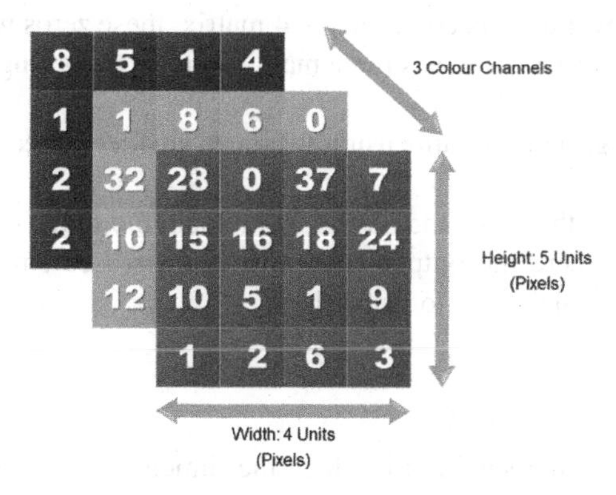

Fig. 3. Input image RGB

We can see in Fig. 4 that is not necessary to have each output value of the activation map connected to every single pixel belonging to the input data. The only method needed is to be connected to the responsive field, where the previous filter was applied.

Now, let's see the hyperparameter which interferes or affects the dimension of the output data, here is the list below.

1. Different dimensions of filters will distinguish different features in the input data which means that the final output will end up in different activation map, therefore is recommended to use filters such as 3 × 3, 5 × 5 or 7 × 7 for larger input data [2].
2. Stride is the number of movements made by the filters to the input images, these movements go across the whole images from left to right, top to bottom and most of time is thoroughly in height and width size. A higher stride triggers a reduced output. [2].

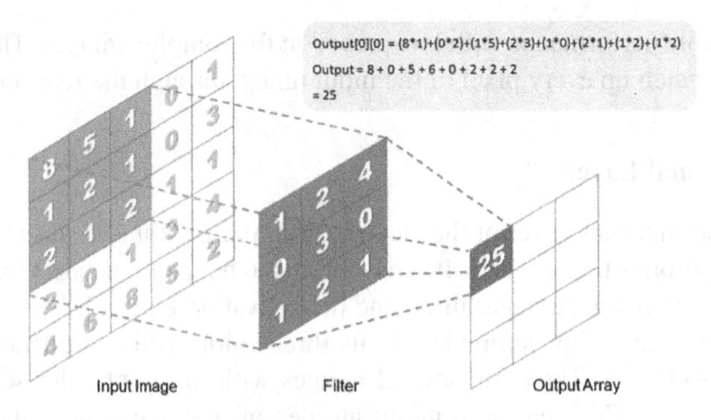

Fig. 4. Activation map.

3. Zero-padding is basically when we add additional filters since the input image doesn't fit. These values added to the input image are zeros, this produces a larger image, for example our original image is a 3 × 3 matrix, when we add an additional filter with zeros, the new image becomes a 4 × 4 matrix, these zeros will be added to the original image's borders. It exists three main paddings according to [3]:

   A. Valid padding, here the dimension becomes smaller or reduced where there's not padding.
   B. Same padding, the input and output will remain the same, no changes applied.
   C. Full padding, here the output sized will be increased since the input image received additional zeros to its borders.

2.2 Pooling Layer

Pooling layers as its name says is to reduce the dimensions of the activation maps. What it basically does is reducing the number of parameters to be learned and numerous computational processes during the training. It sweeps the whole input image, but there's a small change here, this filter doesn't include any weights [3].

Types of pooling

A. Max pooling: is the operation which will select the maximum elements from the area of the activation map which is enclosed by the filter. This operation contains the higher values features of the previous activation map.
B. Average pooling: this procedure involves calculating the average for each area of the activation map and will send it to the output array.

Neurons in a completely connected layer as shown is Fig. 5, connect to the entire activations coming from the prior layer which normal in common neural networks. Those activations can consequently be under a computing process using a matrix and go after a bias offset [3].

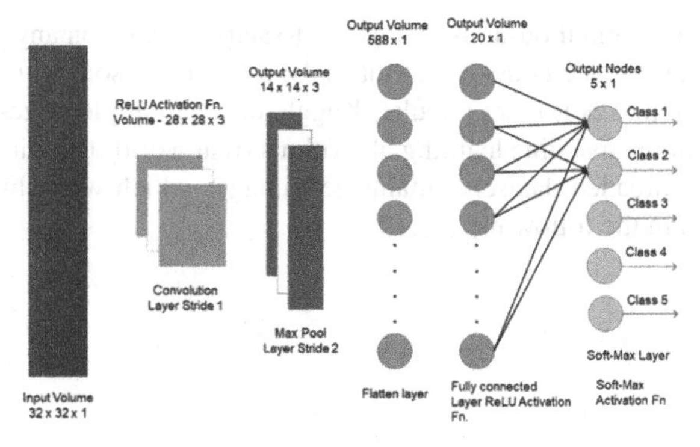

Fig. 5. Fully connected layer

3 Methodology

The whole project has been designed on a Raspberry Pi board and a Raspberry Pi camera since these were the most suitable equipment to execute the algorithm, the board to mount the Raspbian Operating System where we also installed other packages and libraries and the camera for visualizing (detection). There were some steps and processes to follow in order to achieve our goal as shown in Fig. 6.

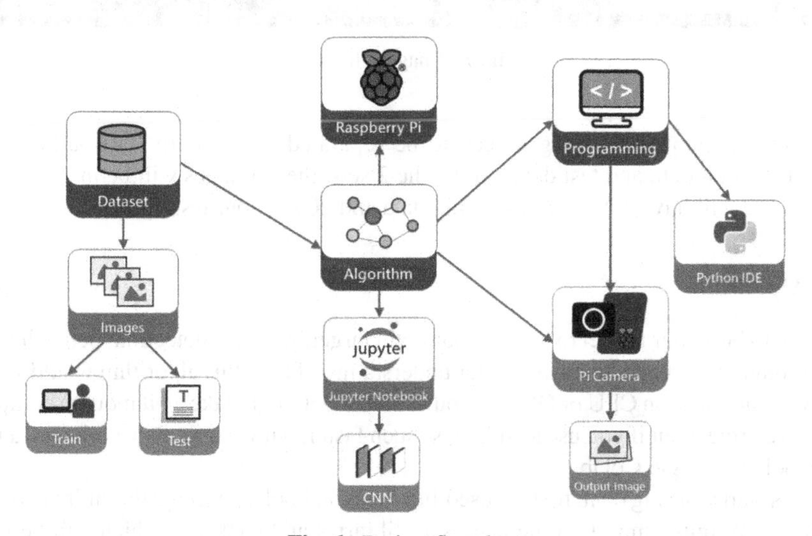

Fig. 6. Project flowchart

3.1 Dataset

This step was basically essential since we needed to start feeding the algorithm, so which was downloaded from Kaggle [4]. Kaggle is a website related to the whole Artificial Intelligence fields and subfields such us Machine Learning, Deep Learning and Computer

Vision. Here you can find thousands of datasets to start working on any project besides that has some data science courses and tutorials, and there's something important on this website which makes it more popular, Kaggle offers multiple prizes to solve easy, medium, and complex machine learning algorithms (real world scenarios).

Now, this downloaded dataset contains 853 images which were divided in into 3 classes, and they are the following: (Fig. 7)

- Mask
- No Mask
- Wear Improperly

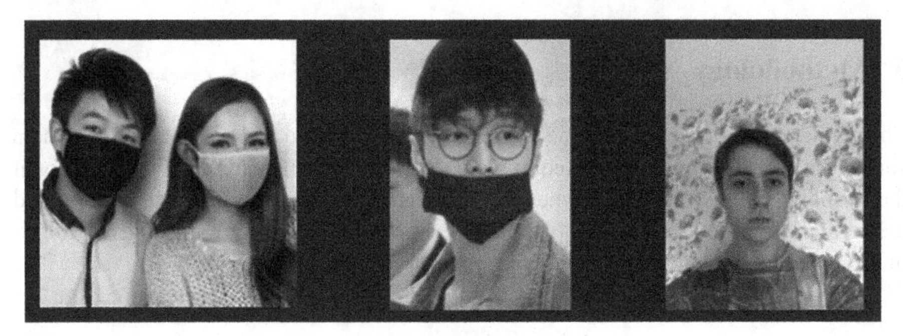

Fig. 7. Images classes

Now, this number of images need to be separated in 2 sets to start building the algorithm, train data and test data will be the 2 sets, these images will be in 2 different folders. We will have 80% as our training data and 20% as our testing data.

3.2 Algorithm

It's time to build our model to achieve our main target, which is detection, here I'll walk you through the process to have a better understating of how this algorithm would work, now you can run it on CPU or GPU, if you use a CPU it might take a bit more to compute the entire project but if you use a GPU it is much faster, but you can also work on Kaggle Notebook or Google Colab.

Let's start working on it, first we used Jupyter Notebook to execute the entire network (project). We must import all the necessary libraries and packages which will be used later as shown in Fig. 8.

```
#import the Libraries
import numpy as np
import pandas as pd
from bs4 import BeautifulSoup
import matplotlib.pyplot as plt
import os
from tensorflow.keras.preprocessing.image import ImageDataGenerator
from tensorflow.keras.applications import MobileNetV2
from tensorflow.keras.layers import AveragePooling2D
from tensorflow.keras.layers import Dropout,BatchNormalization
from tensorflow.keras.layers import Flatten
from tensorflow.keras.layers import Dense
from tensorflow.keras.layers import Input
from tensorflow.keras.models import Model
from tensorflow.keras.optimizers import Adam
from tensorflow.keras.applications.mobilenet_v2 import preprocess_input
from tensorflow.keras.preprocessing.image import img_to_array
from tensorflow.keras.preprocessing.image import load_img
from tensorflow.keras.utils import to_categorical
from sklearn.preprocessing import LabelEncoder
from sklearn.model_selection import train_test_split
from sklearn.metrics import classification_report
import matplotlib.pyplot as plt
import cv2
import random as rand
```

Fig. 8. Import libraries.

After we imported our libraries, we proceed defining the functions to perform multiple tasks, functions where will get the coordinates of the input images given in images file, those coordinates are lower left and upper right corner. Then images need to be imported in a list so the algorithm starts counting with them since they will be trained and tested, as shown in Fig. 9.

```
for i in range(853):
    file_image = 'maksssksksss'+ str(i) + '.png'
    file_label = 'maksssksksss'+ str(i) + '.xml'
    img_path = os.path.join("C:\\Users\cjarq\Desktop\Cagebot\images", file_image
    label_path = os.path.join("C://Users/cjarq/Desktop/Cagebot/annotations/", fi
    #Generate Label
    target,numobj = generate_target(i, label_path)
    targets.append(target)
    numobjs.append(numobj)
```

Fig. 9. Importing and listing images.

Once we finished the below procedure, we move the next part which is encoding and applying data augmentation on our images, the reason why we do it, is because the amount of images we have is minimum and we have to enlarge our dataset to train it better and have a better learning rate and make good prediction, now data augmentation is a method to increase the training data by applying certain amount of transformation to the original image, such us rotation, flip, translation, padding, scaling, cropping, etc. [5]. You can see how it works in a piece of code shown in Fig. 10.

The most beautiful part is coming, and here is where everything starts happening, all the convolution and computational processes, this is called CNN. Now, it might be a little bit confusing but let's see how we can break it down to make it understandable, let's start building our model. The programming in Fig. 11 shows how to build the CNN model.

```
#Encode the Labels in one hot encode form
lb = LabelEncoder()
labels = lb.fit_transform(face_labels)
labels = to_categorical(labels)
labels
```

```
#Perform data augmentation.
aug = ImageDataGenerator(
    zoom_range=0.1,
    rotation_range=25,
    width_shift_range=0.1,
    height_shift_range=0.1,
    shear_range=0.15,
    horizontal_flip=True,
    fill_mode="nearest"
    )
```

Fig. 10. Encoding and augmentation.

```
#define the model
baseModel = MobileNetV2(weights="imagenet", include_top=False,
    input_shape=(224, 224, 3))

# construct the head of the model that will be placed on top of the the base mod
headModel = baseModel.output
headModel = AveragePooling2D(pool_size=(7, 7))(headModel)
headModel = Flatten(name="flatten")(headModel)
headModel = Dense(256, activation="relu")(headModel)
headModel = Dropout(0.25)(headModel)
headModel = Dense(3, activation="softmax")(headModel)

# place the head FC model on top of the base model (this will become the actual
model = Model(inputs=baseModel.input, outputs=headModel)

# loop over all layers in the base model and freeze them so they will *not* be u
for layer in baseModel.layers:
    layer.trainable = False
```

Fig. 11. Building CNN model

MobileNetV2 returns an image classification model which has been loaded by the number of weights pretrained on the ImageNet, [6]. The input shape is basically data format which are width, height, and three input channels, then we move to the next part which is constructing the head of the model, applying max pooling, Flatten, Dense and Dropout. Flatten layers are used when a multidimensional comes up and you make to transform it to linear so Dense layer can receive it, a Dense layer exists when correlation between any feature to any random data feature in data point and finally Dropout oversees cutting the correlation between features by dropping the weights at a probability, [7].

We also needed to divide the training data and the testing data on the algorithm to avoid overfitting or underfitting, once this gets done, we proceed testing the model build to evaluate the network which are: the training loss and accuracy as shown in Fig. 12.

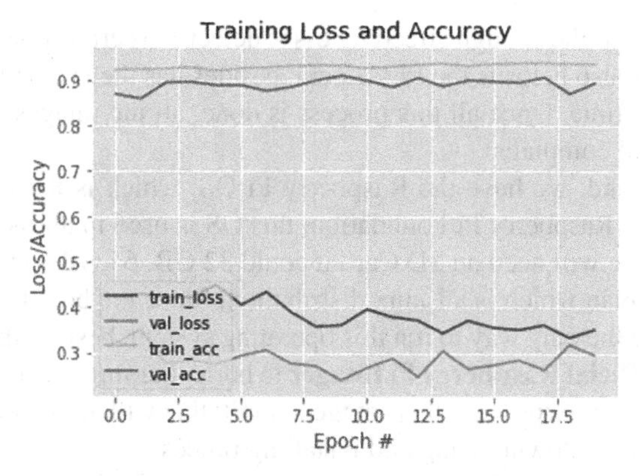

Fig. 12. Training loss and accuracy

Now, to finish up with the algorithm we finally save it.

3.3 Raspberry Pi

Raspberry Pi4 model B as shown in Fig. 13 was the board we used to execute software, this board is basically a desktop computer since the speed and performance is incredible amazing and step from the prior versions. With this new board you can perform a huge number of tasks, such as editing documents, using multiple browsers and open many tabs at the same time, juggling spreadsheet, execute programming software where demonstrates an outstanding performance. We had a great experience using this board because our tasks were done flawlessly, smoothly, and very recognizable. The best part of this board is that smaller, it ties anywhere, easy to use, more energy-efficient and much more affordable than any other boards in the market.

The specifications of this board are silent, energy-efficient which that it doesn't require too much energy as other boards do, the board has a fast networking since it comes with Gigabit Ethernet, Wi-Fi and Bluetooth, besides that it is also coming 4 USB ports two USB2 and two USB3, which has an exceptional transfer data ten times faster, and finally it has 8 GB of RAM [8].

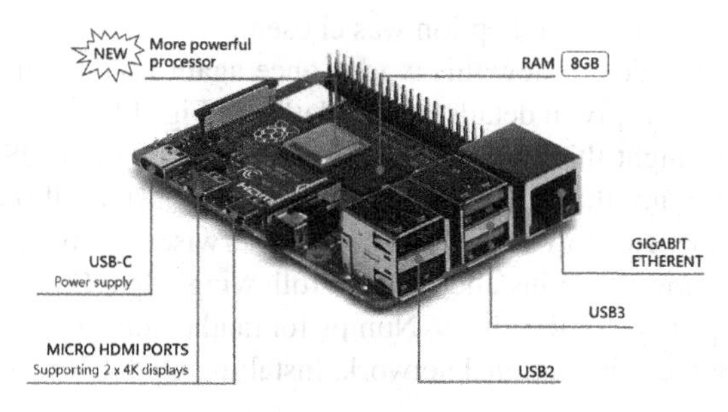

Fig. 13. Raspberry Pi board.

This algorithm has the enough effectiveness to detect to a certain distance and capture the images, what it also helps is the PI Camera, besides has the capacity to detect many people at the same time. Once all this process is done, all the images will be stored in our server (personal computer).

On the other hand, we have the Raspberry Pi OS, which is the official operating system provided by Raspberry Pi Foundation, this OS is used in all its boards. In order to install this OS, we will need an SD Card it could 32 GB, 64 GB, etc. By the way this OS running on Debian which is a Linux distribution. Since the board has a slot for this card, that would be the only way to run this operating system, besides that you will have to download the **Official Raspberry Pi Imager** to burn your.img file, it is recommended to use since it will skip steps that other imagers don't, this will automatically connect to the server and start the downloading and installing process.

Now that we have everything installed, we must proceed with the system configuration, such as date and time, Wi-Fi, username, password, adjust your desktop resolution, etc.

It is recommended to do some tests, to make sure your board is up and running, those tests could be simple tasks, one of them would be restarting your board, open browsers, check the CPU performance, check is the board is overheated or keeps its normal temperature.

If you have done everything, all the necessary tests on your board, so let me tell you that it would time to work on any project you want. This board well with computer vision projects, it has an incredible performance, and highly praised.

3.4 Programming

The most interesting part is coming, which is focus on our final output (detection), we might say, this was done already which is true but not completely, it was done but not on our Raspberry Pi board, since Jupyter Notebook is not supported by Raspberry PI and besides that because what we did was just trained our neural network and saved the model but now the thing is a bit different and I'll tell you why, after the model trained is built, we will use it for computer vision, to start detecting faces which is also made by Jupyter Notebook but at this time is not supported by the board OS so, therefore we will use Python programming language to use the model trained and work on our detection software, here you can choose your preferred IDE, could be PyCharm, Visual Studio Code, etc. In our case, the second option was chosen.

Let's have a better idea of how this works, once again I will walk you through the process but not going deeply in details, alright let's see Fig. 14.

In this part you might think that is just open vs code on your OS and keep going but let me tell that is not the case since we will have to install all these libraries and each of their dependencies to work on the board otherwise, we might encounter many errors, the main libraries to be installed are the following: OpenCV, Numpy, Keras and TensorFlow, etc. OpenCV for the vision, Numpy for mathematical procedures and Keras and TensorFlow for the convolutional network. Installing OpenCV takes around 2 to 3 h normally as shown is Fig. 15.

```
1   import cv2
2   import os
3   import sys
4   import logging as log
5   import datetime as dt
6   from time import sleep
7   from tensorflow.keras.preprocessing.image import img_to_array
8   from tensorflow.keras.models import load_model
9   from tensorflow.keras.applications.mobilenet_v2 import preprocess_input
10  import numpy as np
```

Fig. 14. Import libraries

```
[ 99%] Building CXX object modules/gapi/CMakeFiles/opencv_test_gapi.dir/test/own/conc_queue_tests.cpp.o
[ 99%] Building CXX object modules/gapi/CMakeFiles/opencv_test_gapi.dir/test/own/gapi_types_tests.cpp.o
[ 99%] Building CXX object modules/gapi/CMakeFiles/opencv_test_gapi.dir/test/own/last_written_value_tests.cpp.o
[ 99%] Building CXX object modules/gapi/CMakeFiles/opencv_test_gapi.dir/test/own/mat_tests.cpp.o
[ 99%] Building CXX object modules/gapi/CMakeFiles/opencv_test_gapi.dir/test/own/scalar_tests.cpp.o
[ 99%] Building CXX object modules/gapi/CMakeFiles/opencv_test_gapi.dir/test/render/ftp_render_test.cpp.o
[ 99%] Building CXX object modules/gapi/CMakeFiles/opencv_test_gapi.dir/test/render/gapi_render_tests_ocv.cpp.o
[ 99%] Building CXX object modules/gapi/CMakeFiles/opencv_test_gapi.dir/test/rmat/rmat_integration_tests.cpp.o
[ 99%] Building CXX object modules/gapi/CMakeFiles/opencv_test_gapi.dir/test/rmat/rmat_tests.cpp.o
[ 99%] Building CXX object modules/gapi/CMakeFiles/opencv_test_gapi.dir/test/rmat/rmat_view_tests.cpp.o
[ 99%] Building CXX object modules/gapi/CMakeFiles/opencv_test_gapi.dir/test/s11n/gapi_s11n_tests.cpp.o
[ 99%] Building CXX object modules/gapi/CMakeFiles/opencv_test_gapi.dir/test/s11n/gapi_sample_pipelines_s11n.cpp
.o
[100%] Building CXX object modules/gapi/CMakeFiles/opencv_test_gapi.dir/test/streaming/gapi_streaming_sync_tests
.cpp.o
```

Fig. 15. OpenCV installation.

After installing the necessary libraries and dependencies, we move to the next step. The next move would be capture the image, by activating the camera of the board, we will get there, no rush, after this we will have to capture the input image frame by frame for scaling the color, later on we define a variable to save the output image, once this is done rapidly step into the phase which is label the images with a bounding box, colors will vary depending on the task, if they wear mask, improperly wearing or not wearing at all, I'll show it in Fig. 16.

We finally got to the last step which throwing the result output, where we expect a person always wearing a mask. This is basically the whole code for this result. In order to execute the program, we must do it from the same IDE VS Code, or we can do it from the terminal.

3.5 Pi Camera

This package is used to enable to the usage of the Raspberry Pi Camera and provides a better interface to its module for Python 2.7 or Python 3.8 (or above).

All current Pi boards have a port for connecting the Camera Module, in order to connect the camera to the board, your board must be off, then you connect the pi camera to the board's slot, once is done, the next step would be to turn on the board and check the configuration system, later in preferences click on Raspberry Pi Configuration go to interfaces and enable Camera, you finally need to reboot the board.

```
# determine the class label and color we'll use to draw
# the bounding box and text
if (mask > withoutMask and mask > notproper):
    label = "Without Mask"

elif ( withoutMask > notproper and withoutMask > mask):
    label = "Mask"

else:
    label = "Wear Mask Properly"
if label == "Mask":
    color = (0, 255, 0)
if label=="Without Mask":
    color = (0, 0, 255)
else:
    color = (255, 140, 0)

label = "{}: {:.2f}%".format(label, max(mask, withoutMask, notproper) * 100)
cv2.putText(frame, label, (x, y- 10),
            cv2.FONT_HERSHEY_SIMPLEX, 0.45, color, 2)

cv2.rectangle(frame, (x, y), (x + w, y + h),color, 2)
```

Fig. 16. Labeling images with bounding box.

Now if you want to test the camera is simple, just open the terminal and type the following command:

– **raspistill -o Desktop/image.jpg**

On the other hand, you might want to check the camera using your Python IDE, so follow these basic lines of codes as describe below:

```
from picamera import PiCamera
from time import sleep

cam = PiCamera()

cam.start_preview()
sleep(2)
cam.stop_preview()
```

Next action will be running the program and see the output.

This works the same as it does in our project, where we capture images and having a live video for detection, this camera has been exceptionally useful for this project because helped us with a good image definition and great performance. Next, you'll see the results.

4 Results and Discussion

It's time to see how well the algorithm or software works in real life, ever since this is our main goal on this project. Let's in the following figures the performance accuracy

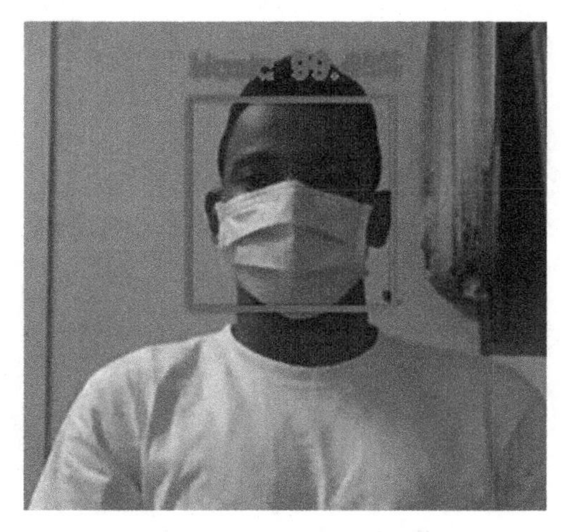

Fig. 17. Mask on

of every action (wearing mask, improperly wearing, and not wearing at all) as shown in Fig. 17.

As you can see the accuracy is well performed, even though, we could say this is overfitting but depending on the movements and distance the accuracy might change and could be lower than the one above.

Now, we will see Fig. 18, where a person is improperly wearing mask, let's look on the performance.

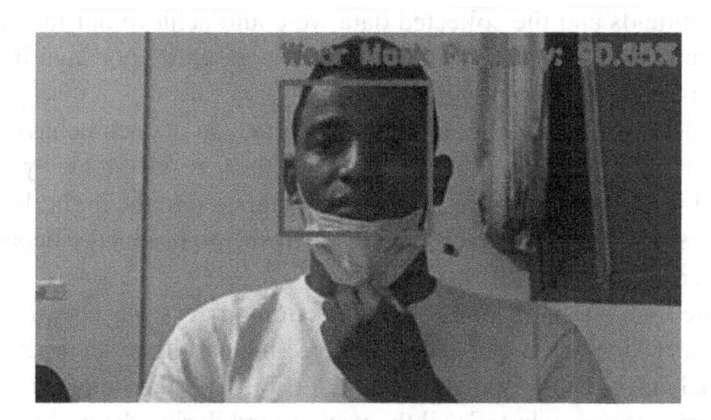

Fig. 18. Improperly wearing mask

In this figure we can see that the accuracy changes and is a bit lower than the one from Fig. 17, so once again, this accuracy might vary due to the movements and distance of the people and the darkness or the environment.

Finally, not wearing mask behavior, which is determinant prohibited in many countries and Taiwan is not the exception, you have to wear masks at all times, no matter where you are, even though there are few scenarios where this rule doesn't fit, alright, let's take a look on Fig. 19.

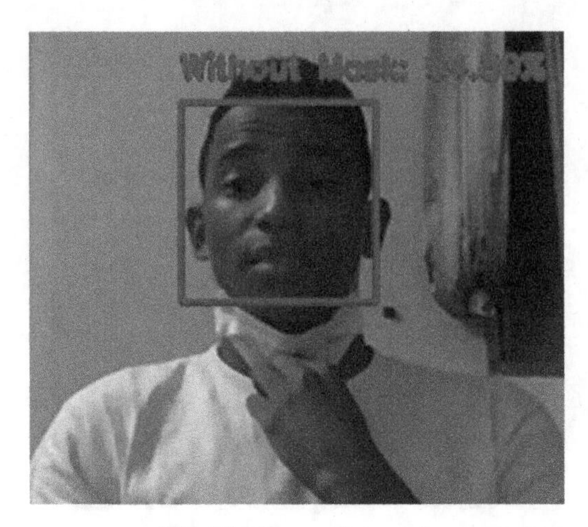

Fig. 19. No wearing mask

This picture shows an accuracy of 94.59% which I consider good enough. It might also change as the others since if there's a person too far the camera will not be able to detect it and if it's the case the accuracy will be low or even confuses, what I mean is it might detect one of the 3 actions or scenarios (wearing mask, improperly wearing, and not wearing at all).

After we performed all these tasks by our software, we found out how important technology has changed in a positive way, with this Facemask detection algorithm we built, we can avoid the spreading of this ongoing pandemic, thanks to Artificial Intelligence and its subfields and the collected data we could achieve our target. In fact, the software performs excellent when we have public area with good lighting, it can also detect multiple people at the same time.

Contrariwise, if we carefully look on the performance of each behavior there is not a big difference between them, and it is evident that these results made by our face mask detection was significantly better during the day compared to the night. However, all of them have almost the same accuracy and all of them might change depending on the situation, place, day, night, etc., [9].

Although the final output results on the detection process was adequate and reasonable, it was observed that the accuracy on the detecting part loses accuracy over the time and became unclear particularly during the night. Initially the images detected during the day had a great accuracy but it's all the way around during the night. Therefore, the color somehow interferes some loss or change during the ongoing detection [9].

5 Conclusion

The purpose of this paper was to identify efficient approaches to fight against COVID 19 pandemic and absolutely with individuals not following the rules established by the government. Based on the research analysis conveyed there were many people not doing the right usage of the masks, therefore it can be concluded that there are different ways to approach and improve this behavior. Nevertheless, future investigations on this matter with more data and techniques could end up as the most convenient way to tackle this ongoing situation. This could improve and save many lives if we all start cooperating and doing much more research.

References

1. Saha, S.: A Comprehensive Guide to Convolutional Neural Networks—the ELI5 way. Towards Data Science, 16 December 2018. https://towardsdatascience.com/a-comprehensive-guide-to-convolutional-neural-networks-the-eli5-way-3bd2b1164a53
2. IBM Cloud Education: Convolutional Neural Networks. IBM, 20 October 2020. https://www.ibm.com/cloud/learn/convolutional-neural-networks
3. Github: Convolutional Neural Networks for Visual Recognition (n.d.). https://cs231n.github.io/convolutional-networks/. Retrieved 11 Feb 2022
4. Larxel: Face Mask Detection. Kaggle, 22 May 2020. https://www.kaggle.com/andrewmvd/face-mask-detection
5. Data Augmentation. Tensorflow Core. TensorFlow. https://www.tensorflow.org/tutorials/images/data_augmentation. Accessed 15 Feb 2022
6. Team, Keras: Keras Documentation: Mobilenet and mobilenetv2. Keras. https://keras.io/api/applications/mobilenet/. Accessed 15 Feb 2022
7. Gak, Kiritee: When to Use Dense, Conv1/2D, Dropout, Flatten, and All the Other Layers? Data Science (blog), 17 Jan 2019. https://datascience.stackexchange.com/questions/44124/when-to-use-dense-conv1-2d-dropout-flatten-and-all-the-other-layers
8. Raspberry Pi: Buy A Raspberry Pi 4 Model B. Raspberry Pi. https://www.raspberrypi.com/products/raspberry-pi-4-model-b/?variant=raspberry-pi-4-model-b-8gb. Accessed 15 Feb 2022
9. McMorrow, M.: Writing Results and Discussion Chapters for Quantitative Research. SlideShare, 11 Sept 2018. https://www.slideshare.net/martinmcmorrow/writing-results-and-discussion-chapters-for-quantitative-research

Leveraging Human and Machine Capabilities for Analyzing Citizen Contributions in Participatory Urban Planning and Development: A Design-Oriented Approach

Gerrit C. Küstermann(✉) and Eva A. C. Bittner

Universität Hamburg, Hamburg, Germany
{gerrit.kuestermann,eva.bittner}@uni-hamburg.de

Abstract. Local authorities are increasingly using online platforms as a means of involving citizens in urban planning and development. However, they encounter several challenges in the analysis and further processing of citizen-generated data when participation occurs on a large scale. Manual evaluation in particular takes individuals to their limits and hampers value extraction, e.g., for better planning and decision-making. To address these challenges, this paper presents a concept and design guidance based on elicited design requirements of an intelligent system that augments officials in data analytical tasks. Therefore, we focus on a hybrid solution to combine the respective strengths of humans and machines in data processing to mutually overcome their respective weaknesses. Particularly, our solution approach integrates human knowledge in the training process for machine learning via a Human-in-the-Loop strategy and simultaneously facilitates participation data analysis performed by officials. Promising initial evaluation results of the prototype indicate the usefulness of the approach.

Keywords: Citizen participation · Urban planning · HITL · Design science

1 Introduction

Citizen participation is today understood as an essential component of sustainable urban planning and development (UPD). As key actors in urban areas, citizens can and should participate in the development of their living environment for several reasons, including legitimacy and learning from local knowledge [1, 2]. To this end, local authorities are increasingly using digital participatory platforms [3] as a means of involving citizens in UPD [4–6]. Thousands of contributions from citizens are no exception when user-friendly participatory platforms using technological and social tools are applied for engagement [7]. However, authorities face several challenges when participation reaches such a scale [3]. In practice, the processing of citizen-generated data, usually contributions in text form, which vary greatly in quality, expression, and lengths, is becoming increasingly cumbersome for the responsible public officials, employees, and planners, as they usually have to analyze the contributions manually [8, 9] (hereafter, we use only

F. Fui-Hoon Nah and K. Siau (Eds.): HCII 2022, LNCS 13327, pp. 56–72, 2022.
https://doi.org/10.1007/978-3-031-05544-7_5

the term "officials", but we explicitly refer to all persons who otherwise carry out public official functions and participate in corresponding tasks, which are detailed below). Searching for meaningful information and patterns to gain insights from these data is a key task that requires organizing contributions logically depending on the process, its objective, and topic. As the amount of information proliferates, the effort required for an individual to accomplish such analytical tasks without assistance quickly becomes onerous or simply unfeasible [10]. However, research acknowledges that individuals are willing to adopt information technologies (IT) to reduce their cognitive workload and enhance their task-related efficiency [11].

In this context, text mining techniques have been proposed to facilitate the processing of citizen-generated data [12], but have hardly been used in practice in the public sphere so far [13], as it has been shown that exclusively machine-based approaches do not yield entirely satisfactory results [14]. This is partly because algorithmic systems require a great amount of high-quality training data [15]. Citizen contributions, however, are often textually flawed [16], inherently unstructured, and multifaceted as they are influenced by a variety of features as a function of the participation process and particular planning phase [2]. It requires profound contextual experiential knowledge and competencies to organize these contributions and extract the information relevant to further decision-making and planning. Algorithms do not yet provide such cognitive human capabilities.

Against this backdrop, the combination of human knowledge and superior cognition (e.g., common sense, flexibility, and transferability) with the performance and precision advantages of machine processing of large data sets and pattern recognition promises to mutually overcome the respective weaknesses [15, 17]. Following that rationale, the underlying idea of this paper is to design an intelligent system that officials can interact with to combine their respective strengths with the capabilities of artificial intelligence (AI), i.e., in the context of this study natural language processing (NLP) and machine learning (ML) technology. Therefore, we ask the question: *How should a system be designed and instantiated that officials can use to effectively analyze participation data in UPD, and that leverages both human and machine intelligence?* By addressing this research question, this study aims to contribute prescriptive knowledge to research and bridge the gap to the outlined practical needs [18]. As such, scholars stressed the need for design guidance considering new means and affordances created by digital technologies in the practice of citizen participation in institutional UPD [19, 20].

In the following, after outlining the background of the study, the applied research approach is described. Then, the concept of the system design is introduced. The subsequent section presents the prototypical instantiation of the system and its evaluation. The ensuing discussion of the study outcomes ends with an outlook for future research and a conclusion.

2 Study Background

In UPD, public authorities use online participation platforms to gather citizen knowledge and incorporate it into government enabled urban planning and decision-making processes that aim to involve a large and diverse group of people [1, 4–6]. The overall surge in popularity of such platforms, also referred to as digital participatory platforms

[3] or digital co-production platforms [5], can be attributed to technological advances and cultural shifts [21–24] and has received a further boost not least because of the COVID-19 pandemic. On such online platforms, citizens can engage in discourse and a co-productive process among each other and authorities to create sustainable and beneficial solutions for their living and working environment in which they have a stake [2, 25]. In particular, this enhances opportunities for citizens to contribute their general and local knowledge to concrete urban development projects and issues based on their interests and dispositions, e.g., in the form of ideas, criticism and suggestions for improvement [4, 26, 27].

While the proliferation of online participation can generally be considered a goal under the e-government umbrella concept [23], it also places new demands on government and administrative work and processes related to citizen-generated data in participatory UPD. That is, these inputs from citizen are usually obtained in large quantities as raw textual data with varying quality and need to be preprocessed before they can be used further [28]. More specifically, participation data are often unstructured and multifaceted as they are influenced by a variety of features depending on the participatory UPD process. In fact, UPD projects are collaborative undertakings [29], which vary widely in terms of their objective and underlying planning data, time span, scope, complexity, and level of participation (cf. e.g., [30]). Simultaneously, the local institutional and non-institutional actors involved vary greatly [5, 19], e.g., the elected officials, district administrators, professional experts, such as urban planners, designers, and architects, as well as citizen participants, and influence the type and quality of the evolving participation data.

At the administrative level, one of the key activities related to this participation data in UPD processes is to extract value from it based on a comprehensive analysis usually drawing on the professional and experiential knowledge of officials. Such convergent tasks have been thoroughly studied in research within different contexts, but are consistent in terms of problem definition, i.e., they are characterized by increasing complexity and time consumption, and the required effort to complete them may exceed the cognitive abilities of individuals [31–35]. These tasks usually necessitate that data is organized, e.g., into logical clusters and class labels are assigned, to enable further decision-making [10, 36]. As the amount of data and information increases greatly, however, the effort for an individual to execute this task without assistance quickly becomes costly or simply infeasible [10]. This can lead to individuals perceiving the task as overly complex, as they are unable to relate all the information to make an informed decision. They consequently, accomplish the task less efficiently and exhibit frustration [11, 33, 37, 38].

Scholars found, nonetheless, that individuals are willing to adopt IT systems in decision contexts to lower their cognitive workload and enhance their task-related efficiency [11]. Others highlighted that the realization of efficiencies associated with the application of algorithmic decision support has led users to prefer algorithmic guidance over human guidance [39, 40]. Generally, advanced machine-based approaches from the field of AI are becoming more prevalent in various data analytic tasks [41]. In this light, intelligent, i.e., learning, algorithms that "rely on large data sets to generate responses, classifications, or dynamic predictions" are increasingly applied to tasks that were previously done primarily by humans [42, p. 62, as quoted in 36]. However, contrary to what the terminology might suggest, AI/ML technology is often not completely independent of human interventions. That is, most of today's ML applications are supervised processes that rely on human feedback and need training beforehand. Therefore, strategic approaches, such as „Human-in-the-Loop“ (HITL)-ML, are adopted to integrate human and machine intelligence in applications that utilize AI [43].

The focus of HITL is to involve humans in the iterative training process of the machines and allow them to interactively engage with their human knowledge in model building [44], e.g., by assigning labels to unmarked data and correct the model's predictions (e.g., by tagging correct and incorrect labels). Regardless of the ML technique applied, a HITL process includes three essential features: (1) a data based prediction, which can be evaluated by a human, (2) validation of the prediction, which serves as feedback, and (3) backpropagation of the feedback into the model. Accordingly, HITL ML processes are especially relevant when the data is unlabeled, full automation is difficult (e.g., requires contextual knowledge), or the data is constantly changing and evolving, as is the case, e.g., in the environment of participatory UPD [43]. The objective of HITL is thus to build a ML model to accomplish future tasks. Simultaneously, such HITL-based interactive system can support humans in the current task completion [44]. This reciprocal process leads to additional benefits arising from combining the complementary strengths of humans and machines (cf., e.g., [15]), "enabling mutual learning and multiplying their capabilities" [17, p. 193].

Considering the task of analyzing complex participation data to extract the information relevant for further processing, i.e., in order to be able to integrate these into the next participation and planning phases in UPD, which is, as previously stated, a cognitively demanding effort for officials, an interactive system design approach based on HITL ML and NLP is reasonable. This will allow officials to contribute their competencies to the data analytic process while relying on the intelligence of the machine.

3 Research Approach

We follow the information systems (IS) DSR approach [45, 46]. With regard to IS DSR, the focus is on the development of novel and useful IT artifacts. In IS DSR, novel IT artifacts include constructs, models, methods, and instantiations [47]. Novelty lies at the juncture of artifact design as well as its application [48]. Specifically, we aim to develop a novel and useful IT artifact in the application domain of IT-enabled participatory UPD, which is also referred to as e-planning and e-participation [46, 49–51].

In our research approach, we take the DSR three-cycle view, which comprises the relevance, design, and rigor cycle [45]. In the relevance cycle, we identified a set of unsolved problems regarding the effective analysis of citizens' textual input from participation processes in UPD. This problem identification is grounded on extant research and interaction with domain experts. In the design cycle, these real-world problems drive the development of our IT artifact. Our artifact is the concept and a prototypical instantiation of an intelligent system that assists officials in analyzing citizen contributions from participatory UPD and integrates human knowledge into the ML process through human-system interaction. For artifact construction, we first derive explicit prescriptions (e.g., principles of form and function [52, 53]) in the mode of design requirements (R) and then derive design principles (P) to provide knowledge for creating instances of artifacts in the same class [54–56]. For this purpose, we follow a scheme capturing the sides of "the implementer, who instantiates the abstract specification of the design principle, and the user, who enacts that instantiation to bring about a goal" [56, p. 1637].

To ensure a design that meets user needs and objective requirements, early involvement of experts from public sector agencies with several years of professional experience was sought (see Table 1). During the conceptualization phase, these requirements for the development of the system were elicited and organized through an iterative reasoning process of deduction and induction based on literature from the applicable knowledge base and domain knowledge conveyed through informal discussion-based expert interviews [57]. Throughout the design process, the formative evaluation of the implemented design features together with the expert group focused on artifact improvement [58]. Further, to evaluate the prototypical instantiation of our artifact, we conducted three online exploratory focus groups (EFG) following [59]. A total of 18 students (10 female, 8 male participants) enrolled in master's degree programs participated in the EFG. Since all participants had basic knowledge of the application domain, they were apt in our context to provide informed feedback from a potential user's perspective. Each focus group was protocolled and subsequently the qualitative evaluations were synthesized. Ensuing the design activities, we take the process step in expanding the applicable knowledge base via the rigor cycle by adding prescriptive knowledge to literature and laying the foundation for further development possible in future research projects [48, 59]. Furthermore, the developed artifact contributes utility for the class of identified problems in the practical environment via the relevance cycle [48].

Table 1. Domain experts involved.

Expert group with brief description of their expertise and background
Official and product owner for digital participation tools. Responsible for the design with technology and integration standards and alignment with internal as well as inter-agency processes and the standardization of participation data
Data scientist and software engineer specializing in urban data applications, employed by a public institution. Involved in several projects with public authorities on data management in participation
Official with engineering background. In the department primarily concerned with the exploitation of new technologies in UPD
Official and citizen participation specialist with a background in UPD Responsible for implementing participatory processes and advising public agencies on participation initiatives on various issues
Public project administrator focusing on the development, implementation, and evaluation of concepts and methods for involving the public in planning and decision processes

4 Conceptualization

4.1 Design Requirements and Principles

The elicited system design requirements are described below and summarized together with the derived design principles in Fig. 1. The purpose of the system is to support officials in complex data analysis, i.e., logically cluster large amounts of text data from citizen e-participation processes in UPD, assign the unlabeled data sets (citizen contributions) to the formed clusters and automatically suggest labels for the data sets (*R2, 3*). Concurrently, officials should engage in the ML process, e.g., at instances where contextualization is needed. This combined clustering and classification approach should be designed to be usefully applied to distinct participation processes that may have different topics and foci in the future (*R1*).

Since the system is supposed to effectively support officials in performing analytical tasks, ease of use and usability are of great importance so that users are cognitively relieved and not additionally burdened when interacting with the system (*R6.1*). That is, by design, officials should be supported in data injection and processing workflows and enabled to provide feedback based on their experiential knowledge and make adjustments according to their needs to yield satisfactory analytic results. That is, since data processing tasks are iterative and the "optimal" number of clusters for processing participation data is context-sensitive [10], officials must be able to change parameters, i.e., adjust the number of clusters and change the predicted label assigned by the system's model, i.e., assign the correct label (*R4.1*). These changes effected by users should be applied when training the model to process new unlabeled data. When a new classification is performed based on the trained model, the user should be able to adjust the labels to induce a new training iteration, and so forth (*R4.2*).

Officials should be able to understand and abstract the meaning of the output of prediction and ML and content algorithmic decisions on the premise that they do not have very detailed knowledge of machine function (*R6.2*) [60–62]. This is important, first, because they have to use the system in their daily work and trust it accordingly [63] and, second, because they have to legitimize their work to the citizens in participatory UPD process [19]. Namely, citizen participation aims to generate collective, broadly accepted decisions, ideally found through consensus among all actors involved, which requires mutual understanding [20]. Against this background, the clustering and classification results must be interpretable and, in particular, the effects of user-system interaction, must be displayed clearly and as immediately as possible (*R6.3*). Therefore, to allow interactive execution, it must be able to deliver results in a short time on demand upon a user's prompt, which can be achieved through efficient algorithms and performant hardware (*R5*). In a broader sense, it must be present to the user that the current effort of training the machine, i.e., providing feedback on the model's performance, is rewarding in the sense that it helps the machine to learn for future domain-related tasks, thus progressively easing the human's prospective tasks. Officials must be able to conceive the rationales that the system offers them, i.e., how the functionality of the system helps them to achieve their goals [64].

R1. The system has to be applicable to current and future analysis tasks, i.e., it must be able to learn based on the users' knowledge and applicable to a variety of textual participation data in the UPD domain.

R2. The system must inductively group and logically name data sets that share common characteristics (clustering function).

R3. The system must classify data sets according to the clusters formed (classification function).

R4.1. The system has to be interactive, which means that the user can influence the clustering and classification process, i.e., set start parameters of the algorithm, correct, and improve results.

R4.2. The system needs to automatically apply user adjustments to subsequent processing iterations. It must be possible to continue these training loops until the user is satisfied with the result.

R5. The system must be performant enough to allow interactive execution, i.e., it must be able to deliver results in a short time.

R6.1. The system must allow a meaningful and productive interaction, i.e., users must be able to intuitively use the system.

R6.2. The system must be operable by users without special technical knowledge, i.e., considering their professional background.

R6.3. The system must clearly display the clustering and classification results. Outputs have to be interpretable in order that the user can make informed changes and understand their effects.

R7.1. The objectives and specifications of administrative processes and system environments must be addressed by the system to ensure interoperability, i.e. using standardized interfaces for the integration and exchange of participation and planning data.

R7.2. For system development, suitable solutions needs to be chosen taking into account that public resources are limited and, in this regard, considering maintainability and extensibility.

R8. For system development, only open source software and algorithms must be used to ensure transparency and transferability to other agencies and authorities and generate public value.

P1. To enable officials responsible for citizen participation processes in public agencies to efficiently analyze citizen textual contributions from current and future participatory UPD endeavors and, thus, gain meaningful insights to improve planning processes, use an interactive system architecture to triangulate officials' related knowledge with NLP and ML techniques, creating an opportunity to engage in such facilitated analytical activities in their agency environment.

P2. To enable domain users from the public sector to understand the process and results of the algorithmic data analysis, these must be presented in a comprehensible and transparent manner. To this end, provide that the parameters that lead to the results are self-explanatory and easily adjustable. The results are to be clearly presented and contestable in the sense that users can make the final decision on algorithmic outcomes without constraints. Thus, allowing officials to develop and sustain trust in the system and legitimize their conclusions to stakeholders in the participatory UPD process.

P3. To address authorities' and overall participation and development goals, consider in the design the organizational and socio-technical requirements for the interplay between the various public institutions and their human and machine actors, as well as public resources, creating and providing public value continuously through system implementation across institutions.

Fig. 1. Artifact design requirements and design principles.

The system should be developed on the premise that it can be implemented in existing process and system environments of administrations to meet data integration interoperability requirements (*R7.1*) [3]. In addition, since administrative resources for the development and operation of advanced IT systems are rather limited [13], easy maintainability and extensibility for future operations are important (*R7.2*). Since the system is intended to provide value to the public domain, it is preferable that the software used is open source and accessible to other public administrations and institutions ensuring transferability (*R8*).

4.2 Concept

Building on the elaborated design requirements and principles, the construction of our IT artifact is structured into three steps: (1) Conceptualization of a clustering and classification process considering a HITL approach, (2) prototypical implementation of a semi-automated HITL-based classification of new data based on the clustering, (3) and the integration into a web-based user interface (UI) for intuitive human-system interaction.

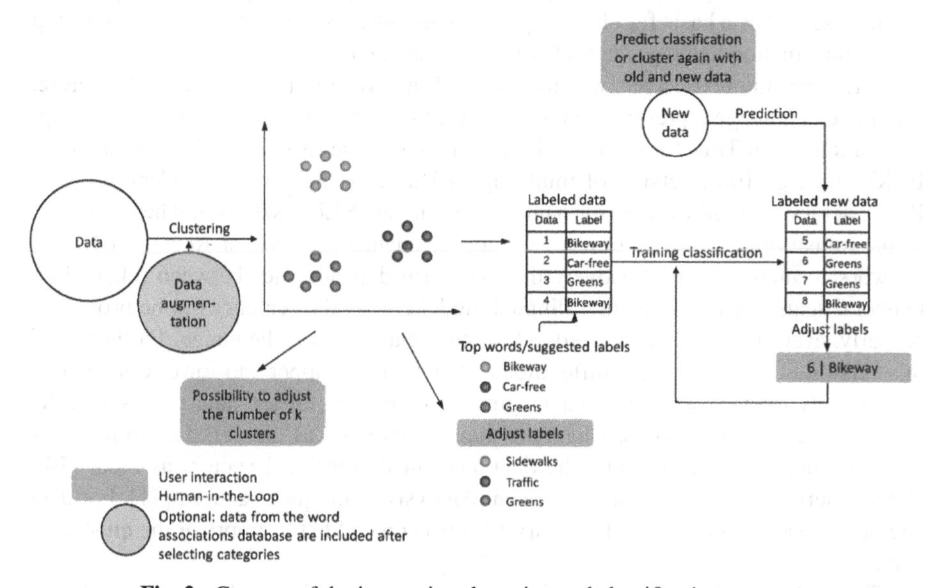

Fig. 2. Concept of the interactive clustering and classification process.

Figure 2 illustrates the conceptualization, which is outlined below. The subsequent steps are presented in the following section.

Clustering is performed depending on the parameter k, which determines the number of clusters to be formed. The user can set this parameter. Using a transformer model, the embedding is done first. Optionally, the user has the possibility to specify words in order to consider additional data, i.e., corresponding word associations, for clustering. For this purpose, data from an associative dictionary is augmented to the clustering process to improve clustering based on these specifications. Afterwards, a clustering model is deployed to group the vectors. The names of the clusters are derived from the top words and then used as label suggestions (cf., e.g., [65]). These labels can also be modified by the user. The data with the corresponding labels are saved. In a next step, these are used as training data to create a classification model. If new data is to be classified, the trained model is deployed and the corresponding class of the individual datasets is displayed. Here the user again has the possibility to adjust the labels and then perform a new training with this data. Alternatively, the user can initiate a completely new clustering with the new and old data in order to re-cluster the data and assign labels to it.

5 Implementation

5.1 Clustering and Classification

The clustering process can be divided into the two steps of data preconditioning and clustering itself. First, participation data from real-world urban development projects was collected. In total, we were able to draw on over 3,500 text contributions from projects in a major city, as provided by the authority. As a first step, this data was prepared and then served as a basis for clustering. Through this cleaning and preprocessing, stop words, punctuation and tags, and other special characters were removed.

After that, the text corpus was transformed into vectors (embeddings). Therefore, pre-trained language representation models were used, i.e., Bidirectional Encoder Representations from Transformer (BERT) and distiluse-base-multilingual-cased (a DistilBERT, cf., e.g., [66], version of multilingual Universal Sentence Encoder) [67, 68]. BERT achieves significant performance across many NLP tasks [67]. The main idea of the transformer model is to use attention mechanism instead of Recurrent Neural Network to model more effectively the long-term dependencies between tokens in a temporal sequence [69]. The multilingual model allows the sentences to be processed properly, even if they contain words of another than the core language. Furthermore, this version of the model also differentiates the words in upper and lower case, which is especially relevant for particular languages. At the end of the data processing, the vectors have 512 dimensions, which are then compressed as the Scikit-learn methods that were used require grouping the data and low-dimensional vectors as input [70]. For this purpose, the Principal Component Analysis technique was used [71]. For data augmentation, an associative dictionary [72] was embedded to improve the clustering results.

The final step of our clustering pipeline is to group the data and generate the labels for the clusters. For clustering, we use a Gaussian mixture model [73]. To control the number of distributions, i.e., clusters, the input field is defined in the UI (see Sect. 5.2). The names of the clusters are derived from top words. For this, the words contained in the texts are first sorted by frequencies and then mapped in the vector space using the BERT model. The first three words that have a maximum cosine similarity between eigenvector and cluster embeddings then serve as label recommendations. The user can customize the names of the clusters as well as the assigned labels in the UI. The data with the associated labels is saved and used as training data in the next step to build a classification model.

After the clustering of the data has been performed, the next step is the classification. For this purpose, two common ML algorithms, Logistic Regression (LR) [74] and Support Vector Machines (SVM) [75], were tested on the data and compared with each other. For this purpose, different parameter settings were tested using a Grid Search [76]. The ML methods were applied to two different ways of transforming the text into vectors with the different parameter settings and compared: Term Frequency-Inverse Document Frequency (TF-IDF) [77] and Sentence-BERT [78]. Sentence-BERT with the LR model yielded the best accuracy and was therefore deployed in our system for training current data and predicting new data.

5.2 User Interface

For the UI we had to combine the interests of developers and end users according to related design requirements. To meet the requirements of the developers for easy maintenance and extensibility, we used up-to-date open source frameworks. Using front-end web development frameworks, we aimed to design an intuitive UI, guiding the user step-by-step through the HITL process. The following describes the final UI of the prototypical instantiation of the system.

The UI is organized into the two areas of clustering and classification. In the first step of clustering, the user initially specifies how many clusters are to be formed and whether additional clustering is to be performed based on predefined categories, e.g., environment and transportation. In the next step, the user has the possibility to adjust and correct the clustering result. Mouse-over on a pie chart shows how many contributions have been assigned to the respective cluster. The user can change the system-proposed label of an entire cluster, navigate through the datasets, and assign individual contributions to a different cluster. Each change is instantly applied and displayed in the UI so that the entire effects of the changes are directly visible to the user, i.e., overall result, labels, and accuracy. Filter options are available to limit the display to contributions that are relevant for the user, e.g., if the system can already label a contribution with high confidence it is probably correct already [43]. As such, only contributions with a certain accuracy can be displayed. For this, a slider can be moved continuously between 0% and 100%. After all changes are made, the user saves them by clicking the corresponding button and automatically starts the training. The results of the training are displayed to the user after successful completion.

After training, the user can perform the classification of data. To do this, the user selects the classification tab in the UI and uploads the new data. The results of the classification are then displayed to the user for adjustment and correction. The interaction through the UI in this step is designed analogously to clustering in order to make the usability as intuitive as possible. As expected, the necessary cognitive transfer effort should thus be low. To complete the classification and start a new training, a click on the respective button is sufficient here as well.

6 Evaluation

Three EFG (*EFG1–3*) that followed a predefined procedure were conducted. First, the context and objective of the study were presented and general issues related to the applied clustering and classification approach were clarified to create a common understanding. Participants then individually performed functional tests based on tasks to be completed to evaluate the system. The researchers moderating the EFG then asked pre-specified open-ended questions to guide the discussion. These questions were devised following the technical functionalities and formulated design requirements to focus on the evaluation of the design knowledge implemented in the instantiated prototype.

For conducting the EFG, the test system was connected to the administrative database to query participation data on the fly and was made available to participants via a web server. In addition, participants received a distinct data set to test the local injection process and new data classification. This allowed the entire system workflow to be modeled, providing an approximation toward the real application context. In the following, we report the synthesis of key insights from the EFG.

In the opening discussions, interesting questions were raised, such as whether the system automatically recognizes and filters out insults or personal data. Further, participants noted that the associative dictionary used was arguably quite general and not context-specific (EFG2). Pros and cons were discussed, whether a subject-specific augmentation of the clustering would lead to different results, without finding consensus.

Using the system, displaying accuracy after training for a classification was easy for participants to recognize by clicking the respective button. Yet, they found that the given accuracy varied among them even though they used the same configuration in this task (EFG1). The reasons for this were explained, i.e., that this is considered normal for the model used and that it is due to the noise in data, which is an inevitable problem in data mining.

Most participants were immediately aware of the function of the slider that specifies the range of probabilities to which the labels in the list should correspond and one participant highlighted its intuitive use (EFG1). However, for some participants it was not so apparent what exactly the probabilities indicated. In this context, the question emerged, whether contributions can correspond to multiple clusters (EFG3). Thus, it was not immediately evident that the cluster assignment was not absolute but based on probabilities, i.e., that several probabilities can be displayed for each data set accordingly and the user makes the final decision of the assignment. On this subject, one participant suggested, and others agreed (EFG3), that it would be better to be able to assign more than one label, which was not feasible with the prototype system. The actual changing of the labels did not pose a challenge to any participants. However, it was proposed to adjust the fields for the label suggestions, as they were considered somewhat too small (*EFG1*). Further desirable features noted by two participants would be the ability to re-sort contributions by filters beyond probabilities and to cache models after training (*EFG2*).

Participants throughout all groups agreed that the system was intuitive and relatively easy to use without expert knowledge. The UI was perceived as very clear and engaging with logical actions and behaviors (*EFG1–3*). Conversely, it was noted that some elements could be named differently and described in more detail, as they were not completely self-explanatory (*EFG1*). At some points, explanation of the data processing workflow was needed, e.g., that new data can be labeled simply by re-executing the classification. It was positively received that the interface of the system is web-based and thus very practical and universally applicable (*EFG2*). Notably, the combined two-stage clustering and classification model was found to be useful (*EFG3*).

7 Discussion and Conclusion

In this paper, we present a concept and prototypical instantiation of an intelligent system that augments officials in analyzing large amounts of citizen-generated data from participatory UPD. Our solution approach focuses on the complementary strengths of human and machines [15, 17]. In particular, our system offers a two-stage combined clustering and classification approach for the analysis of participation data by responsible officials at administrations who can simultaneously engage in the model building based on a HITL ML approach [43]. Therewith, we address specific problems related to the analysis and further processing of citizen contributions when participation occurs on a large scale, i.e., individuals' ability to relate all relevant information in order to make informed decisions. By integrating humans with their knowledge into the ML process, we further address algorithmic learning challenges related to data quality. That is, participation data is complex since it is shaped by the interplay of various features of the dynamic participation landscape as it evolves, i.e., process-, planning-, and actor-specific. In this context, officials can leverage their experiential contextual knowledge to support ML and train the machine for future tasks, which in turn benefits them in task completion. For this, the system displays its current best classification and, after each user interaction, updates the current classification to reflect human input and displays the classification for further review and correction, if necessary [44].

We followed the DSR cycles [45] to achieve our research objective. System design requirements were iteratively elaborated grounded on theoretical and practical domain knowledge and formatively evaluated together with domain experts for artifact conceptualization. Based on these requirements, we abstract design principles, built a prototype and, after an evaluation phase with prospective users, stopped at the proof-of-concept with the possibility of further development in future research projects. Therewith, we expand the applicable knowledge base by adding prescriptive knowledge to literature [54, 55]. In particular, we capture abstract design knowledge through the formulation of design principles which provide "prescriptions for design and action" comprising the perspectives of implementers and users [48, p. 339, 56]. Furthermore, these principles can be utilized in practice to create purposeful IS artifacts in related contexts [54] and present design knowledge embodied with the instantiated and evaluated artifact. Specifically, the assessment of the artifact through EFG focused on design-related aspects as well as technical features and suggested the usefulness of our designed solution approach. The EFG participants positively noted, e.g., that the system's user interface was easy to use and web-based, thus universally applicable. The combined two-stage clustering and classification process was considered useful and comprehensible. In addition, potential for improvement was discovered. Caching of trained models was identified as a possible suitable feature, as were more options for labeling and sorting data sets. This should be considered in the future design. Furthermore, automatic detection for, e.g., insulting texts and third party personal data was suggested. Besides the fact that this was not our focus, the machine training would not have been feasible as the data provided contained only two offending contributions. However, it would have been possible for the user to manually remove such texts in the HITL process. Nonetheless, consistent with research in NLP related to detecting offensive text in various application contexts (e.g., [79]) this topic should not be overlooked in the future with the increasing number of contributions on online participation platforms. Compliance with and integration of, e.g., country-specific privacy requirements into citizen participations platforms and automation techniques is a relevant topic that deserves further attention as well. Another point is data augmentation in clustering to improve results. We embedded general word association data from a publicly available dictionary. In the future, it should be explored, whether a subject-specific database leads to different clustering results.

Our study has several limitations. Challenges arose from the real-world quality data we received from the authority. The data was not labeled, which prevented the final evaluation of clustering results. In this context, however, it must be noted that our focus has been on the conceptualization of the system design and we instantiated our artifact using common algorithms. Looking forward, our solution approach can be studied on a comparative basis with different algorithms and refined based on the evaluation of the performance indicators with ground truth participation data. Furthermore, we did not evaluate an instantiation of the system in a real-world setting. Since our artifact is shaped by the interaction of human and machine and is thus sociotechnical in nature, a naturalistic evaluation in the field of the application domain should be considered accordingly in possible further development [58]. Yet, we were able to draw on expert knowledge throughout the design process and we seized the opportunity to assess the functionalities of the system extensively with potential users and explored opportunities for improvement. With the instantiation of a functioning prototype, which can process complex real-world participation data, we have contributed to the solution space of unsolved problems in the practical environment [80]. Finally, human-AI interaction research in UPD with its specific areas of study, e.g., algorithmic data analysis of participation data and ML, is a research field with high practical relevance, which still offers much potential for novel research approaches to address existing and upcoming challenges due to technological and societal developments.

References

1. Fung, A.: Putting the public back into governance: the challenges of citizen participation and its future. Public Adm. Rev. **75**, 513–522 (2015)
2. Grcheva, L., van den Berg, R., Thung, I.: Urban Planning and Design Labs: tools for integrated and participatory urban planning. United Nations Human Settlements Programme (2016)
3. Falco, E., Kleinhans, R.: Beyond technology: identifying local government challenges for using digital platforms for citizen engagement. Int. J. Inf. Manage. **40**, 17–20 (2018). https://doi.org/10.1016/j.ijinfomgt.2018.01.007
4. Afzalan, N., Muller, B.: Online participatory technologies: opportunities and challenges for enriching participatory planning. J. Am. Plann. Assoc. **84**, 162–177 (2018). https://doi.org/10.1080/01944363.2018.1434010
5. Anttiroiko, A.-V.: Digital urban planning platforms. Int. J. E-Plan. Res. **10**, 35–49 (2021). https://doi.org/10.4018/IJEPR.20210701.oa3

6. Münster, S., et al.: How to involve inhabitants in urban design planning by using digital tools? An overview on a state of the art, key challenges and promising approaches. Procedia Comput. Sci. **112**, 2391–2405 (2017)
7. Lorimer, A.: Mass-participation architecture: social media and the decentralisation of architectural agency as a commercial imperative (2016)
8. Elsen, C., Schelings, C.: Citizen participation through digital platforms: the challenging question of data processing for cities (2019)
9. Küstermann, G., Bittner, E.: Developing a GIS-integrated tool to obtain citizens' input in on-site participation—learnings from participatory urban planning of a large city. In: Bui, T. (ed.) Proceedings of the 54th Annual Hawaii International Conference on System Sciences (2021). https://doi.org/10.24251/HICSS.2021.289
10. Coden, A., Danilevsky, M., Gruhl, D., Kato, L., Nagarajan, M.: A method to accelerate human in the loop clustering. In: Proceedings of the 2017 SIAM International Conference on Data Mining, pp. 237–245. SIAM (2017)
11. Kamis, A., Koufaris, M., Stern, T.: Using an attribute-based decision support system for user-customized products online: an experimental investigation. MIS Q. **32**, 159–177 (2008)
12. Teufl, P., Payer, U., Parycek, P.: Automated analysis of e-participation data by utilizing associative networks, spreading activation and unsupervised learning. In: Macintosh, A., Tambouris, E. (eds.) ePart 2009. LNCS, vol. 5694, pp. 139–150. Springer, Heidelberg (2009). https://doi.org/10.1007/978-3-642-03781-8_13
13. Wirtz, B.W., Weyerer, J.C., Sturm, B.J.: The dark sides of artificial intelligence: an integrated AI governance framework for public administration. Int. J. Public Adm. **43**, 818–829 (2020). https://doi.org/10.1080/01900692.2020.1749851
14. Esau, K., Liebeck, M., Eilders, C.: Mining Arguments in Online Participation: Möglichkeiten und Grenzen manueller und automatisierter Inhaltsanalyse zur Erhebung von Argumentkomponenten. Polkomm (2017)
15. Dellermann, D., Ebel, P., Söllner, M., Leimeister, J.M.: Hybrid intelligence. Bus. Inf. Syst. Eng. **61**, 637–643 (2019)
16. Liebeck, M., Esau, K., Conrad, S.: Text Mining für Online-Partizipationsverfahren: Die Notwendigkeit einer maschinell unterstützten Auswertung. HMD Praxis der Wirtschaftsinformatik **54**(4), 544–562 (2017)
17. Raisch, S., Krakowski, S.: Artificial intelligence and management: the automation–augmentation paradox. AMR **46**, 192–210 (2021). https://doi.org/10.5465/amr.2018.0072
18. Hansen, M.R.P., Haj-Bolouri, A.: Design principles exposition: a framework for problematizing knowledge and practice in DSR. In: Hofmann, S., Müller, O., Rossi, M. (eds.) DESRIST 2020. LNCS, vol. 12388, pp. 171–182. Springer, Cham (2020). https://doi.org/10.1007/978-3-030-64823-7_16
19. Douay, N.: Urban Planning in the Digital Age. From Smart City to Open Government? Wiley, Newark (2018)
20. Hofmann, M., Münster, S., Noennig, J.R.: A theoretical framework for the evaluation of massive digital participation systems in urban planning. J. Geovis. Spat. Anal. **4**(1), 1–12 (2019). https://doi.org/10.1007/s41651-019-0040-3
21. Bannister, F., Connolly, R.: Trust and transformational government: a proposed framework for research. Gov. Inf. Q. **28**, 137–147 (2011). https://doi.org/10.1016/j.giq.2010.06.010
22. Bertot, J.C., Jaeger, P.T., Grimes, J.M.: Using ICTs to create a culture of transparency: e-government and social media as openness and anti-corruption tools for societies. Gov. Inf. Q. **27**, 264–271 (2010). https://doi.org/10.1016/j.giq.2010.03.001
23. Janowski, T.: Digital government evolution: from transformation to contextualization. Gov. Inf. Q. **32**, 221–236 (2015). https://doi.org/10.1016/j.giq.2015.07.001
24. Jo, S., Nabatchi, T.: 12.1 Case Study—Co-Producing Recommendations to Reduce Diagnostic Error. Co-Production and Co-Creation, vol. 161 (2018)

25. Hatuka, T., Rosen-Zvi, I., Birnhack, M., Toch, E., Zur, H.: The political premises of contemporary urban concepts: the global city, the sustainable city, the resilient city, the creative city, and the smart city. Plan. Theor. Pract. **19**, 160–179 (2018). https://doi.org/10.1080/14649357.2018.1455216
26. Nabatchi, T., Ertinger, E., Leighninger, M.: The future of public participation: better design, better laws, better systems. Confl. Resolut. Q. **33**, S35–S44 (2015). https://doi.org/10.1002/crq.21142
27. Stelzle, B., Noennig, J.R.: A method for the assessment of public participation in urban development. Urban Dev. Issues **61**, 33–40 (2019). https://doi.org/10.2478/udi-2019-0005
28. Recalde, L., Meza, J., Terán, L.: Cognitive systems for urban planning: a literature review. In: Santos, H., Pereira, G.V., Budde, M., Lopes, S.F., Nikolic, P. (eds.) SmartCity 360 2019. LNICSSITE, vol. 323, pp. 249–270. Springer, Cham (2020). https://doi.org/10.1007/978-3-030-51005-3_22
29. Paulus, P.B., Baruah, J.: Enhancing creativity in e-planning: recommendations from a collaborative creativity perspective. In: New Approaches, Methods, and Tools in Urban E-Planning, pp. 192–222. IGI Global (2018)
30. Nabatchi, T.: Putting the "public" back in public values research: designing participation to identify and respond to values. Public Adm. Rev. **72**, 699–708 (2012)
31. Cheng, X., et al.: Idea convergence quality in open innovation crowdsourcing: a cognitive load perspective. J. Manag. Inf. Syst. **37**, 349–376 (2020)
32. Fu, S., de Vreede, G.-J., Cheng, X., Seeber, I., Maier, R., Weber, B. (eds.): Convergence of crowdsourcing ideas: a cognitive load perspective (2017)
33. Seeber, I., de Vreede, G.-J., Maier, R., Weber, B.: Beyond brainstorming: exploring convergence in teams. J. Manag. Inf. Syst. **34**, 939–969 (2017). https://doi.org/10.1080/07421222.2017.1393303
34. Brabham, D.C.: Crowdsourcing the public participation process for planning projects. Plan. Theor. **8**, 242–262 (2009)
35. de Vreede, G.-J., Briggs, R.O., de Vreede, T.: Exploring a convergence technique on ideation artifacts in crowdsourcing. Inf. Syst. Front. (2021). https://doi.org/10.1007/s10796-021-10120-0
36. Grønsund, T., Aanestad, M.: Augmenting the algorithm: emerging human-in-the-loop work configurations. J. Strateg. Inf. Syst. **29**, 101614 (2020). https://doi.org/10.1016/j.jsis.2020.101614
37. Benz, C., Zierau, N., Satzger, G.: Not all tasks are alike: exploring the effect of task representation on user engagement in crowd-based idea evaluation. In: 27th European Conference on Information Systems (ECIS), Stockholm & Uppsala, Sweden (2019)
38. Eppler, M.J., Mengis, J.: The concept of information overload: a review of literature from organization science, accounting, marketing, MIS, and related disciplines. Inf. Soc. **20**, 325–344 (2004)
39. Logg, J.M., Minson, J.A., Moore, D.A.: Algorithm appreciation: people prefer algorithmic to human judgment. Organ. Behav. Hum. Decis. Process. **151**, 90–103 (2019). https://doi.org/10.1016/j.obhdp.2018.12.005
40. Lohoff, L., Rühr, A.: Introducing (machine) learning ability as antecedent of trust in intelligent systems (2021)
41. Chui, M., Manyika, J., Miremadi, M., Henke, N., Chung, R., Nel, P., Malhotra, S.: Notes from the AI Frontier: Insights from Hundreds of Use Cases. McKinsey Global Institute (2018)
42. Faraj, S., Pachidi, S., Sayegh, K.: Working and organizing in the age of the learning algorithm. Inf. Organ. **28**, 62–70 (2018)
43. Monarch, R.: Human-in-the-Loop Machine Learning: Active Learning and Annotation for Human-Centered AI. Manning Publications (2021)

44. Ramos, G., Meek, C., Simard, P., Suh, J., Ghorashi, S.: Interactive machine teaching: a human-centered approach to building machine-learned models. Hum. Comput. Interact. **35**, 413–451 (2020). https://doi.org/10.1080/07370024.2020.1734931
45. Hevner, A.R.: A three cycle view of design science research. Scand. J. Inf. Syst. **19**, 4 (2007)
46. Sanford, C., Rose, J.: Designing the e-participation artefact. IJEB **6**, 572 (2008). https://doi.org/10.1504/IJEB.2008.021875
47. Hevner, A.R., March, S.T., Park, J., Ram, S.: Design science in information systems research. MIS Q. **28**, 75–105 (2004). https://doi.org/10.2307/25148625
48. Gregor, S., Hevner, A.R.: Positioning and presenting design science research for maximum impact. MIS Q. **37**, 337–355 (2013). https://doi.org/10.25300/MISQ/2013/37.2.01
49. Nummi, P., Eräranta, S., Kahila-Tani, M.: Enhancing e-participation in urban planning competitions. In: Dima, I., Nunes Silva, C. (eds.) New Approaches, Methods, and Tools in Urban E-Planning. Advances in Civil and Industrial Engineering, pp. 60–94. IGI Global (2018). https://doi.org/10.4018/978-1-5225-5999-3.ch003
50. Anttiroiko, A.-V.: Urban planning 2.0. Int. J. E-Plan. Res. **1**, 16–30 (2012). https://doi.org/10.4018/ijepr.2012010103
51. Lodigiani, A.: E-planning: the digital toolbox in participatory urban planning. In: Contin, A., Paolini, P., Salerno, R. (eds.) Innovative Technologies in Urban Mapping. SSSSI, vol. 10, pp. 135–144. Springer, Cham (2014). https://doi.org/10.1007/978-3-319-03798-1_12
52. Gregor, S., Jones, D.: The anatomy of a design theory. JAIS **8**, 312–335 (2007). https://doi.org/10.17705/1jais.00129
53. Kruse, L.C., Seidel, S., Purao, S.: Making use of design principles. In: Parsons, J., Tuunanen, T., Venable, J., Donnellan, B., Helfert, M., Kenneally, J. (eds.) Tackling Society's Grand Challenges with Design Science, pp. 37–51. Springer, Cham (2016). https://doi.org/10.1007/978-3-319-39294-3_3
54. Chandra, L., Seidel, S., Gregor, S.: Prescriptive knowledge in IS research: conceptualizing design principles in terms of materiality, action, and boundary conditions. In: 48th Hawaii International Conference on System Sciences (2015). https://doi.org/10.1109/HICSS.2015.485
55. Gregor, S.: The nature of theory in information systems. MIS Q. **30**, 611 (2006). https://doi.org/10.2307/25148742
56. Gregor, S., Kruse, L., Seidel, S.: Research perspectives: the anatomy of a design principle. JAIS **21**, 1622–1652 (2020). https://doi.org/10.17705/1jais.00649
57. Dicicco-Bloom, B., Crabtree, B.F.: The qualitative research interview. Med. Educ. **40**, 314–321 (2006). https://doi.org/10.1111/j.1365-2929.2006.02418.x
58. Venable, J., Pries-Heje, J., Baskerville, R.: FEDS: a framework for evaluation in design science research. Eur. J. Inf. Syst. **25**, 77–89 (2016)
59. Tremblay, M.C., Hevner, A.R., Berndt, D.J.: Focus groups for artifact refinement and evaluation in design research. CAIS **26** (2010).https://doi.org/10.17705/1CAIS.02627
60. Nagbøl, P.R., Müller, O.: X-RAI: a framework for the transparent, responsible, and accurate use of machine learning in the public sector. In: EGOV-CeDEM-ePart 2020, p. 259 (2020)
61. Lyons, H., Velloso, E., Miller, T.: Conceptualising contestability: perspectives on contesting algorithmic decisions. Proc. ACM Hum. Comput. Interact. **5**, 1–25 (2021)
62. Asatiani, A., Malo, P., Nagbøl, P.R., Penttinen, E., Rinta-Kahila, T., Salovaara, A.: Challenges of explaining the behavior of black-box AI systems. MIS Q. Exec. **19**, 259–278 (2020)
63. Clarke, M.F., Gonzales, J., Harper, R., Randall, D., Ludwig, T., Ikeya, N.: Better supporting workers in ML workplaces. In: Gilbert, E., Karahalios, K. (eds.) Conference Companion Publication of the 2019 on Computer Supported Cooperative Work and Social Computing, pp. 443–448. ACM, New York (2019). 11092019. https://doi.org/10.1145/3311957.3359429
64. Harper, R.H.R.: The role of HCI in the age of AI. Int. J. Hum. Comput. Interact. **35**, 1331–1344 (2019). https://doi.org/10.1080/10447318.2019.1631527

65. Peikari, M., Salama, S., Nofech-Mozes, S., Martel, A.L.: A cluster-then-label semi-supervised learning approach for pathology image classification. Sci. Rep. **8**, 1–13 (2018)
66. Sanh, V., Debut, L., Chaumond, J., Wolf, T.: DistilBERT, a distilled version of BERT: smaller, faster, cheaper and lighter. arXiv preprint arXiv:1910.01108 (2019)
67. Devlin, J., Chang, M.-W., Lee, K., Toutanova, K.: BERT: pre-training of deep bidirectional transformers for language understanding. In: Burstein, J., Doran, C., Solorio, T. (eds.) Proceedings of the 2019 Conference of the North, pp. 4171–4186. Association for Computational Linguistics, Stroudsburg, PA, USA (2019). https://doi.org/10.18653/v1/N19-1423
68. Reimers, N., Gurevych, I.: Making monolingual sentence embeddings multilingual using knowledge distillation. arXiv preprint arXiv:2004.09813 (2020)
69. Vaswani, A., et al.: Attention is all you need. arXiv preprint arXiv:1706.03762 (2017)
70. Pedregosa, F., et al.: Scikit-learn: machine learning in Python. J. Mach. Learn. Res. **12**, 2825–2830 (2011)
71. Jolliffe, I.T., Cadima, J.: Principal component analysis: a review and recent developments. Philos. Trans. R. Soc. A Math. Phys. Eng. Sci. **374**, 20150202 (2016)
72. Word Associations Network. https://wordassociations.net/
73. Reynolds, D.: Gaussian mixture models. In: Li, S.Z., Jain, A. (eds.) Encyclopedia of Biometrics, pp. 659–663. Springer, Boston, MA (2009). https://doi.org/10.1007/978-0-387-73003-5_196
74. Peng, C.-Y.J., Lee, K.L., Ingersoll, G.M.: An introduction to logistic regression analysis and reporting. J. Educ. Res. **96**, 3–14 (2002). https://doi.org/10.1080/00220670209598786
75. Hearst, M.A., Dumais, S.T., Osuna, E., Platt, J., Scholkopf, B.: Support vector machines. IEEE Intell. Syst. Appl. **13**, 18–28 (1998). https://doi.org/10.1109/5254.708428
76. Liashchynskyi, P., Liashchynskyi, P.: Grid search, random search, genetic algorithm: a big comparison for NAS. arXiv preprint arXiv:1912.06059 (2019)
77. Qaiser, S., Ali, R.: Text mining: use of TF-IDF to examine the relevance of words to documents. IJCA **181**, 25–29 (2018). https://doi.org/10.5120/ijca2018917395
78. Reimers, N., Gurevych, I.: Sentence-BERT: sentence embeddings using siamese BERT-networks. arXiv preprint arXiv:1908.10084 (2019)
79. Sharma, H.K., Kshitiz, K., Shailendra: NLP and machine learning techniques for detecting insulting comments on social networking platforms. In: 2018 International Conference on Advances in Computing and Communication Engineering (ICACCE), pp. 265–272 (2018). https://doi.org/10.1109/ICACCE.2018.8441728
80. vom Brocke, J., Winter, R., Hevner, A., Maedche, A.: Special issue editorial – accumulation and evolution of design knowledge in design science research: a journey through time and space. J. Assoc. Inf. Syst. **21**(3), 520–544 (2020). https://doi.org/10.17705/1jais.00611

The Increasing e-Competence Gap: Developments over the Past Five Years in the German Public Sector

Michael Koddebusch(✉), Sebastian Halsbenning, Paul Kruse, Michael Räckers, and Jörg Becker

University of Münster, ERCIS, Münster, Germany
{koddebusch,halsbenning,kruse,raeckers,becker}@ercis.de

Abstract. The continuously growing speed of the digital transformation also impacts governments and public administrations worldwide. To keep up with this development, public officials are required to obtain and improve so-called e-competences. In order to determine developments in the perceived relevance of those competences over the past five years, we have set up a comparative survey study investigating (a) how the perceived importance of certain e-competences has developed (b) whether the supply of professionals holding certain e-competences has changed and (c) whether there is a change in public sector employees being sent to relevant training. The survey was filled out in 2016 and 2021. Out of 54 inquired variables, 29 show a statistically significant increase. From the employees' perspective, we found a perceived increase in the importance and demand of e-competences and a decrease in having enough professionals available holding these e-competences.

Keywords: e-Competences · Digital competences · e-Government · Digital government · Digital transformation

1 Introduction

Digitalization largely impacts daily life and the whole economy. Especially the public sector can benefit from those changing opportunities by innovating the way public services are delivered. For instance, the public sector can reorganize and digitize its processes and workflows to act more efficiently [23,32]. The provision of digital public services also comes along with benefits and simplifications for citizens and businesses [30]. One huge consequence is a shift in the required abilities of the workforce and even to those of the usual citizen. This, of course, has long been recognized for the different business domains and also for the citizens' everyday life [17]. In the end, the full potential of digitalization-induced innovations can only be leveraged in the public sector with an adequate, corresponding level of public official's competences [20].

However, endeavors to facilitate public service delivery or to foster e-government, in general, are not always performing with the desired speed. Prominent international indices from the United Nations [31] and the European Union [9] reveal a very heterogeneous digital maturity of the public sector across countries, where, e.g., the highly

F. Fui-Hoon Nah and K. Siau (Eds.): HCII 2022, LNCS 13327, pp. 73–86, 2022.
https://doi.org/10.1007/978-3-031-05544-7_6

developed countries Germany or France are not among the top performers. For example, this led Germany to introduce the Online Access Act (OZG) [10], a law obliging the public administration of all federal levels to make their services also online available by the end of 2022. This example uncovers that the typical problems of public sector digitalization remain unsolved [11,13] and challenging. Among other reasons, federal boundaries [28], and lack of sufficiently (digitally) skilled public officials inhibit a prosperous digitalization of the public sector [7,21].

During the last years, a rising focus has been given to researching the needed competences for digitalizing the public sector and e-government. Reinforced by the COVID-19 pandemic [4], the step beyond developing new approaches for the education and acquisition of skills and competences is also an increasing scientific field [14,26]. With the setup of research projects in that field like *Qualifica Digitalis*, the German government, for instance, also acknowledged the importance [27]. In the light of the growing attention, we aim with this paper to gain further insights into the area of e-government competences of public officials.

Even though there is a strong need for such e-competences in the public sector, a bottleneck of these very same competences is still remaining. We can also determine a shift towards the prioritization of educating digital abilities, but the practical impact has not yet closed the competence gap. This leads us to the guiding research question of this paper:

RQ How has the perceived relevance of e-government competences changed over the last five years in Germany?

To gain a more detailed outcome, we subdivided this general research question into three more specific questions:

(a) How has the perceived importance of certain e-competences developed?
(b) How has the supply of professionals holding certain e-competences changed?
(c) How has the share of public officials being sent to relevant training changed?

We followed a quantitative, comparative survey design to meet our research questions. We conducted a survey (mainly) among public officials at first in 2016 and again in 2021. Thereby, we first focused on determining if the thriving importance of e-government is also accompanied by a rising perceived importance of corresponding e-competences. Secondly, we asked for an assessment of the supply on the labor market to figure out the perceived bottleneck. With the third research question's design, we intend to figure out if there might be observable changes in the use of training.

The paper is structured as follows. Based on the motivation of our research, we first introduce the related literature to e-government competences. After that, we elaborate on our research method, which is followed by the results section. Finally, we discuss the relevance and context of the presented results and give a conclusion.

2 Research Background

For the public administration, digital transformation is currently one of the grand challenges. On the one hand, the digitalization of the public sector comes along with big promises and opportunities. It can contribute to making the administration's work more efficient by, e.g., speeding up processing times and—in the long run—reducing financial spending. Also, citizen satisfaction can be positively influenced by a better, more citizen-centric service delivery [22]. On the other hand, public sector digitalization and the establishment of e-government suffer from many organizational, cultural, legal, or technical obstacles [11]. The COVID-19 pandemic revealed the current digitalization status of the public sector since containment scouting or public service delivery during a lockdown would have benefited from already established digital routines.

One important factor affecting the digitalization efforts leads to the public officials' competence [21]. There is a broad consensus that the role of and the required competences of public officials are changing with the digital transformation, which was already acknowledged twenty years ago in the seminal e-government stage model by Layne and Lee [18]. The research on required competences for e-government emerged later with the identification of certain, relevant competences [7,16], clustering to competence categories [7,15] and the establishment of competence frameworks (e.g., [8,15]). In line with prior research [12,19,24], we define competence as *"the combination of an individual's work-related knowledge, skills and abilities"* [7, p. 287]. Thereby, the concept of competence is a broad capture of what an individual can accomplish based, e.g., on her/his education and experience. E-government competences—in the following: *e-competence*—are those competences with relation to public sector digitalization and go beyond single skills or abilities.

According to Hunnius et al. [15], e-competence can be classified in the five categories *technical*, *socio-technical*, *organizational*, *managerial* and *political administrative*, which are described as follows [2,7,15]: The technical category refers to classical IT abilities, including, for instance, conceptual IT development skills. In the socio-technical category, the focus is given to the intersection and influence of IT and human behavior, e.g., to be able to conduct requirements engineering. Organizational competences refer to handling administrative structures, managerial competences to typical business abilities (e.g., change management and controlling), and political-administrative to the intersection of strategy and law (e.g., privacy).

Each of these categories comprises specific competences that are needed to master the digital transformation of the public sector. Having this set at hand, further possibilities come into play. For example, it provides a basis to shape curricula for education or vocational training in the e-government domain [26] or to develop innovative teaching formats conveying those competences [14,25]. Within the public administration, the established competence (categories) can be used to design reference roles for certain fields of work and specify the requirements provided in job postings. Although we can find viable frameworks for digital competence in general and e-government competence in specific, a focus is merely given to a different importance of the single competences and/or competence categories and their change over time. The established five competence categories by Hunnius et al. [15], for instance, highly differ in terms of the impact on e-government projects, the usually required education or the shortage on the job market.

Recent findings indicate that it is worth having a closer look at this. Auth et al. [1] found in an analysis of more than 21.000 job postings from the German public sector domain that digital competences are rarely demanded for classical positions in the public administration. This not only contradicts the fact that almost every public official has to cope with a rising digitalization of daily routines. Also, job market forecasts show a drastic gap for public sector personnel in the area of information technology and classical administrative work [6].

3 Research Approach

In order to answer the research questions, a comparative online survey study was set up, gathering quantitative data from relevant actors working in or closely related to the public sector. The survey was sent out with a temporal difference of five years: in 2016 and 2021. The same questions were asked in both years. The development within those five years was then tested for statistical significance to provide meaningful insights.

Survey Design: The survey design is based on Creswell [5] and consists of three major parts. The first part asked respondents to provide metadata about their relation to the public sector. This primarily includes locational information of the respondent's organization, its size and sector, and the respondent's position within the organization. The second part of the survey collected information on e-competences based on [15]. In total, the perceived importance of 18 e-competences divided into five categories (technical, socio-technical, managerial, organizational, and political-administrative) was inquired (a). Additionally, for each e-competence, it was asked whether (b) there are enough professionals on the market/in the organization that possess the respective e-competence and (c) whether, within the own organization, employees have been sent or are planned to be sent to training for acquiring the respective e-competence. In order to only collect unbiased answers (b) and (c) of a certain e-competence were only answered by those respondents, who rated the e-competence of the matter to be "important" or "very important" beforehand. The third and last part of the survey asked respondents for any other information on competence requirements in their organization they want to provide and their contact information for potential follow-up questions. The main focus of this study lies on the second part of the survey that (in total) consists of 54 items, corresponding to the three questions (a) (b) and (c) to each of the 18 e-competences. The respondents provided information by choosing from a 5-point Likert scale:

- **E-competence importance**: (1) unimportant (2) of little importance (3) moderately important (4) important (5) very important
- **Enough professionals on the market/in the organization**: (1) strongly disagree (2) disagree (3) neither agree nor disagree (4) agree (5) strongly agree
- **Employees have been/are to be sent to training**: (1) strongly disagree (2) disagree (3) neither agree nor disagree (4) agree (5) strongly agree

Sampling: The survey was sent out Germany-wide to various public sector organizations, higher education institutions, and private organizations working closely together with public sector institutions. The total number of respondents was higher in 2016 (n = 587) than in 2021 (n = 470). After removing incomplete responses, we had an organizational distribution of participants, as shown in Table 1. Private sector organizations are primarily located in the field of public sector consulting. *Other* organizations are mostly higher education institutions (e.g., universities), NGOs, or research institutions. It was a conscious decision to limit the catchment area of survey respondents to Germany, as the organizational structures of national public sector architectures often differ heavily from each other. In order to create a sample as genuine homogeneous as possible, we therefore surveyed solely German institutions for being able to afterward abstract general insight from it.

Table 1. Respondent distribution by organization

	2016	2021
Respondents	**587**	**470**
From public sector organizations	537	419
From private sector organizations	19	11
From other organizations	31	40

Analysis: The main objective of the study is to find out the development of (a) the degree of importance of a certain e-competence (b) whether there are enough professionals who offer this e-competence on the market/in the own organization and (c) whether the own organization sends employees to obtain additional training to acquire this e-competence. First, the percentage share of "very important" and "important" for (a), "strongly disagree" and "disagree" for (b), and "agree" and "strongly agree" for (c) were calculated for each item of the second part of the survey for both years. Then, after having established the shares of interest for each year individually, we compared both percentage shares of the same item against each other. The data sets were subjected to a two-sample t-test assuming equal variances to ensure comparability between the two samples. The variances for all data sets were checked in beforehand to make sure this test is appropriate. Additionally, the p-value for the relation of each data set of the 54 items was calculated to determine statistical significance.

4 Results

The following section describes the findings of the survey results, especially the developments of importance in e-competences, the availability of e-competences, and training provision for e-competences. Due to the amount of data, this description will be limited to developments of strong statistical significance ($p \leq 0.05$). Each item's development, including the ones offering less ($p < 0.1$) or no ($p > 0.1$) statistical significance, is shown in Table 2. The table must be read as follows:

- The competences, printed in bold, asked for their respective importance. The following numbers represent the share of respondents stating that they strongly agree or agree with the statement for a competence being important.
- "Competence on the market/in organization" always refers to the competence printed in bold mentioned above. The numbers following this line represent the share

of respondents who strongly disagree or disagree with the statement, that there are enough professionals in the organization/on the job market who offer the respective competence.
- "Employees sent for training" also refers to the previously named, boldly printed competence. The numbers represent the share of respondents agreeing or strongly agreeing to the statement that employees in the organization have been/are planned to be sent for additional training to obtain the respective competence.

For readability purposes, the questions asked in the survey were abbreviated in the table below.

Technical E-competences: Five out of nine items in this category show a strong significant increase from 2016 to 2021. Whereas *IT-competences* were regarded to be important by 72.7% of respondents in 2016, this has risen to 77.5% in 2021. An even more drastic rise can be registered for *IS Design-competences*: only around one-third (34.4%) of all respondents found this to be important in 2016, but the value increased by +13% to 47.5% in 2021. For both of these competences, the share of respondents stating that they disagree with the statement of having sufficient talent on the market/in the organization holding these competences rose as well, greatly. While in 2016, 48.5% did not find that to be true for *IT-competences*, in 2021, it was 69%, indicating a leap of almost +21%. The growth is also strong for *IS Design-competences* on the market/in the organization: 59.4% already disagreed with sufficient talent provision in 2016, which has jumped to 73.1% in 2021, which is almost three-quarters of all respondents. Additionally, it was found that even though *Information Systems competences* did not show any significant development regarding their importance during the years. Their availability on the market/in the organization has decreased as the share of respondents stating to disagree with a sufficient provision of talent holding these competences has leaped by more than +18%, from 46.9% to 65.7%.

Socio-Technical E-competences: All three competences in this category have gained more perceived importance during the past five years. The importance of *Expertise in e-government impact* has increased by more than +15% from 45.7% in 2016 to 61.1% in 2021. The growth of the perceived importance of *Expertise in technology and e-government adoption* has risen from 45.3% to 60%, showing a rise of +14.7%. In complementarity, the importance of *Expertise in politics of e-government* has climbed by +10.8% to 50.9% in 2021, sitting at 40% in 2016. Surprisingly though, despite the growth in perceived importance, the developments of sufficient talent being on the market/in the organization cannot be confirmed to be significant, which might indicate measures being taken by responsible actors to train staff according to the developments. This is also supported by the development of employees being sent for training more often for two of the e-competences: *Expertise in e-Government impact*, +7.7%, and *Expertise in politics of e-Government*, +9.3%.

Organizational E-competences: In this category, four items show significant development, three of which are the importance measurements. For all three e-competences, respondents regard them as increasingly important. The share of respondents deeming *Expertise in e-government structures* important has jumped by +12.4% to 58.9% from

46.5% in 2016. Similarly, while 53.5% realized the importance of *Expertise in organizational design* in 2016, the share has climbed by +11.2% to 64.7% in 2021. *Process management competences* were already perceived to be important by 58.6% in 2016. However, this value elevated by more than +10% to 68.9% in 2021. Regarding the rise of importance of all three competences' importance, the non-significance of all developments around the availability of skilled staff and additional training to be offered comes unexpectedly.

Managerial E-competences: A slightly different picture shows the category of managerial e-competences. Only three of the five competences seem to have gained in significant importance. That being said, it is necessary noting that for this group of competences, the importance was already deemed relatively high in 2016 compared to the other categories. *Business/Public management competences* are perceived to be important in 2021 by 76.8% of respondents, while it was 71.2% in 2016, showing an increase of +5.6%. A similar growth rate can be observed for *Project management competences*: with 73.8% of respondents agreeing on the importance of this competence, a jump of +5% can be ascertained for this item. *Change management competences* record a stronger leap (+10.4%) mounting to 63.2% in 2021 while being at 52.8% in 2016. The other two competences of this category, *Financial management competences* and *Performance management competences* do not show significant developments. It is noticeable that a smaller amount of respondents seems to vouch for these competences' importance; however, this result should not be overrated due to the lack of significance. Moreover, only one of the competences in question, *Business/Public management competences*, show a significant rise in employees being sent for training with an increase of +8.7%, mounting in 65.4% in 2021. The rest of the items in question does not offer any strongly significant developments.

Political-Administrative E-competences: Two of the four competences of this category have grown in their perceived importance: *E-Policy competences* with an increase of +9.4% from 45.5% in 2016 to 54.9% in 2021, and *Expertise in administrative workflows* with an increase of +8% from 76.5% in 2016 to 84.5% in 2021. For the latter, *Expertise in administrative workflows*, respondents ascertain for fewer professionals bearing these competences (+7% disagreement), but, on the other hand, attest for more employees being sent for additional training to obtain those (+4.8% agreement). In general, the respondents agree for overall more training to be obliged to employees: for *E-Policy competences*, an incline of +13.4% can be observed, reaching its peak at 54.3% in 2021. This trend is being continued for the remaining two competences because, in 2021, 64.1% (+8.2%) of respondents agreed for this to be true for *Expertise in legal framework* and 66.9% (+8.8%) for *Expertise in public policy*.

To summarize, out of 18 queried e-competences, 13 have gained in statistically significant importance between 2016 and 2021. Eight of those, *Information Systems Design-competences*, *Expertise in e-government impact*, *Expertise in technology and e-government adoption*, *Expertise in politics of e-government*, *Expertise of e-government structures*, *Expertise in organizational design*, *Process management competences* and *Change management competences* show an increase in perceived importance of at least 10%.

For five out of the 18 competences, we observe a statistically significant increase of respondents stating that they disagree with enough professionals being on the market or in the organization offering those competences. Three out of those, *Information Technology competences*, *Information Systems Design competences* and *Information Systems competences* offer an increase of more than 13% in 2021 compared to 2016, *Information Technology competences* even an increase of more than 20%. All three originate from the category of technical competences. Furthermore, *Expertise in administrative workflows* and *Expertise in public policy* show an increase of individuals disagreeing with the statement as well. It must be mentioned at this point, that even though significance could not be proven for the other competences' market availability, they all tend towards a trend of increasingly not having enough skilled professionals available.

On a positive note, it must be certified that there is a major increase of employees being sent for training in 2021 compared to 2016 across all five competence categories. Out of the 18 competences in question, 11 register for a statistically relevant jump in this question. *E-Policy competences* have made the strongest leap, followed by *Expertise in politics of e-Government*, *Business/Public management competences* and *Expertise in public policy*.

5 Discussion

The presented study and the statistical analysis clearly indicate that the often discussed topic of missing e-competences and the resulting challenges posed by the digital transformation are also relevant for the public sector. In all five competence categories, not only a huge need can be validated, even more, a significant increase of perceived importance of the different e-competences in the last years can be recognized. Hence, the analysis in this paper indicates acknowledging these trends within public administrations. Especially, the increase for technical and socio-technical e-competences is recognizable. The reasons for the increase are not surveyed in this paper, but in light of the last five years, several societal developments reinforce the importance of digital competences. Relevant societal developments that could impact the development are digitalization, the corona pandemic, demographic reasons, and the German OZG.

Five years is a long time to develop new technologies, products, and services in the digital age. This results in a multitude of new opportunities to use these technologies, products, and services to digitize processes within an organization. However, due to the rapid pace of development, the new technologies, products, and services must be continuously reviewed for their applicability to the organization. Different competences are required since other developments are being forged for different use cases simultaneously. This results in a growing need for competences for digitalization within the organization. Moreover, people need additional e-competences to handle multiple technologies, products, and services simultaneously. Consequently, it is a two-sided demand as organizations, and people need additional e-competences to control the new technology, products, and services.

The challenges mentioned above are even enforced in the public sector context, as many public officials have previously not or only slightly been tangent to digital transformation matters. Due to new processes and technologies in their workplace, they must continuously learn to adapt to digitally transforming organizations.

Table 2. Survey results: development of e-competence importance, e-competence availability and training opportunities between 2016 and 2021

Category	Competence/Competence availability/Training	2016	2021	Change	t-value	p-value
Technical	**Information Technology competences**	72.74%	77.45%	+4.7%	−2.25	0.02**
Technical	Competence on market/in organization	48.48%	68.96%	+20.5%	5.79	0.00***
Technical	Employees sent for training	68.15%	71.98%	+3.8%	−1.87	0.06*
Technical	**Information Systems Design competences**	34.41%	47.45%	+13.0%	−4.36	0.00***
Technical	Competence on market/in organization	59.41%	73.09%	+13.7%	2.89	0.00***
Technical	Employees sent for training	60.40%	66.82%	+6.4%	−0.43	0.67
Technical	**Information Systems competences**	24.70%	22.34%	−2.4%	0.45	0.65
Technical	Competence on market/in organization	46.90%	65.71%	+18.8%	3.12	0.00***
Technical	Employees sent for training	47.59%	43.81%	−3.8%	0.42	0.67
Socio-technical	**Expertise in e-Government impact**	45.66%	61.06%	+15.4%	−4.95	0.00***
Socio-technical	Competence on market/in organization	64.18%	69.69%	+5.5%	1.16	0.25
Socio-technical	Employees sent for training	52.24%	59.93%	+7.7%	−2.02	0.04**
Socio-technical	**Expertise in technology and e-Government adoption**	45.32%	60.00%	+14.7%	−5.04	0.00***
Socio-technical	Competence on market/in organization	61.28%	67.38%	+6.1%	1.17	0.24
Socio-technical	Employees sent for training	47.74%	52.13%	+4.4%	−1.40	0.16
Socio-technical	**Expertise in politics of e-Government**	40.03%	50.85%	+10.8%	−4.14	0.00***
Socio-technical	Competence on market/in organization	63.40%	69.04%	+5.6%	0.81	0.42
Socio-technical	Employees sent for training	43.83%	53.14%	+9.3%	−1.99	0.05**
Organizational	**Expertise in e-Government structures**	46.51%	58.94%	+12.4%	−4.43	0.00***
Organizational	Competence on market/in organization	53.48%	59.93%	+6.4%	0.91	0.36
Organizational	Employees sent for training	50.18%	55.96%	+5.8%	−1.69	0.09*
Organizational	**Expertise in Organizational design**	53.49%	64.68%	+11.2%	−4.15	0.00***
Organizational	Competence on market/in organization	40.45%	47.04%	+6.6%	1.31	0.19
Organizational	Employees sent for training	58.92%	59.87%	+1.0%	−0.31	0.76
Organizational	**Process management competences**	58.60%	68.94%	+10.3%	−4.16	0.00***
Organizational	Competence on market/in organization	43.31%	48.46%	+5.1%	1.12	0.26
Organizational	Employees sent for training	60.47%	63.89%	+3.4%	−1.61	0.11
Managerial	**Business/Public management competences**	71.21%	76.81%	+5.6%	−3.06	0.00***
Managerial	Competence on market/in organization	30.14%	36.57%	+6.4%	0.17	0.87
Managerial	Employees sent for training	56.70%	65.37%	+8.7%	−3.17	0.00***
Managerial	**Project management competences**	68.82%	73.83%	+5.0%	−1.99	0.05**
Managerial	Competence on market/in organization	35.64%	40.63%	+5.0%	1.55	0.12
Managerial	Employees sent for training	59.90%	63.98%	+4.1%	−1.64	0.10*
Managerial	**Financial management competences**	50.60%	47.87%	−2.7%	0.48	0.63
Managerial	Competence on market/in organization	23.91%	27.11%	+3.2%	−0.49	0.63
Managerial	Employees sent for training	58.59%	65.33%	+6.7%	−1.86	0.06*
Managerial	**Performance management competences**	37.82%	33.19%	−4.6%	−0.26	0.80
Managerial	Competence on market/in organization	42.34%	51.28%	+8.9%	1.61	0.11
Managerial	Employees sent for training	40.09%	41.03%	+0.9%	−0.84	0.40
Managerial	**Change management competences**	52.81%	63.19%	+10.4%	−4.43	0.00***
Managerial	Competence on market/in organization	50.65%	52.19%	+1.5%	0.03	0.97
Managerial	Employees sent for training	47.10%	48.48%	+1.4%	−0.33	0.74
Political-admin	**E-Policy competences**	45.49%	54.89%	+9.4%	−4.18	0.00***
Political-admin	Competence on market/in organization	49.06%	53.49%	+4.4%	−0.36	0.72
Political-admin	Employees sent for training	40.45%	54.26%	+13.8%	−3.42	0.00***
Political-admin	**Expertise in legal framework**	66.44%	68.09%	+1.6%	−1.11	0.27
Political-admin	Competence on market/in organization	22.31%	25.94%	+3.6%	0.71	0.48
Political-admin	Employees sent for training	55.90%	64.06%	+8.2%	−2.44	0.01***
Political-admin	**Expertise in administrative workflows**	76.49%	84.47%	+8.0%	−2.96	0.00***
Political-admin	Competence on market/in organization	28.06%	35.01%	+7.0%	2.01	0.04**
Political-admin	Employees sent for training	57.68%	62.47%	+4.8%	−2.04	0.04**
Political-admin	**Expertise in public policy**	59.80%	60.43%	+0.6%	−1.09	0.28
Political-admin	Competence on market/in organization	18.80%	27.46%	+8.7%	1.87	0.06*
Political-admin	Employees sent for training	58.12%	66.90%	+8.8%	−2.87	0.00***

Level of significance: *** 1%, ** 5%, * 10%

Another influencing factor is the COVID-19 pandemic. The participants were located in Germany, which was in a lockdown during the second survey iteration (January 2021). The lockdown restrictions included, among others, regulations governing businesses (employers are urged to provide options for working from home) as well as private gatherings (people are allowed to meet one other person besides their household) [3]. These measures, forcing a majority of Germany's workforce to work from home suddenly, may have very bluntly demonstrated the local digital residue to (not only) the respondents of our survey, resulting in the higher perceived importance of e-competences. After battling Corona for more than a year, many people have personally experienced which processes in their organization run well digitally and which do not.

In particular, processes that were once taken for granted are challenging in the Corona pandemic. Due to the *work from home obligation*, processes must be executable from anywhere and thus represent a challenge for public organizations and public service delivery. Within a short time, processes had to be digitally executable in order to be able to continue the processes without interruption. This sudden change may have created problems for public organizations and thus, increased the need for e-competences. Furthermore, the German administration will face a wave of retirements during the following years [6, 29]. This means a shortage of personnel and the need to find and hire new employees to overcome this shortage. To maintain administrative processes with fewer personnel, processes must be adapted to the new conditions. Due to the lack of digitalization in administration, it could be difficult to compensate for the missing workforce. Besides, the retirement of older employees and the lack of digital knowledge recording will lead to the loss of process-related information.

The OZG has been in force since 2018, obliging the German federal and state governments to offer their administrative services electronically by the end of 2022. With the introduction of the OZG, the digitalization of public administration is once again on the political agenda. As a result of this commitment, there will be more and larger projects promoted and supported in the future to achieve the goals of the OZG [13]. However, to implement the OZG, various services in the administration must be digitized, which in turn requires various competences for its realization. To master the implementation of the OZG, new and skilled personnel are required to fulfill the requirements of future public administrations.

In this regard, the second objective of the survey is considered. Considering the supply of professionals, one finding is that most respondents disagree that there are enough professionals available. According to the analysis, the request for more competences has increased over the last five years, while at the same time, the supply of professionals has deteriorated. Hence, hiring enough professionals to accomplish the upcoming digitalization challenges in public administration is critical. Especially for the IT competences, the perceived importance has significantly increased. Most respondents stated that the supply of professionals with IT competences is not sufficient. One explanation is that jobs in the private sector are more attractive compared to the public administration [6]. The lower financial compensation and less innovative technologies in use in public sector organizations are reasons for this. Consequently, job offers in public sector organizations are not as attractive for young and talented graduates. To work against these issues, public sector organizations need to offer incentives to young talents to make the public sector work more attractive.

As a consequence of the increased demand and the lack of supply, organizations need to invest in further training to counter the actual trend. The third part of the survey is addressed in this direction to figure out if organizations invest in such training. The survey shows that organizations send their employees to training courses more frequently than in the 2016 survey. Organizations have thus apparently recognized the growing need for e-competences and are acting accordingly. However, the proportion of individuals actively being sent to additional training does not appear sufficient compared to the need. Due to the ever-faster pace of digital development, it is foreseeable that the trend toward increased demand for e-competences will tend to intensify. Therefore, those responsible in public administration must take the need for more e-competences more seriously and equip their organizations for a digital future. The digitization of public sector organizations can only be carried out successfully if e-competences are in place. Therefore, one suggestion is to integrate more e-competence-related content into public sector training.

As e-competence has become a success factor for public sector digitalization, a corresponding analysis of its perceived importance over time is required to define adequate policy actions. For instance, without any knowledge about which e-competences are needed the most, public organizations cannot set up suitable programs for education or vocational training. However, only few studies are investigating e-competence development over time. Thus, an accurate statement about the development of e-competences in the public sector is oftentimes not or only vaguely possible. By conducting the same survey with a time interval of five years, our results provide a solid ground for actions to overcome the identified challenges. Additionally, complementary research is necessary to quantify the gap also in terms of job vacancies or—even better—to forecast required competence profiles.

6 Conclusion

With this paper, we investigated the area of e-competence in the public sector and pursued the goal of figuring out how those competences' perceived relevance has changed over time. We conducted a large-scale survey in two iterations with a time interval of five years to answer this question. Based on the large amount of gathered survey data, we analyzed the trends and tested them for statistical significance. The sample group was deliberately limited to Germany in order to draw insights from a group of respondents as homogenous as possible and thus avoid unnecessary distortion of our results.

Overall, our findings reveal that (RQ) *public officials now give more attention to e-competence than five years ago*. Consequently, our research contributes to a clear and very topical image of public officials' stance on e-competences. Specifically, 13 of 18 queried e-competences are perceived to be significantly more important in 2021 than in 2016. Thereby, the (a) *perceived importance of most of the competences increased.* Adding to that, the public officials (b) *rate the market availability of many competences as scarce*, five of which, especially the technology-related e-competences, have significantly grown in their rarity since 2016. Thus, the underlining staff shortage in the public sector is an aggravating topic asking for interventions. Taking the insights gained from (a) and (b) into account, there is an (c) increasing, massive need for education and training as the *currently provided training is not sufficient*. Even though respondents stated for 11 out of 18 e-competences, that employees have been/will be sent for training more often in 2021 than in 2016, the offered training does not appear to cover the requirements for competence development in the public sector.

Therefore, we vouch for current curricula of apprenticeships and university courses to adapt to the shift in requirements to meet the growing need for e-competence development. Moreover, innovative education formats and advanced training courses are necessary to train the officials already working in the field to give them a chance to take an active part in the digital transformation, being able to foster it and not just being a *passenger* of that.

Despite those findings, the research also entails limitations. Although we can see a general increase in perceived relevance and market shortage, our data cannot quantify any shortage. The IT competences, for example, are subject to a drastically increased perceived shortage on the market or within the organization. However, whether this is a result of overall rising attention to those competences pushed by organizations, research and media or whether this also goes along with an absolute increase of open positions, cannot be answered with our data. Regarding our quantitative approach, we can rely on a large sample covering many different public organizations, but it is not a representative sample accounting for different kinds of organizations in a balanced way. The statistical investigations also do not account for any mutual influence of the items since, for example, the perceived importance might also influence the awareness of any staff shortage and vice versa. Finally, despite the intentional decision to limit the survey sample to be located in Germany for the above-explained reasons, we must acknowledge that our results are of limited validity for the public sector of other international contexts, as the organization of governments and administrative structures can sometimes differ greatly from each other.

Our study confirms the rising importance of e-competence that could generally be observed over the last years in research and practice. In the next step, it is essential to figure out the reasons for the found developments, i.e., *why* can we observe such changes in the public officials' perceptions on e-competence. They might, for example, be more often confronted with situations in which e-competence is required. Also, it is fascinating whether our observations could also be part of a cultural change within public sector organizations—a tendency towards a broader positive attitude regarding public sector digitalization. If so, research would need to work closely together with practice to develop appropriate education means to meet the growing demand for e-competences. Lastly, our findings are also a first step to give a more robust educational focus on those competences that mark the highest bottleneck.

Acknowledgements. This paper was developed in the research project eGovCAMPUS, which is funded by the German IT Planning Council. The authors thank the Project Management of the FITKO, promotion sign FI-50/043/001-022022.

References

1. Auth, G., Christ, J., Bensberg, F.: Kompetenzanforderungen zur Digitalisierung der öffentlichen Verwaltung: Eine empirische Analyse auf Basis von Stellenanzeigen. In: Proceedings of the 16. Internationale Tagung Wirtschaftsinformatik, Essen, Germany (2021)
2. Becker, J., et al.: E-Government-Kompetenz. Studie im Auftrag der Arbeitsgruppe "E-Government-Kompetenz" des IT-Planungsrat, Berlin, München, Münster, Siegen (2016)
3. Bundesregierung: Videoschaltkonferenz der Bundeskanzlerin mit den Regierungschefinnen und Regierungschefs der Länder am 19. Januar 2021 [Beschluss] (2021). https://www.bundesregierung.de/resource/blob/997532/1840868/1c68fcd2008b53cf12691162bf20626f/2021-01-19-mpk-data.pdf?download=1
4. Crawford, J., et al.: COVID-19: 20 countries' higher education intra-period digital pedagogy responses. J. Appli. Learn. Teach. **3**(1), 1–20 (2020)
5. Creswell, J.W.: Research Design. Qualitative, Quantitative and Mixed Methods Approaches. Sage (2014)
6. Detemple, P., Höhn, A., Fietz, K., Malikzada, R., Unterhofer, U.: Fachkräftemangel im öffentlichen Dienst. Prognose und Handlungsstrategien bis 2030. PricewaterhouseCoopers GmbH Wirtschaftsprüfungsgesellschaft (2018)
7. Distel, B., Ogonek, N., Becker, J.: eGovernment competences revisited - a literature review on necessary competences in a digitalized public sector. In: Proceedings of the 14th International Conference on Wirtschaftsinformatik, Siegen, pp. 286–300 (2019)
8. European Commission: eGovernment Benchmark 2016. A turning point for eGovernment development in Europe? Tech. rep., Luxembourg (2016)
9. European Commission: Digital Economy and Society Index Report 2020 - Digital Public Services (2020). https://ec.europa.eu/digital-single-market/en/digital-public-services
10. Federal Ministry of the Interior: What is the Online Access Act? (2021)
11. Gil-García, J.R., Pardo, T.A.: E-government success factors: mapping practical tools to theoretical foundations. Gov. Inf. Q. **22**(2), 187–216 (2005). https://doi.org/10.1016/j.giq.2005.02.001
12. Gorbacheva, E., Stein, A., Schmiedel, T., Müller, O.: The role of gender in business process management competence supply. Bus. Inf. Syst. Eng. **58**(3), 213–231 (2016). https://doi.org/10.1007/s12599-016-0428-2
13. Halsbenning, S.: Digitalisierung öffentlicher Dienstleistungen: Herausforderungen und Erfolgsfaktoren der OZG-Umsetzung in der Kommunalverwaltung. HMD Praxis der Wirtschaftsinformatik **58**(5), 1038–1053 (2021). https://doi.org/10.1365/s40702-021-00765-5
14. Halsbenning, S., Niemann, M., Distel, B., Becker, J.: Playing (government) seriously: design principles for e-government simulation game platforms. In: Proceedings of the 16. Internationale Tagung Wirtschaftsinformatik (2021)
15. Hunnius, S., Paulowitsch, B., Schuppan, T.: Does E-Government education meet competency requirements? an analysis of the German university system from international perspective. In: 2015 48th Hawaii International Conference on System Sciences, pp. 2116–2123. IEEE (2015)
16. Hunnius, S., Schuppan, T.: Competency requirements for transformational e-government. In: Proceedings of the 46th Hawaii International Conference on System Sciences, pp. 1664–1673 (2013)
17. Initiative D21: Digital Skills Gap - So (unterschiedlich) digital kompetent ist die deutsche Bevölkerung (2021). https://www.digitale-exzellenz.de/digital-skills-gap-uberbrucken/
18. Layne, K., Lee, J.: Developing fully functional E-government: a four stage model. Gov. Inf. Q. **18**, 122–136 (2001)

19. Le Deist, F.D., Winterton, J.: What is competence? Hum. Resour. Dev. Int. **8**(1), 27–46 (2005). https://doi.org/10.1080/1367886042000338227
20. Lindgren, I., Madsen, C.Ø., Hofmann, S., Melin, U.: Close encounters of the digital kind: a research agenda for the digitalization of public services. Gov. Inf. Q. **36**(3), 427–436 (2019). https://doi.org/10.1016/j.giq.2019.03.002
21. Mergel, I.: Kompetenzen für die digitale Transformation der Verwaltung. Innovative Verwaltung **4**, 34–36 (2020)
22. Mergel, I., Edelmann, N., Haug, N.: Defining digital transformation: results from expert interviews. Gov. Inf. Q. **36**(4), 101385 (2019). https://doi.org/10.1016/j.giq.2019.06.002
23. Niehaves, B., Plattfaut, R., Becker, J.: Business process management capabilities in local governments: a multi-method study. Gov. Inf. Q. **30**(3), 217–225 (2013). https://doi.org/10.1016/j.giq.2013.03.002
24. Nordhaug, O.: Human Capital in Organizations: Competence, Training, and Learning. Scandinavian University Press, Oslo (1993)
25. Ogonek, N., Distel, B., Becker, J.: Let's Play ... eGovernment! A Simulation Game for Competence Development among Public Administration Students. In: Proceedings of the 52nd Hawaii International Conference on System Sciences, pp. 3087–3096 (2019). https://doi.org/10.24251/hicss.2019.373
26. Ogonek, N., Hofmann, S.: Governments' need for digitization skills: understanding and shaping vocational training in the public sector. Int. J. Public Adm. Digital Age **5**(4), 61–75 (2018). https://doi.org/10.4018/IJPADA.2018100105
27. Schmeling, J., Bruns, L.: Kompetenzen. Perspektiven und Lernmethoden im Digitalisierten Öffentlichen Sektor. Tech. rep, Berlin (2020)
28. Scholta, H., Niemann, M., Halsbenning, S., Räckers, M., Becker, J.: Fast and federal - policies for next-generation federalism in Germany. In: Proceedings of the 52nd Hawaii International Conference on System Sciences (HICSS), pp. 3273–3282 (2019)
29. Statistisches Bundesamt: Finanzen und Steuern. Personal des öffentlichen Dienstes - Fachserie 14 Reihe 6–2018. Tech. rep. (2019)
30. Twizeyimana, J.D., Andersson, A.: The public value of E-Government - a literature review. Gov. Inf. Q. **36**(2), 167–178 (2019). https://doi.org/10.1016/J.GIQ.2019.01.001
31. United Nations: E-Government Development Index (2020). https://publicadministration.un.org/egovkb/en-us/Data-Center
32. Weerakkody, V., Baire, S., Choudrie, J.: E-Government: the need for effective process management in the public sector. In: Proceedings of the 39th Hawaii International Conference on System Sciences (2006)

Transforming Cultural Heritage—A Digital Humanity Perspective with Virtual Reality

Ling-Ling Lai(✉), Sinn-Cheng Lin, and Han-Chian Wang

Tamkang University, Taipei, Taiwan
llai@mail.tku.edu.tw

Abstract. The study aims to explore a virtual reality (VR) approach representing a local history in the sense of preserving cultural heritage. Two 400-year fort cities, Tamsui and Keelung, located in northern Taiwan, were chosen and transformed into digital fort cities by the researchers from the fields of history, information sciences, and computer science. The researchers invited participants to the VR lab and experience a VR tour. Research tools used for collecting user data include before and after attitude scales and a post-task interview for each participants. The findings of the research revealed that the senses of joy, presence, and control could critically impact VR user experience. The design of the content needed to be more informative and interactive. Affordances in the immersive environment were key for the user to experience a smooth and positive user journey with less pain points. For future work, the next round of data collection will be carried out. Data will be compared and contrasted further to see the differences.

Keywords: Digital cultural heritage · Digital humanities · User experience · User journey · Virtual reality

1 Introduction

Virtual reality (VR) was first predicted in a science fiction novel in the 1930s by writer Stanley G. Weinbaum (Virtual Reality Society 2017). In this study, virtual reality is defined as a computer-generated environment that creates a space where a user can interact with various stimuli (Limniou et al. 2008). The user experiences "scenery, objects, and sound effects" in the virtual space (Lo and Cheng 2020). The sense of immersion and "being there" are expected. The study aims to explore important elements of user experience (UX) when using VR as a tool to relearn and relive local history. As we embrace technology to engage users through various devices, how do users perceive the experience? Could our legacy pass through future generation that meet users' expectations? From the perspective of digital humanities and based on the assumption that VR could bring the past to life, the research team explored users' presumption, perception, attitude, and behavior of VR in the context of a digital collaboration among experts in information sciences and user research. Upon completion of the study in July 2022, researchers will identify current advantages and challenges of VR in preserving culture

F. Fui-Hoon Nah and K. Siau (Eds.): HCII 2022, LNCS 13327, pp. 87–96, 2022.
https://doi.org/10.1007/978-3-031-05544-7_7

heritage with an emphasis of digital curation. For now, this paper serves an exploratory purpose.

According to UNESCO (2021), heritage is defined as "our legacy from the past, what we live with today, and what we pass on to future generations." For every country there is valuable heritage and should be passed on for future generations to understand their invaluable past. For cultural heritage, it means "those sites, objects and intangible things that have cultural, historical, aesthetic, archaeological, scientific, ethnological or anthropological value to groups and individuals" (UNESCO 2021). In this research, our research team continues its effort in digital humanities by rebuilding local history in northern Taiwan with the combination of artificial intelligence (AI) and VR. A virtual space of the past is set up. This paper focuses on the user aspect and evaluates effects of advanced digital tools. Two 400-year fort cities, Tamsui and Keelung, located in northern Taiwan, were chosen and transformed into digital fort cities by our team researchers from the fields of history, information sciences, and computer science. The research questions are as follows:

Q1: What are the user experience?

(1) How do users perceive before and after reliving history in VR?
(2) How do users feel the sense of presence in the virtual environment?
(3) How do users feel the sense of joy in the virtual environment?

Q2: What are the user journey?

(1) What is the attitude change before and after reliving history in VR?
(2) What are the high points of reliving history in VR?
(3) What are the pain points of reliving history in VR?

2 Related Literature

What is an experience? Before we try to understand how users perceive an object in the digital realm, we first needs to understand what constitutes an experience. According to Li et al. (2001), an experience is "more than simply the passive reception of external sensations or subjective mental interpretation of an event or situation." Mathur (1971) noted that "an experience is the product of an ongoing transaction that gains in quality, intensity, meaning, and value integrating both psychological and emotional conditions."

2.1 User Experience

The term "user experience" is defined in ISO 9241-210. According to the definition, user experience is about "a person's perceptions and responses resulting from the use and/or anticipated use of a product, system or service" (Usability.de 2022). Any of all three aspects of positive user experience means that the product, system of service at least meets or goes beyond users' anticipation and perceptions. According to Norman and Nielsen (2021), the most important requirement for an exemplary UX is to "meet the exact needs of the customer, without fuss or bother." Other factors are joy to own and joy to use, because of the simplify and elegance that the products bring. User experience also include "the overall workflow and the steps before and after the product is actually in use" (Product Design 2021). To sum up, it is believed UX covers all aspects of the user's interaction with the company, the service, and the product. Based on this discussion, we look at the user's expectations and perceptions, more specifically the sense of presence.

Sense of Presence. In the immersive environment, feeling real is a critical aspect for users to have a positive experience. Erickson-Davis et al. (2021) stated that the "experience of presence implies a felt sense of 'realness'—the being is really over there, within reach." Looking into the literature of user experience, Cruz-cunha et al. (2013), the sense of presence means the user's degree of presence, usually being discussed in augmented reality (AR) and immersive environments, reflecting the degree to which an individual feels present in the environment, and a sense of "being in and belonging in." More specifically, the feeling of "involvement, warmth, and immediacy" while interacting with each other in online environments." With advancement of technology for virtual reality, the promises that the "sense of presence" and "being there" help users of all sorts immerse in the created environment and continue with the task with joy and involvement.

2.2 User Journey

Experts in the field of user study explains that a user journey map is "a visualization of the process that a person goes through in order to accomplish a goal" (Gibbons 2018). The journey described in the form of a map is a common UX tool that comes in all shapes and formats. Since the use of the map could highly depends on the context, the user journey map in this study is used as a way for the participants to describe their own feelings throughout the journey, thus the attitude change borrowed from consumer and marketing are used as a scaling tool to see before and after virtually touring historic sites in an immersive environment how participants anticipate before the touring and perceive the actual experience afterwards.

Attitude Change. Attitude change before and after the virtual experience helps us understands the user's attitude and the associated behavior throughout the journey. Originated from customer behavior, attitude change usually involves measuring before and after going through a certain experience, whether it is well-meaning or manipulative. Levine (2003) pointed out that "a well-meaning persuasion is called education; when it is manipulative, it might be called mind control." In a similar sense, the immersive virtual environment persuade customers that what is being experienced is what could be seen in the real world. When it is positively received, the persuasion is successful; thus brings an attitude change.

In the field of marketing, in addition to ask people directly about their thoughts, such as an interview, scales are one of the tools that are used for attitude change measurement (Epstein, n.d.). Thus, to measure attitude change, participants are asked to draw on a numbering scale with arrows indicating users' attitude towards each statement regarding their VR user journey.

(1) A numbering scale

In order for participants to sufficiently express their feelings, a scale with plus and minus 10 is designed to measure frequency, quality, importance, and likelihood.

(2) Figure drawings

The study uses the technique of figure drawings to assess the attitude change (Mcleod 2009). The technique originates from psychology is used to understand an individual's psychological state and unspoken opinions. This projective diagnostic technique works as an individual is instructed to draw arrows toward a positive or a negative state so that his/her cognitive, interpersonal, or psychological state can be assessed.

3 Research Methods

A VR laboratory was set up and research participants were recruited for a sequence of tasks. The research team designed the research tool for collecting data, which included a brief demographic and VR experience survey, a before and an after attitude scale, and post-task interview questions.

More specifically, the process of data collection included the task of having users wearing a VR device for a virtual tour to the two historic fort cities, think-aloud during the virtual tour, and post-task interviews. The virtual environment is where two 400-year old fort cities are built collaboratively with team experts with expertise in 3D technology. The researchers invited participants to the VR lab and experience a VR tour as shown in Fig. 1. Before the tour started, participants were asked to fill in a questionnaire and a before-task attitude scale, expressing his/her prior experience and/or anticipation for VR. The participants were then asked to wear the VR device and begin the tour. During the tour, participants were asked to think aloud; the entire process was recorded for data analysis. Afterwards, a post-task questionnaire was administered including a post-task

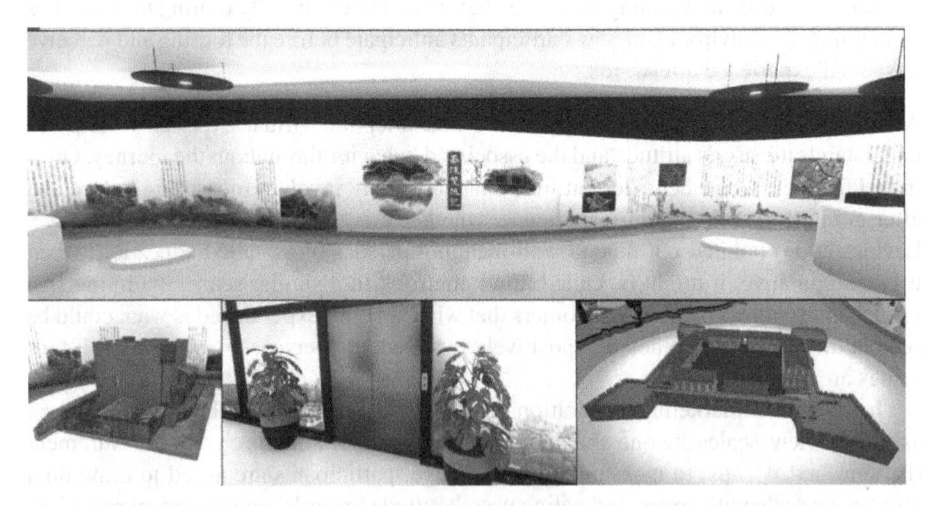

Fig. 1. The virtual environment of historical fort cities, Tamsui and Keelung.

attitude scale. In addition, a short post-task interview were carried out for the researcher to clarify any ambiguities that might occur and served as a follow-up for obtaining insights for attitude change. Figure 2 and 3 below showing participants filling out the before and after questionnaires as well as the lab setting were taken during one of the session for data collection.

Fig. 2. A participant draw on the attitude scale in the lab.

Fig. 3. User explored the historical setting in the VR

3.1 Attitude Statements and Scales

Examining related literature, Besoain et al. (2022) looked into attitude change in their evaluation study on the effects of a virtual museum of cultural heritage. The method used was a factorial design of 2 × 2 × 2 including the (1) direction of thoughts, (2) level of presence, and (3) two test settings, i.e., a virtual museum versus an interactive website. In this study, we designed a scale with positive and negative directions, from 0 to positive 10 and negative 10, which helped participants express their attitude towards 14 descriptions (Table 1) about VR user experiences covering three elements: (1) system, (2) process, and (3) content. The system element covers questions and statements related to behaviors; the process element covers how participants perceive and feel over the course of the experience; the content element covers representations of the local history in its digital form. For instance, in the system element, the statement "I feel comfortable moving around in the VR" is intended to reveal how the participant feels the actual and perceived comfort in the VR tour. The actual comfort might mean the VR headset and device is easy to wear, but it also could mean moving around in the immersive virtual environment is easy and comfortable, with full control of directions. In this case, the perceived comfort is achieved.

Table 1. Attitude statements grouped by elements of user experience.

Elements of user experience	Item #	Statements
System	4	I feel comfortable moving around in the VR
	8	I feel headaches or dizziness during the process
	10	I think I can perfectly control where I want to go through my own actions
	13	I can't control myself in the VR
Process	1	I felt like I was there
	5	I feel happy in the VR
	7	I would like to stay in the VR scenario
	11	I felt like I was experiencing an exciting moment while interacting with the scene
Content	2	I think the scenes are very informative
	3	I (would) have a different feeling about the events before (after) I revisit the history in the VR
	6	I (would) have different thoughts about the events before (after) reliving the history in the VR
	9	I think the VR scenes are beautiful and impressive
	12	I have a deep understanding of the content presented in the VR
	14	I have full interaction with the scenery during the experience

4 Findings and Discussions

The team was able to recruit 17 participants in a data collection period of approximately 3 weeks (December 20th, 2021 to January 14th, 2022), which included 8 (47%) of the participants had zero VR experience and 9 (52.9%) of the participants had previous VR experienced; 12 of the participants were female (70.5%); 5 of the participants were male (29.4%). The participants were composed of mostly undergraduate students (82.3%) and graduate students (17.6%) from mainly Liberal Arts background of a comprehensive university in northern Taiwan as shown in Table 2.

Table 2. Participant demographics.

		Number of participants	Percentage (%)
Gender	Female	5	29.4
	Male	12	70.6
Education level	Undergraduate	9	75
	Graduate school	3	25
Previous VR experience	Yes	4	33.3
	No	8	66.6

4.1 Sense of Joy, Presence, and Control

Examining user experience from the sense of joy, participants showed greatest attitude changes when comparing the before and after attitudes. The statement "I felt like I was experiencing an exciting moment while interacting with the scene," as seen in Table 3, indicated that there was disappointment on the level of emotion. Participants felt a loss when expectations for interactions and emotional high were not met, according to the analysis of interview data. The attitude changes for the statement "I would like to stay in the VR scenario" as well as the interview results both showed that the scenario in the VR tour was not engaging nor inviting. Participants expressed that they expected there would be objects in the room, and possibly the one could use his controller to interact with some of the objects if there were any. There was one exception, though. Since the tour was rather simple and straightforward, participants were pleasantly surprised when they realized that they could fly to the rooftop of the ancient forts, because in the reality it was an impossible scenario. The statement "I feel happy in the VR" received a negative attitude change as well, compared with the before attitude scale. We could say at its current state, the sense of joy was absent in the design of the immersive environment.

The sense of presence and the sense of "being there" are expected in the VR as one of the core elements and an essential criteria for good user experiences. Research results showed that the statement "I felt like I was there" received a reversed change of attitude when participants were asked to evaluate their expectations and perceptions. On the other hand, because of the slight problem of the VR headset, the statement "I feel headaches or dizziness during the process" received the 3rd highest ranking indicating participants were reminded that they were manipulating a VR.

Participants noted that before experiencing the VR tour, it was expected that they would not be able to have a pleasant and smooth movement in the environment. The scale of attitude changes revealed that the statements concerning "sense of control" reversed the attitude directions afterwards, meaning participants were actually able to move pleasantly and smoothly around in the virtual space, feeling the sense of control over what was seemingly an uncontrollable situation. The statement "I think I can perfectly control where I want to go through my own actions" received the 2nd highest ranking comparing with the before attitude also proved that participants were pleased with the sense of control.

Table 3. Rankings for attitude change directions.

Attitude change		
Negative to positive	Ranked by responses	Statements/Elements
	1	I can't control myself in the VR/System
	2	I think I can perfectly control where I want to go through my own actions/System
	3	I have a deep understanding of the content presented in the VR/Content
	4	I have full interaction with the scenery during the experience/Content
	5	I felt like I was there/Process
Positive to negative	Ranked by responses	Statements/Elements
	1	I felt like I was experiencing an exciting moment while interacting with the scene/Process
	2	I think the scenes are very informative/Content
	3	I feel headaches or dizziness during the process/System
	4	I have different thoughts about the events (after) reliving the history in the VR/Content
	5	I would like to stay in the VR scenario/Process
	6	I think the VR scenes are beautiful and impressive/Content
	7	I feel comfortable moving around in VR/System
	8	I (would) have a different feeling about the events before (after) I revisit the history in the VR/Content
	9	I feel happy in the VR/Process

4.2 The Design of VR Content

Research data showed that the most critical element of VR is interaction. Before experiencing the VR, participants expected that there would be more interaction implemented in the tour; participants' after-task attitude changes revealed there were not enough interaction. On the other hand, the post-task interview data also pointed out that the content of VR needed to be more interactive and informative (see Table 3). Furthermore, examining past literature regarding the design of VR, a number of things need to be considered, especially affordance.

According to Norman (1988), the concepts of affordances could be traced back to Gibson's (1966) which studied human perception. The type of expected interaction between users and products is called affordances. Users bring their experiences, assumptions, and anticipations from the physical world into the virtual environment.

For instance, the buttons, levers, handles of products are affordances that suggest behavioral interaction. Similarly, these designs in the VR also evoke users' anticipations and emotions. Participants revealed that they felt disappointed when there were no persons nor objects to be interacted with when opening the door and entering a room in the VR.

The richness of the design is also a problem participants pointed out. Since the content of the VR is related to a period of local history, the research team is curious to find out whether reliving the history changed the cognitive state. Overall, participants gave positive attitude of the purpose, yet felt disappointed that the historical setting was lacking richness and details. Participants noticed a map was present but were surprised that there was no possible interaction with the map.

Another example is the VR controller. Participants expressed that they assumed the controller would serve as their feet and take them to any directions they wish as they moved the controller to the corresponding angle; but in fact the controller could only move forwards and backwards, taking the participant going forward and backward in the VR. These incidents again showed that when designing the content of the VR, it is critical to meet users' expectations, offer affordances accordingly. Norman (1988) stated that "In the design of objects, real affordances are not nearly so important as perceived ones; it is the perceived affordances that tell the user what actions can be performed on an object and, to some extent, how to do them" (p. 123). Norman also stated that "when affordances are taken advantage of, the user knows what to do just by looking: no picture, label, or instruction needed" (p. 9). The lessons learned here is that the design of content should, if at all possible, very closely meet the expectation of affordances that the user perceives to ensure a positive user experience in terms of learning and receiving the message of the VR design team tries to convey.

5 Conclusions

The study took a qualitative approach and the results so far revealed current perceptions and insights from users. Even though the results represented initial findings from a rather small number of participants, the results could serve as a base for the still ongoing process of data collection. A number of problems of user experience for VR were discussed. The analysis revealed that users experienced not enough interaction during the process of the VR tour, which was the most critical problem. Content per se, users expected a rather rich historical content that is more informative from the learning point of view. For the system aspect, users expressed discomfort such as headaches or dizziness when taking the VR tour. Researchers expected that upon completion of the research in the next couple of months, the researchers would be able to build the framework of user experience and identify key points for VR design and experience, especially in the context of digital preservation of cultural heritage.

VR technology already has significant applications and impacts in various fields such as healthcare, entertainment, education, and marketing. From the viewpoint of marketing digital cultural heritage, the use of VR is highly rated as "Business analysts have suggested that the development of VR is comparable in importance with that of social media" (Morris 2016). Barnes (2017) noted, as technologies continue to advance, VR is very likely to become a key medium for marketing. From the user experience and user journey point of view, incorporating VR to promote local history for creating and preserving collective memory as well as e-learning, this study identified with its initial research results the key points that needed to be considered when designing VR content related to representing local histories.

The research project is set out to continue collecting research data for another 6 weeks approximately based on the planned schedule. For the current paper, it serves as a progress report. Researchers are curious to find out whether participants from different academic backgrounds would show differences in attitude changes. For the next round of data collection and analysis, it could be compared and contrasted to see the differences academic backgrounds make and a framework would be proposed.

References

Barnes, S.: Understanding virtual reality in marketing: nature, implications and potential (2017). https://ssrn.com/abstract=2909100

Besoain, F., González-Ortega, J., Gallardo, I.: An evaluation of the effects of a virtual museum on users' attitudes towards cultural heritage. Appl. Sci. **12**, 1341 (2022). https://doi.org/10.3390/app12031341

Cruz-cunha, M.M., Miranda, I., Gonçalves, P.: Handbook of Research on ICTs for Human-Centered Healthcare and Social Care Services (2013)

Epstein, L.: Measuring attitude change, the right way. SurveyMonkey (n.d.). https://www.surveymonkey.com/curiosity/measuring-attitude-changes/

Erickson-Davis, C., et al.: The sense of presence: lessons from virtual reality. Relig. Brain Behav. **11**(3), 335–351 (2021)

Gibson, J.J.: The Senses Considered as Perceptual Systems. Houghton Mifflin (1966)

Gibbons, S.: Journey mapping 101 (2018). https://www.nngroup.com/articles/journey-mapping-101/

Levin, R.: The Power of Persuasion: How We're Bought and Sold. Wiley, Hoboken, NJ (2003)

Li, H., Daugherty, T., Biocca, F.: Characteristics of virtual experience in electronic commerce: a protocol analysis. J. Interact. Mark. **15**(3), 13 (2001)

Limniou, M., Roberts, D., Papadopoulos, N.: Full immersive virtual environment CAVE in chemistry education. Comput. Educ. **51**, 584–593 (2008)

Lo, W.H., Cheng, K.L.B.: Does virtual reality attract visitors? The mediating effect of presence on consumer response in virtual reality tourism advertising. Inf. Technol. Tour. **22**(4), 537–562 (2020). https://doi.org/10.1007/s40558-020-00190-2

Mathur, D.C.: Naturalistic Philosophics of Experience. Warren H. Green Inc., St. Louis, MO (1971)

McLeod, S.A.: Attitude measurement. Simply Psychology, 24 October 2018. https://www.simplypsychology.org/attitude-measurement.html

Morris, C.: Virtual reality and the new sales experience. Campaign, 22 April 2016. http://www.campaignlive.co.uk/article/virtual-reality-new-sales-experience/1392253

Norman, D.: The Psychology of Everyday Things. Basic Books, New York (1988)

Norman, D., Nielsen, J.: The definition of user experience (UX) (2021). https://www.nngroup.com/articles/definition-user-experience/

Product Design. User experience (2021). https://www.productplan.com/glossary/user-experience/

UNESCO: Concept of cultural heritage (2021). https://en.unesco.org/themes/information-preservation/digital-heritage/concept-digital-heritage

Usability.de: It is the user who decides on the success of your products (2022). https://www.usability.de/en/usability-user-experience.html

Virtual Reality Society: History of virtual reality (2017). https://www.vrs.org.uk/virtual-reality/history.html

An Epistemological Analysis of the "Brain in a Vat" Approach for the Philosophy of Artificial Intelligence

Batnasan Luvaanjalba(✉) and Bo-chiuan Su

Department of Information Management, National Dong Hwa University, Shoufeng Township, Taiwan
batnasanlu@gmail.com, bsu@gms.ndhu.edu.tw

Abstract. The development of artificial intelligence (AI) attempts to model the human mind and processes. It has became clear that the interest of scientists in modelling human thought is inextricably linked to the brain, language, and physical body. Hilary Putnam is one of many researchers who have researched this area. His study called The Brain in the Vat has attracted a lot of attention for philosophers. The most important issue he raises in this article is the concept of the Brain in Vat, focusing on the development of brain, body, and language issues. The purpose of this paper is to analyze the causes of the Brain in the vat hypothesis and the interaction between the brain and the physical body for artificial intelligence. Since the argument is whether it is impossible to know if someone is the brain of a vessel. In principle, it is impossible to deny that you are the brain of the vessel, so there is no good reason to believe what you believe. The dubious argument argues that they cannot be known and raises issues related to the definition of knowledge. An example of this is the Brain in Vat hypothesis. It is necessary to study not only artificial intelligence scientists but also philosophers, neuroscientists, psychologists, medical scientists, physicists and many other scientists. It would be ethical to conduct human-centred artificial intelligence research in this way; always develop a link between artificial intelligence and the physical body in order to develop epistemological hypotheses in the philosophy of artificial intelligencee. In order to develop artificial intelligence, it is necessary to combine brain function and physical body interaction together. It is more important to develop and direct the activities of human-centred robots than to promote robots beyond the human mind. Although neuroscience is developing rapidly, not all the secrets of the brain have been revealed yet. Therefore, it is important to develop many interesting hypotheses from the philosophical and linguistic point of view for artificial intelligence. Many scientists have found that collaborative research is more likely to lead to better results. The development of human-centered artificial intelligence can bring positive technological and economic advances to humanity in many ways. The challenge is to direct and use this new evolution in a timely manner. When it comes to artificial intelligence, the question of the physical body is always on the agenda. Hilary Putnam's Brain in A Vat is important not only to show many researchers what is possible and what is not possible but also to remind them what to focus on in the future.

F. Fui-Hoon Nah and K. Siau (Eds.): HCII 2022, LNCS 13327, pp. 97–111, 2022.
https://doi.org/10.1007/978-3-031-05544-7_8

Keywords: artificial intelligence · Human-computer interaction · philosophy · Brain in a vat

1 Introduction

Hilary Whitehall Putnam (1926–2016) was an American philosopher, computer scientist, and mathematician, and a major figure in analytic philosophy in the second half of the 20th century. He made significant contributions to the philosophy of language, the philosophy of mathematics, the philosophy of the mind, the philosophy of science also to mathematics and computer science.

In 1981, he wrote a book, Reason, Truth, and History. The first chapter of this work deals with "Brains in a Vat". This article will analyze this issue.

The Brain in Vat (BIV) is a variant used in the experiment of various thoughts aimed at producing certain notions about human knowledge, reality, truth, intelligence, consciousness, and perception. It identifies the option by removing the human brain from the body, placing it in a life-sustaining fluid container, and connecting the nerve cells with a wire. On the other hand, there is a supercomputer that provides normal electrical impulses to the brain. The computer mimics reality, and the "bodiless" brain transmits these normal functions to the computer.

The simplest use of "Brain in Vat" variants helps to broaden the knowledge of philosophical doubts. The simplest version of this is as follows: The brain in a water tank sends and receives the same impulses as the skull, so these are the only way to interact with the environment. Tell him whether he is in the skull or in the fluid from the point of view of the brain. However, in the first case, most of the person's beliefs may be true (if they say they are walking in the wind or going shopping); In the latter case, their beliefs are false.

Since the argument is that it is impossible to know if someone is the brain of a vessel, it is impossible to know whether most beliefs can be completely false.

In principle, it is impossible to deny that you are the brain of the vessel, so there is no good reason to believe what you believe; The dubious argument argues that they cannot be known and raises issues related to the definition of knowledge.

Recently, many modern philosophers believe that virtual reality has a serious impact on human independence. But another view is that VR (Virtual Reality) does not destroy our cognitive structure or our connection to reality. In contrast, VR allows us to have more new ideas, new insights, and new perspectives on the world.

2 Literature Review

Many theories have been developed on the Input-Process-Output line and it has also been the subject of much speculation by generations of philosophers. There were advanced many methods of the time about what information to include and how to process that information and what the results would be.

Among philosophers is Descartes' "Evil demon hypothesis". In the mid-20th century, the Austro-English philosopher Ludwig Wittgenstein also defined his view of language as "The Way Out of the Fly-Bottle."

So why do they value a container, a box, a room as their object of study?

This is because they are program processing machines that have a different shape and design but are close in content.

As for Descartes, question was, "Where is my body?" Is the question important?

Rene Descartes' words "I think, therefore I am".

So, if something is thought, can it be considered to exist?

Now let's analyze the main topic.

Suppose someone succeeds in inventing a computer which can actually carry on an intelligent conversation with one (on as many subjects as an intelligent person might). How can one decide if the computer is 'conscious'?

The British logician Alan Turing proposed the following test: (2 A. M. Turing, 'Computing Machinery and Intelligence', Mind (1950), reprinted in A. R. Anderson (ed.), Minds and Machines.) let someone carry on a conversation with the computer and a conversation with a person whom he does not know.

If he can-not tell which is the computer and which is the human being, then (assume the test to be repeated a sufficient number of times with different interlocutors) the computer is conscious. In short, a computing machine is conscious if it can pass the 'Turing Test'. (The conversations are not to be carried on face to face, of course, since the interlocutor is not to know the visual appearance of either of his two conversational parmers [1].

Although the Turing test is important, it is more important today to test its significance in a variety of ways.

Nor is voice to be used, since the mechanical voice might simply sound different from a human voice. Imagine, rather, that the conversations are all carried on via electric typewriter. The interlocutor types in his statements, questions, etc., and the two partners—the machine and the person—respond via the electric keyboard.

Also, the machine may lie—asked 'Are you a machine', it might reply, 'No, I'm an assistant in the lab here.') The idea that this test is really a definitive test of consciousness has been criticized by a number of authors (who are by no means hostile in principle to the idea that a machine might be conscious). But this is not our topic at this time. I wish to use the general idea of the Turing test, the general idea of a dialogic test of competence, for a different purpose, the purpose of exploring the notion of reference [1].

Of course, there are many advantages to testing.

Imagine a situation in which the problem is not to determine if the partner is really a person or a machine, but is rather to determine if the partner uses the words to referas we do. The obvious test is, again, to carry on a conversation, and, if no problems arise, if the partner 'passes' in the sense of being indistinguishable from someone who is certified in advance to be speaking the same language, referring' to the usual sorts of objects, etc., to conclude that the partner does refer to objects as we do. When the purpose of the Turing test is as just described, that is, to determine the existence of (shared) reference, I shall refer to the test as the Turing Test for Reference.

And, just as philosophers have discussed the question whether the original Turing test is a definitive test for consciousness, i.e. the question of whether a machine which 'passes' the test not just once but regularly is necessarily conscious, so, in the same way,

I wish to discuss the question of whether the Turing Test for Reference just suggested is a definitive test for shared reference [1].

Is it a definite test? However, after repeated testing, the criteria should be reconsidered.

The answer will turn out to be 'No'. The Turing Test for Reference is not definitive. It is certainly an excellent test in practice; but it is not logically impossible (though it is certainly highly improbable) that someone could pass the Turing Test for Reference and not be referring to anything. It follows from this, as we shall see, that we can extend our observation that words (and whole texts and discourses) do not have a necessary connection to their referents.

Even if we consider not words by themselves but rules deciding what words may appropriately be produced in certain contexts—even if we consider, in computer jargon, programs for using words—unless those programs themselves refer to something extralinguistic there is still no determinate reference that those words possess [1].

The evaluator or examiner should also consider the criteria for testing.

This will be a crucial step in the process of reaching the conclusion that the Brain-in-a-Vat Worlders cannot refer to anything external at all (and hence cannot say that they are Brain-in-a-Vat Worlders).

Suppose, for example, that I am in the Turing situation (playing the 'Imitation Game', in Turing's terminology) and my partner is actually a machine. Suppose this machine is able to win the game ('passes' the test).

Imagine the machine to be programmed to produce beautiful responses in English to statements, questions, remarks, etc. in English, but that it has no sense organs (other than the hookup to my electric typewriter), 'and no motor organs (other than the electric typewriter). (As far as I can make out, Turing does not assume that the possession of either sense organs or motor organs is necessary for consciousness or intelligence) [1].

In this case, it is easier to detect by differences in sensory and foreign language characteristics.

"Assume that not only does the machine lack electronic eyes and ears, etc., but that there are no provisions in the machine's program, the program for playing the Imitation Game, for incorporating inputs from such sense organs, or for controlling a body. What should we say about such a machine?

To me, it seems evident that we cannot and should not attribute reference to such a device. It is true that the machine can discourse beautifully about, say, the scenery in New England. But it could not recognize an apple tree or an apple, a mountain or a cow, afield or a steeple, if it were in front of one. What we have is a device for producing sentences in response to sentences.

But none of these sentences is at all connected to the real world. If one coupled two of these machines and let them play the Imitation Game with each other, then they would go on fooling' each other forever, even if the rest of the world disappeared! There is no more reason to regard the machine's talk of apples as referring to real world apples than there is to regard the ant's 'drawing' as referring to Winston Churchill [1].

In this case, the machine will only be able to perform tasks in data-configured mode. When tested outside of the data, it immediately makes a mistake and proves itself to be a machine.

What produces the illusion of reference, meaning, intelligence, etc., here is the fact that there is a convention of representation which we have under which the machine's discourse refers to apples, steeples, New England, etc. Similarly, there is the illusion that the ant has caricatured Churchill, for the same reason.

But we are able to perceive, handle, deal with apples and fields. Our talk of apples and fields is intimately connected with our nonverbal transactions with apples and fields.

There are 'language entry rules' which take us from experiences of apples to such utterances as 'I see an apple', and 'language exit rules' which take us from decisions expressed in linguistic form ('I am going to buy some apples') to actions other than speaking [1].

There is a need to clarify the difference between real and unreal.

Lacking either language entry rules or language exit rules, there is no reason to regard the conversation of the machine (orof the two machines, in the case we envisaged of two machines playing the Imitation Game with each other) as more than syntactic play. Syntactic play that resembles intelligent discourse, to be sure; but only as (and no more than) the ant's curve resembles a biting caricature.

In the case of the ant, we could have argued that the ant would have drawn the same curve even if Winston Churchill had never existed. In the case of the machine, we cannot quite make the parallel argument; if apples, trees, steeples and fields had not existed, then, presumably, the programmers would not have produced that same program [1].

Therefore, it is more important to look for similarities in all interrelated things based on logic.

Although the machine does not perceive apples, fields, or steeples, its creator-designers did. There is some causal connection between the machine and the real world apples, etc., via the perceptual experience and knowledge of the creator-designers. But such a weak connection can hardly suffice for reference.

Not only is it logically possible, though fantastically improbable, that the same machine could have existed even if apples, fields, and steeples had not existed; more important, the machine is utterly insensitive to the continued existence of apples, fields, steeples, etc. Even if all these things ceased to exist, the machine would still discourse just as happily in the same way. That is why the machine cannot be regarded as referring at all.

The point that is relevant for our discussion is that there is nothing in Turing's Test to rule out a machine which is programmed to do nothing but play the Imitation Game, and that a machine which can do nothing but play the Imitation Game is clearly not referring any more than a record player is [1].

Here is a science fiction possibility discussed by philosophers: imagine that a human being (you can imagine this to be yourself) has been subjected to an operation by an evil scientist.

The person's brain (your brain) has been removed from the body and placed in a vat of nutrients which keeps the brain alive. The nerve endings have been connected to a super-scientific computer which causes the person whose brain it is to have the illusion that everything is perfectly normal.

There seem to be people, objects, the sky, etc.; but really all the person (you) is experiencing is the result of electronic impulses travelling from the computer to the nerve endings. The computer is so clever that if the person tries to raise his hand, the feedback from the computer will cause him to 'see' and 'feel' the hand being raised [1].

There have been many experiments in the history of mankind. Thanks to those experiments, science is correcting its mistakes and moving forward. Moreover, by varying the program, the evil scientist can cause the victim to 'experience' (or hallucinate) any situation or environment the evil scientist wishes.

He can also obliterate the memory of the brain operation, so that the victim will seem to himself to have always been in this environment. It can even seem to the victim that he is sitting and reading these very words about the amusing but quite absurd supposition that there is an evil scientist who removes people's brains from their bodies and places them in a vat of nutrients which keep the brains alive [1].

Many things change when you erase all memories of the brain and begin to feel like you've always been there. The nerve endings are supposed to be connected to a super-scientific computer which causes the person whose brain it is to have the illusion that... When this sort of possibility is mentioned in a lecture on the Theory of Knowledge, the purpose, of course, is to raise the classical problem of scepticism with respect to the external world in a modern way. (How do you know you aren't in this predicament?) But this predicament is also a useful device for raising issues about the mind/world relationship [1].

Unravelling the secrets of the nervous system can lead to many improvements. The argument I am going to present is an unusual one, and it took me several years to convince myself that it is really right. But it is a correct argument.

What makes it seem so strange is that it is connected with some of the very deepest issues in philosophy. (It first occurred to me when I was thinking about a theorem in modern logic, the 'Skolem-Löwenheim Theorem', and I suddenly saw a connection between this theorem and some arguments in Wittgenstein's Philosophical Investigations.)

A 'self-refuting supposition' is one whose truth implies its own falsiry. For example, consider the thesis that all general statements are false. This is a general statement. So if it is true, then it must be false. Hence, it is false. Sometimes a thesis is called 'self-refuting' if it is the supposition that the thesis is entertained or enunciated that implies its falsity. For example, 'I do not exist' is selfrefuting if thought by me (for any 'me'). So one can be certain that one oneself exists, if one thinks about it (as Descartes argued) [1].

So there is an urgent need to be interested in how one's own existence relates to one's own thinking. What I shall show is that the supposition that we are brains in a vat has just this property. If we can consider whether it is true or false, then it is not true (I shall show). Hence it is not true. Before I give the argument, let us consider why it seems so strange that such an argument can be given (at least to philosophers who subscribe to a 'copy' conception of truth).

We conceded that it is compatible with physicallaw that there should be a world in which all sentient beings are brains in a vat. As philosophers say, there is a 'possible world' in which all sentient beings are brains in a vat. (This 'possible world' talk makes it sound as if there is a place where any absurd supposition is true, which is why it can be very misleading in philosophy) [1].

Putnam argues to the following effect. In order to possess the concepts brain, in, and vat, a thinker must somehow be informationally linked to brains, instances of the spatial relation of containment, and vats [2].

Communication is a denial of some things, but a confirmation of some things.

I want now to ask a question which will seem very silly and obvious (at least to some people, including some very sophisticated philosophers), but which will take us to real philosophical depths rather quickly. Suppose this whole story were actually 'true'. Could we, if we were brains in a vat in this way, say or think that we were?

I am going to argue that the answer is 'No, we couldn't.' In fact, I am going to argue that the supposition that we are actually brains in a vat, although it violates no physical law, and is perfectly consistent with everything we have experienced, cannot possibly be true. It cannot possibly be true, because it is, in a certain way, self-refuting [1].

We can find something more interesting if we focus on how powerful the brain is, rather than arguing that it is a bodyless body. I believe that Warfield's efforts are unsuccessful. Here is how Warfield sets up the Cartesian argument for skepticism regarding knowledge of the external world: I do not know that I am not a brain in a vat in an otherwise empty world.

If I do not know that I am not a brain in a vat in an otherwise empty world, then I do not know that I am currently drinking water. So, I do not know that I am currently drinking water. This argument generalizes to all claims to external-world knowledge.

Warfield attacks the argument's first premise. He presents the following anti-skeptical argument (hereafter AS):

I think that water is wet. No brain in a vat in an otherwise empty world can think that water is wet. So, I am not a brain in a vat in an otherwise empty world [3].

The fact that someone can think that water is wet in an empty world is inextricably linked to his mind that knows water.

Schematically, the argument (which I will call the Skeptical Argument) looks like this:

P1. I do not know that I am not a brain in a vat in an empty world
P2. If I do not know that I am not a brain in a vat in an empty world then I do not know that I am currently water.
Cl. So, I do not know that I am currently drinking. This argument is valid and its premises seem at least initially plausible [4].

When there is a physical body, there is a brain. When there is a brain, it has some kind of thinking process. However, the potential of the brain varies. The question arises as to whether its capacity depends on the physical body.

The humans in that possible world have exactly the same experiences that we do. They think the same thoughts we do (at least, the same words, images, thought-forms, etc., go through their minds). Yet, I am claiming that there is an argument we can give that shows we are not brains in a vat.

How can there be? And why couldn't the people in the possible world who really are brains in a vat give it too?

The answer is going to be (basically) this: although the people in that possible world can think and 'say' any words we can think and say, they cannot (I claim) refer to what we can refer to. In particular, they cannot think or say that they are brains in a vat (even by thinking 'we are brains in a vat') [1].

Therefore, it is important to study the body and the mind in a mutually beneficial way and draw the right conclusions from them. We argue that the minimal biological requirements for consciousness include a living body, not just neuronal processes in the skull.

Careful examination of this thought experiment indicates that the null hypothesis is that any adequately functional "vat" would be a surrogate body, that is, that the so-called vat would be no vat at all, but rather an embodied agent in the world.

Thus, what the thought experiment actually shows is that the brain and body are so deeply entangled, structurally and dynamically, that they are explanatorily inseparable. Such entanglement implies that we cannot understand consciousness by considering only the activity of neurons apart from the body, and hence we have good explanatory grounds for supposing that the minimal realizing system for consciousness includes the body and not just the brain [5].

If we want to complete consciousness in a machine, we need consciousness in the whole body, not just the brain. The machine will only be able to perform tasks in data-configured mode. When tested outside of the data, it immediately makes a mistake and proves itself to be a machine.

I am not here interested in a skeptic who says that merely because there is a nonzero (perhaps incredibly tiny) probability that one is a brain in a vat, one does not "know" facts about the external world.

The skepticism with which I am concerned is external-world, justification skepticism.

My skeptic holds that because we have no reason to reject the brain-in a-vat scenario, or at least no strong reason to reject it, our beliefs about the external world are not even justified [6].

Is there any reason to believe that his work is complete?

Putnam appears to take his argument to have the form 'p implies not-p, therefore not-p'. However, this is not actually the structure of the argument as stated.

Putnam's argument could then be stated:

(1) If we are brains in a vat. Then the sentence 'we are brains in a vat' says something false.
(2) If the sentence 'we are brains in a vat' says something false then we are not brains in a vat (implicit assumption).
(3) Therefore, if we are brains in a vat, then we are not brains in a vat (from (1) and (2)).
(4) Therefore, we are not brains in a vat (from (3)) (Jane McIntyre, 1984).

Epistemologically, there is a need to clarify the explicit and implicit meanings of each word, term, concept, and category. In order to clarify the meaning, it is important to clearly distinguish the feature and nature of the phenomenon.

Let me give you an example: The term tree is related to the tree, but it is not the tree itself. Understanding this is the reason for the development of automatic machines. If you enter the word tree in an automatic machine and run a mode that reminds you of the definition of a tree, the machine will have more detailed information about the tree.

They would still use the word 'tree' just as they do, think just the thoughts they do, have just the images they have, even if there were no actual trees. Their images, words, etc., are qualitatively identical with images, words, etc., which do represent trees in our world; but we have already seen that qualitative similarity to something which represents an object does not make a thing a representation all by itself. In short, the brains in a vat are not thinking about real trees when they think 'there is a tree in front of me' because there is nothing by virtue of which their thought 'tree' represents actual trees [1].

All they think about is data about something called a tree.

If this seems hasty, reflect on the following: we have seen that the words do not necessarily refer to trees even if they are arranged in a sequence which is identical with a discourse which (were it to occur in one of our minds) would unquestionably be about trees in the actual world. Nor does the 'program', in the sense of the rules, practices, dispositions of the brains to verbal behavior, necessarily refer to trees or bring about reference to trees through the connections it establishes between words and words, or linguistic cues and linguistic responses. If these brains think about, refer to, represent trees (real trees, outside the vat), then it must be because of the way the 'program' connects the system of language to non-verbal input and outputs [1].

In such cases, the order in which the meanings of the words in the linguistics are entered into the machine should be considered. There are indeed such non-verbal inputs and outputs in the Brainin- a-Vat world (those efferent and afferent nerve endings again!), but we also saw that the 'sense-data' produced by the automatic machinery do not represent trees (or anything external) even when they resemble our treeimages exactly. Just as a splash of paint might resemble a tree picture without being a tree picture, so, we saw, a 'sense datum' might be qualitatively identical with an 'image of a tree' without being an image of a tree.

How can the fact that, in the case of the brains in a vat, the language is connected by the program with sensory inputs which do not intrinsically or extrinsically represent trees (or anything external) possibly bring it about that the whole system of representations, the language-in-use, does refer to or represent trees or anything external? [1].

The program is not only a representative program but also an imaginative one. Once we see that the qualitative similarity (amounting, if you like, to qualitative identity) between the thoughts of the brains in a vat and the thoughts of someone in the actual world by no means implies sameness of reference, it is not hard to see that there is no basis at all for regarding the brain in a vat as referring to external things [1].

I have now given the argument promised to show that the brains in a vat cannot think or say that they are brains in a vat. It remains only to make it explicit and to examine its structure. By what was just said, when the brain in a vat (in the world where every sentient being is and always was a brain in a vat) thinks 'There is a tree in front of me', his thought does not refer to actual trees.

On some theories that we shall discuss it might refer to trees in the image, or to the electronic impulses that cause tree experiences, or to the features of the program that are responsible for those electronic impulses [1].

In this case, some things depend on where the brain is located. These theories are not ruled out by what was just said, for there is a close causal connection between the use of the word 'tree' in vat-English and the presence of trees in the image, the presence of electronic impulses of a certain kind, and the presence of certain features in the machine's program.

On these theories the brain is right, not wrong in thinking 'There is a tree in front of me.' Given what 'tree' refers to in vat-English and what 'in front of' refers to, assuming one of these theories is correct, then the truthconditions for 'There is a tree in front of me' when it occurs in vat-English are simply that a tree in the image be 'in front of' the 'me' in question—in the image—or, perhaps, that the kind of electronic impulse that normally produces this experience be coming from the automatic machinery, or, perhaps, that the feature of the machinery that is supposed to produce the 'tree in front of one' experience be operating [1].

So do you give the meaning of the word with an image? From a linguistic point of view, one must also consider whether to give only the meaning of the word. Many neuroscientists and philosophers would say that your brain directly determines what you experience, but your body affects what you experience only via its influence on your brain [5].

So, if we are brains in a vat, then the sentence 'We are brains in a vat' says something false (if it says anything). In short, if we are brains in a vat, then 'We are brains in a vat' is false. So it is (necessarily) false [7].

Similarly, 'nutrient fluid' refers to a liquid in the image in vat-English, or something related (electronic impulses or program features). It follows that if their 'possible world' is really the actual one, and we are really the brains in a vat, then what we now mean by 'we are brains in a vat' is that we are brains in a vat in the image or something of that kind (if we mean anything at all)" [1].

So what do we mean by images?

But part of the hypothesis that we are brains in a vat is that we aren't brains in a vat in the image (i.e. what we are 'hallucinating' isn't that we are brains in a vat). So, if we are brains ina vat, then the sentence 'We are brains in a vat' says something false (if it says anything). In short, if we are brains in a vat, then 'We are brains in a vat' is false. So it is (necessarily) false [1].

A closer look at physical capabilities reveals in more detail how the brain interacts with the body's powers. Then the brain can work more closely with which organs of the body.

This kind of knowledge requires only true belief that meets a safety condition: belief that remains true in most or nearly all nearby possible worlds. Assuming that the world in which we are brains in vats is distant, our belief that we are not such remains true in nearby possible worlds [8].

It is important to examine more closely those preconditions based on the nature of the mind.

It follows that brains in a vat, or at least those which completely lack the right sort of causal interactions with the rest of the world, may employ the representations we use but they 'cannot refer to what we refer to.' Combine this with the claim that thinking about Xs involves the use of representations which refer to Xs, and we can conclude that brains in a vat cannot even think about what we think about.

In particular, they cannot even think about such things as brains and vats: if they employ the very sentence 'We are brains in a vat,' they still cannot be meaning what we mean - i.e. they cannot use the sentence to say or think that they are brains in a vat. So, if we are brains in a vat, then the sentence 'We are brains in a vat' says something false (if it says anything). In short, if we are brains in a vat, then 'We are brains in a vat' is false. So it is (necessarily) false [9].

Differences in such perceptions create differences in classification and species and require a more detailed understanding of them. The words, images, and senses that we perceive in our minds are part of some complex element of thought. In such cases, correct understanding is not only an expression used in accordance with the correct grammar of the language but also a correct concept.

In Chapter, I (entitled 'Brains in Vats') of Reason, Truth and History Professor Putnam has held that the hypothesis that everyone is, like Ludwig, a brain in a vat, though physically possible, is conceptually impossible. Though brains in vats will be able to combine words with other words to form correct English sentences, they will not be able to correlate the words in these sentences with actual things, because they have never come across any.

Hence they will not know what these words mean. Since they have seen only image brains and image vats, their word 'brain' will mean 'image brain', and their word 'vat' will mean 'image vat'. But they are not image brains in image vats; they are real brains in real vats.

So what they all believe (that they are image brains in image vats) will be false. Hence, in a world in which everyone is a brain in a vat, it will be false that anyone is a brain in a vat [10].

In such a case, the question of where and in what space the word is used will inevitably raise the question of meaning. We have seen that possessing a concept is not a matter of possessing images (say, of trees—or even images, 'visual' or 'acouslic', of sentences, or whole discourses, for that matter) since one could possess any system of images you please and not possess the ability to use the sentences in situationally appropriate ways (considering both linguistic factors—what has been said before—and non-linguistic factors as determining 'situational appropriateness') [1].

It is clear that human perception depends on many things, including phenomena, images, feelings, thoughts, and imaginations. When all this is combined, the correct understanding will increase.

The doctrine that there are mental presentations which necessarily refer to external things is not only bad natural science; it is also bad phenomenology and conceptual confusion [1].

Early Putnam may have intended to overestimate the functional role. Thus in short, 'functionalism' may be defined as the theory that explains mental phenomena in terms of the external input and the observable output. It explains the mind as a complicated machine [11].

Thus according to him, the functionalist is wrong in saying that semantic and propositional attitude predicates are semantically reducible to computational predicates, which can be realized in a physical system like the human brain. There is no reason why the study of human cognition requires that we try to reduce cognition either to computations or to brain processes. The reductionist approach to functionalism gives an inadequate picture of the human mind [11].

Early Putnam argued for the possibility of robotic consciousness. As a functionalist, early Putnam shows that a human being is an automation: that is, the human mind is a computing machine. The later Putnam, however, has found that his earlier thesis was wrong as mind can never be reduced to a machine [11].

It indicates that bodies and intelligence are not distinct things. The claim that the body is fundamental to all facets of intelligent life is not merely the claim that bodies and intelligence are co-extensive, that wherever intelligence is found so too is there a body. Rather, it is the much stronger claim that bodies are intelligent.

The more or less discrete physical systems we call bodies are just the sort of physical systems with the capacity to interact skillfully with their environments. The distinction between bodies and intelligence is an analytical distinction—it refers to two aspects of the same phenomenon (its physical properties and its skills or capacities).

Thus one finds AI researchers attempting to strap humanoid robot "bodies" onto complex computers, or conversely, trying to capture the dynamics of embodiment in complex digital models. In both cases, the body is understood as something that intelligence requires, a necessary feature which must be supplied or involved or made reference to, instead of being understood as what intelligence is.

But the point that Dreyfus is making in his work is precisely that such a conception is misguided, that intelligence and the body are inseparable, that they are two sides of the same coin, that they develop together in the world, that intelligent creatures are intelligent because they are embodied, and as a result, that intelligence must be understood in terms of embodied activity in the world.

For AI researchers going forward, then, the first point to consider is that artificial intelligence and artificial embodiment must be developed in tandem. "Hardware" and "software" cannot be understood as fundamentally distinct. Instead, the very organizational structure of physical systems must be designed to produce intelligent behavior. In order to develop truly intelligent systems, we must design physical systems whose raisons d'être are to cope with their environments [12].

So we need to understand intelligence behaviour. The development of intelligent behaviour is mutually correct developmental behaviour.

Understanding intelligence and the body in the way I've described suggests that AI researchers ought to be thinking not only about how intelligent creatures are intelligent, but also about why they are intelligent. As Dreyfus has shown, following Heidegger, meaningfulness and intelligence arise in the pursuit of interests, in relation to a world in which one is inexorably embedded—a world about which one has no choice but to care.

Bickhard's conception of recursively self-maintaining systems brings this notion into even sharper relief: building beings that understand the world—in whatever way they do—and that are able, therefore, to behave intelligently in the world, means building beings that need to be intelligent in order to successfully function. This constitutive need for intelligence is crucial to understanding intelligence as such [12].

3 Methodology

Artificial intelligence research is, in a way, a methodology. Many scientists seem to be studying artificial intelligence, but on the other hand, they are discovering many new things through artificial intelligence. Each of these new approaches is the basis of new cognition.

Through this new cognitive approach and methodology, many new studies can be developed. We all know that in recent years, not only in the field of technology but also in the field of production, various methods, techniques and research of artificial intelligence have been widely studied and its results have been tested. One of them is this study by Hilary Putnam. Qualitative approaches and descriptive methods were used in this study.

Design/methodology/approach – Acquainted with the works of Hilary Putnam. In particular, I studied the article in Brain in Watt in detail. To do this, a complete search of a large number of LIS magazine databases was conducted electronically.

Findings – Researchers have done a lot of research, either from the brain or from the physical side. However, it is rare to study the relationship between the brain and the physical body and how it works.

Research limitations/implications – This is a limitation because there is still a lack of detailed research on the structure of the human brain and how it works.

Originality/value – An in-depth study of this research topic will be useful in making some assumptions.

Research Design and Data Analysis – In this paper, we analyze a collection of epistemological hypotheses.

4 Result

- It is true that the artificial intelligence sector is developing at an appropriate level. However, in doing so, I found that research links between sub-sectors were better and more likely to develop. An example of this is the "brain in soap" hypothesis we studied. I think it is necessary to study not only artificial intelligence scientists but also philosophers, neuroscientists, psychologists, medical scientists, physicists and many other scientists. It would be ethical to conduct human-centred artificial intelligence research in this way.
- Always develop a link between artificial intelligence and the physical body in order to develop epistemological hypotheses in the philosophy of artificial intelligence and artificial intelligence.
- In order to develop artificial intelligence, it is necessary to develop brain function and physical body interaction together.
- It is more important to develop and direct the activities of human-centred robots than to promote robots beyond the human mind.
- Although neuroscience is developing rapidly, not all the secrets of the brain have been revealed yet. Therefore, it is important to develop many interesting hypotheses from the philosophical and linguistic point of view. So while the secrets of the brain have not yet been fully understood, many scientists have found that collaborative research is more likely to lead to better results.

5 Discussion

In recent years, artificial intelligence research has been developing rapidly. As the sector develops, sub-research is progressing. Many scholars are paying close attention to the development of these disciplines and are working hard to develop them in the right direction.

Whatever sub-field that researchers study, that sub-field is inextricably linked to the cognition of the world. We all know that there is a new field of research, a new cognitive industry. Therefore, it would be more effective to study epistemology, a branch of philosophy, in connection with one's new research.

Hillary Patnam's Brain in Vat Research is very interesting and I hope it will be of more interest to many researchers in the future. Balancing the findings of this study and improving it will undoubtedly have a positive effect on many aspects of the study, including the human brain, cognitive limits, and the interrelationships between the body and the brain.

Therefore, I would like to say that it would be more effective to study this research in the field of neuroscience, in the field of philosophy and AI.

6 Conclusions

The development of artificial intelligence is viewed negatively by many people, but the development of human-centered artificial intelligence can bring positive technological and economic advances to humanity in many ways. The challenge is to direct and use this new evolution in a timely manner.

In such cases, it is important to study the research topic little by little, even if it is controversial and unresolved. Therefore, Patnam's hypothesis about the relationship between the brain and the physical body was chosen.

When it comes to artificial intelligence, the question of the physical body is always on the agenda. Putnam's idea of this experiment is important not only to show many researchers what is possible and what is not possible but also to remind them what to focus on in the future.

References

1. Putnam, H.: Brains in a vat, chap. 1. In: Reason, Truth, and History, pp. 1–21. Cambridge University Press (1981)
2. Forbes, G.: Realism and skepticism: brains in a vat revisited. J. Philos. **92**(4), 205–222 (1995)
3. Brueckner, A.: A priori knowledge of the world not easily available. Philos. Stud. **104**, 109–114 (2001). https://doi.org/10.1023/A:1010374904795
4. Warfield, T.A.: A priori knowledge of the world: knowing the world by knowing our minds. Philos. Stud. Int. J. Philos. Anal. Tradit. **92**(1/2, A), 127–147 (1998)
5. Thompson, E., Cosmelli, D.: Brain in a vat or body in a world? Brainbound versus enactive views of experience. Philos. Top. **39**(1), 163–180 (2011). Embodiment
6. Huemer, M.: Serious theories and skeptical theories: why you are probably not a brain in a vat. Philos. Stud. Int. J. Philos. Anal. Tradit. **173**(4), 1031–1052 (2016). Special Issue: Perceptual Evidence

7. Hetherington, S.: Re: brains in a vat. Dialectica **54**(4), 307–311 (2000)
8. Goldman, A.H.: The under determination argument for brain-in-the-vat scepticism. Analysis **67**(1), 32–36 (2007)
9. Smith, P.: Could we be brains in a vat? Can. J. Philos. **14**(1), 115–123 (1984)
10. Harrison, J.: Professor Putnam on brains in vats. Erkenntnis **23**(1), 55–57 (1985)
11. Nath, R.: Philosophy of Artificial Intelligence: A Critique of the Mechanistic Theory of Mind. Universal-Publishers, Boca Raton, Florida, USA (2009). Chapter 3, pp. 96–100
12. Susser, D.: Artificial intelligence and the body: Dreyfus, Bickhard, and the future of AI. In: Müller, V. (eds.) Philosophy and Theory of Artificial Intelligence. Studies in Applied Philosophy, Epistemology and Rational Ethics, vol. 5, p. 279. Springer, Heidelberg (2013). https://doi.org/10.1007/978-3-642-31674-6_21

Fans with Benefits - Posting User-Generated Content on Brand-Owned Social Media Channels

Jawin Schell and Christopher Zerres(✉)

Offenburg University, Badstraße 24, 77652 Offenburg, Germany
christopher.zerres@hs-offenburg.de

Abstract. Brand-related-user-generated-content allows companies to achieve several important objectives, such as increasing sales and creating higher user engagement. In this paper a research framework is developed that provides an overview of the necessary processes to successfully use brand-related-user-generated-content. The framework also helps managers to understand the main motives of users when posting brand-related-user-generated-content. Expert interviews were carried out to validate the research framework. The results from the interviews support the proposed framework. Brand-related-user-generated-content can increase purchase intention and the community engagement. From a user's perspective the opportunity to interact with a brand and be featured on official brand channels could be seen as the main motivation for creating brand-related-user-generated-content.

Keywords: User-generated content · Social media · Brand-related-user-generated content

1 Introduction

Social media has changed the communication and interaction behavior of consumers and organizations. One central reason is the fact that users can create and publish content themselves (user-generated content - UGC). In this paper UGC is defined as "…original contributions that are created by users, are expressed in a number of different media […] and are widely shared with other users and/or with firms." [24, p. 23]. UGC has a significant impact on the communication activities of organizations. In social media, brand-related content and messages are no longer provided exclusively by organizations, but also by users. This "noticeable shift in power and control" [14, p. 405] means a loss of control for marketers and communications specialists: "Consumers-as-Creators can post either positive or negative content about a brand, product, or service on social media […] With more empowered consumers, advertisers have less control over what kind of brand-related content they can convey and how they are perceived on social media than in traditional forms of advertising" [32, p. 8]. Brand-related-user-generated-content (BRUGC) has a major influence on the purchase decision of consumers and the image of

F. Fui-Hoon Nah and K. Siau (Eds.): HCII 2022, LNCS 13327, pp. 112–126, 2022.
https://doi.org/10.1007/978-3-031-05544-7_9

a brand [18, 23, 34]. This form of content is therefore becoming increasingly important for the organizations' communication [24].

One of the crucial questions is therefore how BRUGC can be integrated into marketing communication and successfully used. The aim of this article is to create a framework that helps companies to understand 1) why users create and post content about companies; 2) the (marketing) objectives that can be achieved in connection with BRUGC; 3) the strategies and implementation possibilities. This framework is developed, in a first step, on existing research and theories. In a second step, interviews with experts from different companies are conducted to verify and improve the framework.

2 Research Framework

2.1 Overview Framework

To better understand the possibilities of BRUGC we developed a research framework that consists of three main areas (Fig. 1).

To successfully integrate user-generated content into the companies' communication activities it is important to understand the reasons why users create and publish BRUGC in social media. The main reasons can be summarized with the categories Impression Management and User Engagement [6, 8]. Impression Management summarizes the basic idea, that users who share BRUGC in social media are primarily concerned with optimizing and shaping their own image in the eyes of others [3, 6]. It includes the concepts of self-enhancement and identity-signaling [6]. User Engagement relates to user motives to actively interact and participate in a specific brand community [11].

The second area of the research framework deals with the different aims that can be achieved by using BRUGC. From a company's perspective BRUGC can directly influence purchase intention and product purchases [6, 12]. BRUGC can also lead to a more interactive and therefore more attractive brand community [11, 27]. Furthermore, BRUGC can help to reduce own media spending and is an important source for market research (e.g. regarding trends) [8, 26].

The third area summarizes aspects regarding BRUGC strategies and their implementation from a company perspective.

2.2 User Perspective: Motives and Reasons to Create and Publish BRUGC

Organizations that want to integrate BRUGC into their marketing communication need to understand the reasons and motivations of users to create and publish BRUGC.

According to Daugherty et al. [10], users who share user-generated content in social networks are primarily concerned with optimizing and shaping their own image in the eyes of others. Self-concepts and values that are personally considered important play an important role, as well as the feeling of being a member of a community that shares similar values and self-concepts. Both aspects can also be applied to the creation of BRUGC. In order to shed light on which motives can motivate customers and fans of the brand to create BRUGC, two key motivations will be analyzed in the following. Expressing one's own self-concepts is one of the main motives for the creation of user-generated content

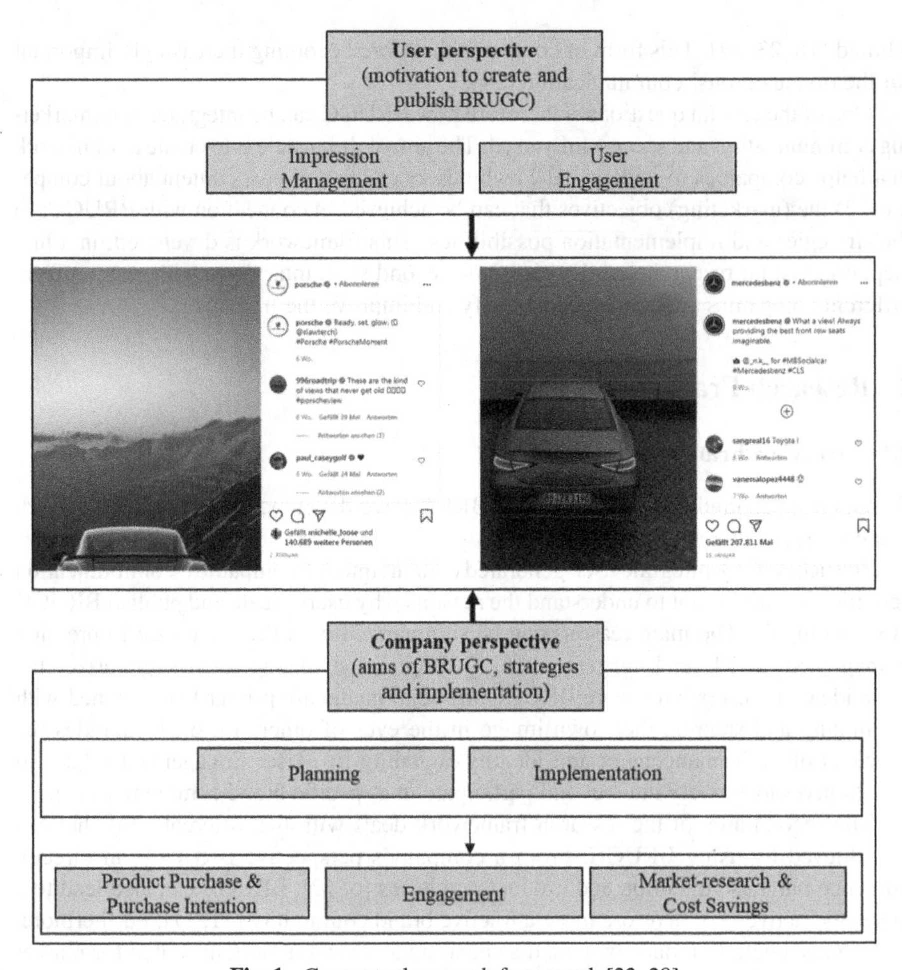

Fig. 1. Conceptual research framework [22, 28]

[10]. According to Ind and Iglesias [17, p. 64] brands deliver "functional, emotional and self-expressive benefits to customers." Therefore, it can be assumed that "impression management" [6, p. 588], i.e., shaping the impression that other people get of a person, finds an ideal form of expression in BRUGC. This is also the view of Seong and Kyung [30, p. 1289], who note: "Consumers find a brand attractive when they believe the brand matches or improves their self-image. Therefore, consumers want to identify with a brand perceived as attractive and are amenable to having an ongoing relationship with the brand." Berger [6] refers to this form of impression management as self-enhancement. He writes, "People like to be perceived positively and present themselves in ways that garner such impressions" [6, p. 588]. For this reason, they primarily share content that makes them look good in the eyes of others [9, 31]. For example, their own distinctiveness, connoisseurship regarding certain products, and social status should be highlighted [29]. Customers could thus share BRUGC - under the premise of self-enhancement - primarily

to signal to others that they are closely in tune with the times in terms of their product selection.

Berger [6] identifies a second form of expression within impression management in the area of identity-signaling. "Beyond generally looking good, people also share things to communicate specific identities, both to themselves and others" [6, p. 589]. This is not only to improve one's own image, but also to clarify personal interests and character traits [9]. For this reason, users communicate desired identity characteristics and avoid revealing undesired ones [4, 12]. Publicly displayed product and brand choices play a crucial role in this process [6]. In particular, cars, clothes, and other publicly highly visible products are often used to signal one's identity [4]. BRUGC is thus also likely to be published by users to express their own identity - or at least an idealized version of it.

Based on these main motives, Berger [6] suggests that further underlying sub-motives can be identified. For example, users share entertaining content with their network in order to appear more interesting themselves. In the case of entertaining content with brand references, interesting products are more likely to be mentioned than mundane products [5].

A second subcategory can be identified with regard to content that Berger describes as "self-concept relevant things" [6, p. 590]. Here, users publish content with brand references in which products are discussed that are supposed to provide information about their (desired) identity [6]. These are therefore often used as identity markers [4]. For this reason, content is more often shared with symbolic products than with less symbolic items [9].

Also relevant for image and identity optimization is the presentation of products that enjoy high prestige and promise status gains [6]. Luxury and premium brands might therefore come up more often than less high-priced alternatives [20].

Another potential motive for the creation of BRUGC can be summarized under the umbrella term 'customer or consumer engagement'. This can be defined as a "psychological state that occurs by virtue of interactive, co-creative customer experiences with a focal agent/object (e.g. a brand)" [8, p. 260]. The creation of BRUGC in particular plays a central role here. This is because although social media communication without UGC also allows people to interact with the brand - for example, by commenting on the brand channel or sharing content with friends - BRUGC gives customer engagement an even more interactive component [32].

The possibility for users to initiate social media communication with the brand on their own initiative can thus definitely be understood as an important motive for creating brand-related user-generated content. Customer or consumer engagement is thus supplemented by a component that is likely to make it more relevant in terms of its position within marketing communication.

Participation in so-called online brand communities is also likely to be one of the motives for the creation of BRUGC. In today's society, social interactions are increasingly taking place online, with social networks playing a dominant role [11]. Consequently, brand-related social interactions are also shifting to social media. The sense of belonging to a particular brand community is of central importance here [6]. For companies, these communities offer potential for customer loyalty.

Social media provides the ideal platform for this specific form of social bonding [6]. One of the forms of expression to position oneself as part of an online brand community is the creation of BRUGC [10]. Regarding users who create user-generated content, Daugherty et al. [10, p. 18] state: "[They] become members of an online community that shares the principles they consider important. It validates and helps them feel good about who they are and what they believe about the world." Ultimately, this form of UGC creation would provide "a sense of belonging" [10, p. 18].

2.3 Company Perspective: Aims of BRUGC

From a business perspective, BRUGC can be used for numerous objectives. To derive some of the possible objectives of BRUGC, the stimulus-organism-response model (S-O-R model) is used. Kim and Johnson [18] propose an adapted S-O-R model for this purpose. The initial stimulus, i.e., viewing BRUGC, is followed by cognitive and emotional processing, such as perceiving information and evaluating its quality and feeling pleasure or arousal [18]. These internal processes can subsequently lead to behaviors such as product purchases or purchase intentions, sharing of product information, and active brand engagement [18].

Product Purchase and Purchase Intention
Following the concept of the S-O-R model an unmediated behavioral response can be so-called impulse buying [18, 33]. A study by Adelaar et al. [1] demonstrated a positive association between emotional response to sensory stimuli such as audio, text, and pictures, and impulse buying. Kim and Johnson's [18] work shows the link between impulse purchases and viewing BRUGC that features specific products. In particular, the opportunities for direct linking to the respective brand's online stores reinforce this effect.

In addition, future purchase intentions and purchase decisions can also be influenced by receiving BRUGC [18]. For example, recipients might expand their knowledge about certain brands and products when viewing BRUGC, and use this information, when making future purchase decisions [16].

Product purchases are indirectly influenced by the sharing of product information. The sharing of product information (information pass-along) can clearly be identified as a phenomenon related to eWord-of-Mouth [18]. This is because BRUGC often conveys "opinions, facts or user experiences with brands or products" [18, p. 101], which are then shared again by the recipient with other users. Pleasure and excitement when viewing content can be seen here as an important reason for sharing and the word-of-mouth communication [18, 19].

Kim and Johnson's [18] modified S-O-R model was tested on a group of 533 subjects. For this purpose, postings containing BRUGC for the shoe brand Sperry Top-Sider were created and embedded in a brand channel, in this case a mock-up Facebook page [18]. Here, the postings were modeled using real BRUGC on brand pages, and included product descriptions, product usage information, and product recommendations [18]. Both emotional and more informative messages were used [18]. The results of the study show that cognitive and emotional responses to the BRUGC stimulus influenced all previously hypothesized behavioral responses [18].

Another important concept explaining the influence of BRUGC on product purchases and purchase intention is social proof. Social proof is defined as "demonstration that other people have made a choice or partaken in a product/service - such as reviews, testimonials or social shares - thereby encouraging others to do so" [7]. The persuasive power here stems from the assumption that products that are liked by others could also be liked by the viewer himself [3]. Social proof is particularly effective when content is published that shows people who are similar to the intended target group in certain parameters [3]. Especially by using BRUGC, such effects could also be achieved. Since BRUGC comes directly from the brand community - i.e., from customers, fans of the brand, or even brand evangelists - images or videos are likely to be shown primarily to those people who are similar to the intended target group. The use of BRUGC in the sense of co-communicating could therefore prove to be a way of incorporating social proof concepts into marketing communications. In particular, the fact that BRUGC creators - in contrast to paid influencers or advertisers - do not normally receive any financial compensation is likely to contribute more to their credibility and the authenticity of the brand message. Interestingly, social proof effects can also be achieved with products that are unaffordable or unavailable to many people [3]. For this reason, social proof is also likely to be considered suitable for brands in the high-priced or luxury segment, as it appeals not only to potential customers but also to fans of the brand who currently cannot afford the products or may never be able to.

Engagement

A central objective of many social media activities is to increase engagement [35]. BRUGC can help companies to increase user engagement and active participation [2]. Customer participation, i.e., the customer's involvement in the brand's communication process, can be understood as a sign of appreciation for the brand [2]. Users are featured on brand-owned channels, gaining visibility and reach, and are allowed to feel part of the brand. Kim and Johnson [18] identified an increase of brand engagement as a latent behavioral response to BRUGC. Brand engagement can be understood as an emotional bond that connects the customer with the brand [15]. Brand-customer relationships that enable the participation of users allow them to become more actively involved in the co-creation of the brand [17]. By publishing on the brand's own channels, customers and fans of the brand become active partners in the co-creation of the brand experience. Companies can use this to give customers and fans of the brand a sense of belonging and make them aware that they are contributing to the development of the brand and the brand experience through BRUGC [17]. The use of BRUGC can therefore strengthen customer-brand loyalty and intensify the brand experience for individuals [17].

Cost Savings

By outsourcing the production of content to the users, agency and creation costs can be saved. Earned media with a positive brand reference becomes, in a sense, owned media through publication on the brand channel. Since content marketing requires a high level of constant communication output, brand-owned channels can be operated much more cost-effectively with the use of BRUGC. O'Hern and Kahle [24, p. 25] argue that "Much of the attractiveness of a co-communication campaign may lie in its ability to generate engaging new promotional content while greatly reducing the production costs associated with creating and in-house advertising." Since BRUGC is available almost free of charge to companies, there are clear advantages here for the allocation of communication budgets [26].

However, no major sacrifices should be made in terms of aesthetic quality, brand adequacy, and communication message. Due to high-quality image editing tools and powerful cameras/cell phone cameras, restrictions in terms of image quality are not necessarily to be expected.

Market Research
The fourth objective involves gathering insights about users. This includes in particular the brand perception of the customers and fans. These insights subsequently allow the identification of particularly relevant topic areas. The market research findings obtained can be divided into two subcategories. Insights into what customers or fans associate with the brand and how they express their personal brand experience through BRUGC, as well as insights that reveal which BRUGC is received by other users of the channel. By analyzing the performance of BRUGC, companies can gain access "to a rich stream of consumer input that enable[s] [them] to monitor which of these user-generated ads [are] especially appealing to other customers." [24, p. 25] Both analysis options promise to identify topics that are particularly relevant to customers and fans of the brand. In this way, insights can be gained which are helpful for optimizing the content marketing process - especially content collection and filtering.

2.4 Company Perspective: BRUGC Planning, Initiation and Production

In order to achieve the BRUGC company aims identified in the theoretical framework, it is very important to integrate them into a planning process. Therefore, a suitable content marketing process is essential. Thus, to put BRUGC into practice and achieve the discussed objectives, certain steps are necessary.

To integrate BRUGC into existing content management or marketing processes, established approaches such as that of O'Callaghan and Smits [25] are adjusted. In this paper we propose the following model: it includes, the planning phase, the initiation phase and the production phase (Fig. 2). These are framed by the risks and challenges which might occur when using BRUGC.

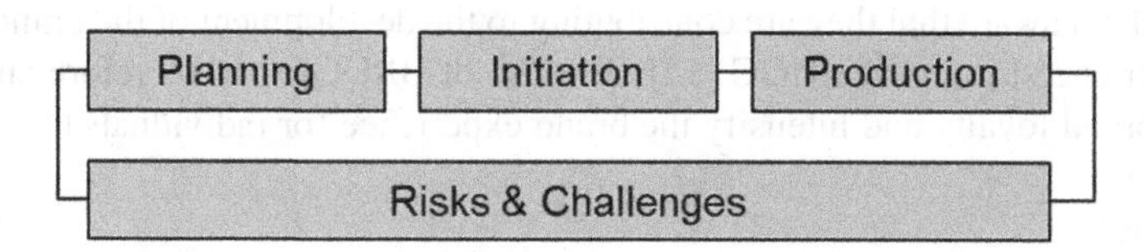

Fig. 2. BRUGC planning model

The content planning phase should be understood as preparation for content collection and filtering. For example, questions regarding the desired content (tonality and visual aesthetics) and format (photo, video, audio, text) of the content to be generated by the user must be clarified. This allows an efficient content initiation, collection and filtering. In addition, it should be determined on which brand-owned channels BRUGC is to be requested and distributed. With regard to the specific promotion and initiation of BRUGC, fixed concepts should therefore be developed which, above all, clarify the question of how customers and fans of the brand are to be motivated to create BRUGC with regard to the previously listed motivations.

In order to be able to collect BRUGC and distribute it, its creation must first be initiated by the company. Consumers and users must be motivated to create BRUGC messages and thus contribute positively to the brand. For campaigns and challenges, social media appeals are typical in this regard: "Co-communication initiatives typically begin with an open call or contest that encourages consumers to [develop] promotional collateral that the firm can utilize in a future advertising campaign." [24, p. 24]. The establishment of permanent, visible and clearly structured BRUGC communication should lead to acceptance and knowledge of the process by users. Users who recognize that brands or companies regularly publish user images and videos on their channels could become aware of the process required for this and subsequently are willing to contribute BRUGC.

In the content production phase, the outsourcing of production to users is taking place. Therefore, the target group-appropriate initiation of BRUGC must be communicated on brand-owned channels to trigger content collection. With regard to this, it should be clarified whether automated or semi-automated tools can be used, or whether the collection is realized manually and organically via the respective channel. Questions regarding the use of automated or semi-automated tools are also important with regard to content filtering, which should take place on the basis of the guidelines formulated in the content planning phase. Without clear guidelines the collected content flows manually and organically through the respective media channel. Finally, content consolidation, and thus ultimately the decision on whether BRUGC appears on brand-owned channels, requires detailed concepts regarding responsibilities and approval processes.

Risks and Challenges

The use of externally produced content and materials in corporate communications always raises legal questions, for example regarding copyrights. It is necessary for companies to carefully examine the copyright and personal rights situation before publishing BRUGC on their own brand channels. Consent for further use by users plays an important role here. To avoid problems or even legal issues prophylactically, it is therefore advisable to obtain written consent for further use from the BRUGC creators. The form in which this should take place in order to guarantee legal validity should be clarified with the company's legal department. The possibility of more complicated cases, such as erroneous consent to the further use of BRUGC contributions by persons who do not in fact hold the personal rights or copyrights to the content, should also be clarified with the legal department as a precautionary measure so that appropriate countermeasures can be initiated immediately.

Besides legal issues there are other challenges and risks that need to be considered before implementing a BRUGC strategy. "Although co-communicating offers firms a number of benefits, there are certain challenges and possible side effects that firms wishing to employ this type of strategy should anticipate" [24, p. 25]. One risk factor concerns the lack of user participation: "Firms that overestimate consumers' involvement in their brand may unintentionally create their own negative publicity by launching contests that attract only a limited number of submissions" [24, p. 25]. Furthermore, BRUGC posts could be of poor quality, portray the brand or products in an unflattering light, or simply not elicit the desired responses from other customers [24]. In addition, social networks are highly volatile environments where negative reactions can occur. For example, calls for BRUGC could be met with content highlighting perceived misconduct by the company. While these posts would likely not be published on brand-owned channels, they could still attract unwanted attention in the public and social networks. "Firms that produce controversial products may be especially prone to attracting a negative public backlash to their co-communication campaigns" [24, p. 25]. Coherence with regard to the general brand positioning can be considered as a general risk factor.

3 Empirical Research

3.1 Research Design and Methodology

To validate the theoretical framework, we conducted expert interviews. The first step was to identify companies and brands that use BRUGC as part of their communication policy. The 72 international companies and brands identified in this way were then assigned to different industries and were asked to participate in the research. Ten experts from different industries agreed to participate. A qualitative research approach according to Mayring [21] and a structured interviewing process according to Fontana and Frey [13] were used to determine the expert opinions. For this purpose, before conducting the qualitative expert interviews, significant upper categories for the creation of an interview guideline were deductively worked out from the developed research framework. The interviews were conducted according to the developed interview guide, recorded and later transcribed. The transcribed material was condensed, structured, and examined for specifics, differences, and connections in a qualitative content analysis [13, 21]. The coding and identification of the relevant text passages were done in relation to the research framework.

	Company	Occupational title
Expert 1	Luxury car manufacturer	Manager Content & Channels
Expert 2	Premium car manufacturer	Social Media Manager
Expert 3	Premium car manufacturer	Content Management Consultant
Expert 4	Beverage manufacturer	Digital Marketing Manager
Expert 5	Consumer goods manufacturer	Digital Marketing Strategist
Expert 6	Professional sports club	Social Media & Content Manager
Expert 7	Game and toy company	International Social Media Manager
Expert 8	Fashion company	Manager PR & Social Media
Expert 9	Bicycle helmet manufacturer	Growth Marketing Manager
Expert 10	Backpack manufacturer	Head of Performance Marketing

3.2 Results

User Perspective: Motives and Reasons to Create and Publish BRUGC

The main motivation for creating BRUGC is to improve customer engagement, according to the opinion of the experts. Thus, above all, the opportunity to interact with a loved brand and the possibility to be featured on an official brand channel is the main user side motivation behind BRUGC creation. Expert 4 says that: *"People want to be part of it, want to be taken along and also want to see their own posts get recognition."* Expert 8 mentions: *"I believe that people like to see themselves on the brand channels and are proud that their picture has made it onto our website, for example."* Less frequently, experts cited some form of self-expression as a motivator for BRUGC creation. Expert 5 thinks *"…in general, the reason that drives people to share things publicly on social media is a certain aspect of self-expression that probably everyone who is active in this way enjoys to some extent."* This can be related to the behavioral experiences in which the brand is used as an instrument for self-presentation and a psychological projection surface, and according to Berger [6], corresponds with impression management. In summary all experts agreed on the identified motivational factors in the research framework.

3.2.1 Company Perspective: Aims of BRUGC

Product Purchase and Purchase Intention

Almost all experts stated that social proof effects and the hope of associated sales play a strategic role in the use of BRUGC on brand-owned channels or could do so in the future. In particular, the possibility of showing authentic content and customer images and thus promoting trust, thereby increasing purchase intention, and positively influencing the purchase decision, is seen as very promising. Expert 6 says: *"This is the best form of advertising a company can get. People who are not influencers, pay money for the products and later also deliver advertising for the products. That's worth much, much more than running a big advertising campaign. Because that's the most effective advertising."* To increase product purchase and purchase intention, a sociodemographic match between the company's target group and the BRUGC creators is described as a key factor by experts. Expert 7 affirms: *"We notice that customers like to look at these pictures and are then perhaps also influenced in their purchase decision. After all, this is the product in action, so to speak, and you don't normally see that. We definitely notice that this also supports sales on our website."* These findings from corporate practice are in line with the theories regarding social proof and the S-O-R model and seem to correspond with the basic theoretical assumptions of Armstrong [3] and Kim and Johnson [18]. Thus direct and indirect effects on purchases, purchase intentions and recommendations are seen as the most important aims of BRUGC implementation by corporate experts.

Engagement

Strengthening the brand-customer relationship in online brand communities and increasing interactivity in these communities are important strategic reasons for the use of

BRUGC for all ten experts. Their aim is to convey a feeling of appreciation and participation to customers and fans of the brand and, as a result, promote brand commitment and identification. Expert 3 says: *"It is of course also an appreciation matter. We are very close to our community and it´s always a special highlight for the fans when they actually see their own pictures on the channel."* Expert 5 confirms: *"By communicating directly with consumers and also using consumer content we are able to show our brand and strengthen our brand image. Of course, we want to reward this on this stage via BRUGC and thus appear close and approachable as a brand. So it's all about appreciation and community."* BRUGC appears to be a very suitable instrument, as each of the companies surveyed uses such content to strengthen the brand-customer relationship. The findings from the expert interviews are in line with the ideas identified in the research framework.

Cost Savings

Almost all of the experts mentioned the possibility of using BRUGC to communicate content on their own brand channels in a more resource-efficient and cost-effective way. Companies with a high social media output and high posting frequency in particular seem to use BRUGC to reduce costs. Expert 3 mentioned in that context: *"As a brand, you don't manage to produce things yourself with the high frequency with which you have to post content. The use of user-generated content is therefore very efficient. You really have the opportunity to keep this flow of content going through your relationship with the community, which you would otherwise not be able to do with your own resources."* Expert 5 adds: *"Because you have additional content for your channels but of course also lower production costs, this is a great method to use, especially for brands with a small production budget."* Implications of the research framework regarding the strategic component of cost savings can therefore be confirmed. Cost savings should therefore be regarded as an important reason for the use of BRUGC on a brand's owned channels, as stated in the research framework.

Market Research

Due to its potential to provide customer insights regarding brand perception and potentially relevant brand topics, market research was identified in the research framework as a potential objective to be considered by companies. However, according to the expert interviews this aim is only important to some companies. Many of the companies are aware of this strategic opportunity, but value other aspects as more important. It is interesting to note, however, that some companies actually include learning from the performance of BRUGC and the underlying user preferences in their future content planning and filtering, in order to tailor BRUGC. According to the expert interviews, the way in which customers present and visualize their brand experiences makes it possible to adapt future content even more precisely. Expert 1 says: *"With such findings, we then of course continue to work on content that will be successful in the future."* This seems to confirm the assumptions formulated in the research framework regarding the BRUGC content management process and the resulting opportunities to better understand customers' content preferences and brand experiences. Nevertheless, only a few company experts identify market research as the primary aim of using BRUGC in marketing communication.

Company Perspective: BRUGC Planning, Initiation and Production

The management of BRUGC communication is handled in very different ways across the different companies. While content collection and filtering are handled with powerful tools by some brands, they are operated manually/editorially by other brands. The tools range from fully automated to semi-automated approaches. The experts describe the simplification of day-to-day work with large volumes of data, the creation of transparency and legal certainty, and the uncomplicated integration into various channels as particularly useful. Expert 1 explains the benefits: *"For example, when we use the tool to search for specific car models, we can filter directly whether the image is indoors or outdoors, whether a woman or a man is the protagonist, and also which car it is."* Expert 2 adds: *"Our tool automatically finds when someone uses the hashtag of our brand and then automatically asks if we can use it. And if the user then gives his or her consent, the image is automatically saved in a content store. You haven't lifted a finger yet, but you have a folder that you know you can use."* Expert 9 takes a different position: *"We've looked at some of the BRUGC programs and tools and they were really expensive. And doing it yourself you could, even though it entails a bit more manual work, find better quality and maybe even 'crack the code' with a better understanding."* With regard to the BRUGC content management process developed in the research framework, it could be assumed that for a certain company size - and the associated very high volume of BRUGC - this can only be implemented in a resource-friendly way by using tools. For smaller companies, on the other hand, the acquisition costs of specialized BRUGC tools seem to be a deterrent or simply not necessary.

Larger and better-known brands do not seem to have to implement sophisticated concepts when activating BRUGC, since the process is already known to the brand's users and fans and is self-reinforcing due to the brand's appeal. Expert 4 explains: *"We're very fortunate that people already send us an insane amount of content without being asked or tag us in pictures. People know that we post BRUGC and that it's well received. I can only confirm that this is an inexhaustible source."* Expert 3 mentions in that context: *"When people see pictures of other users on the channel, the next photos keep coming in. That somehow feeds on itself. People realize that there is a realistic chance of being featured on the brand channel and that you can submit your contributions accordingly."* Nevertheless, constant and active reminders of the existence of BRUGC submission opportunities also seem common. Often, BRUGC in this context is stimulated by the brand's own photo or video content.

Risks and Challenges

In almost all expert interviews, challenges are seen primarily in the area of brand fit or brand adequacy. Aesthetics, tonality and image quality in particular are named as important factors. Expert 9 says: *"I think for a lot of marketers the challenge with BRUGC, especially when you are a brand that is investing a lot in design, is the quality of BRUGC. It needs to represent the customer but also the brand and the look."* In the case of smaller companies, concerns about content submissions of inferior quality and quantity are also evident. Expert 10 mentions: *"There's always the risk that content has low quality, but people still expect you to post or publish it."* These findings are consistent with the risk factors identified earlier. However, all experts agree that these challenges can be overcome through thorough content filtering. Expert 1 clarifies: *"It's not a risk because you have that under control. You can filter which contributions you use and which you don't use. We don't select contributions that don't meet our visual standards."*

Legal risks, especially in the area of copyright and image rights, are considered to be the greatest risks. This corresponds with the position developed in the research framework. Legal risks are countered in practice by the use of tools that promise transparent processes and written declarations of consent, as well as by consultation with the respective legal departments.

4 Conclusions

An understanding of the mechanisms of BRUGC and its integration into the marketing communications strategy allows companies to achieve several important objectives, such as increasing sales and creating a higher user engagement. The research framework developed in this paper provides an understanding of the necessary processes and aspects that need to be considered when integrating BRUGC into the communication strategy. All interviewed experts underlined the high importance of BRUGC in the communication mix. They stated that the opportunity to interact with a brand and be featured on its official brand channels could be seen as the main motivation for users to create BRUGC. The greatest risk of using BRUGC on brand channels was identified in the area of legal issues and uncertainties. Main goals were related to sales improvements and conveying to customers a sense of belonging and appreciation. Strengthening the interactivity and intensity of the customer-brand relationship and brand community was mentioned by the experts as a very important goal. However, sales promotion through authentic and therefore credible customer content was also seen as a key objective. In addition, cost savings as well as insights into the brand perception of customers and fans can be realized. In summary, engagement, market research, cost savings and impact on sales can be considered as the main categories of our framework. In summary, BRUGC is an important influencing factor for consumer's decision making process and their perception of brands.

Certainly a key element of success for companies is to better analyze, systematize and understand the processes on the user side which affect attitudes toward BRUGC and thus influence the consumption and creation of BRUGC. For this purpose, our developed research framework offers a comprehensive overview of the most important aspects that should be considered when using BRUGC. It can be understood as a guide to unleashing BRUGC's potential in marketing communications. This paper argued primarily from the marketer's point of view, since the study exclusively featured experts from the industry. Therefore, it would be interesting for future research to also investigate the view of the users, with qualitative and quantitative research. Furthermore, research could also be conducted involving quantitative studies.

References

1. Adelaar, T., Chang, S., Lancendorfer, K.M., Lee, B., Morimoto, M.: Effects of media formats on emotions and impulse buying intent. J. Inf. Technol. **18**(4), 247–266 (2003)

2. Algesheimer, R., Borle, S., Dholakia, U.M., Singh, S.S.: The impact of customer community participation on customer behaviors: an empirical investigation. Mark. Sci. **29**(4), 711–726 (2010)
3. Armstrong, J.: Persuasive Advertising: Evidence-based Principles. Palgrave Macmillan, Basingstoke (2014)
4. Berger, J., Heath, C.: Where consumers diverge from others: identity signaling and product domains. J. Consum. Res. **34**(2), 121–134 (2007)
5. Berger, J., Iyengar, R.: Communication channels and word of mouth: how the medium shapes the message. J. Consum. Res. **40**(3), 567–579 (2013)
6. Berger, J.: Word of mouth and interpersonal communication: a review and directions for future research. J. Consum. Psychol. **24**(4), 586–607 (2014)
7. BigCommerce: What is social proof and why it's necessary for ecommerce success (2021). https://www.bigcommerce.com/ecommerce-answers/what-is-social-proof/. Accessed 22 Dec 2021
8. Brodie, R.J., Hollebeek, L.D., Jurić, B., Ilić, A.: Customer engagement: conceptual domain, fundamental propositions, and implications for research. J. Serv. Res. **14**(3), 252–271 (2011)
9. Chung, C., Darke, P.: The consumer as advocate: self-relevance, culture, and word of-mouth. Mark. Lett. **17**(4), 269–279 (2006)
10. Daugherty, T., Eastin, M.S., Bright, L.: Exploring consumer motivations for creating user-generated content. J. Interact. Advert. **8**(2), 16–25 (2008)
11. Dennhardt, S.: User-Generated Content and its Impact on Branding: How Users and Communities Create and Manage Brands in Social Media. Springer Gabler, Wiesbaden (2014)
12. Escalas, J.E., Bettman, J.R.: You are what they eat: the influence of reference groups on consumer connections to brands. J. Consum. Psychol. **13**(3), 339–348 (2003)
13. Fontana, A., Frey, J.H.: Interviewing: the art of science. In: Denzin, N.K., Lincoln, Y.S., (eds.) Handbook of Qualitative Research, pp. 361–376. Sage Publications, Inc., Thousand Oaks (1994)
14. Gangadharbatla, H.: Social media and advertising theory. Advert. Theory, 402–416 (2012)
15. Goldsmith, R.E.: Brand engagement and brand loyalty. In: Kapoor, A., Kulshrestha, C. (ed.) Branding and Sustainable Competitive Advantage: Building Virtual Presence, pp. 121–135). IGI Global, Hershey (2012)
16. Hung, K.H., Li, S.Y.: The influence of eWOM on virtual consumer communities: social capital, consumer learning, and behavioral outcomes. J. Advert. Res. **47**(4), 485–495 (2007)
17. Ind, N., Iglesias, O.: Brand Desire: How to Create Consumer Involvement and Inspiration. Bloomsbury Business, London (2016)
18. Kim, A.J., Johnson, K.K.P.: Power of consumers using social media: examining the influences of brand-related user-generated content on Facebook. Comput. Hum. Behav. **58**, 98–108 (2016)
19. Ladhari, R.: The effect of consumption emotions on satisfaction and word-of-mouth communications. Psychol. Mark. **24**(12), 1085–1108 (2007)
20. Lovett, M.J., Peres, R., Shachar, R.: On brands and word of mouth. J. Mark. Res. **50**, 427–444 (2013)
21. Mayring, P.: Qualitative content analysis. Forum: Qualitat. Soc. Res. **1**(2) (2000). Art. 20
22. Mercedes-Benz Instagram (2021). https://www.instagram.com/mercedesbenz/. Accessed 22 Dec 2021
23. Nusairat, N.M., et al.: User-generated content – consumer buying intentions nexus: the mediating role of brand image. Acad. Strat. Manage. J. **20** (2021)
24. O'Hern, M.S., Kahle, L.R.: The empowered customer: user-generated content and the future of marketing. Global Econ. Manage. Rev. **18**, 22–30 (2013)

25. O'Callaghan, R., Smits, M.T.: A strategy development process for enterprise content management. In: Bartmann, D. (eds.) Proceedings of the 13th European Conference on Information Systems (ECIS 2005) ECIS (2005)
26. Onishi, H., Manchanda, P.: Marketing activity, blogging and sales. Int. J. Res. Mark. **29**, 221–234 (2012)
27. Park, H., Kim, Y.K.: The role of social network websites in the consumer–brand relationship. J. Retail. Consum. Serv. **21**(4), 460–467 (2014)
28. Porsche Instagram (2021). https://www.instagram.com/porsche/. Accessed 23 Dec 2021
29. Rimé, B.: Emotion elicits the social sharing of emotion: theory and empirical review. Emot. Rev. **1**, 60–85 (2009)
30. Seong, H.L., Kyung, S.J.: Loyal customer behaviors: identifying brand fans. Soc. Behav. Pers. **46**(8), 1285–1304 (2018)
31. Sundaram, D.S., Mitra, K., Webster, C.: Word-of-mouth communications: a motivational analysis. Adv. Consum. Res. **25**, 527–531 (1998)
32. Wan, A.: Effectiveness of a Brand's paid, owned, and earned media in a social media environment (2019). https://scholarcommons.sc.edu/etd/5262. Accessed 29 Jun 2021
33. Weun, S., Jones, M.A., Beatty, S.E.: Development and validation of the impulse buying tendency scale. Psychol. Rep. **82**, 1123–1133 (1998)
34. Xie, K., Lee, Y.J.: Social media and brand purchase: quantifying the effects of exposures to earned and owned social media activities in a two-stage decision making model. J. Manag. Inf. Syst. **32**(2), 204–238 (2015)
35. Zerres, C.: Too hard to measure. Measurement of social media activities. an objective-based process. Manage. Market. **19**(2), 201–211 (2021)

The Economic Theoretical Implications of Blockchain and Its Application in Marine Debris Removal

Ting Jung Tsao(✉)

Department of Marine Environment and Engineering, National Sun Yat-Sen University, Kaohsiung 80424, Taiwan
dominbelle2@gmail.com

Abstract. Following the success of digital coins, such as Bitcoin, blockchain technology, on which digital coins are based, has attracted much attention. One of the main capabilities of blockchain is "to ensure any saved data is exactly the same as when it was generated" and that such data cannot be tampered with or destroyed. Therefore, the information written on blockchain is undeniable and absolutely credible. The cover story of the Economist in its October 2015 issue calls blockchain the "trust machine". Related theories about the core capabilities of blockchain, i.e., "increases in information symmetry " and "increases in mutual trust", are used to construct the basis of an economic theory of blockchain to explore the potential application of blockchain and to consider incentives for the application of blockchain in marine debris removal in this research.

Plastic debris leaks into the ocean and cause an unprecedented ecological catastrophe. This has become a hot topic for discussion amongst all disciplines. In an article published in 2016 by the Ellen MacArthur Foundation, which is committed to promoting the global circular economy, it stated that, at present, as much as 8 million metric tons of plastic debris leaked into the ocean each year, and if this continued, there would be more plastic than fish in the ocean around the globe by 2050. Following this shocking prediction, it was firmly declared in the G20 Osaka Summit, consisting of representatives from 20 nations, to reduce the environmental impacts caused by marine plastic debris "with a goal to end marine plastic pollution by 2050" and to collectively join actions in marine plastic removal.

There are also many issues of information asymmetry and mutual trust in the governance of marine resource protection, for example, within a marine debris recycling system, the relationship between the manufacturing industry and marine debris, the increased cost of petrochemical plastic manufacturers in reducing marine debris, issues of trust in the government in promoting plastic reduction, the lack of the credit information of recyclers by cooperating financial institutions, the low willingness to participate by environmental recycling agencies in marine protection due to the high cost of marine debris removal, and issues of cooperation between the industry and the government in decision-making processes. These issues cause difficulties in introducing reform policies, such as, in respect of adjustments of guaranteed prices for marine debris recycling, and income insurance (assurance) policies for fishermen for their assistance in marine debris recycling. Promoting policies can encounter great difficulties as well, such

F. Fui-Hoon Nah and K. Siau (Eds.): HCII 2022, LNCS 13327, pp. 127–138, 2022.
https://doi.org/10.1007/978-3-031-05544-7_10

as promoting policies in respect of adjustments of marine debris removal policies, classifications of marine debris recycling, marking systems for qualified recyclers for product conversions, and non-project based fishermen loans for marine debris recycling, etc. If blockchain technology can be used to solve the issues of information asymmetry and mutual trust among people, between manufacturers and consumers, among recycling agencies, manufacturers and fund providers, and between the people and the government, it will play a crucial role in introducing and implementing reform policies in marine resource conservation development. To demonstrate its determination in marine debris removal, the Ocean Affairs Council in Taiwan R.O.C. participated in the Presidential Hackathon 2020. With the theme of "Ocean Blockchain, Garbage Turning into Gold" and through the public-private collaboration, it proposed to use blockchain technology to connect volunteers and enterprises that would be willing to sponsor marine debris removal. In addition, based on the concept of a recycling economy of marine debris traceability, it was proposed that the collected information could be used in the future to establish hot spots for marine debris to call on all people for its removal, and, as a result, the quality of marine resource conservation could be improved to deal with the increasingly severe marine ecological destruction.

Keywords: Blockchain · Marine debris

1 Introduction

Following the success of digital coins, such as Bitcoin, blockchain technology, on which digital coins are based, has attracted much attention. One of the main capabilities of blockchain is "to ensure any saved data is exactly the same as when it was generated" and that such data cannot be tampered with or destroyed. Therefore, the information written on blockchain is undeniable and absolutely credible. The cover story of the Economist in its October 2015 issue calls blockchain the "trust machine" [1].

This capacity of blockchain has a profound significance in economic theories, for example, in economic phenomena, transaction costs are often increased due to an asymmetric information relationship between a principal and an agent and the existence of adverse selection and moral hazard [2], and as a result, optimal allocation (Pareto optimality) decisions [3] are not being made, reducing the overall social well-being, such as financial credit rationing phenomena. If blockchain technology can be used to enhance mutual information symmetry and mutual trust, transaction costs can be reduced, and as described in the Coarse theory (1960), the decision-making processes of both parties in transactions can be improved toward Pareto optimality, and the overall social well-being can therefore be improved.

There are also many issues of information asymmetry and mutual trust in the governance of marine resource protection, for example, within a marine debris recycling system, the relationship between the manufacturing industry and marine debris, the increased cost of petrochemical plastic manufacturers in reducing marine debris, issues of trust in the government in promoting plastic reduction, the lack of the credit information of recyclers by cooperating financial institutions, the low willingness to participate by environmental recycling agencies in marine protection due to the high cost of marine debris removal, and issues of cooperation between the industry and the government in decision-making processes. These issues cause difficulties in introducing reform policies, such as, in respect of adjustments of guaranteed prices for marine debris recycling, and income insurance (assurance) policies for fishermen for their assistance in marine debris recycling. Promoting policies can encounter great difficulties as well, such as promoting policies in respect of adjustments of marine debris removal policies, classifications of marine debris recycling, marking systems for qualified recyclers for product conversions, and non-project based fishermen loans for marine debris recycling, etc. If blockchain technology can be used to solve the issues of information asymmetry and mutual trust among people, between manufacturers and consumers, among recycling agencies, manufacturers and fund providers, and between the people and the government, it will play a crucial role in introducing and implementing reform policies in marine resource conservation development, and, as a result, the quality of marine resource conservation can be improved to deal with the increasingly severe destruction of marine ecology.

In this research, the relevant economic theory literature is discussed, and theories related to the core capabilities of blockchain, i.e., "increases in information symmetry" and "increases in mutual trust", are systematically discussed to construct the basis of an economic theory of blockchain to explore its potential application. Examples of the application of blockchain in agriculture and fishery are used as supporting evidence.

The principles and technology of blockchain are described in the next section to help understand the examples of its application. In the section under the heading of "Methodology - Systematic Literature Review", theories related to the core capabilities of blockchain, i.e., "increases in information symmetry" and "increases in mutual trust", are systematically organized to highlight the functions and importance of the "trust machine" in improving resource allocation decisions and the overall social well-being. In the section under the heading of "Examples of Application of Blockchain in Agriculture and Fishery", several application examples of blockchain in agriculture and fishery are introduced, and other potential economic application of blockchain is also discussed. In the last section, conclusion and some suggestions are provided.

2 Basic Principles and Technology of Blockchain

In the cyber network, at least several requirements must be met to "ensure any saved data is exactly the same as when it was generated" and that such data cannot be tampered with or destroyed:

1. There is a robust and durable data storage mechanism.
2. There is sufficient manpower to maintain this network information system.
3. There is a mechanism to verify the account which sends off the data.
4. There is a mechanism to ensure that the data will not be tampered with during network transmission processes.
5. It needs to be ensured that the data will not be tampered with after being stored in a public ledger (i.e., blockchain).
6. It needs to be ensured that the public ledger storing the data is not compromised.

If all the above requirements are met, it is a set of blockchain technology. More than one blockchain technology has been widely used at present. The more well-known ones are Bitcoin, Ethereum, and Hyperledger, etc., while others call themselves or are called Blockchain 2.0 or 3.0. They are mainly to increase the capabilities of smart contracts, to increase the chain speed, or to extend the capacity of ledgers.

Blockchain meets the above requirements with the following creative ideas and information technology:

a. an ingenious design digital coin mining rewards;
b. three underlying information theories - the Hash method, Elliptic Curve Cryptography, and peer to peer broadcasting; and
c. a computer program that implements the rules set by it (the blockchain policy white paper).

A brief introduction of related technologies is listed in Appendix 1. The technical points are summarized below.

1. Digital coin mining rewards: This ingenious design can be said to be the most critical invention of blockchain. A problem-solving competition rule has been designed into the blockchain system (the most used method is the proof of work method). The first person to solve the problem and provide the answer will be rewarded with a certain number of digital coins (for example, 12.5 coins to be rewarded to the winner, and the number of digital coins rewarded can also be adjusted according to rules). The person who solves the problem first can link the collected data to be billed (a candidate block) to the blockchain. This process is called mining. The difficulty of competition questions (more on this later) is adjusted according to the number of miners, so that the time taken to solve problems is about the same for every mining competition (for example, 10 min per competition on average). Taking the Bitcoin blockchain as an example, at present, 12.5 digital coins can be mined (issued) every 10 min on average. The market price in January 2022 was above USD 35,000 per Bitcoin, meaning that the reward for winning miners was around USD 437,500 every 10 min, as a result, tens of thousands of Internet hosts ("miners") around the world joined in the Bitcoin mining. A miner needs to save a copy of a public ledger (that is, the blockchain) when mining, which provides the information and communication equipment and professional manpower required to maintain this system at the same time. Because there are tens of thousands of copies of the public ledgers, there is little possibility to tamper with so many copies or destroy all of them at once. The digital coin mining rewards allow the first, second and sixth blockchain requirements to be met.
2. The public and private key pairing mechanism: A user (the person who sends off the data) in a blockchain system uses an "electronic wallet" (i.e., a computer program) to set a password (with a length requirement, such as 256 binary digits) and keep it safe, called a private key. The elliptic curve algorithm is used in the electronic wallet to generate a public key from the private key and present it publicly in the form of an account URL. This algorithm ensures that the private key cannot be deduced from the public key. When a user sends the data (such as a payee account number and the

amount of a transaction), the electronic wallet will lock the sent data with the private key. When any miner in the network intercepts this information, it can be unlocked with the sender's account URL (i.e., the public key), verifying that the information is indeed sent by the account of the private key holder. The public and private key pairing mechanism allows the third blockchain requirement to be met.

3. The electronic signature mechanism: In fact, when a user sends the data, the electronic wallet will first compress the data into a Hash code to form an electronic signature and transmit it together with the plaintext data. The person who receives the data can also compress the plaintext data and compare the Hash code obtained from the compressed plaintext data with the electronic signature. If they are exactly the same, it can be certain that the data has not been changed during the transmission process. The electronic signature mechanism allows the fourth blockchain requirement to be met.
4. The mechanism to add blocks to blockchain: Miners of a blockchain system send (thousands of pieces of) the data collected from many users through peer-to-peer broadcasting within the solving time (around 10 min) in the last block. The data is consolidated into a candidate block and compressed into a Hash code.[1] At this time, a problem-solving argument (called "nonce") and a difficulty target will be set in the latest block of the blockchain system. The earliest miner, who can use the nonce to calculate the Hash value that meets the set difficulty target, will receive digital coin rewards and also chain its candidate block to the blockchain system. Such a mechanism interlocks and chains the blocks together, so if any information written on blockchain is tampered with, even just a little, will be discovered. This mechanism allows the fifth blockchain requirement to be met.

The capabilities of blockchain and the way the system operates have made redundant the roles of central institutions (such as governmental certification bodies) or intermediaries (such as financial institutions) that used to play a role in strengthening the trust between strangers in transactions. This is called "decentralization". However, decentralization is not a necessary condition for blockchain.

2.1 Methodology - Systematic Literature Review

Definitions and theories explored by some key research using economic theoretical implications of blockchain are discussed below.

The famous Coarse theory of economics (1960) states that "if trade in an externality is possible and there are sufficiently low transaction costs, bargaining will lead to a Pareto efficient outcome regardless of the initial allocation of property." This theory is widely used in many policies or analyses, such as negative externalities of environmental pollution or property rights disputes [4, 5]. In fact, not only is there no circumstance that warrants "no transaction costs", but there are huge transaction costs incurred. For

[1] The Hash method is widely used in the blockchain system, and is used in the compression of individual data (i.e., electronic signatures), the compression of full-page block data, the proof of work (that is, finding the Hash value that meets the target), etc. But the three Hash values are not related.

example, in the prevention and control of soil and water pollution, the administrative costs required are so high as to how to monitor who emits polluted water and how to collect credible evidence to make polluters admit to their polluting acts, that it is almost unaffordable [6]. Taking another example, although manufacturers can use a production and marketing traceability marking system to prove to consumers that their products are safe, the cost is prohibitive for manufacturers. Therefore, optimal allocation decisions are not reached in many transactions and the overall social welfare has not been improved. What remains is so called economic problems, or environmental problems, etc.

Transaction costs generally refer to "all costs other than transformation costs (i.e., production costs)" [7], including any cost, time and effort required for finding and matching transaction objects, negotiating transactions, entering into transaction contracts, monitoring the performance of contracts, remedying breaches, and mediating disputes and litigation. The main source of these costs is "insufficient information" and "information asymmetry" between transaction parties [2]. Information asymmetry leads to problems of adverse election, that is, hiding information before a transaction, and problems of moral hazard, that is, hiding actions (breaches of contracts) after a transaction [2, 8].

The adverse selection and moral hazard problems caused by information asymmetry are detrimental to human well-being. The economic credit rationing model [9] has done a very typical demonstration. Due to insufficient credit information and trust problems for borrowers, financial institutions are concerned of providing funds to borrowers because of adverse selection and moral hazard issues, while borrowers are unable to carry out their business plans due to the lack of sufficient funds. Under the market equilibrium interest rates, there is a surplus demand of loans in the lending market. This happens in the agricultural financial market in Taiwan as well. For example, capital investments are urgently needed to transform the agriculture industry in Taiwan. Many farmers reported that funds were insufficient, but in 2018, the agricultural financial market had surplus funds of approximately NT$300 billion [10].

To improve the credit rationing phenomenon, it has been recommended in past theories and literature that collateral should be increased and the verification carried out on the implementation of borrower's plans should be strengthened [11]. The former asserts and infers that sufficient collateral enables borrowers to obtain required loans at the market equilibrium interest rates. However, borrowers often do not have enough to offer as loan collateral. For example, farmers in Taiwan do not have many assets to offer as collateral, especially those new farmers who do not own land. The problem with the latter method is how to find a lower cost and efficient method of verification, as verification processes (or signoffs, certifications, assessments, etc.) require additional costs, just like the analyses in the principal and agent theory [12]. Moreover, the professional knowledge of the financial institution personnel about the personal career of the borrower is far less than that of the borrower himself/herself.

The emergence of blockchain and the "trust machine" is the solution to the information asymmetry, its derived adverse selection or moral hazard problems, because a party or parties to a transaction can disclose their information through blockchain and gain trust with each other. A self-authentication (also known as "the second party authentication") through blockchain is also more reliable and cheaper than a third party authentication [13].

The studies reviewed in this research are selected using four main criteria: (a) only studies published in English, (b) studies that have been published in peer reviewed academic journals and can be openly accessed online, (c) studies that analyze and explore information asymmetry between transaction costs and between transaction parties, and (d) the application of blockchain on a recycling economy of marine debris traceability jointly developed by Ocean Affairs Council, the Ocean Conservation Administration and the National Academy of Marine Research etc. in Presidential Hackathon 2020 in Taiwan R.O.C.[2]

The main focus of the author of this research is to analyze and evaluate the usability of blockchain on marine debris removal and recycling through its application on agriculture and fishery, which pioneers the use of blockchain on the verification of marine debris recycling.

2.2 Examples of Application of Blockchain in Marine Debris Removal

There is still much to be explored in terms of the potential blockchain application. In addition to its original application of issuing digital coins and conducting digital coin transfers and transactions, the most basic application is proof of existence. Any information that may require verification, such as degree certificates, certificates of rights, property titles, wills, production and sales traceability certificates, inspection data, instrument data, geolocation, timestamps, etc. (not limited to text, graphic, video and other electronic files), can be written on blockchain to gain trust of stakeholders. The authority or person that originally issues the certificate or grants the right only needs to write its information on blockchain, and the person who wants to verify the certificate can check the information directly on blockchain without the need for the original right holder or the certified person to provide the original or photocopy of the certificate or document. The authenticity of the right or certificate no longer needs to be verified with the original issuing authority or person.

A further example of the application of blockchain is smart contracts. A smart contract is an executable code written on blockchain that automatically determines whether contract conditions are fulfilled, and automatically executes the contract when the conditions are fulfilled. For example, an online fundraising contract, which takes effect automatically, otherwise the funds will be refunded; or a flight delay insurance contract, which automatically settles claims when a flight delay occurs.

Below illustrates how blockchain is applied in Taiwan in marine debris removal for the conservation of marine resources. 1. A system is designed to upload the quantity, collector, timestamp and location information of marine debris onto blockchain. 2. Those who are willing to help remove marine debris will obtain points converted by

[2] The participating teams in the Presidential Hackathon 2020 included the Ocean Conservation Administration and the National Academy of Marine Research. The graduate school of Department of Computer Science & Information Engineering of National Taiwan University, Taiwan Cement Corporation, LITE-ON Technology Corporation, The Society of Wilderness and Taiwan Marine Social Enterprise Co., Ltd were also invited to participate.

uploaded record on blockchain. 3. Those who remove marine debris (such as fleets of environmental protection ships and submarines) can exchange points for merchandises donated by CSR enterprises. 4. CSR enterprises can verify marine debris removal data from web dashboards and provide merchandises for exchanges in stores to those who help remove marine debris. 5. CSR enterprises use the points as CSR evidence and can use them to obtain tax-deductions, loan subsidies or to become award recipients in the future (Figs. 1, 2, and 3).

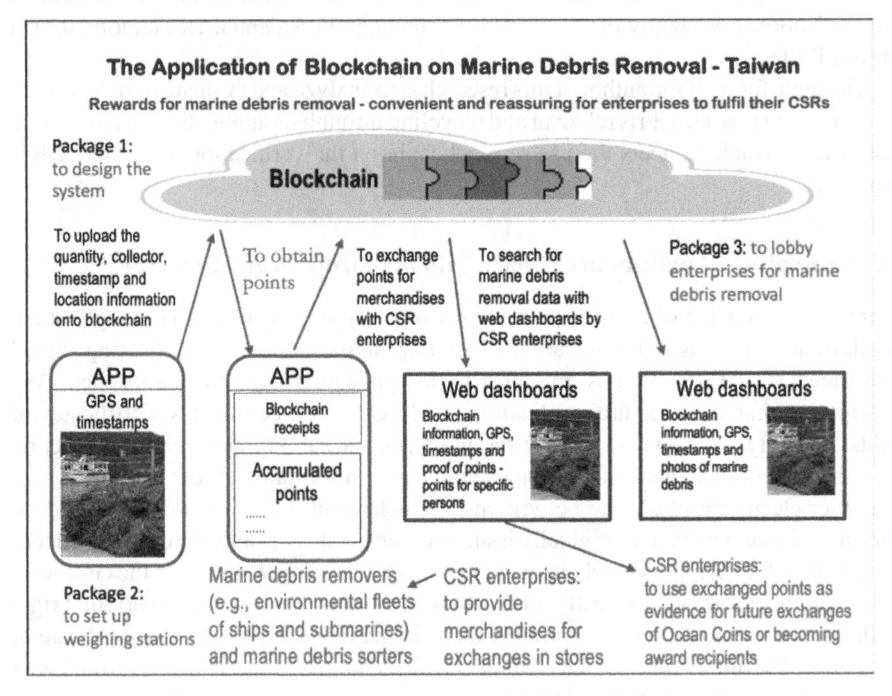

Fig. 1. Concept map of marine debris planning in Taiwan (1)

The incentives of the whole mechanism first come from the CSR of enterprises. Enterprises provide their products as incentives for marine debris removal. In return, enterprises can use points received from offering their merchandises for exchanges for marine debris removal as CSR evidence and the basis for becoming praise recipients (such as the Ocean Contribution Award). Enterprises can also accumulate the points first, which can then be exchanged for "Ocean Coins" or used for cashing in in the long run. In the early stage, those who clean up marine debris include fleets of environmental protection ships, various boats or ships that can bring back marine debris, those who dive to the seabed to collect marine debris, and those who recycle and reuse marine debris. Weighing stations for recycling marine debris are set up mainly at marine debris temporary storage areas near where ships are ashore. Information and communication technology supported weighing scales are used to reduce requirements for manpower and to improve the objectivity of weighing results. Required manpower are recruited from port management authorities or private entities or supported by enterprises. Smart

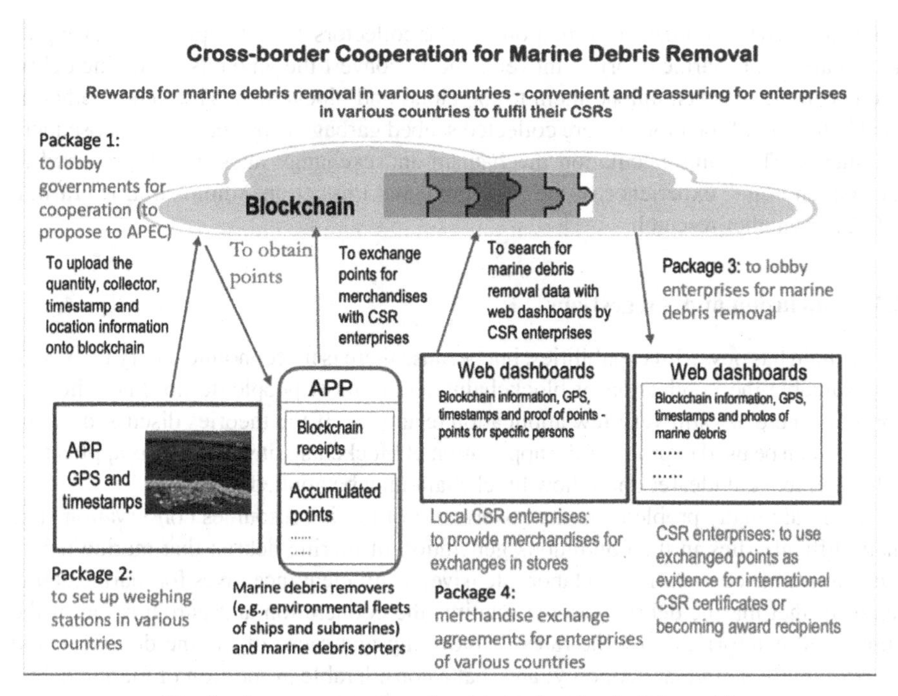

Fig. 2. Concept map of marine debris planning in Taiwan (2)

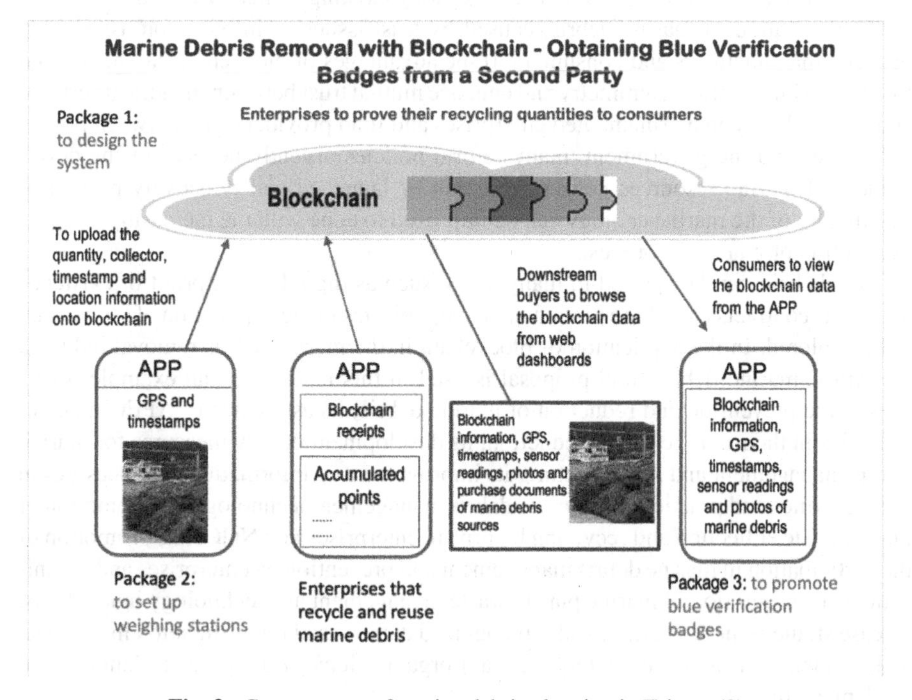

Fig. 3. Concept map of marine debris planning in Taiwan (3)

scales are used to record the information, such as collectors, timestamps, objects, weights and locations of marine debris removers, and to convert the amounts of marine debris into ocean blue points, upload them to the cloud and blockchain. The same method is used in respect of locations where collected seabed garbage is placed and its subsequent treatments. The point calculation mechanism and exchange rates may be referred to the "Green Point" experience of the Environmental Protection Administration, Taiwan R.O.C. or another research.

2.3 Conclusion and Suggestions

Blockchain has powerful capabilities, but because there is no economic theory to explain how valuable the application of blockchain can be, most people do not know how to apply it. There are still very few application examples. If the theories discussed in this research can be used to promote the application of blockchain, it will show its application value and make it clearer about how blockchain may be applied.

There are many problems in the governance of marine resources conservation. The main difficulty lies in the continuous generation of marine debris; that marine debris removal is time consuming and labor intensive; insufficient incentives for marine debris removal; that marine debris removal requires the cooperation and participation of the public and enterprises; that the rates of recycling and reuse of marine debris are too low to achieve a circular economy; and that a considerable proportion of marine debris drifts across national borders. It therefore requires multinational cooperation in the governance of marine debris removal. The other long existing difficulty is with the regard of the generation of marine debris, caused by trust issues or information asymmetry between manufacturers and consumers. If the advantages of blockchain can be utilized to improve information asymmetry and enhance mutual trust between manufacturers and consumers, between marine related enterprises and fund providers, and between private enterprises and the government, many reform policies urgently needed for the development of marine resources conservation can be launched and effectively promoted. The quality of the marine ecology can be improved to cope with the increasingly severe destruction of marine resources.

Blockchain can be applied in many areas, such as digital coins, proof of existence and smart contracts etc. There remain many possibilities in the application of blockchain to be explored. In the application of blockchain in the marine debris removal industry, the APEC marine debris fund proposal is used in this research as an example for its uses in the prevention and reduction of marine debris, its assistance of APEC member countries in their marine debris reduction, its development of new measures for marine debris management and infrastructure, its improvement of information and measures in policies, rules and practices for marine debris management technology, its promotion of plastic waste collection and recycling by private enterprises and NGOs, its promotion of the participation in marine debris management and prevention by enterprises and organizations, its promotion of marine plastic waste management and technology innovations, its case studies where internationally recognized data is used for verification in practice, and its inclusion of private enterprises and organizations in the project planning and implementation etc.

Looking to the future, it is urged to support the 1982 United Nations Convention on the Law of the Sea and the establishment of an open data platform for the information of marine ecology and biodiversity.

Appendix 1: Description of main blockchain technology

1. **Hash**

Hash is a key technology of blockchain. Many links in blockchain rely on Hash technology, such as:

I. Public keys: to generate public keys from private keys through Hash.
II. Electronic signatures: to compress any digital data (numbers, symbols, letters, words, sentences, articles, charts, graphics, photos, etc.) to be sent by blockchain end users into electronic signatures.
III. Hash codes for blocks: Multiple compressed hash codes can be further Hashed in pairs to make into a Merkle root (that is, a Hash code for a block).
IV. Mining: A problemsolving argument (nonce) provided by the latest block of the blockchain and a set difficulty target are used for miners to compete to find the earliest miner, who can use the nonce to calculate the Hash value that meets the set difficulty target.

Hash is a complicated mathematical conversion program (for details, see: https://zh.wikipedia.org/wiki/SHA-2). The following characteristics can be found in Hash codes obtained by Hash transformation: [14].

a. The length of a Hash code is fixed, no matter how long or short it is before Hashing.
b. The Hash code for the same data is the same after Hashing.
c. If there is a slight difference in two pieces of data, or if one piece of data is modified a little, the Hash code after Hashing will be very different.
d. The data before Hashing cannot be deduced from a Hash code.

2. **Elliptic Curve Cryptography (EEC)**

When a private key is used to generate a public key for the "electronic wallet" of a blockchain end user, in addition to using Hash, Elliptic Curve Cryptography may be used (for details, see: [14]). The formula of EEC is similar to (a1-1):

$$Y^2 = x^3 + aX + b \quad \text{(a1-1)}$$

The graph of formula (a1-1) is shown in Fig. 4. EEC is an algorithm with its own unique definition for four arithmetic operations. For example: According to the EEC definition, the addition of p1 + p2 in Fig. 4 equals p3 in the Figure. Under this uniquely defined operation, a public key can be generated from a private key, but a private key cannot be derived from a public key. If a piece of data is locked with a private key, only its paired public key can be used to open it. Likewise, if a piece of data is locked with a public key, only its paired private key can be used to open it.

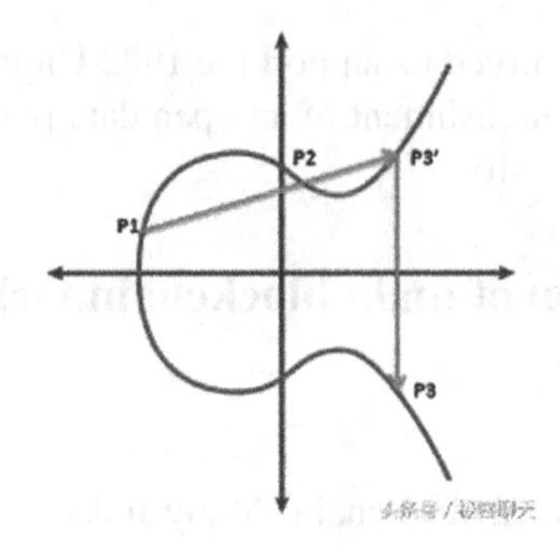

Fig. 4. Elliptic curve cryptography. (Figure source: https://kknews.cc/tech/y6je4yn.html)

References

1. The promise of the blockchain-the trust machine-the technology behind bitcoin could transform how the economy works. The Economist (2015)
2. Akerlof, G.A.: Market for lemons - quality uncertainty and market mechanism. Quart. J. Econ. **84**(3), 488–500 (1970)
3. Coase, R.: The problem of social cost. J. Law Econ. **3**(Oct), 1–44, (1960). 1992, WE und OATES, Hrsg: The Economics of the Environment
4. McCann, L., Easter, K.W.: Transaction costs of policies to reduce agricultural phosphorous pollution in the Minnesota River. Land Econ., 402–414 (1999)
5. Rees, G., Stephenson, K.: Transaction cost of nonpoint source water quality credits: implications for trading programs in the Chesapeake Bay Watershed (2014)
6. Chang, T.K.: The establishment of automated irrigation water quality monitoring system and classified management of water and soil resources. National Taiwan University, Taiwan R.O.C. (2016). (funded research by Council of Agriculture, Executive Yuan)
7. Allen, D.W.: What are transaction costs? Res. Law Econ. **14**, 1–18 (1991)
8. Stiglitz, J.E., Weiss, A.: Credit rationing in markets with imperfect information. Am. Econ. Rev. **71**(3), 393–410 (1981)
9. Jaffee, D., Stiglitz, J.: Credit rationing. In: Handbook of Monetary Economics, vol. 2, pp. 837–888 (1990)
10. Hille, K.: Taiwan's rice farmers use big data to cope with climate change. Financial Times (2018)
11. Holmstrom, B., Tirole, J.: Financial intermediation, loanable funds, and the real sector. Quart. J. Econ. **112**(3), 663–691 (1997)
12. Jensen, M.C., Meckling, W.H.: Theory of the firm: managerial behavior, agency costs and ownership structure. J. Financ. Econ. **3**(4), 305–360 (1976)
13. Kaal, W.: Blockchain Solutions for Agency Problems in Corporate Governance (2017). https://medium.com/semadaresearch/blockchain-solutions-for-agency-problems-in-corporate-governance-a83aae03b846
14. Tagomori, T.: Developing Blockchain Smart Contract and Its Security Practice. NRI, Gijyutsu-Hyoron Co., Ltd., Tokyo (2017)

Intelligent Data Analysis and Business Analytics

The Corpus of Emotional Valences for 33,669 Chinese Words Based on Big Data

Chia-Yueh Chang[1], Yen-Cheng Chen[1], Meng-Ning Tsai[1], Yao-Ting Sung[1], Yu-Lin Chang[1], Shu-Yen Lin[1], Shu-Ling Cho[2], Tao-Hsing Chang[3], and Hsueh-Chih Chen[1,4,5,6(✉)]

[1] Department of Educational Psychology and Counseling, National Taiwan Normal University, Taipei, Taiwan
chcjyh@gmail.com

[2] Department of Clinical Psychology, Fu-Jen Catholic University, Taipei, Taiwan

[3] Department of Computer Science and Information Engineering, National Kaohsiung, University of Science and Technology, Kaohsiung, Taiwan

[4] Institute for Research Excellence in Learning Sciences, National Taiwan Normal University, Taipei, Taiwan

[5] Chinese Language and Technology Center, National Taiwan Normal University, Taipei, Taiwan

[6] MOST AI Biomedical Research Center, Taipei, Taiwan

Abstract. Emotion theories are mainly classified as categorical or dimensional approaches. Given the importance of emotional words in emotion research, researchers have constructed a co-occurrence corpus of 7 types of emotion words through word co-occurrence and big data corpora. However, in addition to the categorical approach, the dimensional approach plays an important role in natural language processing. In particular, valence has an important influence on the study of emotion and language. In this study, the co-occurrence corpus of 7 types of emotion words constructed by Chen et al. [1] was expanded to create a corpus of emotional valences. Then, stepwise multiple regression analysis was performed with the predicted criterion variables and 15 predictor variables. The criterion variables were the emotional valences of 553 frequently occurring stimulus words included in the Chinese Word Association Norms [2]. The predictor variables included the emotion co-occurrences scores for 2 clusters (a cluster of literal emotion words and a cluster of metaphorical emotion words) and 7 types of emotions (happiness, love, surprise, sadness, anger, disgust, and fear) [the emotional words were common words from both the co-occurrence corpus of 7 types of emotion words constructed by Chen et al. [1] and the Chinese Word Association Norms established by Hu et al. [2]] and the virtue word co-occurrences score. The results showed that the scores for literal happiness word co-occurrences, metaphorical happiness word co-occurrences, literal disgust word co-occurrences, literal fear word co-occurrences, and virtue word co-occurrences could predict the valence values of emotion words, with the multiple correlation coefficients of multiple regression analyses reaching .729. Subsequently, the valence values of 33,669 words were established using the formula obtained from the multiple regression analysis of the 553 words. Next, the correlation between the actual valence values and the predicted valence values was analyzed to test the cross-validity of the

F. Fui-Hoon Nah and K. Siau (Eds.): HCII 2022, LNCS 13327, pp. 141–152, 2022.
https://doi.org/10.1007/978-3-031-05544-7_11

established valences using the common words in the norm established by Lee and Lee [3] for the emotionality ratings and free associations of 267 common Chinese words. The results showed that the correlation between the 2 was .755, indicating that the predicted values generated by the big data corpora and word co-occurrence had a degree of similarity with the manually determined values. Based on theories and tests, this study used the co-occurrence data of 7 emotions and virtue to construct the corpus of emotional valences for 33,669 Chinese words. The results showed that the combined use of big data corpora and word co-occurrence can effectively expand existing corpora that were established based on emotional categories, improve the efficiency of manual construction of corpora, and establish a larger corpus of emotional words.

Keywords: Emotion · Valence · Word co-occurrence · Big data · Chinese

1 Introduction and Literature Review

Emotion theory falls within 2 approaches, namely, the categorical approach [4] and the dimensional approach [5]. Researchers supporting the categorical approach argue that certain facial expressions can be accurately recognized by people of different cultures and ethnic groups, which indicates that certain emotional categories have cross-cultural consistency; these emotions are called basic emotions. Basic emotions exist in the early stages of human life and have specific categories, and different basic emotions are accompanied by different physiological responses or representations to help people adapt to the environment [6, 7]. For example, Ekman and Friesen [8] classified human emotions into 6 categories: Happiness, sadness, fear, anger, disgust, and surprise. Each of these emotions has a corresponding facial expression and may be accompanied by related physiological responses [9]. Later, Shaver et al. [10] classified the words for emotions into 6 basic categories: Love, joy, surprise, anger, sadness, and fear.

Researchers supporting the dimensional approach argue that human emotions are characterized by dimensions, and they classify emotions based on their intensity within different dimensions and regard these dimensions as independent of one another [5, 11–15]. The 2-dimensional theory of emotion is the most common dimensional classification of emotions. For example, Russell and Pratt [5] believed that there are 2 main dimensions of emotion: Pleasure (pleasant–unpleasant) and arousal (high arousal–low arousal). Pleasure refers to the degree of comfort or discomfort felt by a participant after exposure to the stimulus. The greater the pleasure is, the more the perceived comfort, that is, the more positive the participant's feeling, and vice versa. Valence is commonly used to refer to this dimension. Arousal refers to the intensity of the emotion aroused in a participant after exposure to the stimulus, that is, the emotional high or low that the stimulus generates in the participant, and the higher the degree of arousal is, the stronger the emotional feeling. In addition to the 2-dimensional classification of emotions, Osgood et al. [14] categorized emotions into three dimensions: valence, arousal, and dominance.

Previous studies on emotion and language often explored the effects of valence and arousal on personal cognition and behavior from different perspectives, of which valence has the greater influence. In terms of a more detailed argument, some studies explored the relationship between emotion and language in terms of words and found that valence has a stronger effect on word processing than arousal does, and it can explain a higher proportion of variance in word recognition latencies than arousal does. Furthermore, both valence and arousal affect word recognition, but valence has a greater influence. In lexical decision and naming tasks, valence can explain approximately 2% of the variance in word recognition latencies and approximately 0.2% of the variance in naming times, while arousal can only explain approximately 0.1% of the variance in word recognition latencies and approximately 0.1% of the variance in naming times, indicating that valence has a stronger effect on word processing than arousal does [16]. This shows the importance of the valence dimension in emotion and language research. In addition, the macro effect of valence on the processing of language and facial stimuli was explored from the perspective of personal development [17].

In emotion research, emotional words have an especially important position and wide applications. Emotional words are words that people use to represent or convey emotions, and they contain both semantic information and emotional connotations. In empirical studies, emotional words have been used to promote the emotions of participants and to further understand the impact of emotions on cognitive function [18]. In language and psychological research, personality and psychological tendencies have been analyzed using the free association of emotional words [19]. To further study emotions, countries around the world have established several emotional word norms or corpora, including English emotional word norms or corpora, such as the Affective Norms for English Words (ANEW) [20–23], and Chinese emotional word norms or corpora, such as the Taiwan emotional word corpus and the Taiwan implicit emotion corpus [3, 24–27].

Past studies on the use of emotional words show that there are 2 major categories of emotional words: literal emotional words and metaphorical emotional words. Literal emotional words refer to words that directly describe emotional experiences, such as happiness and sadness, or words that directly evoke emotions, such as success and failure. Cho et al. [25] categorized literal emotion words into emotion-describing words and emotion-inducing words and established the Taiwan emotional word corpus. Metaphorical emotional words refer to words that use a known concrete object to describe a complex and abstract emotion; these serve as indirect and implicit stimuli to trigger emotions. Chen et al. [26] established the Taiwan implicit emotion corpus. Emotional metaphors can be classified into 4 categories, namely, sensation metaphors, event metaphors, object metaphors, and personality metaphors. These metaphor categories, combined with positive or negative emotional categories, can trigger emotions. For example, "cheerfulness" is the personality metaphor with the highest number of responses indicating happiness, and "ghost" is the object metaphor with the highest number of responses indicating fear.

Among the categories of emotional metaphors, personality metaphors are metaphors of emotions based on a person's personality, style, and attitude. Virtue is a personality trait. The relationship between morality and emotion has been widely discussed since ancient times (e.g., in The Nicomachean Ethics). Previous literature also pointed out that the emotional response to an event is affected by whether the associated behavior conforms to social and cultural behavioral norms [28–30]. In addition, some researchers believe that some high-standard moral behaviors (such as virtue) are also emotions [31]. It is undeniable that emotion and morality have a considerable degree of correlation [32].

Emotional word norms and corpora are mostly established according to the dimensional approach [33], and most of them dimensionally contain positive and negative emotion valence data. At the same time, in natural language processing, sentiment analysis techniques primarily label texts according to the positivity and negativity of the words in the dictionary or in the corpus based on the valence dimension of the dimensional approach and then determine the number of positive and negative words to infer the emotional tendency of the text. These techniques have been applied to English corpora, such as SentiWordNet [34] and SO-CAL [35], and to Chinese corpora, such as the NTUSD [36] and E-HowNet. The importance of emotional valence data in emotional research can be summarized by combining the establishment of previous norms and corpora and the results of emotion and language research.

In addition to corpora in which emotional words comprise the main body, corpora of non-emotional words have also been built and developed into software. A representative of such corpora is the Linguistic Inquiry and Word Count (LIWC) software developed by Pennebaker et al. in early 1990, which is a program for the linguistic and word analysis of written texts. The software makes research more efficient [37–39].

However, most of the abovementioned emotional word norms were manually constructed and have the disadvantage that they include too few to be applied to text analysis. In addition, manual construction is time-consuming and labor-intensive, making manually constructed word norms difficult to expand. In addition, the reliability and validity of these norms are likely to be affected by the life experience and domain knowledge of the research participants [40]. Recently, Chen et al. [1] constructed a co-occurrence corpus of 7 types of emotion words based on big data corpora and word co-occurrence, which not only circumvented the previously mentioned shortcomings of manual construction but included a large number of words. More importantly, the corpus can be continuously expanded as long as its text corpus is continuously updated, indicating that the combined use of big data corpora and word co-occurrence has the potential to be used to expand manually created corpora or norms.

In summary, emotional words have already been applied at many levels. Countries around the world have also established emotional word corpora or norms for research purposes. Most emotional word corpora and norms are established using the dimensional approach, and positive and negative emotional valence is a particularly important aspect of the dimensional approach. However, the methods that were used in the past are time-consuming and expensive. With the use of big data, co-occurrence measurements and large corpora may become effective tools for generating corpora that contain a large number of emotional words.

Therefore, this study aimed to establish the valence values of a large number of words through word co-occurrence and big data corpora. If emotions are classified based on the intensity changes of different dimensions according to the dimensional approach, then the intensity of a word in a specific dimension can be anchored by the difference in intensity between different emotion categories and words. The most appropriate formula can be generated by using the emotion co-occurrence scores of different emotion categories in a large number of words as predictor variables and the manually assessed valence values of words as criterion variables. Then, the valence value of another emotion norm can be used to determine whether the valence values predicted by the formula are correlated with the manually assessed valence values.

In this study, stepwise multiple regression analysis was performed with the predicted criterion variables and 15 predictor variables. The criterion variables were the emotional valences of words in the Chinese Word Association Norms established by Hu et al. [2]. The predictor variables included the emotion co-occurrence scores of 2 types of clusters (the cluster of literal emotion words and the cluster of metaphorical emotion words) of 7 types of emotions (happiness, love, surprise, sadness, anger, disgust, and fear) and the virtue co-occurrence scores. Then, the formula generated by the stepwise multiple regression analysis was used to obtain the valence values of words in the co-occurrence corpus of 7 types of emotion words. Finally, the correlation between the actual valence values and the predicted valence values was analyzed using the common words in the norm established by Lee and Lee [3] for the emotionality ratings and free associations of 267 common Chinese words.

2 Method

2.1 Material

Co-occurrence Corpus of 7 Types of Emotion Words. This study used the co-occurrence corpus of 7 types of emotion words constructed by Chen et al. [1]. This corpus contains the emotion co-occurrence scores of 33,669 words in 2 types of clusters (a literal emotion words cluster and a metaphorical emotion words cluster) of 7 types of emotions (happiness, love, surprise, sadness, anger, disgust, and fear).

Virtue Emotion Co-occurrence Score. This study used virtue as the personality metaphor emotion in the metaphorical emotions cluster. With reference to the method that Chen et al. [1] used to establish the co-occurrence corpus of 7 types of emotion words, a total of 124 virtue words (e.g., honesty, integrity, righteousness, frugality, and gentleness) were selected from the thesis of Chang Chien [41] to comprise the virtue words cluster. Second, the word co-occurrence was calculated using the window size between words as the chunk, 2 punctuation marks as the window size, and the Baroni-Urbani measure as the measure [42] to calculate the co-occurrence values of the words. The Baroni-Urbani measure is a commonly used co-occurrence measure [43] and has a minimum value of 0 and a maximum value of 1. A larger value indicates a higher likelihood of the co-occurrence of 2 words in the same clause, and a value of 0 indicates no co-occurrence of 2 words in the same clause. Finally, from the co-occurrence matrix of the Chinese text corpora, the co-occurrence vectors of all words in the cluster of virtue words were selected, and then the average co-occurrence score for each word and the cluster of virtue words was calculated. This average emotion co-occurrence score was regarded as the virtue co-occurrence score of the word. The results showed that a total of 35,905 words had no co-occurrence with the cluster of virtue words. For example, if a word w1 obtains k co-occurrence scores (E1–Ek) in the E category, then the average co-occurrence of the word w1 in the E category (CoE) is as follows:

$$Co_E = \frac{\sum_k^1 E_i}{k} \tag{1}$$

Chinese Word Association Norms. This study used the Chinese Word Association Norms established by Hu et al. [2]. The norm contains the valence values of 1,200 words, with a minimum value of 1 and a maximum value of 1. The valence values of 553 frequently occurring 2-character Chinese words in the norm were used as the dependent variables for calculating the regression-predicted valence.

Emotionality Ratings and Free Association Norms of 2-Character Chinese Words. This study used the valence values in the norm established by Lee and Lee [3] for the emotionality ratings and free associations of 267 common Chinese 2-character words as the indexes to validate the valence values predicted by the formula.

2.2 Procedure

In this study, we first selected the words that were common to both the co-occurrence corpus of 7 types of emotion words constructed by Chen et al. [1] and the Chinese Word Association Norms established by Hu et al. [2]. Then, stepwise multiple regression analysis was performed with the predicted criterion variables and the 15 predictor variables using the listwise method. The criterion variables were the emotional valences of words in the Chinese Word Association Norms established by Hu et al. [2]. The predictor variables included the emotion co-occurrence scores of 2 types of clusters (the literal emotion words cluster and the metaphorical emotion words cluster) of 7 types of emotions (happiness, love, surprise, sadness, anger, disgust, and fear) and the virtue co-occurrence score. Then, the formula generated by the stepwise multiple regression analysis was used to predict the valence values of all words in the co-occurrence corpus of 7 types of emotion words. Finally, the correlation between the actual valence values and the valence values predicted by the formula was validated using the common words in the norm established by Lee and Lee [3] for the emotionality ratings and free associations of 267 common Chinese words.

3 Results

3.1 Predicting the Valence Value with the Emotion Co-occurrence Scores of Words

A stepwise multiple regression analysis of the predicted criterion variables and the 15 predictor variables was performed using the listwise method. The criterion variables were the emotional valences of 553 frequently occurring stimulus words in the Chinese Word Association Norms [2]. The predictor variables included the emotion co-occurrence scores of 2 types of clusters (literal emotion words and metaphorical emotion words) of the 7 types of emotions (happiness, love, surprise, sadness, anger, disgust, and fear) and the virtue co-occurrence scores. As Table 1 shows, the multiple correlation coefficient between the 2 reached $R = .729$ (adjusted $R^2 = .531$), indicating a significantly high correlation. The maximum positive standardized β coefficient corresponded to the variable "score of virtue co-occurrences"; in other words, the virtue co-occurrence score could effectively predict the valence scores of words, indicating that whether a word's connotation is consistent with the society's moral standards will indeed affect the judgment of the emotion associated with the word. In addition, Table 1 shows that for positive emotions, only the emotion co-occurrence score of the happiness category could predict the valence of the words, which is consistent with the fact that happiness is the only positive emotion among the 6 basic emotions defined by Ekman and Friesen [8]. The variables with negative standardized β coefficients included the literal fear co-occurrence score and the literal disgust co-occurrence score. Finally, by multiplying the variables in Table 1 by the β coefficient and adding a constant term of 3.957, the following formula was obtained: the valence value of a word = the literal happiness co-occurrence score × 6.504 + the literal fear co-occurrence score × (−10.642) + the virtue co-occurrence score × 8.102 + the literal disgust co-occurrence score × (−7.881) + the metaphorical happiness co-occurrence score × 3.465 + 3.957.

Table 1. Summary of the multiple regression analysis.

Variable	Multiple correlation coefficient R	Coefficient of determination R^2	β coefficient	Standardized β coefficient	t-value	F-value
	.73	.53				123.82***
LH			6.50	.33	4.49	
LF			−10.64	−.46	−8.35	
VO			8.10	.35	9.09	
LD			−7.88	−.28	−4.50	
MH			3.47	.19	2.78	

Note. LH = Literal happiness co-occurrence score, LF = Literal fear co-occurrence score, VO = Virtue co-occurrence score, LD = Literal disgust co-occurrence score, MH = Metaphorical happiness co-occurrence score

3.2 Generating the Valence Values of 33,669 Chinese Words

In this study, the valence values of 33,669 Chinese words were obtained using the formula generated by the stepwise multiple regression analysis. The valence values of these words had an average of 4.098, a standard deviation of 0.363, Q1 of 3.937, Q2 of 4.033, Q3 of 4.191, kurtosis of 6.307, and skewness of 1.258. Table 2 shows the 10 examples highest and lowest valence values in the co-occurrence corpus of 7 types of emotion words. Because the present study used big data corpora and word co-occurrence as the tool for generating the valence values of words, words that cannot form a sentence independently (e.g., **難免 (inevitably), 招致 (cause), 導致 (lead to)** are inevitably assigned valence values due to their high frequency of co-occurrence with specific types of words. The percentage of such words among the 200 words with the lowest valence values was 18%, and the percentage of such words among the 200 words with the highest valence values was 15%.

Table 2. The 10 Chinese words with the highest and lowest valence values in the co-occurrence corpus of 7 types of emotion words.

The 10 words with the highest valence values	Valence value	The 10 words with the lowest valence values	Valence value
熱情 (Enthusiasm)	7.00	遭受 (Suffer)	1.25
感恩 (Gratitude)	6.93	恐懼 (Terror)	1.50
真誠 (Sincerity)	6.91	憤怒 (Anger)	1.78
培養 (Cultivation)	6.83	暴力 (Violence)	1.88
展現 (Show)	6.76	不堪 (Unbearable)	2.03
分享 (Share)	6.64	難免 (Inevitably)	2.08
親切 (Cordial)	6.59	忍受 (Endure)	2.10
贏得 (Win)	6.56	免於 (Avoid)	2.17
美德 (Virtue)	6.54	害怕 (Fear)	2.17
愛心 (Love)	6.50	虐待 (Abuse)	2.18

3.3 Validating the Correlation Between the Predicted Valence Values and the Actual Valence Values

To understand whether the valence values generated by the formula obtained in this study are consistent with the manually assessed valence values, this study analyzed the correlations between the emotional valences of 265 common words in the norm established by Lee and Lee [3] and the valence values predicted in this study. The results showed a high correlation between the 2 values ($r = .76, p < .001$). That is, the emotional valences generated by the co-occurrence-based indicators were in high agreement with the manually determined values and have the potential to replace human ratings.

4 Discussion

This study aimed to establish the emotional valences of a large number of words through word co-occurrence and a big data corpus. First, stepwise multiple regression analysis was performed with the predicted criterion variables and 15 predictor variables. The predictor variables included the emotion co-occurrence scores of 2 types of clusters (literal emotion words and metaphorical emotion words) of 7 types of emotions (happiness, love, surprise, sadness, anger, disgust, and fear) and the virtue co-occurrence score (obtained according to the method used by Chen et al. [1]). The criterion variables were the emotional valences of the frequently occurring stimulus words in the Chinese Word Association Norms established by Hu et al. [2]. The multiple correlation coefficient of multiple regression analysis reached .729. The variables literal happiness co-occurrence score, metaphorical happiness co-occurrence score, literal disgust co-occurrence score, literal fear co-occurrence score, and virtue co-occurrence score can predict the valence values of words. Then, the formula obtained by the multiple regression analysis was used to generate the valence values of 33,669 words. Finally, the words that were also found in the norm compiled by Lee and Lee [3] were analyzed to determine the correlation between the predicted valence values and the actual valence values, and the coefficient of correlation between the two was .755.

For the corpus of emotional words included in this study, the predictive variables included not only the existing literal emotion word and metaphorical emotion word clusters but also the data for the corpus of virtue words in the personality metaphor category. In this study, stepwise multiple regression analysis was performed using the same criterion variables that were used in past studies. After the inclusion of virtue (specifically, the use of the virtue co-occurrence score as a predictive variable), the multiple correlation coefficient increased from .698 to .729, and the variable with the largest standardized β coefficient changed from the literal happiness co-occurrence score to the virtue co-occurrence score. The results are consistent with the view of previous studies that the emotional response to an event is affected by whether the behavior conforms to social and cultural behavioral norms [28–30], a factor that may be manipulated in future experiments. The relationship between virtue and the valence values of words can be further explored with experimental manipulations, such as evaluating the valence values of the words that do not conform to expected behavior after the subjects are prompted to pay more attention to virtues.

In recent years, an increasing number of researchers have focused on emotion research and have established many corpora or norms of emotional words for emotion research. However, considerable time and money are required to establish corpora or norms of emotional words. In addition, these data cannot be retained for long periods, and the corpora must be rebuilt over time. It was not until recently that Chen et al. [1] established a co-occurrence corpus of 7 types of emotion words based on co-occurrence and big data corpora. This study further used big data corpora and word co-occurrence to automatically generate the valence values of words. Hence, the corpus constructed in this study has the advantages of the co-occurrence corpus of 7 types of emotion words established by Chen et al. [1], including a large number of words, low money and time costs, and the ability to update the data in real time. More importantly, the correlation analysis based on the valence values of the emotional words in the norm established by Lee and Lee [3] found that the valence values of the words generated in this study were highly correlated with the manually assessed valence values, indicating that the valence values predicted by the norms for emotional words (established with the method used in this study) are highly similar to the valence values derived from subjects in actual experiments. Therefore, the combined use of big data corpora and word co-occurrence measures can effectively expand past corpora that were established based on emotional categories and can potentially improve the efficiency of manually constructing corpora and establish a larger corpus of emotional words.

At the application level, the results of this study can be used in selecting material for emotion research. Material selection has a profound impact on the results of emotion research. The virtue co-occurrence scores and valence values of the 33,669 words generated in this study can provide subsequent researchers with more options for material selection than traditional corpora or norms, and this corpus has been authorized for free use by researchers.

The contribution of this study is the establishment of the emotional valences and virtue co-occurrence scores of 33,669 words through big data and word co-occurrence based on theory and experimental findings. In addition, the high similarity between the data generated by this method and the manually constructed norms or corpora was experimentally validated. The limitation of this study is that it was impossible to include all co-occurrence measures. According to the collation of Pecina (2009), there are more than 80 co-occurrence measures. Hence, the use of other co-occurrence measures may further improve the similarity between automatically assessed results and manually assessed results.

Subsequent researchers may use different co-occurrence measures to automatically generate emotional valences and determine the degree of similarity between automatically and manually assessed values. They may also use the method proposed in this study to predict other dimensions (e.g., arousal, continuance, and concreteness) of emotions in the dimensional approach in an effort to make the corpus of emotional words more complete and to help achieve a better understanding of emotions.

Acknowledgements. This work was financially supported by the grant MOST-111-2634-F-002-004 from Ministry of Science and Technology (MOST) of Taiwan, the MOST AI Biomedical Research Center, and the "Institute for Research Excellence in Learning Sciences" and "Chinese Language and Technology Center" of National Taiwan Normal University from The Featured Areas Research Center Program within the framework of the Higher Education Sprout Project by the Ministry of Education in Taiwan.

References

1. Chen, C.H., et al.: Building a "Corpus of 7 types emotion co-occurrences words" of Chinese emotional words with Big Data Corpus (in Chinese) [Paper presentation]. In: 24th International Conference on Human-Computer Interaction (HCII2022), Virtual Only Conference, June 2022 (2022)
2. Hu, J.-F., Chen, Y.-C., Zhuo, S.-L., Chen, H.-C., Chang, Y.-L., Sung, Y.-T.: Word "association" and "associated" norms for 1200 Chinese two-character words (in Chinese). Bull. Educ. Psychol. **49**(1), 137–161 (2017). https://doi.org/10.6251/BEP.20161111
3. Lee, H.M., Lee, Y.S.: Emotionality ratings and free association of 267 common Chinese words (in Chinese). Formosa J. Mental Health **24**(4), 495–524 (2011)
4. Darwin, C., Ekman, P., Prodger, P.: The Expression of the Emotions in Man and Animals. Oxford University Press, New York (1998)
5. Russell, J.A., Pratt, G.: A description of the affective quality attributed to environments. J. Pers. Soc. Psychol. **38**(2), 311–322 (1980). https://doi.org/10.1037/0022-3514.38.2.311
6. Eibl-Eibesfeldt, I.: The Expressive Behavior of the Deaf-and-Blind Born Social Comm. and Mov., Ed. I., pp. 163–193. Academic Press (1973)
7. Plutchik, R.: A general psychoevolutionary theory of emotion. Theor. Emotion **1**, 3–31 (1980)
8. Ekman, P., Friesen, W.V.: Measuring facial movement. Environ. Psychol. Nonverb. Behav. **1**(1), 56–75 (1976). https://doi.org/10.1007/bf01115465

9. Ekman, P., Friesen, W.V., Ellsworth, P.: Emotion in the Human Face: Guidelines for Research and an Integration of Findings. Pergamon Press, Oxford (1972)
10. Shaver, P., Schwartz, J., Kirson, D., O'Connor, C.: Emotion knowledge: Further exploration of a prototype approach. J. Pers. Soc. Psychol. **52**(6), 1061–1086 (1987). https://doi.org/10.1037/0022-3514.52.6.1061
11. Fontaine, J.R., Scherer, K.R., Roesch, E.B., Ellsworth, P.C.: The world of emotions is not two-dimensional. Psychol. Sci. **18**(12), 1050–1057 (2007). https://doi.org/10.1111/j.1467-9280.2007.02024.x
12. Lang, P.J., Bradley, M.M., Cuthbert, B.N.: Emotion, attention, and the startle reflex. Psychol. Rev. **97**(3), 377–395 (1990). https://doi.org/10.1037/0033-295X.97.3.377
13. Larsen, R.J., Diener, E.: Promises and Problems with the Circumplex Model of Emotion, pp. 25–59. Sage Publications Inc., Thousand Oaks (1992)
14. Osgood, C.E., Suci, G.J., Tannenbaum, P.H.: The Measurement of Meaning. University of Illinois Press (1957)
15. Thayer, R.E.: Activation-deactivation adjective check list: current overview and structural analysis. Psychol. Rep. **58**(2), 607–614 (1986). https://doi.org/10.2466/pr0.1986.58.2.607
16. Kuperman, V., Estes, Z., Brysbaert, M., Warriner, A.B.: Emotion and language: valence and arousal affect word recognition. J. Exp. Psychol. Gen. **143**(3), 1065 (2014)
17. Kauschke, C., Bahn, D., Vesker, M., Schwarzer, G.: The role of emotional valence for the processing of facial and verbal stimuli—positivity or negativity bias? Front. Psychol. **10**, 1654 (2019)
18. Gendron, M., Lindquist, K.A., Barsalou, L., Barrett, L.F.: Emotion words shape emotion percepts. Emotion **12**(2), 314–325 (2012). https://doi.org/10.1037/a0026007
19. Ko, Y.H., Cho, S.L.: The relationship between proneness to borderline personality and suicide ideation based on the analysis of mental function and free association of emotion words (in Chinese). Fu-Jen Journal of Medicine **11**(2), 59–71 (2013)
20. Bradley, M.M., Lang, P.J.: Affective Norms for English Words (ANEW): Instruction Manual and Affective Ratings, pp. 1–45. The Center for Research in Psychophysiology (1999)
21. Mukherjee, S., Heise, D.R.: Affective meanings of 1,469 Bengali concepts. Behav. Res. Methods **49**(1), 184–197 (2016). https://doi.org/10.3758/s13428-016-0704-6
22. Stadthagen-Gonzalez, H., Imbault, C., Pérez Sánchez, M.A., Brysbaert, M.: Norms of valence and arousal for 14,031 Spanish words. Behav. Res. Methods **49**(1), 111–123 (2016). https://doi.org/10.3758/s13428-015-0700-2
23. Warriner, A.B., Kuperman, V., Brysbaert, M.: Norms of valence, arousal, and dominance for 13,915 English lemmas. Behav. Res. Methods **45**(4), 1191–1207 (2013). https://doi.org/10.3758/s13428-012-0314-x
24. Wang, Y.N., Zhou, L.M., Luo, Y.J.: The pilot establishment and evaluation of Chinese affective words system (in Chinese). Chin. Ment. Health J. **22**(8), 608–612 (2008)
25. Cho, S.L., Chen, H.C., Cheng, C.M.: Taiwan corpora of Chinese emotions and relevant psychophysiological data-a study on the norms of Chinese emotional words (in Chinese). Chinese J. Psychol. **55**(4), 493–523 (2013). https://doi.org/10.6129/cjp.20131026
26. Chen, H.C., Chan, Y.C., Feng, Y.J.: Taiwan corpora of Chinese emotions and relevant psychophysiological data-a norm of emotion metaphors in Chinese (in Chinese). Chinese J. Psychol. **55**(4), 525–553 (2013). https://doi.org/10.6129/CJP.20130112b
27. Yao, Z., Wu, J., Zhang, Y., Wang, Z.: Norms of valence, arousal, concreteness, familiarity, imageability, and context availability for 1,100 Chinese words. Behav. Res. Methods **49**(4), 1374–1385 (2016). https://doi.org/10.3758/s13428-016-0793-2
28. Haidt, J.: The Moral Emotions Handbook of Affective Sciences, pp. 852–870. Oxford University Press, New York (2003)
29. Rich, J.M.: Moral education and the emotions. J. Moral Educ. **9**(2), 81–87 (1980). https://doi.org/10.1080/0305724800090202

30. Rudolph, U., Tscharaktschiew, N.: An attributional analysis of moral emotions: naïve scientists and everyday judges. Emot. Rev. **6**(4), 344–352 (2014). https://doi.org/10.1177/1754073914534507
31. Jackson, M.: Emotion and Psyche. John Hunt Publishing Limited (2010)
32. Tangney, J.P., Stuewig, J., Mashek, D.J.: Moral emotions and moral behavior. Annu. Rev. Psychol. **58**, 345–372 (2007). https://doi.org/10.1146/annurev.psych.56.091103.070145
33. Hinojosa, J.A., et al.: Affective norms of 875 Spanish words for five discrete emotional categories and two emotional dimensions. Behav. Res. Methods **48**(1), 272–284 (2015). https://doi.org/10.3758/s13428-015-0572-5
34. Baccianella, S., Esuli, A., Sebastiani, F.: Sentiwordnet 3.0: an enhanced lexical resource for sentiment analysis and opinion mining. Paper Presented at the Lrec (2010)
35. Taboada, M., Brooke, J., Tofiloski, M., Voll, K., Stede, M.: Lexicon-based methods for sentiment analysis. Comput. Linguist. **37**(2), 267–307 (2011). https://doi.org/10.1162/COLI_a_00049
36. Ku, L.-W., Chen, H.-H.: Mining opinions from the web: beyond relevance retrieval. J. Am. Soc. Inf. Sci. Technol. **58**(12), 1838–1850 (2007). https://doi.org/10.1002/asi.20630
37. Pennebaker, J.W., Booth, R.J., Francis, M.E.: Linguistic Inquiry and Word Count (LIWC2007) (2007)
38. Pennebaker, J.W., Booth, R.J., Francis, M.E.: Operator's manual: linguistic inquiry and word count: LIWC2007, Austin: LIWC. net (2007). http://homepage.psy.utexas.edu/HomePage/Faculty/Pennebaker/Reprints/LIWC2007_OperatorManual.pdf. Accessed 1 Oct 2013
39. Pennebaker, J.W., Francis, M.E., Booth, R.J.: Linguistic Inquiry and Word Count: LIWC 2001, vol. 71. Lawrence Erlbaum Associates, Mahway (2001)
40. Jaffe, E.: What Big Data Means for Psychological Science. Observer **27**(6) (2014)
41. Chang Chien, C.Y.: Analysis of Life Curriculum Textbooks in Elementary Schools with Regard to Character Building Education (in Chinese). Master's thesis in the summer social education program of the Department of Further Education, National Taitung University (2010)
42. Baroni-Urbani, C., Buser, M.W.: Similarity of binary data. Syst. Biol. **25**(3), 251–259 (1976). https://doi.org/10.2307/2412493
43. Pecina, P.: Lexical association measures and collocation extraction. Lang. Resour. Eval. **44**(1–2), 137–158 (2009). https://doi.org/10.1007/s10579-009-9101-4

Predicting the Usefulness of Questions in Q&A Communities: A Comparison of Classical Machine Learning and Deep Learning Approaches

Langtao Chen(✉)

Department of Business and Information Technology, Missouri University of Science and Technology, Rolla, MO 65409, USA
chenla@mst.edu

Abstract. Questioning and answering (Q&A) communities have become an important platform for online knowledge exchange. With a vast number of questions posted to elicit high-quality solutions as well as a large number of participants engaged in online knowledge sharing, a grand challenge for Q&A communities is thus to effectively and efficiently identify and rank useful questions. The current approach to solving this problem is either through user voting or by community moderators. However, such manual processes are limited in terms of efficiency and scalability, especially for large Q&A communities. Thus, automatically predicting the usefulness of questions has significant implications for the management of online Q&A communities. To provide guidelines for assessing the quality of online questions, this research investigates and compares various classical machine learning and deep learning methods for predicting question usefulness. A dataset collected from a large Q&A community was used to train and test those machine learning methods. The findings of this research provide important implications for both the research and practice of online Q&A communities.

Keywords: Q&A communities · Question usefulness · Machine learning · Deep learning

1 Introduction

Users are increasingly participating in online questioning and answering (Q&A) communities such as Yahoo! Answers, Reddit, and forums hosted on Stack Exchange to seek answers to their questions and/or provide solutions to solve others' problems [1, 2]. In 2021 alone, Stack Exchange network had 3.2 million questions posted[1]. That means there were on average 365 questions asked on the platform in every single hour. The efficiency and effectiveness of problem-solving in Q&A communities, however, depends on how quickly the submitted questions are made noticeable to experts with

[1] https://stackexchange.com/about (accessed on February 13, 2022).

F. Fui-Hoon Nah and K. Siau (Eds.): HCII 2022, LNCS 13327, pp. 153–162, 2022.
https://doi.org/10.1007/978-3-031-05544-7_12

relevant knowledge as well as how potential answer providers perceive the usefulness of the questions. Accordingly, large online Q&A platforms such as Reddit and Stack Exchange have adopted the mechanism of user voting to filter/rank questions submitted to the community. Users can voluntarily and anonymously vote up or vote down questions submitted. Questions with the highest user votes are displayed on the top of the question list or recommended to potential problem solvers with the highest priority.

However, user voting of questions is not efficient especially in large online communities, since it requires a significant amount of cognitive effort spent in assessing various quality aspects of the content submitted. Furthermore, the voluntary nature of user voting in most online communities may lead to a systemic problem due to the error of omission [3]. Studies have shown that the percentage of users participating in content voting is relatively low in various online settings [3, 4]. In addition, user voting may also be seriously biased under certain conditions [5]. Thus, to facilitate effective and efficient knowledge exchange, an imperative task for Q&A communities is to automatically predict the usefulness of questions by using machine learning methods.

Machine learning is to learn patterns from data without explicit programming. There are two broad approaches to machine learning: classical machine learning and the recently developed deep learning methods. Although deep learning methods have shown prospects in various applications especially when large amounts of training data are available, the classical machine learning methods are still popularly applied in numerous scenarios. In the context of online Q&A communities, questions remain as to: (1) how classical machine learning and deep learning methods can be implemented to assess the usefulness of questions, (2) what are the design principles that can guide the implementation of machine learning methods, and (3) under what conditions deep learning methods would perform better than the classical methods.

To provide guidelines for research and practice, this research investigates the application of a set of classical machine learning and deep learning methods for predicting the usefulness rating of questions in online Q&A communities. A large dataset collected from a Q&A platform was used to train those machine learning methods and compare their predicting performance. The results of this research provide important implications for both the research and practice of online Q&A communities.

This paper is organized as follows. The next section reviews work related to the prediction of question usefulness, machine learning, deep learning, and word embedding methods for machine learning. Then, research method is explained in Sect. 3. Section 4 presents preliminary results. The Sect. 5 discusses the current work and future directions for improving the performance of predictive models.

2 Related Work

2.1 Usefulness of Questions

Rating the usefulness of user-generated content is a common mechanism on online platforms. For example, consumers can rate the usefulness of customer reviews posted by others [6, 7]. In Q&A communities, not all questions posted to the communities have an equal opportunity of being solved. Those questions that are perceived useful are deemed to receive more attentions from potential experts who have sufficient knowledge and experience to solve the problems. Thus, appropriately composing a question can often determine whether and how long the question will be solved. This can be comprehended from the perspective of signaling theory. Signaling theory suggests that people assess the quality of content through a variety of cues or signals that can help reduce information asymmetry between the information signaler and recipient [8]. Thus, knowledge seekers purposively include important information in their questions such that the questions could attract attention and interest from other peers in the community. Guided by the theoretical framework of signaling theory, this research proposes that a set of important cues can signal the usefulness of questions.

Specifically, there is an abundance of basic linguistic cues that can be used to transfer purposive information from one party to another. As presented in Table 1, a set of important cues such as informativeness, diversity, media richness, readability, spelling, and sentiment can be used to explain or predict the usefulness of questions in Q&A communities. In addition, features of Linguistic Inquiry and Word Count (LIWC) can also be used to predict the usefulness of questions. The validity and reliability of LIWC features have been verified by previous studies [9–11].

Table 1. Description of basic linguistic features.

Usefulness Cues	Definitions	Sample studies
Informativeness	The amount of information embedded in the content	[12–14]
Diversity	The extent to which diverse topics are discussed in the content	[15, 16]
Media richness	The extent to which visual information (e.g., images) is included in the content	[15, 17]
Readability	The ease of reading the content by others	[13, 18]
Spelling	The level of correct spelling in the content	[6, 13]
Sentiment	The strength of opinion expressed in the content	[13]

2.2 Machine Learning and Feature Engineering

Machine learning methods can automatically learn structural patterns from data. In various application scenarios where analytical solutions are not possible and a dataset is accessible, machine learning methods are often preferred to construct empirical solutions such as spam filtering, credit scoring, product recommendation, and image recognition. The well-known no-free-lunch (NFL) theorem proposed by Wolpert [19] suggests that there is not such a single machine learning algorithm that performs best for all learning tasks. In other words, a comparison of machine learning methods (both classical and deep learning approaches) is needed for a specific domain task. A typical machine learning process includes data processing, feature extraction, feature selection, model training, model evaluation, and implementation.

A key factor for the success of machine learning projects is feature engineering that generates and prepares a set of important features from the raw data [20]. The process of feature engineering is also the key difference between classical machine learning methods (such as Linear Regression, Decision Trees, Support Vector Machines, Random Forests, and AdaBoost) and the recently developed deep learning methods. Classical machine learning methods rely on a manual process of feature engineering in which a set of important features need to be extracted from the raw data by experts, while deep learning methods have the capability of automatically extracting multiple levels of features from raw data [21].

2.3 Deep Learning

The recent advances in deep learning methods have motivated researchers and practitioners to apply deep neural networks to predict outcomes in numerous applications. Compared to classical machine learning methods, deep learning methods are more computationally expensive. Interestingly, deep learning methods tend to have good performance even when models overfit data [22], a phenomenon generally called benign overfitting [23]. With recent advances in algorithms and hardware, deep learning has emerged as an attractive learning algorithm for various applications including the classification or prediction of user-generated content on social media [24, 25]. Specifically, convolutional neural network (CNN) and recurrent neural network (RNN), the two major types of deep learning algorithms, have been used for various natural language processing and text mining tasks [26]. CNN was originally developed for image recognition by using convolution layers to automatically extract important features. RNN processes sequential data by using a loop structure to connect early state information back to the current state. Long-short term memory (LSTM) is a specific RNN model that was originally developed to learn long-term dependencies in the data [27].

2.4 Word Embedding

Machine learning methods applied for text mining usually require a specific type of embedding methods that map the raw data (characters, words, documents, etc.) to vectors that can be further fed into the machine learning models. The word2vec model [28] and doc2vec model [29] are two popular wording embedding methods for text mining such as sentiment analysis [30], online content quality assessment [31], and news classification [32]. Both the word2vec and doc2vec embedding methods can be used as an alternative to the traditional bag-of-words (BOW) approaches such as TF-IDF (term frequency-inverse document frequency) matrices.

Since the word2vec method only supports vector representation for words, the vector representation cannot be directly used for predictive analytics at document level. In practice, word2vec representations need to be aggregated to document level for document classification. Being an extension of the word2vec model, the doc2vec method directly learns the continuous representation of documents. Doc2vec is particularly attractive for various text mining tasks given its capability in capturing semantic meanings from textual data. Thus, this research applies the doc2vec embedding method. Specifically, two variants of doc2vec including distributed memory (DM) and distributed bag-of-words (DBOW) models are used to extract vector representations of online questions.

3 Research Method

An experiment was conducted to implement various classical machine learning methods and deep learning approaches to predict the usefulness of questions. Then those predictive models were compared. The following subsections explain the details of research method used in this study.

3.1 Data

The dataset was collected from a community-based open Q&A website for user experience designers and professionals. In the community, users can ask questions related to the design of user interfaces and answer questions posted by other peers. After a user submits a question to the community, other users can vote up or vote down the usefulness of the question. Those questions with the highest net votes (i.e., positive votes – negative votes) are displayed on the top of the question list so that all community users can first view them when looking at the question list. Figure 1 shows a sample question with usefulness votes.

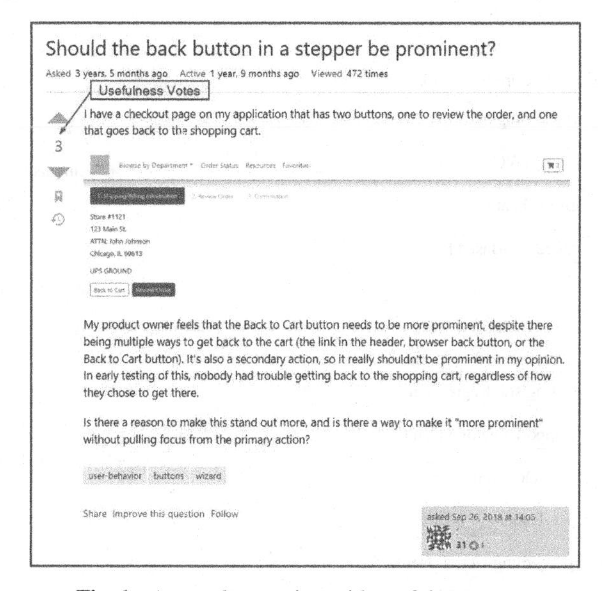

Fig. 1. A sample question with usefulness votes.

The dataset contains 30,718 questions posted from January 2010 to November 2021. The whole dataset was split into a training set of 24,574 questions (80%) and a test set of 6,144 questions (20%). The training set was used to train machine learning models, with the test set used to test the performance of these models.

3.2 Predictive Modeling

Given that a question posted to the community can be voted up and down, usefulness of the question is dichotomized as a binary variable.

$$Usefulness = \begin{cases} 1, \text{ if } up\ votes - down\ votes \geq 1 \\ 0, \text{ if } up\ votes - down\ votes \leq 0 \end{cases}$$

Figure 2 presents the overall predictive modeling procedure. After the dataset was collected from the online Q&A community, important features were extracted from the raw data. Specifically, the feature set includes basic linguistic features (explained in Table 1), LIWC features calculated by using the software tool LIWC [10], TF-IDF matrix as BOW features, and doc2vec features (using both DM and DBOW models) trained by utilizing the Gensim package [33]. In total, 1,216 features were extracted. Then, classical machine learning methods including logistic regression, support vector machines, decisions trees, and random forests were applied to classify usefulness based on features extracted. In addition, a CNN deep learning model was directly applied to

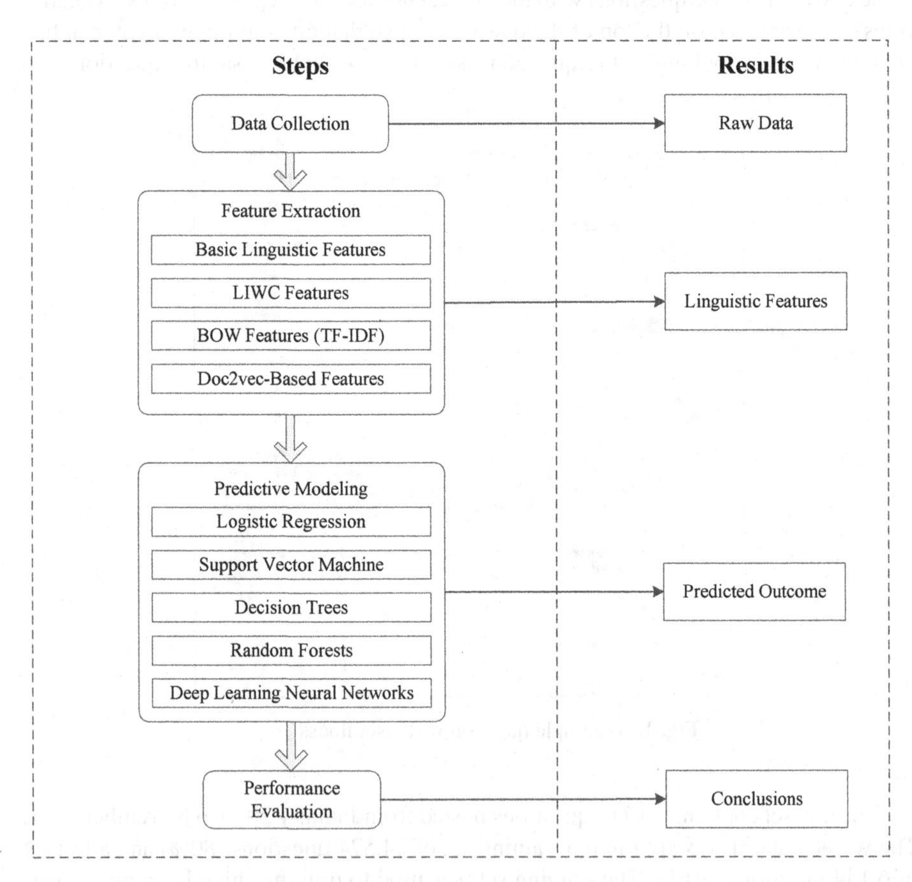

Fig. 2. Predictive modeling procedure.

the textual data to classify usefulness of questions. Finally, all predictive models were compared in terms of their predictive performance.

3.3 Feature Selection

The importance of all features was evaluated by applying a random forests algorithm. Figure 3 presents the importance scores of all 1,216 features.

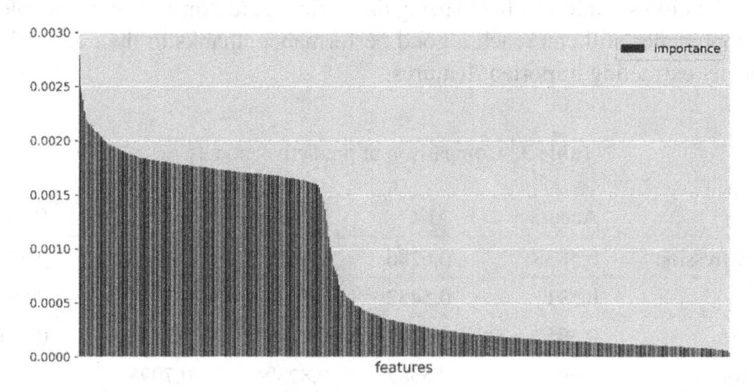

Fig. 3. Feature importance.

To reduce the dimensionality of predictive models, only the 600 most important features were selected for classical machine learning modeling. Table 2 presents a summary of those most important features with their average importance scores. Among all 600 important features, 400 features are trained from doc2vec models (i.e., 200 features from doc2vec DBOW, and 200 features from doc2vec DM). This clearly shows the capability of doc2vec models in deriving important features.

Table 2. Summary of top 600 most important features.

Feature category	Number of features	Mean importance
Doc2vec DBOW	200	0.0020
Doc2vec DM	200	0.0016
LIWC	84	0.0016
Basic linguistic feature	12	0.0014
BOW (TF-IDF)	104	0.0005

4 Preliminary Results

Table 3 summarizes the preliminary comparison of both classical and deep learning models. Among all machine learning models compared, random forest has the highest level of accuracy (0.6918), F1 score (0.8139) and recall (0.9544), whereas logistic regression has the highest level of AUC (area under the curve of ROC, 0.6286). The CNN model that directly learns word embeddings from the textual data achieves a mediate performance. This result indeed shows the need for theoretical guidance for classical machine learning modeling. With strong theoretical bases (such as signaling theory in this study) guiding feature engineering, classical machine learning methods could outperform deep learning methods. The result also shows the prospect of deep learning methods in automatically extracting important features for textual content classification. In application situations where strong theoretical guidelines are not possible, deep learning approaches still can reach a good performance, thanks to their capabilities of automatically extracting important features.

Table 3. Comparison of predictive models.

Method	Accuracy	AUC	F1 score	Precision	Recall
Logistic regression	0.5838	0.6286	0.6629	0.7743	0.5795
SVM	0.5911	0.5452	0.7074	0.7150	0.6999
Decision tree	0.5953	0.5139	0.7129	0.7144	0.7114
Random forest	0.6918	0.5382	0.8139	0.7095	0.9544
CNN	0.6234	0.5330	0.7420	0.7211	0.7641

5 Discussion

Online Q&A communities have offered an excellent opportunity for people to solve their problems without temporal and spatial constraints. To effectively seek answers, questions need to be composed in a way that can reduce the information asymmetry between knowledge seekers and potential knowledge providers. Informed by signaling theory, this research suggests that a variety of linguistic features can be used to predict the usefulness of questions submitted to Q&A communities. Specifically, this research has explored various classical machine learning and deep learning methods for predicting question usefulness.

As demonstrated in the preliminary results in Sect. 4, this study has evaluated a set of classical machine learning methods in classifying usefulness of questions. However, only a specific CNN model was evaluated in this study. For the future work, more deep learning neural network structures (such as a simple RNN and an LSTM) will be thoroughly evaluated. Features manually extracted from textual content can also be fed to deep learning structures to test how the deep learning methods perform with those manual features. An ensemble of both classical machine learning and deep learning methods can also be further evaluated. Importantly, grid search strategy will be used to tune numerous hyper-parameters in deep learning models.

Future work can also model the prediction of question usefulness as a regression problem by applying a variety of regression models to predict the natural count of usefulness votes. Findings of this research will provide practical and theoretical implications for improving the effectiveness and efficiency of knowledge exchange in online Q&A communities. Machine learning algorithms provide a technical approach to automatically filter/rank questions submitted to online Q&A communities, without the need for usefulness voting by users. This brings rich opportunities for designing new online community features or mechanisms that can address the grand challenge of supporting effective online knowledge exchange.

References

1. Chen, L., Baird, A., Straub, D.: Why do participants continue to contribute? Evaluation of usefulness voting and commenting motivational affordances within an online knowledge community. Decis. Support Syst. **118**, 21–32 (2019)
2. Chen, L., Baird, A., Straub, D.: The impact of hierarchical privilege levels and non-hierarchical incentives on continued contribution in online Q&A communities: a motivational model of gamification goals. Decis. Support Syst. **153**, 113667 (2022)
3. Liu, X., Wang, G.A., Fan, W., Zhang, Z.: Finding useful solutions in online knowledge communities: A theory-driven design and multilevel analysis. Inf. Syst. Res. **31**, 731–752 (2020)
4. Cao, Q., Duan, W., Gan, Q.: Exploring determinants of voting for the "helpfulness" of online user reviews: a text mining approach. Decis. Support Syst. **50**, 511–521 (2011)
5. Ochi, M., Matsuo, Y., Okabe, M., Onai, R.: Rating prediction by correcting user rating bias. In: 2012 IEEE/WIC/ACM International Conferences on Web Intelligence and Intelligent Agent Technology, pp. 452–456 (2012)
6. Ghose, A., Ipeirotis, P.G.: Estimating the helpfulness and economic impact of product reviews: mining text and reviewer characteristics. IEEE Trans. Knowl. Data Eng. **23**, 1498–1512 (2011)
7. Chen, L.: The impact of the content of online customer reviews on customer satisfaction: Evidence from yelp reviews. In: Proceedings of the ACM Conference on Computer-Supported Cooperative Work and Social Computing 2019, Austin, TX, USA (2019)
8. Spence, M.: Job market signaling. Quart. J. Econ. **87**, 355–374 (1973)
9. Tausczik, Y.R., Pennebaker, J.W.: The psychological meaning of words: LIWC and computerized text analysis methods. J. Lang. Soc. Psychol. **29**, 24–54 (2010)
10. Pennebaker, J.W., Boyd, R.L., Jordan, K., Blackburn, K.: The Development and Psychometric Properties of LIWC2015. University of Texas at Austin, Austin (2015)
11. Pennebaker, J.W., Francis, M.E.: Cognitive, emotional, and language processes in disclosure. Cogn. Emot. **10**, 601–626 (1996)
12. Huang, K.-Y., Long, Y.: Fighting together: Discovering the antecedents of social support and helpful discussion threads in online support forums for cannabis quitters. In: Proceedings of the 52nd Hawaii International Conference on System Sciences, pp. 4319–4328 (2019)
13. Chen, L., Baird, A., Straub, D.: A linguistic signaling model of social support exchange in online health communities. Decis. Support Syst. **130**, 113233 (2020)
14. Mudambi, S.M., Schuff, D.: Research note: what makes a helpful online review? A study of customer reviews on Amazon.com. MIS Q. **34**, 185–200 (2010)
15. Wu, L.: Social network effects on productivity and job security: evidence from the adoption of a social networking tool. Inf. Syst. Res. **24**, 30–51 (2013)
16. Bechmann, A., Nielbo, K.L.: Are we exposed to the same "news" in the news feed? An empirical analysis of filter bubbles as information similarity for Danish Facebook users. Digit. J. **6**, 990–1002 (2018)
17. Hlee, S., Lee, J., Yang, S.-B., Koo, C.: The moderating effect of restaurant type on hedonic versus utilitarian review evaluations. Int. J. Hosp. Manag. **77**, 195–206 (2019)

18. Yin, D., Bond, S., Zhang, H.: Anxious or angry? Effects of discrete emotions on the perceived helpfulness of online reviews. MIS Q. **38**, 539–560 (2014)
19. Wolpert, D.H.: The lack of a priori distinctions between learning algorithms. Neural Comput. **8**, 1341–1390 (1996)
20. Domingos, P.: A few useful things to know about machine learning. Commun. ACM **55**, 78–87 (2012)
21. LeCun, Y., Bengio, Y., Hinton, G.: Deep learning. Nature **521**, 436–444 (2015)
22. Belkin, M., Hsu, D., Ma, S., Mandal, S.: Reconciling modern machine-learning practice and the classical bias–variance trade-off. Proc. Natl. Acad. Sci. **116**, 15849–15854 (2019)
23. Bartlett, P.L., Long, P.M., Lugosi, G., Tsigler, A.: Benign overfitting in linear regression. Proc. Natl. Acad. Sci. **117**, 30063–30070 (2020)
24. Chen, L.: A classification framework for online social support using deep learning. In: Nah, FH., Siau, K. (eds.) HCI in Business, Government and Organizations. Information Systems and Analytics. HCII 2019. LNCS, vol. 11589. Springer, Cham (2019). https://doi.org/10.1007/978-3-030-22338-0_14
25. Haralabopoulos, G., Anagnostopoulos, I., McAuley, D.: Ensemble deep learning for multilabel binary classification of user-generated content. Algorithms **13**, 83 (2020)
26. Chai, J., Li, A.: Deep learning in natural language processing: a state-of-the-art survey. In: 2019 International Conference on Machine Learning and Cybernetics (ICMLC), pp. 1–6 (2019)
27. Hochreiter, S., Schmidhuber, J.: Long short-term memory. Neural Comput. **9**, 1735–1780 (1997)
28. Mikolov, T., Chen, K., Corrado, G., Dean, J.: Efficient estimation of word representations in vector space. In: Proceedings of Workshop at the International Conference on Learning Representations (2013)
29. Le, Q., Mikolov, T.: Distributed representations of sentences and documents. In: International Conference on Machine Learning, pp. 1188–1196 (2014)
30. Liang, H., Fothergill, R., Baldwin, T.: Rosemerry: a baseline message-level sentiment classification system. In: Proceedings of the 9th International Workshop on Semantic Evaluation (SemEval 2015), pp. 551–555 (2015)
31. Dang, Q.V., Ignat, C.-L.: Quality assessment of Wikipedia articles without feature engineering. In: Proceedings of the 16th ACM/IEEE-CS on Joint Conference on Digital Libraries, pp. 27–30. ACM (2016)
32. Trieu, L.Q., Tran, H.Q., Tran, M.-T.: News classification from social media using twitter-based doc2vec model and automatic query expansion. In: Proceedings of the Eighth International Symposium on Information and Communication Technology, pp. 460–467. ACM (2017)
33. Rehurek, R., Sojka, P.: Software framework for topic modelling with large corpora. In: Proceedings of the LREC 2010 Workshop on New Challenges for NLP Frameworks. Citeseer (2010)

Building a "Corpus of 7 Types Emotion Co-occurrences Words" of Chinese Emotional Words with Big Data Corpus

Ching-Hui Chen[1], Yu-Lin Chang[1], Yen-Cheng Chen[1], Meng-Ning Tsai[1], Yao-Ting Sung[1], Shu-Yen Lin[1], Shu-Ling Cho[2], Tao-Hsing Chang[3], and Hsueh-Chih Chen[1,4,5](✉)

[1] Department of Educational Psychology and Counseling, National Taiwan Normal University, No. 162, Sec. 1, Heping E. Road, Taipei 106, Taiwan
Chcjyh@gmail.com

[2] Department of Clinical Psychology, Fu-Jen University, Xinbei, Taiwan

[3] Department of Computer Science and Information Engineering, National Kaohsiung University of Science and Technology, Kaohsiung, Taiwan

[4] Institute for Research Excellence in Learning Sciences, National Taiwan Normal University, No. 162, Sec. 1, Heping E. Road, Taipei 106, Taiwan

[5] Chinese Language and Technology Center, National Taiwan Normal University, No. 162, Sec. 1, Heping E. Road, Taipei 106, Taiwan

Abstract. Past studies used human rated as the way of establishing a corpus which costs a lot of time and money but contains insufficient words, also the Categorical Approach was seldom used for building corpus, which may also lead to study bias. Therefore, study 1 of present study has used the Spreading Activation Model as the structure, and used big data of text corpus and word co-occurrences to build a corpus that contains more categories of emotions and much more words. First, study 1 selected the words that can clearly describe the meanings or can effectively evoke the feeling of its emotion category for seven emotions, including Happiness, Surprise, Sadness, Anger, Disgust, Fear, and Love. Then study 1 calculated the averages of co-occurrences for selected words and text corpora by seven emotions categories (measure is Baroni-Urbani, unit is chunk), it computes the averages of co-occurrences by emotional categories for 33669 words, it represents the conceptual consonance of words and the emotions. Study 2 has investigated the practical use of the corpus built in study 1, and used C-LIWC dictionary which was built by human rated as a comparison, taking the posts of Happy Board, Sad Board, Hate Board of PTT Bulletin Board System into the analyses of emotions recognition, result showed that Corpus of 7 Types Emotion Co-occurrences Words" built in study 1 had higher correct rate than human rated corpus. Present study has also compared the correct rates between the Corpus of 7 Types Emotion Co-occurrences Words and CLIWC (Chinese Linguistic Inquiry and Word Count), result showed correct rates of two databases were significant different, the corpus of present study has higher correct rate. Present study has built a text corpus for the material of emotion research, and the results also supports a potential of building the corpora of emotional words with big data measures.

Keywords: Emotional words · Co-occurrence · Chinese · Big data · Corpus

F. Fui-Hoon Nah and K. Siau (Eds.): HCII 2022, LNCS 13327, pp. 163–181, 2022.
https://doi.org/10.1007/978-3-031-05544-7_13

1 Introduction

Human beings conceptualize their mind in three levels, cognition, emotion, and intention. In particular, emotion is most important, because it was found to be strongly associated to cognition as well as to philological and psychological health [1]. The "Linguistic Analysis" plays a key role on exploring the psychological characteristics and processes, for human beings' thoughts, inner process and emotional traits can be reflected by elaborating the verbal texts existed naturally among daily life. For "linguistic analysis", an important way of understanding emotional states and mental health was the analysis of emotional words [2, 3].

For the linguistic analysis of emotion words, the research approaches are different due to theoretical interpretation of emotions, two major approaches are dimensional approach [4] and categorical approach [5]. The categorical approach claims that some facial expressions are commonly identified among the people from different cultures and races, these emotions are agreeably identified across cultures and thus are defined as "basic emotions". Basic emotions developed in early stage of life, each basic emotion represents its respective category, and has its specific physiological responses and expressions, which is in order to help individuals to adapt to the environment [6, 7]. For example, Ekman and Friesen [8] have classified human emotions into 6 categories, among which, each emotion had its corresponding facial expression and was accompanied with related body reactions [9]. Dimensional approach claims that human emotions have several dimensions, the classifications of emotions are defined by the strengths of these dimensions, and each dimension was distinguishing from others [4, 10–14].

For the purpose of emotional words research, the norms or corpora of emotion words have been generally built [15–19]; Researchers have also built several emotion words corpora of Chinese language [1, 20–23]. And those emotion words corpora constitute of the materials in many cognitive experiments [24, 25].

Not only the emotion words but also the relevant non-emotion words have been included, researchers have integrated and developed these results into computer programs, which made the research of emotion words more effectively [26, 27]. Among these computer programs, the "Linguistic Inquiry and Word Count" (LIWC) developed by Pennebaker and the colleagues in early 90's is the most generally used [27–29]. After more than a decade, LIWC is a program which contains 80 words categories and has good reliability and fine validity, it was also generally used in the studies that aimed at analysis of the linguistic characteristics of written texts and verbal communication [30]. For example, Rude, Gortner, and Pennebaker [31] have found that depressed participants used more negative emotion words than control group; people with extraversion used more positive emotion words and less negative emotion words than average people [32].

However, previous studies results have relied on human judgement work for establishing the emotional words corpora, and didn't include enough emotional words for linguistic analysis use. Also, human judgement was time-consuming and required a large amount of manpower, and further database expanding would be hard. Furthermore, the reliability and validity of corpora were easily influenced by participants' life experience and domain knowledge [33]. Another question of human-rated corpus was that most of the norms or corpora of Chinese emotional words were based on dimensional approach instead of categorical approach, because the norms or corpus based on dimensional approach were easy to apply in experiments.

With the innovation of technology and the coming of internet era, it is common for people to write down their life experience, daily events, and creation in the internet, eventually these mega digital text data becomes the mind treasure of exploring human behaviors and psychological characteristics. Certain big data is natural, updated with the time and comprehensive. Without any drawbacks of human-rated corpus, it is prefect for building the emotion words corpus.

It remains to be seen how the emotion words corpora will be built by using the naturalistic database of massive text. According to Association Theory [34], people can bring words with similar concepts gradually in the way of associative thinking. The stronger the connection between the word and the concept, the faster people bring it to mind. One solution to analyze big text data is from the perspective of co-occurrence of words, which means the frequency that two words occur together in a general corpus or daily conversations. Many researchers have indicated that every word activated the linking of related concepts, words association would also link related concepts or reflect the collocation of relevant concepts of daily life experience, and the concepts linking sequentially influenced the performance of related behaviors or attitudes [35, 36]. Besides, many studies have claimed the significant relation between the level of lexical co-occurrence and the association strength of words [37–39]. Based on the analyses of the massive text corpus with measure of word co-occurrences, we could build a simulated concept network of human's mind and mark out the emotional characteristics of words.

In the massive text corpus, however, a word can co-occur with many words and different attributes, the emotional characteristics of word is just one of them; mining the word co-occurrences with single word, as a result, the outcome would be a mess that included too many words but most of them were not significantly relevant to the emotional characteristics. For the sake of generating a systematic outcome, the idea of the Spreading Activation Model [40] is used in the present study. The idea of the Spreading Activation Model is that as the node of the concept of the sematic network is active, the node relative to the concept will be active one after another, and the one with similar concept or has stronger connection is active in priority. Thus, present study suggests to use the emotion words collections which mean a group of words with similar meanings as the data mining keys of massive text corpora, then to compute the word co-occurrences of every collection, even the outcome might contain functional words which were there for idiomatic expressions, because certain solution will gain more effective output.

In brief, many studies from different language have built the norms or corpora of emotion words for research purposes, and those norms or corpora were used not only in cognitive experiments, but also in others field, for example, text analysis. However, previous research methods were time consuming, expensive, less vocabulary and may cause the doubt of reliability and validity. Also, less corpus based on categorical approach were available. With the aid of big data, using the measure of word co-occurrence and the massive text corpus together would effectively develop the corpus of emotion words.

Therefore, present study aims at building a corpus of emotional words which doesn't have the short comings of human-rated database, using the "Spreading Activation Model" as structure, to build a corpus of emotional words for categorical approach by word co-occurrences and text corpora of big data. Last, in order to verify the practical use of this corpus of emotional words, we also implemented a comparison of correct rates of emotional attributes of written text between the corpora of emotional words by present study and an emotional words corpus by human-rated, and provide the supporting data that corpora of emotional words by present study is possible to replace the human-rated ones.

2 Study 1

Present study aimed at building an emotional words database by using Spreading Activation Model as the framework and analyzing the massive text corpus with measure of word co-occurrences.

First, present study used the six basic emotions proposed by Ekman and Friesen [8], which are "happiness", "surprise", "sadness", "anger", "disgust", and "fear"; since the six basic emotions only had one positive emotion, present study also included "love" [41] into the research.

Second, present study selected the words that could clearly describe the emotional content of its basic emotion categories, and the words with meanings could sufficiently activate the corresponding emotions. Besides, past studies have indicated that emotion had two styles of expression, direct/explicit and indirect/implicit. Therefore, present study selected emotional words of direct/explicit expression and, named them as literal emotion words, and the collection of literal emotion words as the cluster of literal emotion words. Present study also selected other emotion words of indirect/implicit expression and named them to be metaphorical emotion words, and the collection of metaphorical emotion words as the cluster of metaphorical emotional words. Then, according to the analyses of the massive text corpus, results presented the scores of word co-occurrences by seven emotion categories and two clusters of expressions. Last, present study integrated the emotion words, the scores of co-occurrences for seven emotion categories and two expression clusters, and built a Chinese emotion words data base.

2.1 Method

Material. The Cluster of Literal Emotion Words.

Present study used 2016 emotion-describing words and emotion-inducing words from the "Norm of Chinese emotional words" [20] and the emotion words in the "Chinese linguistic inquiry and word count dictionary" [42] as the literal emotion words. Then, a total of 1,317 literal emotion words were chosen in order to generate the clusters of literal emotion words for each emotion, and computed the word co-occurrences. Table 1 lists the amounts and some examples for the clusters of literal emotion words by categories of emotions. In particular, among the literal emotion words of seven emotions, amount of "sadness" is largest, and amount of "surprise" is smallest.

Table 1. The amount and example of literal emotion words by seven emotions

Emotion category	Happiness	Love	Surprise	Anger	Sadness	Disgust	Fear
Amount	220	79	51	224	314	281	148
Examples	快活 (Merry)	熱愛 (Like)	驚奇 (Amazed)	暴怒 (Rage)	難過 (Sad)	憎恨 (Hate)	懼怕 (Terrified)
	歡喜 (Gladness)	摯愛 (Beloved)	驚訝 (Astonish)	震怒 (Fury)	遺憾 (Sorry)	齷齪 (Filthy)	顫抖 (Tremble)
	歡呼 (Cheer)	深愛 (Deep love)	震驚 (Shocked)	激怒 (Irritated)	傷感 (Sorrow)	噁心 (disgusting)	怕生 (Shy)
	熱情 (Passion)	愛心 (Love)	受驚 (Frightened)	義憤 (Righteous anger)	愁悵 (Disconsolate)	輕蔑 (Contempt)	心急 (Impatient)
	豪爽 (Forthright)	甜蜜 (Sweet)	詫異 (Inquiring)	怒氣 (Resentment)	悲慘 (Miserable)	輕視 (Belittle)	恐懼 (Dread)

The Cluster of Metaphorical Emotion Words

Present study selected the metaphorical words from the "Norm of emotion metaphors in Chinese" [1] as the metaphorical emotion words; and in order to decrease the bias of infrequent used metaphor words, present study deleted the words which had response times less than 10. Then, 1,559 words from the norm were chosen as metaphorical emotion words for generating the cluster of metaphorical emotion words for each emotion and computed the co-occurrences of words. Table 2 lists the amounts and some examples of metaphorical emotion words by emotions. In particular, among the metaphorical words of seven emotions, the amount of "sadness" is largest, and the amount of "anger" is smallest.

Chinese Lexical Association Database

In order to compute the co-occurrence of emotion words, present study used the "Chinese Lexical Association Database" [43] as for the source of the text corpus which has over 400 million words, including the "UDN corpus" and "Sinica corpus".

Table 2. The amount and example of metaphorical emotion words by seven emotions

Emotion category	Happiness	Love	Surprise	Anger	Sadness	Disgust	Fear
Amount	270	219	181	175	314	185	215
Example	朋友 (Friend)	巧克力 (Chocolate)	禮物 (Gift)	火山 (Volcano)	眼淚 (Tear)	蟑螂 (cockroach)	鬼 (ghost)
	小鳥 (Bird)	玫瑰 (Rose)	中獎 (Win lottery)	火熱 (Heat)	墳墓 (Tomb)	腐臭 (Stinky)	蜘蛛 (Spider)
	糖果 (Candy)	太陽 (Sunny)	懷孕 (Pregnant)	恐龍 (Dinosaur)	孤兒 (Orphan)	黏黏 (Sticky)	老鼠 (Mouse)
	唱歌 (Sing)	約會 (Date)	外星人 (Alien)	背叛 (Betray)	喪禮 (Funeral)	嘔吐 (throw up)	黑暗 (Darkness)
	旅行 (Travel)	悅耳 (Pleasing sound)	刺 (Sting)	火藥味 (Smell of gunpowder, tension)	冰冷 (Gold)	垃圾 (Garbage)	無聲 (Noiseless)

2.2 Procedure

Present study computed the scores of co-occurrences of emotion words in the Chinese Lexical Association Database in order to establish the emotion words database. To compute the scores of co-occurrences, the window size (the distance between two computed words) is important; if window size was too big, the database would include many words with unqualified collocations; if windows size was too small, the database would include fewer words but be too conservative; both would result in bias. Henceforth, present study used "clause" as the unit for computing the words co-occurrences. Present study defined a "clause" to be a sentence between two punctuation marks. As the measure for computing the words co-occurrences, present study used the measure "Baroni-Urbani" [44], the minimum is 0, the maximum is 1, bigger value indicates a higher possibility of words co-occurred in a sentence, value 0 indicates no word co-occurrences. Previous research showed that the co-occurrences analysis by using "Baroni-Urbani" have higher predictive power of priming effects than association norms [43]. Then, in the co-occurrences matrices of Chinese Lexical Association Database, present study picked up the vectors value of co-occurrences by seven emotions and two clusters of expression for each word, and computed the mean of all values of co-occurrences of this word by emotion category and cluster, the mean represented "score of word co-occurrences". The formula of "score of word co-occurrences" is explained as above: the word (w1) has k scores of co-occurrences (E1~Ek) in emotion E, the mean of co-occurrences (CoE) of w1 for emotion E is:

$$\mathrm{Co_E} = \frac{\sum_k^1 E_i}{k}$$

2.3 Result

Not all of the Chinese words can co-occur in a sentence due to the logics of grammar, thus, the amounts of co-occurrences words of each emotion were different form each other. Table 3 lists the amounts of literal emotion co-occurrences words and metaphorical emotion co-occurrences words by seven emotions categories. For the literal emotion co-occurrences words, amount of happiness was largest, amount of surprises cluster was smallest; for metaphorical emotion co-occurrences words, amount of happiness cluster was largest but amount of anger was smallest.

Table 3. The amounts of literal and metaphorical emotion co-occurrences words in seven emotions

Emotion category	Happiness	Love	Surprise	Anger	Sadness	Disgust	Fear
Literal emotion words	49,704	35,021	19,016	36,430	36,755	35,184	37,160
Metaphorical emotion words	69,084	62,463	57,551	38,833	47,790	44,368	52,957

After deleting the function words, result had 79,008 co-occurrences words for seven emotions and two clusters of expression. Every word had its scores of word co-occurrences for seven emotions and two clusters. Tables 4 and 5 list the descriptive statistics of the scores of literal emotion co-occurrences and the scores of metaphorical emotion co-occurrences.

Table 4. The descriptive statistics of literal emotion co-occurrences by seven emotions

	Happiness	Love	Surprise	Sadness	Anger	Disgust	Fear
Maximum	.298	.395	.304	.269	.231	.270	.289
Mean	.011	.012	.005	.006	.006	.006	.008
S.D.	.027	.030	.017	.017	.016	.016	.020
Q3	.008	.007	0	.003	.003	.003	.005
Skewness	3.948	4.342	5.800	5.448	5.050	5.322	4.396
Kurtosis	18.112	23.001	47.794	37.31	32.041	36.081	23.779

Table 5. The descriptive statistics of metaphorical emotion co-occurrences by seven emotions

	Happiness	Love	Surprise	Sadness	Anger	Disgust	Fear
Maximum	.317	.321	.293	.217	.236	.290	.260
Mean	.017	.016	.013	.009	.010	.009	.010
S.D.	.036	.035	.028	.021	.022	.021	.023
Q3	.014	.013	.011	.007	.008	.007	.008
Skewness	3.416	3.564	3.620	4.010	3.908	4.281	4.088
Kurtosis	12.853	14.310	14.930	18.968	17.937	22.480	19.848

Tables 4 and 5 show that both the scores of co-occurrences of literal and metaphorical words had high kurtosis and positive skewness; also, among the "scores of emotion co-occurrences words", at least 75% of words had the score of co-occurrences less than mean. Therefore, present study deleted the words that had low scores of co-occurrences, considering some words had high co-occurrences only with one emotion, present study further deleted the words which had the scores of co-occurrences less than 0.1, and finally results had 33,669 words. Tables 6 and 7 list the descriptive statistics of scores of literal and metaphorical emotion co-occurrences of these 33,669 words by seven emotions, and it found that the kurtosis and skewness of "emotion co-occurrences scores" decreased. Tables 8 and 9 list the percentile of scores of literal and metaphorical emotion co-occurrences words as the reference of choosing research material for further studies.

Table 6. The descriptive statistics of scores of literal emotion co-occurrences by seven emotions after deleting the words with low scores of co-occurrences

	Happiness	Love	Surprise	Sadness	Anger	Disgust	Fear
Maximum	.298	.395	.304	.269	.231	.27	.289
Mean	.025	.026	.012	.013	.012	.013	.018
S.D.	.037	.042	.024	.025	.022	.023	.028
Skewness	2.447	2.733	3.79	3.503	3.223	3.423	2.768
Kurtosis	6.468	8.799	20.718	15.123	12.765	14.686	9.134

Table 7. The descriptive statistics of scores of metaphorical emotion co-occurrences by seven emotions after deleting the words with low scores of co-occurrences

	Happiness	Love	Surprise	Sadness	Anger	Disgust	Fear
Maximum	.317	.321	.293	.217	.236	.290	.260
Mean	.038	.036	.029	.020	.021	.019	.022
S.D.	.047	.047	.038	.028	.030	.029	.031
Skewness	2.084	2.197	2.251	2.526	2.457	2.732	2.598
Kurtosis	4.141	4.863	5.215	7.046	6.599	8.802	7.534

Table 8. The percentile of scores of literal emotion co-occurrences by seven emotions

	Happiness	Love	Surprise	Sadness	Anger	Disgust	Fear
75	.029	.030	.014	.012	.013	.013	.020
80	.038	.039	.018	.017	.017	.017	.026
85	.051	.054	.025	.024	.024	.024	.036
90	.074	.078	.036	.037	.036	.036	.052
95	.111	.121	.059	.064	.060	.062	.080

Table 9. The percentile of scores of metaphorical co-occurrences by seven emotions

	Happiness	Love	Surprise	Sadness	Anger	Disgust	Fear
75	.045	.043	.034	.023	.025	.022	.025
80	.058	.055	.044	.029	.033	.028	.032
85	.076	.073	.058	.039	.043	.038	.043
90	.106	.102	.080	.055	.060	.054	.060
95	.151	.147	.118	.085	.091	.084	.093

Tables 10 and 11 are the examples of words and its scores of co-occurrences by seven emotions. Present study has built a corpus of emotion words by the measure of word co-occurrences with the Chinese Lexical Association Database. The researchers can use our corpus to indicate the scores of co-occurrences of positive emotions and negative emotions, the most common classification of emotions, and also can find the scores of co-occurrences of particles (empty words). For examples, "Sharing" has nearer scores of emotion co-occurrences to the maximum of happiness and love, however, has farther scores of emotion co-occurrences from the maximum of other categories; "Violence" has farther scores of emotion co-occurrences from the maximum of happiness and love but nearer scores of emotion co-occurrences to the maximum of anger and fear. As for the examples of particles (empty words), "Such" has high scores of emotion co-occurrences with all seven emotions.

Present study has established an emotion words corpus of categorical approach. For application of the results, researchers can select the words that have higher scores for particular emotion, for example, "First love" to "Love", and "Indeed" for "surprise"; also, researchers can select the words that have high scores with two or more emotions, even with both positive and negative emotions at the same time, such as "memory" has higher scores with both positive emotion "love" and also with negative emotion "sadness". Besides, present study also builds the database of scores of emotions co-occurrences for literal and metaphorical words, researchers can select the words that have bigger differences of scores of emotion co-occurrences between literal words and metaphorical words, for example, "elf" and "picture" have higher scores of metaphorical emotion co-occurrences than scores of literal emotion co-occurrences in all seven emotions.

Table 10. The example of words of literal emotion words and its scores of co-occurrences by seven emotions

Words	Happiness	Love	Surprise	Sadness	Anger	Disgust	Fear
Sharing	.267	.275	.075	.087	.059	.059	.062
Violence	.054	.118	.077	.125	.171	.146	.215
Such	.252	.321	.216	.208	.147	.169	.193
First love	.078	.222	.024	.046	.020	.015	.079
Indeed	.056	.031	.239	.046	.012	.010	.066
Memory	.202	.300	.110	.166	.081	.094	.135
Elf	.107	.105	.048	.028	.022	.019	.052
Picture	.094	.091	.006	.004	.007	.007	.009

Table 11. The example of words of metaphorical emotion words and its scores of co-occurrences by seven emotions

Words	Happiness	Love	Surprise	Sadness	Anger	Disgust	Fear
Sharing	.262	.270	.157	.113	.083	.091	.112
violence	.099	.114	.094	.092	.118	.121	.141
Such	.248	.269	.203	.170	.142	.167	.195
First love	.079	.126	.067	.058	.042	.033	.038
Indeed	.028	.030	.048	.029	.030	.018	.018
Memory	.238	.276	.182	.171	.127	.111	.148
Elf	.189	.152	.139	.063	.075	.061	.102
Picture	.188	.185	.192	.056	.075	.067	.059

Present study has built an emotion words corpus which has 33,669 words by seven emotions and two cluster of expression styles, and their scores of emotion co-occurrences. Among these 33,669 words, 20,045 words are nouns, 12,619 words are verbs, and 1,332 words are adverbs.

The purpose of study 1 was to build a corpus of emotional words without the shortcomings of human-rated ones, present study constructed it with the frame of the Spreading Activation Model and calculated the co-occurrences of words and text corpora of big data. Study 1 developed a corpus of emotional words which contains 33669 Chinese words, every word has its scores of co-occurrences in seven emotions categories and two sources (literal and metaphorical). This corpus achieves the goal of study 1, it costs less expense and takes less time, but contains more words and presents the results closer to actual situation in daily use.

Study 2 examines the practical using of the corpus of emotional words built in study 1, and recognizes the emotional attributes of article by using this corpus of emotional words, and compared the results of human rated corpus, in order to support the possibility of replacing human rated corpora with program calculated one for establishing the corpus of emotional words.

3 Study 2

The purpose of study 2 was to examine the recognition ability of emotion categories of corpus developed in study 1 and compared the result with human rated corpus or dictionary. In order to compare the results, study 2 took Chinese LIWC, the "C-LIWC" as the representative of human rated corpus, C-LIWC is well known and generally used. Present study calculated the category of emotional words in given material, weighted the numbers of each category to output the emotion scores, and classified the emotion of an article, then compared the results between using C-LIWC and using the corpus of present study, in order to investigate the possibility of using corpus of present study to replace using the human rated one for the research of big data.

3.1 Method

Material
The Posts on the PTT (One bulletin board system).
Present study sampled 553 posts from the Happy Board, 2076 posts from the Hate Board, 2601 posts from the Sad Board in the popular bulletin board system named PTT in Taiwan, 5230 posts in total.

The Corpus of 7 Types Emotion Co-occurrences Words.
It was built in study 1.

C-LIWC
It was established by Huang et al. [44], and is embodied in the dictionary section of "Chinese Linguistic Inquiry and Word Count", in order to weight the emotional words by category.

Procedure
Choose the Text Material by Emotions

Present study chose the posts by the themes of boards from PTT, one of well know Bulletin Board System (BBS) in Taiwan, and used these posts as the text material of emotions. Present study chose the length of text from 200 to 400 words per post, and the content was about the themes of boards. In total, 5230 posts were selected, 553 posts from Happy Board, 2076 posts from Hate Board, and 2601 posts from Sad Board.

Calculate the Emotional Words by Category and Weighted

Since the text material of present study was the posts selected from the Happy Board, Hate Board, Sad Board, these posts respectively represent the emotions: Happiness, Anger, and Sadness, it has to consider the weight and amounts of these three emotional words by the "corpus of 7 types emotion co-occurrences words" and "C-LIWC". The C-LIWC doesn't provide words of happiness but named it "positive words", also, C-LIWC doesn't provide weights by emotional category, therefore, present study gave the weights of all kinds of emotion categories in C-LIWC to be 1. The way of calculation the emotional score, take Happiness for explanation, was to count the number of emotional words of happiness in one post based on the "corpus of 7 types emotion co-occurrences words" and "C-LIWC" respectively, and multiplies the number by weight of emotion, which is Happiness in this example, and then divided this weighted number by total words of this post, the quotient was the emotional score of Happiness. The emotional scores of Sadness and Anger used the same way as for Happiness. For example, the total words of a post were 200, contains 2 categories of emotion Anger based on the C-LIWC, the emotional score of Anger of this post was 2 × 1 (weight) ÷ 200 (Total words) = 0.01, and so on for other emotions.

Discriminant Analysis

Present study used emotional scores of Happiness, Anger, and Sadness computed based on the "corpus of 7 types emotion co-occurrences words" as the independent variables, and the source of board of each post as dependent variable, implemented the discriminant analysis to identify the emotions categories of text and get the correct rate of categorization of emotions based on the "corpus of 7 types emotion co-occurrences words", then the same method but using the emotional scores based on the C-LIWC and got the correct rate of categorization of emotions based on C-LIWC.

The Test of Significance

At last, present study implemented a test of significance between the correct rates of C-LIWC and the "corpus of 7 types emotion co-occurrences words", in order to know the difference of emotion categorizations based on these two corpora.

3.2 Result

Descriptive Statistics

Table 12 lists the percentages of emotional words by category among the selected posts of Sad Board, Happy Board, and Hate Board, and compares the results by C-LIWC and by the "corpus of 7 types emotion co-occurrences words". As Table 12 shown, among three boards, based on C-LIWC, the emotional words were at most 3.1%, however, based on the "corpus of 7 types emotion co-occurrences words", the emotional words were more than 36.4%. Result indicated that C-LIWC has less emotional words.

Table 12. Percentages of emotional words by category among the selected posts of Sad, Happy, and Hate boards

Board	Selected posts	C-LIWC			Corpus of 7 types emotion co-occurrences words		
		Happiness	Anger	Sadness	Happiness	Anger	Sadness
Sad	2601	2.40%	0.50%	0.90%	36.80%	36.50%	36.40%
Happy	553	3.10%	0.20%	0.30%	41.20%	40.00%	40.30%
Hate	2076	1.60%	0.90%	0.40%	37.80%	37.20%	37.50%

Because the percentages of emotional words categorized by C-LIWC among the selected posts was too low, present study further looked into the posts, based on C-LIWC, that had zero percentage of emotional words in its corresponding boards (emotional words of Happiness in Happy Board, emotional words of Anger in Hate Board, emotional words of Sadness in Sad Board), and found that based on C-LIWC, among the posts from Sad Board, 537 posts (21%) had no emotional words of Sadness; among the Happy Board, 8 posts (1%) had no emotional words of Happiness; among the Hate Board, 499 posts (24%) had no emotional words of Anger. However, when calculating based on the "corpus of 7 types emotion co-occurrences words", every post had its board-corresponding emotional words. Since the emotional scores of were computed according to the emotional words within an article, it would bias the analysis result if no emotional words were embodied in the article.

Analysis of Variance (ANOVA)
It has to implement analysis of variance (ANOVA) before discriminant analysis, if emotional scores are different between the boards then present study continues the discriminant analysis. Table 13 lists the means, standard deviations (S.D.), and the F value of ANOVA for scores of Happiness, Sadness, Anger of the posts in Happy Board, Hate Board, and Sad Board based on C-LIWC and on the "corpus of 7 types emotion co-occurrences words". Whether it calculated based on C-LIWC or the "corpus of 7 types emotion co-occurrences words", the scores of Happiness, Sadness, Anger were different between boards ($p < .001$). Then present study can implement the discriminant analysis.

Discriminant Analysis
Table 14 shows result of discriminant analysis, and the correct rates of C-LIWC and the "corpus of 7 types emotion co-occurrences words" are listed respectively as Tables 15 and 16. For C-LIWC, Happiness had the biggest coefficient 1, Sadness had the smallest coefficient 1; Sadness had the biggest coefficient 2, Happiness had the smallest coefficient 2. For the "corpus of 7 types emotion co-occurrences words", Anger had the biggest coefficient 1, Sadness had the smallest coefficient 1; Happiness had the biggest coefficient

Table 13. The means, standard deviations (S.D.), and the F value of ANOVA for scores of Happiness, Sadness, and Anger of the posts in Happy Board, Hate Board, and Sad Board based on C-LIWC and on the "corpus of 7 types emotion co-occurrences words"

		Happy board	Sad board	Hate board	$F(2, 5227)$
		Mean (S.D.)	Mean (S.D.)	Mean (S.D.)	
CLIWC Dictionary	Happiness	31.39(15.23)	23.53(12.02)	16.44(10.46)	421.70
	Sadness	3.15(4.71)	9.33 (7.74)	4.01 (5.41)	455.61
	Anger	2.32 (3.92)	5.34 (6.05)	8.87 (7.91)	277.22
Corpus of 7 types emotion co-occurrences words	Happiness	143.62(18.59)	129.86(16.57)	132.3(49.01)	38.32
	Sadness	89.29(12.08)	92.51(12.85)	89.44(38.71)	9.11
	Anger	81.23(10.87)	82.04(11.24)	86.65(45.06)	16.19

Note: base unit is 1/1000, except F value

2, Anger had the smallest coefficient 2. Table 15 shows the correct rates of Happiness, Sadness, and Anger identified by C-LIWC, the correct rate was 68.2% for the posts of Happy Board, the correct rate was 50.8% for the posts of Sad Board, the correct rate was 69.3% for the posts of Hate Board, overall correct rate was 60.0%.

Table 16 shows the correct rates of Happiness, Sadness, and Anger identified by the "corpus of 7 types emotion co-occurrences words", the correct rate was 71.8% for the posts of Happy Board, 73.3% for the posts of Sad Board, 70.0% for the posts of Hate Board, overall correct rate was 71.8%. Compare the results of Tables 15 and 16, present study finds that corrects rate were most different between two corpora in Sad Board.

Table 14. The discriminant analysis result of Happiness, Sadness, and Anger based on C-LIWC and the "corpus of 7 types emotion co-occurrences words"

		Canonical discriminant function Standardized coefficients		Discriminant function Structure coefficients		Canonical discriminant function Unstandardized coefficients	
		Coefficient 1	Coefficient 2	Coefficient 1	Coefficient 2	Coefficient 1	Coefficient 2
C-LIWC	Happiness	0.58	−0.55	0.64	−0.47	0.05	−0.05
	Sadness	−0.63	0.85	0.52	0.82	0.09	0.13
	Anger	0.57	0.16	−0.54	0.31	−0.09	0.02
Corpus of 7 types emotion co-occurrences words	Happiness	2.11	3.29	0.10	0.185	0.06	0.10
	Sadness	−5.82	0.50	−0.08	0	−0.22	0.02
	Anger	3.77	−3.74	0.08	−0.10	0.12	−0.13

The test of significance

Last, Table 17 lists the results of test of significance, the cross tabulation of correct rates of categorization for discriminant analysis based on C-LIWC and the "corpus of 7 types emotion co-occurrences word", test of significance by percentage. Results showed there was significantly different ($z = 24.79$, $p < .001$) of categorization of emotions between using C-LIWC and using the "corpus of 7 types emotion co-occurrences word".

4 General Discussion

Emotions closely connect to individual's cognition system and psychological health; many studies have built the emotional words corpora or norms for the material of emotion research. However, establishing an emotion norm or corpus is time-consuming and expensive. Besides, updating a norm or corpus is also important for database maintenance. Present study has found the ways to overcome these disadvantages in the study 1. Study 1 used the big data texts corpus and co-occurrences to build an emotion words database which has more words of seven emotions than past studies, it included 33,669

Table 15. Correct rates of discriminant analysis by taking three categorical scores as independent variables based on C-LIWC

		Predicted			Total	Overall correct rate
		Happy board	Sad board	Hate board		
Actual	Happy board	377(68.2%)	81(14.6%)	95(17.2%)	533	60.0%
	Sad board	630(24.2%)	1321(50.8%)	650(25.0%)	2601	
	Hate board	309(14.9%)	328(15.8%)	1439(69.3%)	2076	

Table 16. Correct rates of discriminant analysis by taking three categorical scores as independent variables based on the "corpus of 7 types emotion co-occurrences words"

		Predicted			Total	Overall correct rate
		Happy board	Sad board	Hate board		
Actual	Happy board	397(71.8%)	81(14.6%)	75(13.6%)	533	71.8%
	Sad board	233(9.0%)	1906(73.3%)	462(17.8%)	2601	
	Hate board	279(13.4%)	344(16.6%)	1453(70.0%)	2076	

Table 17. The percentage of correct rate change of discriminant analysis

		Corpus of 7 types emotion co-occurrences words		Total of C-LIWC
		Correct	Wrong	
C-LIWC	Correct	0%	60%	60%
	Wrong	28.2%	11.8%	40%
Total of corpus of 7 types emotion co-occurrences words		28.2%	71.8%	100%

words categorized by seven emotions and two clusters of expression styles, and their scores of emotion co-occurrences. Additionally, the corpus is built with the massive text corpus by the measure of co-occurrences, which makes the maintenance of corpus easier since successors only need to update or supply the content of the massive text corpus, then the scores of co-occurrences of emotion words will update itself accordingly.

However, is it possible to replace traditional method of building corpus with the method of building "corpus of 7 types emotion co-occurrences words" established in study 1? To answer this question, study 2 was implemented. Study 2 has used Chinese LIWC "C-LIWC" to be the representative of human rated corpus, compared the results of categorization of emotion between these two corpora, and carried on a discriminant analysis. Results show that "corpus of 7 types emotion co-occurrences words" has more emotional words and also higher correct rate than C-LIWC.

Moreover, the "corpus of 7 types emotion co-occurrences words" was designed based on concept of the Spreading Activation Model, therefore can avoid the bias of having too many irrelevant emotional words due to co-occurrence. However, there are still some exceptions that some words cannot be used alone, and are used together with special words to evoke emotions, for example, " "多麼"" (such), " 著實" (indeed). Past studies took these words as meaningless words or as the filler for the experiment, and would delete these words from the corpora and norms eventually. For daily communication, however, these words can clearly make native language speakers feeling differently, taking aforementioned "著實" (indeed) for example, the score of surprise co-occurrence of "著實" (indeed) is closer to max value of emotion "surprise" than its score of disgust co-occurrence to max value of emotion "disgust", which shows that " 著實" (indeed) makes people stronger feeling when it used together with surprise. Even it is not a grammar mistake, for native Chinese language speakers, " 著實厭惡" (indeed dislike) is not commonly seen or heard and is rather weird when you heard it. It suggests further research for the function of these words which have no emotional meanings but have influence on feeling, such as comparing the effect of emotional trigger with typical emotional words.

In general, the "corpus of 7 types emotion co-occurrences words" established by present study has the advantages of prompt update, bigger capacity of emotional words, less time and expense for establishment than traditional ways of building corpora. As for the categories of emotional words, the "corpus of 7 types emotion co-occurrences words" also includes some words which were deleted from traditional corpora but have certain attributes of triggering emotions. Moreover, study 2 compared the correct rates of emotions recognition between one human-rated corpus and the "corpus of 7 types emotion co-occurrences words", results also show that the way of establishing emotional norms or corpora by big data text corpora and the co-occurrence of words is able to replace the traditional ways.

Present study contributes to find a novel way of corpora building. By the method of present study, researchers can save time and money, and also have a better outcome which includes more and various words, besides, to update the database is effective and easy. Present study generates the scores of emotion co-occurrences of 33,669 words in a frame contained seven emotions and two clusters (expression styles). In particular, seven emotions are happiness, love, surprise, sadness, anger, disgust, and fear, and two clusters are literal emotion words cluster and metaphorical emotion words cluster.

The measure "Baroni-Urbani" was used in the present study to compute the words co-occurrences and to build the corpus. There are still a lot of measures available and might be other measures which have the results closer to the human judgement about emotional words. Present study suggests further research to use different measures and verify the results with the corpora built by human judgement. Present study also suggests investigating the aspects of dimensional approach of emotion, such as the valence, arousal, continuance, concreteness of emotion, by using the scores of words in seven emotions, for expanding the emotion words database and helping further emotion research. It is also worthwhile to discuss using emotional words corpus on more special text material, such as Chinese poetry or Chinese classical, in order to help people have further acquisition for this research topic.

Acknowledgements. This work was financially supported by the grant MOST-111-2634-F-002-004 from Ministry of Science and Technology (MOST) of Taiwan, the MOST AI Biomedical Research Center, and the "Institute for Research Excellence in Learning Sciences" and "Chinese Language and Technology Center" of National Taiwan Normal University from The Featured Areas Research Center Program within the framework of the Higher Education Sprout Project by the Ministry of Education in Taiwan.

References

1. Chen, H.-C., Chan, Y.-C., Feng, Y.-J.: Taiwan corpora of Chinese emotions and relevant psychophysiological data-a norm of emotion metaphors in Chinese. Chin. J. Psychol. **55**(4), 525–553 (2013). https://doi.org/10.6129/CJP.20130112b
2. Kiefer, M., Schuch, S., Schenck, W., Fiedler, K.: Mood states modulate activity in semantic brain areas during emotional word encoding. Cereb. Cortex **17**(7), 1516–1530 (2007). https://doi.org/10.1093/cercor/bhl062
3. St-Hilaire, A., Cohen, A.S., Docherty, N.M.: Emotion word use in the conversational speech of schizophrenia patients. Cogn. Neuropsychiatry **13**(4), 343–356 (2008). https://doi.org/10.1080/13546800802250560
4. Russell, J.A., Pratt, G.: A description of the affective quality attributed to environments. J. Pers. Soc. Psychol. **38**(2), 311–322 (1980). https://doi.org/10.1037/0022-3514.38.2.311
5. Darwin, C., Ekman, P., Prodger, P.: The Expression of the Emotions in Man and Animals. Oxford University Press, Oxford (1998)
6. Eibl-Eibesfeldt, I.: The expressive behavior of the deaf-and-blind born. In: Social Communication and Movement, pp. 163–193. Academic Press (1973)
7. Plutchik, R.: A general psychoevolutionary theory of emotion. Theories Emotion **1**, 3–31 (1980)
8. Ekman, P., Friesen, W.V.: Measuring facial movement. Environ. Psychol. Nonverbal Behav. **1**(1), 56–75 (1976). https://doi.org/10.1007/bf01115465
9. Ekman, P., Friesen, W.V., Ellsworth, P.: Emotion in the Human Face: Guidelines for Research and an Integration of Findings. Pergamon Press, Oxford (1972)
10. Fontaine, J.R., Scherer, K.R., Roesch, E.B., Ellsworth, P.C.: The world of emotions is not two-dimensional. Psychol. Sci. **18**(12), 1050–1057 (2007). https://doi.org/10.1111/j.1467-9280.2007.02024.x
11. Lang, P.J., Bradley, M.M., Cuthbert, B.N.: Emotion, attention, and the startle reflex. Psychol. Rev. **97**(3), 377–395 (1990). https://doi.org/10.1037/0033-295X.97.3.377
12. Larsen, R.J., Diener, E.: Promises and Problems with the Circumplex Model of Emotion. In: Emotion, pp. 25–59. Sage Publications Inc, Thousand Oaks (1992)
13. Osgood, C.E., Suci, G.J., Tannenbaum, P.H.: The Measurement of Meaning. University of Illinois Press, Illinois (1957)
14. Thayer, R.E.: Activation-deactivation adjective check list: current overview and structural analysis. Psychol. Rep. **58**(2), 607–614 (1986). https://doi.org/10.2466/pr0.1986.58.2.607
15. Bradley, M.M., Lang, P.J.: Affective norms for English Words (ANEW): instruction manual and affective ratings. Technical report C-1, The Center for Research in Psychophysiology, University of Florida (1999). https://pdodds.w3.uvm.edu/teaching/courses/2009-08UVM-300/docs/others/everything/bradley1999a.pdf
16. Hinojosa, J.A., et al.: Affective norms of 875 Spanish words for five discrete emotional categories and two emotional dimensions. Behav. Res. Methods **48**(1), 272–284 (2015). https://doi.org/10.3758/s13428-015-0572-5

17. Mukherjee, S., Heise, D.R.: Affective meanings of 1,469 Bengali concepts. Behav. Res. Methods **49**(1), 184–197 (2016). https://doi.org/10.3758/s13428-016-0704-6
18. Stadthagen-Gonzalez, H., Imbault, C., Pérez Sánchez, M.A., Brysbaert, M.: Norms of valence and arousal for 14,031 Spanish words. Behav. Res. Methods **49**(1), 111–123 (2016). https://doi.org/10.3758/s13428-015-0700-2
19. Warriner, A.B., Kuperman, V., Brysbaert, M.: Norms of valence, arousal, and dominance for 13,915 English lemmas. Behav. Res. Methods **45**(4), 1191–1207 (2013). https://doi.org/10.3758/s13428-012-0314-x
20. Cho, S.-L., Chen, H.-C., Cheng, C.-M.: Taiwan Corpora of Chinese emotions and relevant psychophysiological data-a study on the norm of Chinese emotional words. [Taiwan Corpora of Chinese Emotions and Relevant Psychophysiological Data-A Study on the Norm of Chinese Emotional Words]. Chin. J. Psychol. **55**(4), 493–523 (2013). https://doi.org/10.6129/cjp.20131026
21. Lee, H.-M., Lee, Y.-S.: Emotionality ratings and free association of 267 common Chinese words. Formosa J. Ment. Health **24**(4), 495–524 (2011)
22. Wang, Y.-N., Zhou, L.-M., Luo, Y.-J.: The pilot establishment and evaluation of Chinese affective words system. Chin. Ment. Health J. **22**(8), 608–612 (2008)
23. Yao, Z., Wu, J., Zhang, Y., Wang, Z.: Norms of valence, arousal, concreteness, familiarity, imageability, and context availability for 1,100 Chinese words. Behav. Res. Methods **49**(4), 1374–1385 (2016). https://doi.org/10.3758/s13428-016-0793-2
24. Brainerd, C.J., Holliday, R.E., Reyna, V.F., Yang, Y., Toglia, M.P.: Developmental reversals in false memory: effects of emotional valence and arousal. J. Exp. Child Psychol. **107**(2), 137–154 (2010). https://doi.org/10.1016/j.jecp.2010.04.013
25. Casasanto, D., de Bruin, A.: Metaphors we learn by: directed motor action improves word learning. Cognition **182**, 177–183 (2019). https://doi.org/10.1016/j.cognition.2018.09.015
26. Pennebaker, J.W., Chung, C.K., Ireland, M., Gonzales, A., Booth, R.J.: The development and psychometric properties of LIWC2007. Austin, TX. LIWC.net (2007)
27. Pennebaker, J.W., Francis, M.E., Booth, R.J.: Linguistic Inquiry and Word Count: LIWC 2001, vol. 71. Lawrence Erlbaum Associates, Mahway (2001)
28. Pennebaker, J.W., Booth, R.J., Francis, M.E.: Linguistic Inquiry and Word Count (LIWC 2007) (2007a)
29. Pennebaker, J.W., Booth, R.J., Francis, M.E.: Operator's manual: linguistic inquiry and word count: LIWC2007, Austin, Texas. LIWC.net (2007b). http://www.gruberpeplab.com/teaching/psych231_fall2013/documents/231_Pennebaker2007.pdf
30. Tausczik, Y.R., Pennebaker, J.W.: The psychological meaning of words: LIWC and computerized text analysis methods. J. Lang. Soc. Psychol. **29**(1), 24–54 (2009). https://doi.org/10.1177/0261927x09351676
31. Rude, S., Gortner, E.-M., Pennebaker, J.W.: Language use of depressed and depression-vulnerable college students. Cogn. Emot. **18**(8), 1121–1133 (2004). https://doi.org/10.1080/02699930441000030
32. Mehl, M.R., Gosling, S.D., Pennebaker, J.W.: Personality in its natural habitat: manifestations and implicit folk theories of personality in daily life. J. Pers. Soc. Psychol. **90**(5), 862–877 (2006). https://doi.org/10.1037/0022-3514.90.5.862
33. Jaffe, E.: What big data means for psychological science. APS Observer **27** (2014). https://www.psychologicalscience.org/observer/what-big-data-means-for-psychological-science
34. Mednick, S.: The associative basis of the creative process. Psychol. Rev. **69**(3), 220–232 (1962). https://doi.org/10.1037/h0048850
35. Bargh, J.A., Chen, M., Burrows, L.: Automaticity of social behavior: direct effects of trait construct and stereotype activation on action. J. Pers. Soc. Psychol. **71**(2), 230–244 (1996). https://doi.org/10.1037/0022-3514.71.2.230

36. Fazio, R.H.: On the automatic activation of associated evaluations: an overview. Cogn. Emot. **15**(2), 115–141 (2001). https://doi.org/10.1080/02699930125908
37. Charles, W.G., Miller, G.A.: Contexts of antonymous adjectives. Appl. Psycholinguist. **10**(3), 357–375 (1989). https://doi.org/10.1017/S0142716400008675
38. Justeson, J.S., Katz, S.M.: Co-occurrences of antonymous adjectives and their contexts. Comput. Linguist. **17**(1), 1–19 (1991)
39. Spence, D.P., Owens, K.C.: Lexical co-occurrence and association strength. J. Psycholinguist. Res. **19**(5), 317–330 (1990). https://doi.org/10.1007/bf01074363
40. Collins, A.M., Loftus, E.F.: A spreading-activation theory of semantic processing. Psychol. Rev. **82**(6), 407–428 (1975). https://doi.org/10.1037/0033-295X.82.6.407
41. Shaver, P., Schwartz, J., Kirson, D., O'Connor, C.: Emotion knowledge: further exploration of a prototype approach. J. Pers. Soc. Psychol. **52**(6), 1061–1086 (1987). https://doi.org/10.1037/0022-3514.52.6.1061
42. Huang, C.-L., et al.: The development of the Chinese linguistic inquiry and word count dictionary. Chin. J. Psychol. **54**(2), 185–201 (2012)
43. Lin, S.-Y., Chen, H.-C., Chang, T.-H., Lee, W.-E., Sung, Y.-T.: CLAD: a corpus-derived Chinese lexical association database. Behav. Res. Methods **51**(5), 2310–2336 (2019). https://doi.org/10.3758/s13428-019-01208-2
44. Baroni-Urbani, C., Buser, M.W.: Similarity of binary data. Syst. Biol. **25**(3), 251–259 (1976). https://doi.org/10.2307/2412493

China's CO_2 Emissions Interval Forecasting Based on an Improved Nonlinear Fractional-Order Grey Multivariable Model

Hang Jiang[1], Xijie Zhang[1], and Peiyi Kong[2,3](✉)

[1] School of Business Administration, Jimei University, Xiamen 361021, China
[2] Ph.D. Program in Business, Chung Yuan Christian University, Taoyuan 320314, Taiwan
10804606@ccycu.edu.tw
[3] School of Economics and Management, Xiamen Nanyang University, Xiamen 361012, China

Abstract. Accurately predicting carbon emissions and mastering the law of carbon emissions are the premise for effective energy saving and emission reduction and realizing the goal of "carbon peaking and carbon neutrality". This paper takes foreign direct investment and environmental regulation as the influencing factors, and uses the nonlinear fractional-order grey multivariable model to predict carbon emissions interval. The results showed that foreign direct investment intensifies carbon emissions, while environmental regulation contributes to carbon emissions, with total carbon emissions still on the rise in the next few years. Paying great importance to the quality of "bring in" and making good use of environmental regulation is an important way to achieve sustainable development.

Keywords: Carbon dioxide emissions · FDI · Environmental regulations · Fractional-order · Grey multivariable model

1 Introduction

In recent years, excessive carbon dioxide emissions (CO_2) have caused a series of severe environmental problems, such as global warming, the melting of glaciers, and the frequent occurrence of extreme weather, and they also have had a harmful influence on human health. For those reasons, countries around the world have put their great efforts into energy-saving and emission-reduction. China, as the largest developing country with the largest CO_2 in the world, promised at the UN Climate Change Summit in 2009 (COP15) that the CO_2 per unit of GDP will be reduced by 40%–45% at the end of 2020 compared to 2005. At the Paris Climate Change Conference in 2015 (COP21), China further committed to reduce its CO_2 by 60%–65% by 2030 compared to 2005. Since then, President Xi proposed that China will intend nationally determined contributions to have carbon emissions peak before 2030 and achieve carbon neutrality before 2060. Therefore, it is meaningful for the government to accurately forecast China's CO_2 to know about the emission trend and to realize reduction commitment.

F. Fui-Hoon Nah and K. Siau (Eds.): HCII 2022, LNCS 13327, pp. 182–193, 2022.
https://doi.org/10.1007/978-3-031-05544-7_14

The issues regarding the impact factor analysis and prediction of CO_2 have received great attention from academics and practitioners. Previous studies have found that economic growth, energy consumption, urban population, and research and development were important factors affecting CO_2, predicting emissions by considering the effects of these factors. General speaking, reducing CO_2 mainly depends on technological advances, which come from the R&D undertaken and foreign direct investments (FDI). In the past few decades, many academics have focused on analyzing the relationship between inward FDI and CO_2, and they concluded that FDI could affect CO_2.

Low standards of environmental regulations, as well as inexpensive energy and labor provided by countries with emerging economies like China's, are attractive for FDI, which also confirms the existence of the pollution haven hypothesis (PHH). To solve this problem, China has introduced a series of relevant environmental protection policies and increased environmental expenditures to address the increased impact of FDI on the environment. The improvement of environmental regulation intensity and investment in environmental expenditures were bound to restrict the entry of high-carbon emission industries and promote the carbon emission reduction effect. Thus, it is very important to analyze the impact of FDI on carbon emissions.

Due to the importance of FDI and environmental regulations, forecasting China's CO_2 by considering the impact of FDI and environmental expenditure help to improve the prediction accuracy, which also makes the grey multivariable GM(1,N) model stand out from the numerous grey models. The GM(1,N) model is a prediction model with one system characteristics variable and $N - 1$ relevant variables. More specifically, the GM(1,N) model is a typical causal prediction model because it takes full account of the effect of the relevant factors on the system behavior change in the whole modeling process. At present, the traditional GM(1,N) model uses the first-order accumulated generating operation (1-AGO) to generate a sequence; however, 1-AGO is not consistent with the principle of the new information priority in the grey system theory. Therefore, the proposed fractional-order accumulation gives separate weights to older and newer data. In this study, the fractional-order accumulated generating operation is combined with the traditional GM(1,N) model (the FGM(1,N) model) to further improve the prediction accuracy. Furthermore, the GM(1,N) is a linear prediction model; in other words, with this model, it is not appropriate to reflect the nonlinear relationship between system characteristics variable and relevant variables in reality. Referring to the nonlinear grey model used in previous studies, this study proposes the nonlinear fractional grey multivariable model, abbreviated as the NFGM(1,N), to predict China's CO_2 by considering the nonlinear relationship between the relevant variables and emissions. The fractional-order, power, and coefficients in modeling are the crucial issues that influence the prediction accuracy of the proposed model. Therefore, finally, this study applies the optimization algorithm to determine the values of the parameters.

As carbon emissions are affected by various variables, the series prediction could generate the bias results, and could not truly reflect the future changes in CO_2. Therefore, this paper extends the original data to intervals by applying the method of Limit error, and then interval forecasting China's CO_2 by using the proposed NFGM(1,N) model.

2 Literature Review

2.1 Impact of FDI on Carbon Emissions

The impact of foreign direct investment on the carbon emissions of the host country can be summarized into the following three kinds of view. First, the pollution haven hypothesis (PHH) believed that as developed countries pay more and more attention to environmental protection, environmental treatment costs for pollution-intensive industries will increase, thus forcing enterprises to move out to developing countries with weak environmental regulation, leading to an increase in carbon emissions from host countries. Although the research subjects were different, it was found that foreign direct investment and carbon emissions were in causal relations, which will cause the intensified carbon emissions of the host country [1, 2]. Secondly, the pollution halo effects (PHE) is that green production technology, advanced management level, and strict environmental standards will be transferred with foreign direct investment, forming a spillover effect on the green production technology of the host country, and then have a significant effect in promoting carbon emission reduction [3, 4]. Finally, it is also believed that the impact of foreign direct investment on the carbon emissions of host countries depends on their urbanization rate, environmental regulation, financial development, and other factors. Meanwhile, there is a threshold effect or a non-linear relationship between the two [5–7].

2.2 Impact of Environmental Regulations on Carbon Emissions

The intensity of the environmental regulation is also one of the important factors affecting carbon emissions. It is similar to the impact of FDI on carbon emissions, and the conclusions of the impact of environmental regulation on carbon emissions are also different. Due to the strict environmental regulation in the home country, the industries with high pollution, high energy consumption, and high emission are transferred from the home country, and as FDI flows to the host country with a low level of environmental regulation, which is the pollution haven hypothesis. With the implementation of environmental regulation measures and intensity in host countries, its impact on carbon emissions is also different. On the one hand, Yin et al. [8] believed that implementation of environmental regulation had an emission reduction effect on host carbon emissions. On the other hand, Gao and Li [9] believed that the environmental regulation did not reduce carbon emission, instead increased the burden on enterprises, which was called the "green paradox" effect.

2.3 Interval Grey Prediction on Carbon Emissions

Accurately predicting the total carbon emissions and grasping the future development trend of carbon emissions are the guarantee for realizing the "double carbon" goal and implementing the new development concept. Past studies have mainly used measurement models including regression analysis, time series analysis, and intelligent computing technology to predict carbon emissions, but they need to grasp the sample distribution and a large amount of data as the basis for accurate prediction. Grey system theory was proposed by Professor Deng in 1982, which is characterized by the fact that it does not need to meet the premise assumptions of statistics, and is suitable for the analysis of uncertain systems, small samples, and poor information [10]. Among them, the grey univariate prediction GM(1,1) model and the grey variable collar prediction GM(1,N) model and their improved models have been widely used in the field of carbon emission prediction, and have achieved fruitful results [11, 12]. But the past research is mainly for real prediction of carbon emissions, although through the background value correction, optimization algorithm, residual correction to improve the prediction accuracy, but because the data in the system cannot be expressed with exact values, can provide information is limited, so the real prediction to interval prediction is helpful to the cognition and grasp of data, also can provide rich information [13]. The existing research of interval grey prediction mainly focuses on two aspects. Some scholars apply grey prediction models to interval sequences with upper and lower bounds for prediction [14]. Some scholars use the grey number and neural network interval expansion method to expand the real number sequence into the interval sequence and combined it with the grey prediction model [15, 16].

3 Model Setting and Data Collection

3.1 Interval Expansion

Expanding the real number sequence into the interval sequence is the key to make the interval prediction, and to ensure the prediction accuracy. The previous studies used different interval expansion methods to expand real numbers sequences into an interval sequence, and we choose to expand real numbers into interval sequences through limit error. Limit error is the maximum range of errors under a certain probability guarantee, the critical value of a confidence multiplied by the sampling average error, the limit error of real data to form upper and lower bounds of the interval, which can be expressed as:

$$x_i \pm t_{\frac{\alpha}{2}} \times \sigma \tag{1}$$

where α is the standard deviation, and the confidence level is 95%.

3.2 Nonlinear Fractional Grey Multivariable Model

The computational steps to construct an NFGM(1,N) model are demonstrated as follows.

The computational steps to construct an NFGM(1,N) model are demonstrated as follows.

Step 1: Present the data sequences. The system characteristics sequence $X_1^{(0)} = \left(x_1^{(0)}(1), x_1^{(0)}(2), \ldots, x_1^{(0)}(n)\right)$ and relevant factors sequence $X_i^{(0)} = \left(x_i^{(0)}(1), x_i^{(0)}(2), \ldots, x_i^{(0)}(n)\right)$ are presented, where $i = 2, 3, \ldots, N$.

Step 2: Perform the accumulated generating operation. The accumulative sequence $X_i^{(r_i)} = \left(x_i^{(r_i)}(1), x_i^{(r_i)}(2), \ldots, x_i^{(r_i)}(n)\right)$ can be obtained from $X_i^{(0)}$ by the r_i-order accumulated generating operation (r_i-AGO) as follows:

$$x_i^{(r_i)}(k) = \sum_{j=1}^{k} \binom{k-j+r_i-1}{k-j} x_i^{(0)}(j),\ k = 1,\ 2,\ \ldots,\ n \tag{2}$$

where $\binom{k-j+r_i-1}{k-j} = \frac{(k-j+r_i-1)(k-j+r_i-2)\ldots(r_i+1)r_i}{(k-j)!}$. Based on the study of Wu and Liu [17], r should be taken in the interval (0,1).

Step 3: Construct the NFGM(1,N) model. Drawing on the concept from the GM(1,1) model, a grey control parameter term is added to the model. Therefore, the NFGM(1,N) model is given as

$$\left(x_1^{(r_i)}(k) - x_1^{(r_i)}(k-1)\right) + az_1^{(r_i)}(k) = \sum_{i=2}^{n} b_i\left(x_i^{(r_i)}(k)\right)^{\gamma_i} + u \tag{3}$$

where a, b_i, and u are the coefficients of the modeling, which represent the development term, the driving coefficients, and grey control parameter, respectively. Moreover, r_i and γ_i are the order of the fractional for different variables and the power, respectively. The NFGM(1,N) model reverts to the FGM(1,N) model when $\gamma_i = 1$, and it reverts to the traditional GM(1,N) model when $r_i = 1$ and $\gamma_i = 1$.

Step 4: Estimate the coefficients. The parameters a, b_i, and u can be estimated by using the OLS method

$$[a,\ b_i,\ u]^T = (\mathbf{B}^T\mathbf{B})^{-1}\mathbf{B}^T\mathbf{Y} \tag{4}$$

where

$$\mathbf{B} = \begin{bmatrix} -z_1^{(r_1)}(2) & \left(x_2^{(r_2)}(2)\right)^{\gamma_2} & \cdots & \left(x_N^{(r_2)}(2)\right)^{\gamma_N} & 1 \\ -z_1^{(r_1)}(3) & \left(x_2^{(r_2)}(3)\right)^{\gamma_2} & \cdots & \left(x_N^{(r_2)}(3)\right)^{\gamma_N} & 1 \\ \vdots & \vdots & \ddots & \vdots & 1 \\ -z_1^{(r_1)}(n) & \left(x_2^{(r_2)}(n)\right)^{\gamma_2} & \cdots & \left(x_N^{(r_2)}(n)\right)^{\gamma_N} & 1 \end{bmatrix},\ \mathbf{Y} = \begin{bmatrix} x_1^{(r_1)}(2) - x_1^{(r_1)}(1) \\ x_1^{(r_1)}(3) - x_1^{(r_1)}(2) \\ \vdots \\ x_1^{(r_1)}(n) - x_1^{(r_1)}(n-1) \end{bmatrix} \tag{5}$$

where the background values of the system characteristics variable $z_1^{(r_i)}$ is adjoining mean generated sequence as follows:

$$z_1^{(r_1)}(k) = 0.5 \times \left(x_1^{(r_1)}(k) + x_1^{(r_1)}(k-1)\right) \tag{6}$$

Then, the time response function can be represented as

$$\begin{aligned} \hat{x}_1^{(r)}(k) = & \left[x_1^{(r)}(1) - \frac{1}{a}\left(\sum_{i=2}^{n} b_i\left(x_i^{(r_i)}(k)\right)^{\gamma_i} + u\right)\right]e^{-a(k-1)} \\ & + \left[\frac{1}{a}\left(\sum_{i=2}^{n} b_i\left(x_i^{(r_i)}(k)\right)^{\gamma_i} + u\right)\right] \end{aligned} \tag{7}$$

Step 5: Perform the inverse accumulated generating operation. The r-order inverse fractional accumulation can be defined as

$$x^{(-r)}(k) = \sum_{i=1}^{k} \binom{k-i-r-1}{k-i} x^{(0)}(i) \tag{8}$$

Therefore, the predicted values $\hat{X}_1^{(0)}$ can be calculated by using the inverse ri-order accumulated generating operation (r_i-IAGO) as follows

$$\begin{aligned} \hat{x}_1^{(0)}(k) &= \left(\hat{x}_1^{(r_i)}(k)\right)^{(-r_i)} = \sum_{j=1}^{k} \binom{k-j-r_i-1}{k-j} \hat{x}_1^{(r)}(j) \\ &- \sum_{j=1}^{k-1} \binom{k-1-j-r_i-1}{k-1-j} \hat{x}_1^{(r)}(j),\ k = 2,\ 3,\ \ldots,\ n \end{aligned} \tag{9}$$

where $\hat{x}_1^{(r)}(1) = x_1^{(0)}(1)$.

3.3 Evaluating Prediction Accuracy

Using the mean absolute percentage error (MAPE) to evaluate the prediction accuracy of the prediction model.

$$MAPE = \frac{1}{n} \sum_{i=1}^{n} \left| \frac{x^{(0)}(k) - \hat{x}^{(0)}(k)}{x^{(0)}(k)} \right| \times 100\% \tag{10}$$

If the smaller the MAPE value is, the smaller the prediction error is, the higher the prediction accuracy. According to Lewis (1982), MAPE 10% indicates the high-quality forecast; 10% $<$ MAPE $\leq$ 20%, indicates the good accuracy; 20% $<$ MAPE $\leq$ 50%, indicates the reasonable forecast; MAPE $>$ 50% means the weak forecast [18]. On the other hand, to solve the shortage that the point prediction cannot provide rich information, the interval prediction determines whether the real value and the out-of-sample predicted value can fall within the predicted interval by constructing the upper and lower bounds [19]. Therefore, whether the prediction interval can cover the actual value and the out-of-sample prediction is the key to evaluating the accuracy and coverage of the prediction interval.

3.4 Data Description

To analyze the impact of foreign direct investment and environmental regulation on carbon emissions, this paper selects the total carbon emissions as the prediction variable and selects the actual foreign direct investment and environmental expenditure to represent foreign direct investment and environmental regulation as the influencing variables, respectively. Carbon emissions are derived from the International Energy Agency (IEA), and foreign direct investment and environmental regulation data are derived from the China Statistical Yearbook. Since the statistics of environmental regulation data in China began in 2007 and was limited by the data availability, this paper selected the relevant data from 2007–2019 for analysis. Descriptive statistics of carbon emissions, FDI, and environmental regulation data are shown in Table 1.

Table 1. Variable descriptive statistics.

Variables	Unit	N	Mean	S.D.	Min	Max
CO_2	Million tons	11	82.90	3.23	64.73	98.09
FDI	Billions of dollars	11	1101.01	53.25	747.70	1443.69
ER	Billions of dollars	11	3166.67	443.66	995.82	6317.04

4 Empirical Study

4.1 Data Interval Expansion

To obtain more abundant data information, the real number sequence is expanded into the interval sequence according to Eq. (1) by using the limit error, and the interval sequence of each variable is shown in Table 2. As can be seen from Table 2, China's carbon emissions showed volatility during the sample period, but the overall carbon emissions showed an upward trend.

4.2 Carbon Emissions Forecasting

According to the steps of interval expansion and computation above, this paper takes FDI and environmental regulation as the relevant variables to predict carbon emissions. To analyze the predictive effect of the model, data from 2007 to 2017 were used for model fitting, and data from 2018 to 2019 were reserved for ex-post testing. Then, extrapolation predictions for the period of 2020–2025 were made using the proposed model. In this paper, the NFGM(1,N) model is used to make point predictions of carbon emissions, which are shown in Table 3. As can be seen from Table 3, the MPAE of the GM(1,N) model and NFGM(1,N) model is 4.77% and 2.96%, respectively, and the MAPE of the ex-post testing is 3.53% and 1.38%, respectively. The prediction accuracy of NFGM(1,N) model is superior to that of the original GM(1,N) model.

Table 2. Sequence of the variable intervals.

Year	CO_2			FDI			ER		
	Lower bound	Actual value	Upper bound	Lower bound	Actual value	Upper bound	Lower bound	Actual value	Upper bound
2007	57.62	64.73	71.84	638.06	747.70	857.34	19.31	995.82	1972.33
2008	59.58	66.69	73.80	814.31	923.95	1033.59	474.85	1451.36	2427.87
2009	64.21	71.32	78.42	790.69	900.33	1009.97	957.53	1934.04	2910.55
2010	71.20	78.31	85.42	947.66	1057.30	1166.94	1465.47	2441.98	3418.49
2011	78.59	85.70	92.80	1050.47	1160.11	1269.75	1664.47	2640.98	3617.49
2012	81.08	88.18	95.29	1007.52	1117.16	1226.80	1986.95	2963.46	3939.97
2013	84.78	91.88	98.99	1066.22	1175.86	1285.50	2458.64	3435.15	4411.66
2014	84.06	91.16	98.27	1085.98	1195.62	1305.26	2839.09	3815.60	4792.11
2015	83.83	90.93	98.04	1153.03	1262.67	1372.31	3826.38	4802.89	5779.40
2016	83.44	90.55	97.65	1150.37	1260.01	1369.65	3758.29	4734.80	5711.31
2017	85.35	92.46	99.56	1200.71	1310.35	1419.99	4640.82	5617.33	6593.84
2018	88.17	95.28	102.39	1240.02	1349.66	1459.30	5321.10	6297.61	7274.12
2019	90.98	98.09	105.20	1271.70	1381.35	1490.99	6413.69	7390.20	8366.71

Table 3. Comparison of predicted values by different models

Year	Actual	GM(1,N)		NFGM(1,N)	APE
		Predicted value	APE	Predicted value	
2007	64.73	64.73	0.00	64.73	1.47
2008	66.69	57.52	13.76	66.69	5.60
2009	71.32	79.48	11.45	75.09	1.47
2010	78.31	86.46	10.40	82.55	5.60
2011	85.70	92.13	7.51	88.77	1.47
2012	88.18	86.36	2.07	83.61	5.60
2013	91.88	89.46	2.64	87.41	1.47
2014	91.16	89.71	1.60	88.44	5.60
2015	90.93	91.86	1.02	91.48	1.47
2016	90.55	91.88	1.47	92.50	5.60
2017	92.46	92.99	0.57	94.73	1.47
MAPE			4.77		2.96
2018	95.28	93.89	1.47	96.90	1.70
2019	98.09	92.60	5.60	97.05	1.06
MAPE			3.53		1.38

Similarly, the NFGM(1,N) model is used to predict the upper and lower bounds to construct the prediction intervals. Meanwhile, given the good prediction accuracy of the NFGM(1,N) model, out-of-sample predictions from 2020 to 2025 were performed. The prediction results are shown in Table 4 and Fig. 1.

Table 4. Prediction of carbon emissions interval for 2020–2025

Year	Predicted lower bound	Predicted value	Predicted upper bound
2007	57.62	64.73	71.84
2008	59.58	66.69	73.80
2009	67.70	75.09	82.38
2010	75.67	82.55	89.41
2011	81.83	88.77	95.68
2012	76.61	83.61	90.57
2013	80.31	87.41	94.45
2014	81.33	88.44	95.50
2015	84.30	91.48	98.61
2016	85.44	92.50	99.56
2017	87.66	94.73	101.81
2018	89.88	96.90	103.96
2019	90.06	97.05	104.09
2020	101.34	108.04	114.90
2021	94.21	107.33	114.89
2022	97.81	110.25	118.05
2023	101.58	113.15	121.27
2024	105.54	116.04	124.55
2025	109.71	118.89	127.90

As can be seen from Fig. 1, the prediction interval formed by the data modeling from 2007–2017 can cover the actual and predicted data from 2018–2019. Furthermore, from the out-of-sample predictions from 2020 to 2025, the prediction interval also covers the prediction data obtained through point predictions. It can be judged that establishing an interval prediction model with the NFGM(1,N) model has a good prediction effect.

According to the forecast results, China's total carbon emissions will have certain fluctuations in the next few years with a rising trend. At 95% confidence level, China's carbon emissions will be within the interval of 109.71–127.90 million tons in 2025. Therefore, China still has a long way to go to achieve the goal of reaching the carbon peak before 2030 and achieving carbon neutrality by 2060.

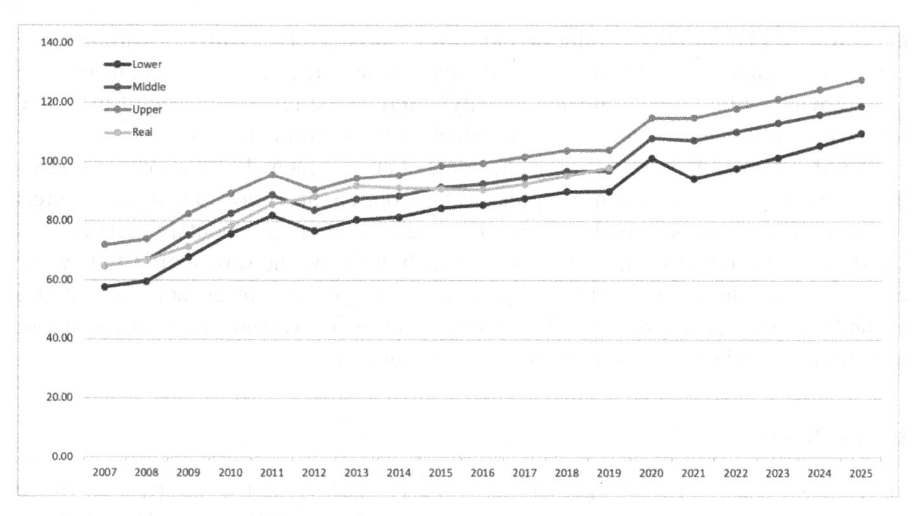

Fig. 1. Carbon emissions and their forecast range for 2007–2025

5 Conclusion and Discussions

Under the goal of "carbon peaking and carbon neutrality", accurately predicting carbon emissions is the key to promoting energy conservation and emission reduction. In this paper, the NFGM(1,N) model is used to predict the interval of Chinese carbon emissions from 2020 to 2025. From the prediction results, the NFGM(1,N) model has high prediction accuracy, while the prediction intervals constructed based on this model are able to cover the actual values and out-of-sample predictions. From the prediction results, China's total carbon emissions will still show a rising trend in the next few years. To ensure the realization of the "double-carbon" goal on schedule and achieve the high-quality development path of coordinated environmental protection and economic development, the paper puts forward the following policy suggestions.

First, look for the optimal environmental regulation intensity. Increasing the intensity of environmental regulation and increasing investment in environmental protection will all contribute to carbon emission reduction, and it will also help to limit the transfer of carbon emissions to China with foreign direct investment. However, it is necessary to pay attention to the implementation of environmental regulations maybe a "green paradox" in the implementation of environmental regulation, which results in rising environmental protection costs of enterprises, but is not conducive to carbon emission reduction. Therefore, reasonable environmental protection measures are formulated and different environmental regulation methods are adopted to find the optimal environmental regulation intensity, which is conducive to the carbon emission reduction effect and realize the carbon emission reduction while promoting domestic economic development.

Second, focus on the quality of introducing foreign investment. We will raise the threshold of environmental control in the process of foreign direct investment, implement stricter environmental regulation strategies, and screen for "green" foreign investment with high technology, high efficiency, and low energy consumption. We will avoid the inflow of FDI into high-pollution industries, introduce FDI with low carbon emissions and cutting-edge technology, give full play to the role of foreign direct investment in improving China's environmental quality, improve the utilization rate of industrial resources, and achieve carbon emission reduction in the production process.

Third, pay attention to the synergistic effect of environmental regulation and foreign direct investment. Environmental regulation and foreign direct investment have a direct impact on carbon emissions. Improving the intensity of environmental regulation can avoid developed countries from transferring high-emission industries into China while emerging technologies brought by the process of foreign direct investment can produce an inhibition on carbon emissions. The synergy between the two under certain conditions can minimize carbon emissions from domestic industries.

References

1. Jiang, S.L., Shao, Y.H.: Whether industrial agglomeration leads to "Pollution Paradise": based on the data analysis of 239 prefecture-level cities in China. Ind. Econ. Rev. **11**(4), 109–118 (2020)
2. Jorgenson, A.K.: Does foreign investment harm the air we breathe and the water we drink? Organ. Environ. **20**(2), 137–156 (2007)
3. Wang, Y.F., Liao, H., Wang, Y.F.: Emission reduction effect of China's two-way FDI coordinated development. Sci. Res. Manage. (1), 1–19 (2021)
4. Xu, Y.D.: FDI, trade openness and CO2 emissions by taking Shandong province as an example. Sci. Res. Manage. **8**, 76–84 (2016)
5. Wang, X.L., Zhang, H.M.: Research on carbon emission effect of FDI in China——based on threshold panel model of urbanization. Forecasting **39**(1), 59–65 (2020)
6. Yin, Q.M., Fan, M.Y.: Threshold effect of two-way FDI on China's carbon emission viewing from environmental regulations. Resour. Ind. **22**(1), 24–31 (2020)
7. Wang, X.H.: Financial development, two-way FDI and carbon emissions: empirical analysis of threshold model based on China's province-level panel data. Value Eng. **38**(26), 110–112 (2019)
8. Yin, J.H., Zhang, M.Z., Chen, J.: The effects of environmental regulation and technical progress on CO2 Kuznets curve: an evidence from China. Energy Policy **77**, 97–108 (2015)
9. Gao, Z.G., Li, M.R.: Spatial-temporal heterogeneity and synergy for the effect of formal and informal environmental regulation on carbon emission reduction: empirical analysis of 14 Prefectures of Xinjiang during 2007–2017. J. Chongqing Technol. Bus. Univ. (West Forum) **30**(6), 84–100 (2020)
10. Deng, J.L.: Control problems of grey systems. Syst. Control Lett. **1**(5), 288–294 (1982)
11. Jiang, H., Yu, J.L.: A predictive analysis of CO2 emissions based on the impact of bilateral FDI and environmental regulation——an evidence from Fujian province. J. Jingdezhen Univ. **36**(4), 24–29 (2021)
12. Bai, Y.X., Wang, L.J., Sheng, M.Y.: Empirical study on carbon emission of agricultural production in Karst region of Guizhou province. Chin. J. Agric. Resour. Reg. Planning **42**(3), 150–157 (2021)
13. Li, Y., Ding, Y.P.: Construction and optimization of interval grey number NGM(1,1) prediction model. Math. Practice Theory **51**(10), 316–322 (2021)
14. Meng, W., Liu, S.F., Zeng, B.: Standardization of interval grey number and research on its prediction modeling and application. Control Decis. **27**(5), 773–776 (2012)
15. Xiong, P.P., Zhang, Y., Yao, T.X., Zeng, B.: Multivariable grey forecasting model based on interval grey number sequence. Math. Practice Theory **48**(9), 181–188 (2018)

16. Jiang, P., Hu, Y.-C., Wang, W.B., Jiang, H., Wu, G.: Interval grey prediction models with forecast combination for energy demand forecasting. Mathematics **8**(6), 1–12 (2020)
17. Wu, L.F., Liu, S.F., Yao, L.G., Yan, S.L., Liu, D.L.: Grey system model with the fractional order accumulation. Commun. Nonlinear Sci. Numer. Simulat. **18**, 1775–1785 (2013)
18. Lewis, C.: Industrial and Business Forecasting Methods. Butterworth Scientific, UK (1982)
19. Quan, H., Srinivasan, D., Khosravi, A.: Uncertainty handling using neural network-based prediction intervals for electrical load forecasting. Energy **73**, 916–925 (2014)

User-Centered Assembly Knowledge Documentation: A Graph-Based Visualization Approach

Christian Kruse(✉), Daniela Becks, and Sebastian Venhuis

University of Applied Sciences Gelsenkirchen, Muensterstr. 265, Bocholt, Recklinghausen, Germany
{christian.kruse,daniela.becks,sebastian.venhuis}@w-hs.de
https://www.w-hs.de

Abstract. Due to the rapid progress of digitization and the increasing importance of modern information technologies like collaboration platforms companies need to find appropriate ways to overcome the looming qualification lag of their employees. Aggravated by the demographic change, they have to ensure that the knowledge of older, experienced employees is made available to others to prevent its loss in case of retirement. This is especially relevant for small and medium-sized companies where deep personal knowledge is crucial to maintain competitiveness [6]. In this context, it is of paramount importance to enable employees to document their knowledge without interrupting their daily work routine. Knowledge acquisition, distribution and utilization have to be seamlessly integrated into the employees' daily operations. This paper presents a graph-based approach to visualize assembly knowledge and its prototypical implementation. The resulting knowledge management system has been evaluated within one medium-sized company in the domain of mechanical engineering. These results as well as future developments are discussed at the end of this case study.

Keywords: Analytics and visualization · Industry 4.0 · User-centered design · Knowledge management · Assembly priority graph

1 Introduction

Imagine the following situation: You are supposed to assemble a very complex, customer-specific machine. As you have never worked on this type of machine before you need to find suitable guidance. In a first step, you search for the current construction drawings in the ERP and PLM system of your company, but you end up with two completely different versions. Because you are not

This research and development project is funded by the European Social Fund (ESF) and the German Federal Ministry of Education and Research (BMBF) within the program "Future of work" (02L18B000) and implemented by the Project Management Agency Karlsruhe (PTKA). The authors are responsible for the content of this publication.

F. Fui-Hoon Nah and K. Siau (Eds.): HCII 2022, LNCS 13327, pp. 194–207, 2022.
https://doi.org/10.1007/978-3-031-05544-7_15

sure which one to use you have to take a look at the annotations left by your fellow colleagues on the department PC. After having searched multiple files, applications and folders you find some five-year old notes that do not contain any information you are looking for. Unfortunately, the colleague who last worked on this machine type just left for a two-week holiday.

Such an scenario describes a typical situation found in numerous small and medium-sized companies. But how to solve this problem? To overcome this deficiency an approach was chosen to develop a system that intuitively allows to acquire, distribute and utilize the predominantly tacit knowledge of and by the employees. This prototypical knowledge management system is designed and tested in cooperation with Spaleck Oberflächentechnik GmbH & Co. KG, a medium-sized company in Germany that specializes in producing complex customer-specific machines.

2 Knowledge Management in Small and Medium-Sized Companies

In a nutshell, knowledge management entails the acquisition, distribution and utilization of a company's knowledge base, which represents the sum of all information, skills and capabilities within a company (organizational knowledge) and its employees (individual knowledge) [5]. For small and medium-sized companies information is a valuable resource because in many cases it ensures economic advances [6]. This is especially the case for product-specific as well as technical knowledge of the employees which is not documented sufficiently in many companies and may become a bottleneck when introducing new employees.

2.1 Knowledge Management: Status Quo

Research and application of knowledge management have a long history. Many companies have already tried to implement such systems, but with mixed success. Some researchers even go as far as declaring knowledge management a dying topic [4]. However, in the context of Industry 4.0 and its increasingly complex manufacturing challenges knowledge management has rebounced [9].

In its most rudimentary form, knowledge management in small and medium-sized companies manifests itself by using office software such as the Microsoft Office Suite with the help of a simple file structure. In more advanced settings, Wikis, Blogs or collaboration platforms such as Microsoft Sharepoint[1] or Atlassian Confluence[2] are utilized. But their inherent software complexity results in additional expenditures for introducing and maintaining these systems. As a result, ease of use and acceptance on the shop floor are often not satisfactory.

---

[1] https://www.microsoft.com/en-us/microsoft-365/sharepoint/collaboration/.

[2] https://www.atlassian.com/software/confluence.

2.2 Transferring Knowledge Within a Company

Knowledge transfer addresses the aspect of sharing and distributing a company's knowledge base among its members. Knowledge typically resides in abstract artifacts such as tools, tasks or information systems as well as in the heads of organizational members. Knowledge transfer can be achieved both by organizational as well as technological means. The former use methods such as job rotation, work shadowing, action learning or internal knowledge fairs to spread knowledge among employees.

According to Renaud and van Biljon [7] knowledge visualization can be described as

> [...] the use of graphical means to communicate experiences, insights and potentially complex knowledge. [...] Such representations facilitate and expedite the creation and transfer of knowledge between people by improving and promoting knowledge processing and comprehension.

This is where graph-based knowledge visualization comes into play. It can be considered an interdisciplinary research field with applications within skill-intensive domains such as medicine [10], bioinformatics, engineering [1] or even crime fighting [3]. There also exists more general research into visualizing semantic graphs like the one proposed by Hirsch [2].

3 Introducing Assembly Priority Graphs as a Visualization Concept for Knowledge Management

Within the research project FlexDeMo[3] a participatory workshop with assembly workers of Spaleck Oberflächentechnik was held to discuss possible ways to document different variations of machines. During this, the assembly workers reacted very positive and liked the visualization of the assembly priority graph because of its simplicity as well as its clearness. Furthermore, they quickly started contributing and sharing own ideas to the workshop and discussed the advantages of this kind of machine presentation. At this point, the idea arose to implement a graph-based software tool to manage and transfer assembly knowledge.

3.1 Fundamentals of Assembly Priority Graphs

An assembly priority graph is utilized to visually present dependencies within an assembly or production process [11]. Thereby, assembly groups are displayed in the relevant order, either as sequences or in parallel [8], so that it immediately becomes clear when to perform which step. An example of such a graph can be seen in Fig. 1.

When producing complex, custom-specific products there may be numerous modifications in the assembly order. Thus, work processes are much more difficult to plan and to structure. In this context, assembly priority graphs are of great importance, because they help to navigate employees through the complex assembly process.

[3] https://www.flexdemo.eu/.

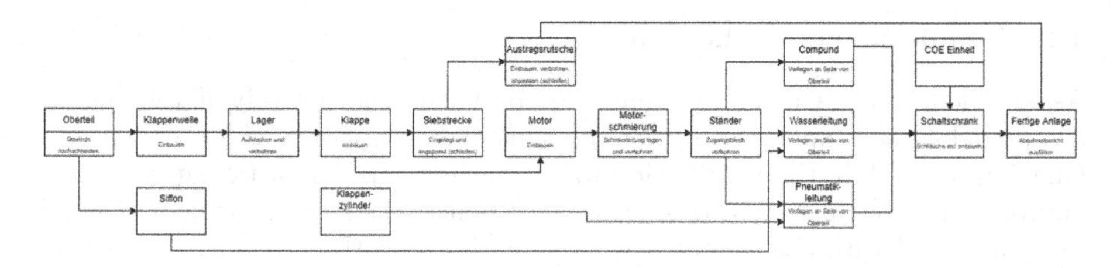

Fig. 1. Example of an assembly priority graph

3.2 Potentials of Assembly Priority Graphs

The usage of assembly priority graphs within knowledge management systems may have further advantages for the assembly and production processes. Some of these potentials are the following:

- They may be used to gather, link and present assembly knowledge in an user-friendly way.
- Furthermore, the structured visualization of the assembly process allows employees to select the next suitable assembly procedure [8].
- The assembly priority graph of a machine, which is created by the assembly employees themselves, may be used for planning further machines. This potentially leads to optimized assembly processes in the future.
- Instead of creating many knowledge bases for variations of a machine, they may be included into one single graph, just visually marked. This results in less redundancy of information.

One of the tasks of the FlexDeMo project is to transfer knowledge of older, experienced assembly workers to the younger ones. Because of the above mentioned potentials, the idea was born to implement and evaluate a knowledge management system which is based upon the assembly priority graph as the navigation and visualization concept.

4 Developing a Graph-Based Tool to Manage Assembly Knowledge

Spaleck Oberflächentechnik GmbH & Co. KG has already tried to establish a platform for knowledge management. However, prior attempts were unsuccessful due to unsuitable technical and methodological solutions.

In the past the documentation and retrieval of assembly information often was too complex and time-consuming, so that employees of Spaleck did not continue to maintain the system. As a consequence, the information landscape of the company has developed into numerous heterogeneous and department-specific applications for knowledge management. To improve the documentation of assembly knowledge and the transfer of information between different departments a smart and user-friendly tool shall be implemented within the FlexDeMo project. Its design and implementation process will be discussed in the next paragraphs.

4.1 Requirements of the Tool

As knowledge management systems will only be used actively if the tool is accepted by the employees, it is important that it satisfies the users' needs. Ongoing research in the FlexDeMo project suggests, that a participatory design approach may be applied to reach this goal. Hence, the requirements of the tool were discussed with the assembly workers within a workshop which resulted in the following prerequisites:

Acceptance and Ease of Use. The developed software needs to be accepted by the users to ensure its sustainable use. Joy of use and intuitive user-interfaces are decisive factors for increasing the acceptance of knowledge management systems.

Graph as Central Navigation Concept. The assembly priority graph should be the main navigation component for the users of the software.

Minimal Effort. The user must be able to graphically document all details of an assembly process with minimal effort.

Detailed Description. It must be possible to specify single steps and intrinsic knowledge of the assembly process with the help of a textual description.

Attachment of External Data. Photos and documents from multiple external sources may be attached or linked to single process steps.

Expert Knowledge. The user needs to be able to store expert knowledge in form of best practices.

Variant Management. Different types of a machine have to be documented in an easy way.

Corporation-Wide Use. The tool is planned to be used across different departments.

4.2 Evaluating Existing Methods and Tools to Create Assembly Priority Graphs

Keeping in mind these requirements different existing tools were evaluated and compared with each other.

Diagrams.net is a diagram tool[4] that is available under the Apache license 2.0 and as such is free to use, even commercially. It supports many different kinds of diagrams and symbols which may be customized to fit the users needs. A quite interesting feature is to add links to symbols which may point to web addresses (see attachment of external data in Sect. 4.1) or to different "pages" of the created diagram allowing the user to add more information to a single assembly step. To be able to effectively view the diagrams they need to be converted into HTML files and can then be opened in the web browser, or hosted within the company's network. To edit the created graph, a copy of the Diagrams.net file needs to be opened in the web editor which then needs to be reexported. An example of the exported diagram can be found in Fig. 2.

[4] https://www.diagrams.net/.

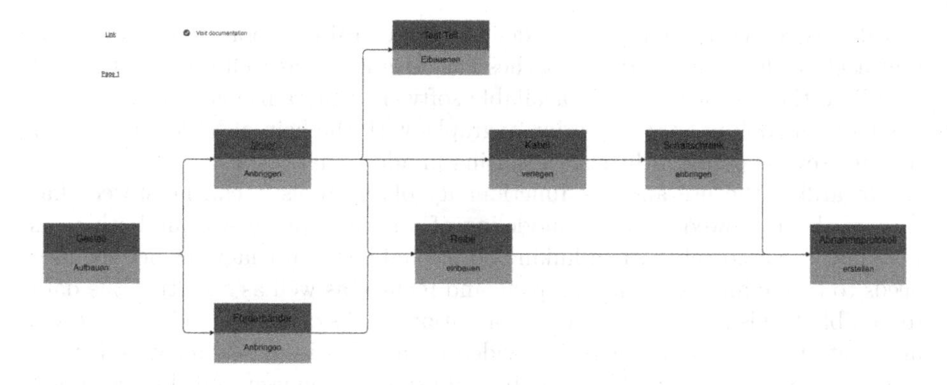

Fig. 2. Example of a Diagrams.net HTML export

Microsoft Visio and ViFlow. Visio[5] is particularly suitable to visualize processes and workflows. Thus, the assembly priority graph could be easily created but linking of files or other types of information was not possible. This missing functionality was added by using the extension ViFlow[6]. Thus, additional data such as descriptions, images and web links could be added to each assembly step. Furthermore, ViFlow facilitates the export of Visio models into a web app allowing to display the model in web browsers as seen in Fig. 3.

Tool Comparison. The alternative options for implementing a knowlegde management system were presented to the employees of Spaleck Oberflächentechnik

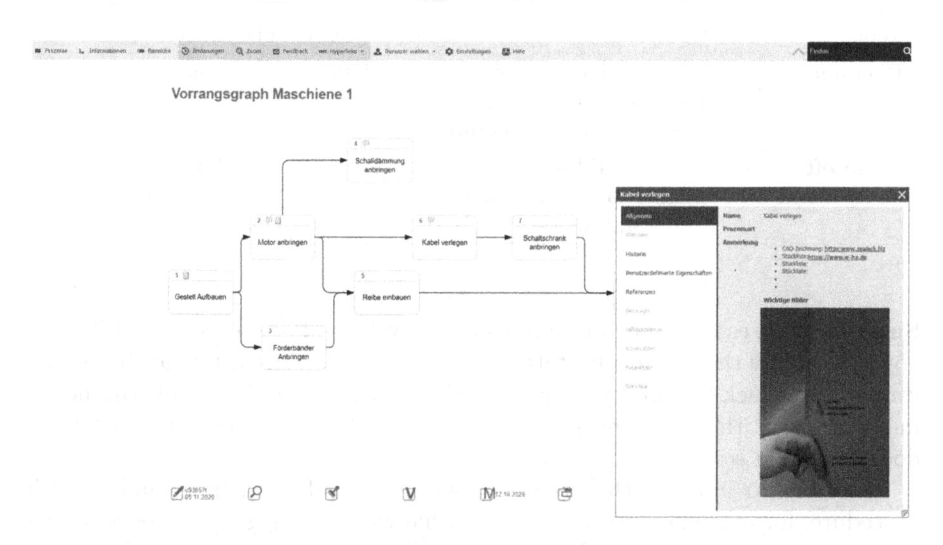

Fig. 3. Example of an exported ViFlow model

[5] https://www.microsoft.com/en/microsoft-365/visio/flowchart-software.

[6] https://www.viflow.de/en/.

and discussed with regard to the defined prerequisites. Soon, it became clear that each of the mentioned approaches has advantages as well disadvantages.

While the use of the freely available software Diagrams.net does not cause any costs, modeling assembly priority graphs with the help of Visio and ViFlow requires several software licenses resulting in additional costs.

Regarding the use and the functionality of the tools it can be stated that each of them allowed intuitive modeling of assembly processes but both tools were not as comfortable when linking additional data. In Diagrams.net the user needs to create an extra diagram page and include as well as structure the data to be able to visualize it. Adding information in Visio models and ViFlow was more intuitive, but a lot of the provided data fields were not relevant for the assembly documentation. This may be confusing for the user and slow down the documentation process considerably.

In both tools it was necessary to export the models so that they could be viewed in the browser. Additionally, the models generated with ViFlow needed to be hosted on a separate web server. Because of this, the graph could no longer be edited by the user easily. Instead, they would need to open the model on their PC, edit the documentation and reexport the model. This can be seen as one of the main hurdles to adjust the documentation and is likely to effect the acceptance[7] of the knowledge management tool.

To conclude, these advantages as well as disadvantages are summarized by the following table (Table 1).

Table 1. Comparison of existing software

Tool	Complexity	Functionality	Cost
Diagrams.net	Very complex editor Fairly simple Linking Complicated to structure	High	none
Microsoft Visio and ViFlow	Complex Editor Simple to add information Many unnecessary fields	High	Microsoft Visio and ViFlow Licences

Summary. In essence, both Diagrams.net as well as Microsoft Visio/ViFlow do not fully satisfy the previously stated user requirements. In particular, both software systems lack the ability to quickly edit existing data. This is of paramount importance, as the daily routine of assembly workers may not allow extended documentation sessions.

One way to improve this, is to automate part of the export and publish procedure, for example with the help of a file watcher triggering export scripts once a file has been changed. But the user still needs to open, edit and save the graph.

[7] See Sect. 4.1.

Another way to improve this workflow is by extending Diagrams.net or writing plugins for Visio. While a potentially viable option, the necessary changes may be fairly substantial. At this point it is not clear whether this complexity could be reduced.

Discussing both options with Spaleck Oberflächentechnik, it was finally decided to pursue a third option: developing a custom tool with a user-specific graphical interface tailor-made for the assembly workers.

4.3 Developing a Custom Tool for Knowledge Management

The development of a custom tool has a couple of advantages. In particular, the target group may define the design of the user interface and the functions needed which strongly influences the acceptance of knowledge management systems. Furthermore, it is possible to consider the existing IT infrastructure allowing integration into other applications.

Designing an Intuitive User Interface. To involve the potential users into the design process, at the beginning different workshops with the assembly workers of Spaleck were held. In this context the technical requirements as well as the structure and the core components of the graphical user interface were discussed and sketched using wire framing. An excerpt of the resulting mock-ups can be seen in Fig. 4. At the end, the following core components were assembled together with the employees.

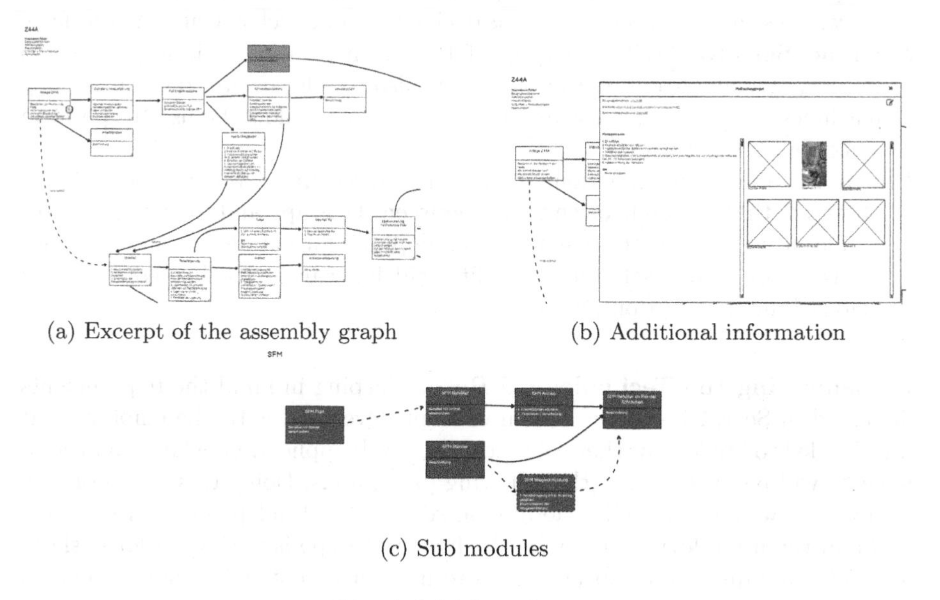

(a) Excerpt of the assembly graph

(b) Additional information

(c) Sub modules

Fig. 4. Initial mock-ups

Assembly Step Graph. This main component is closely related to the assembly priority graph. Each documented step is represented as a square containing the name of the step. The Steps will be connected by arrows that can be drawn between the steps.

Assembly Step Graph Editor. The editor should be as simple and intuitive as possible allowing to document each necessary step of a specific assembly process using drag & drop functionality. Although it has some similarities to Diagrams.net or Visio, the main difference is that there are only two symbols which can be drawn. Because of this, no toolbox component was needed. Instead a new assembly step can be added by dragging a connection spot[8] on an existing step and dropping. This will then create a new empty assembly step and an arrow connecting both.

Information Popup. It is used to display additional information for an assembly step. This popup contains static information for the assembly step, a description of the assembly process and relevant images. It can be edited by the user at any time by toggling the editing mode.

Image Upload. This dialog can be used to upload images or link existing images that are available on the internet or intranet. Linking images should be preferred because it does not create an additional information pool and redundant storage. The choice was made to allow uploads as well, as it better fits the workflow of the assembly workers. Currently, at Spaleck Oberflächentechnik there exists a directory based image archive, which is regularly used by the workers. However, it neither allows access to the images via a web address nor does it provide easy integration into different tools. As the company has already been in the process of implementing a new product lifecycle management system which allows these missing function, it was decided to not develop a custom interface.

Machine Selector. With the help of the machine selector it is possible to select a machine from existing ones, but also to add, delete, rename or copy machines. Additionally, a search bar was added to allow the user to perform a simple keyword search.

Editor Sidebar. This functionality was integrated in a later iteration. It was observed, that the users often toggle between the graph and the popup editor, a quicker way to edit the assembly step popup information was needed. Once an assembly step is selected, the editor sidebar can be used to edit all data fields of the information popup.

Implementing the Technological Base. Keeping in mind the requirements discussed in Sect. 4.1 and the design concept developed with the employees, it was decided to implement the tool in form of a web application, whose main components will be presented in the following paragraphs. Doing this, the tool may be accessed without installing additional software and independent of the location and the used device. However, the first prototype is optimized for desktop use and will require some special views to make it accessible on mobile devices.

[8] Every assembly step has four anchor points, one at each side. To all of them, arrows can be connected.

Frontend. The user interface was designed to be highly interactive with movable components. As such, the interface was developed using React[9], which allows the rapid development of interactive and reactive web applications. To better handle the state of the application, Redux Toolkit[10] was chosen to provide a single state object which all frontend components can access. As the graph editor utilizes multiple drag and drop user interactions to move the graph and its nodes and to create new assembly steps, the React-Draggable[11] library was integrated. Finally, symbols from Font Awesome[12] were used to customize several components like for example buttons.

Backend. Furthermore, a backend was needed to store and retrieve data. For this purpose Python and the Flask[13] web server were selected. SQLAlchemy[14] is used to access an external database that nearly stores all application data. For each image, an additional thumbnail is created by use of the pillow[15] library and both the original and thumbnail are saved on the file system of the server.

Authentication. Authentication has been realized using Keycloak[16], which provides a good and secure user management, authentication and authorization service. It can also be integrated into existing Active Directory or LDAP systems allowing the use of existing user accounts in the developed app.
Currently, a login function has not yet been integrated into the react application. Instead, users must authenticate on a login page provided by the backend over the Keycloak server before being able to access the tool and its API. In the future, the complete integration of Keycloak into the react app is planned.

Deployment. The software is deployed with the help of Docker. Both backend and frontend are hosted in a single Docker container for which the storage of the database and image directory can be configured. Additionally, a Docker Compose script was created to deploy this aforementioned container, as well as a Keycloak container. This script allows easy and fast deployment of the created software components in each company, as long as a Linux server is available.

4.4 Working with SPOTLink

The prototype of SPOTLink[17] was deployed on the servers of Spaleck Oberflächentechnik to be tested by the assembly workers. To give a short overview of how the tools work, the most important functionalities will be highlighted.

9 https://reactjs.org/.
10 https://redux-toolkit.js.org/.
11 https://github.com/react-grid-layout/react-draggable.
12 https://fontawesome.com/.
13 https://flask.palletsprojects.com/.
14 https://www.sqlalchemy.org/.
15 https://pillow.readthedocs.io/en/stable/.
16 https://www.keycloak.org/.
17 SPOT - abbreviation for SPaleck OberflächenTechnik; Link - linking information.

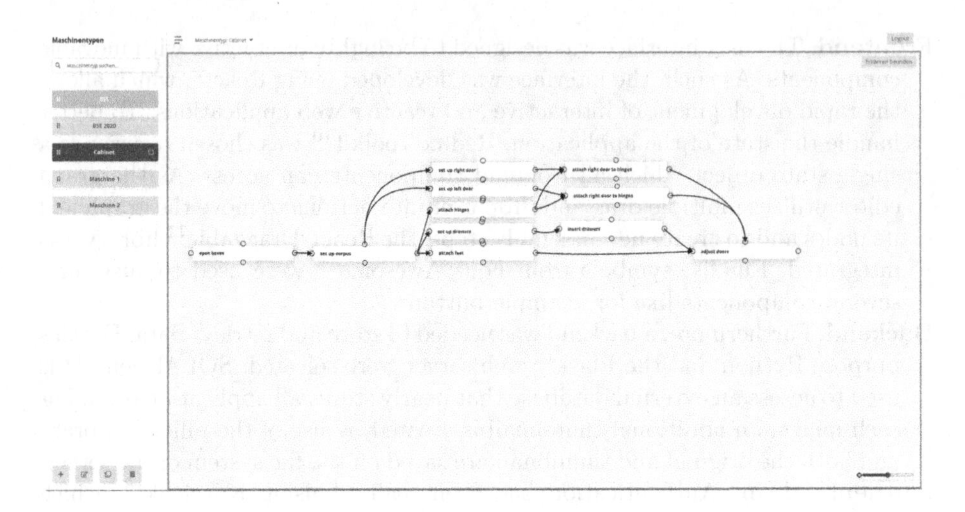

Fig. 5. SPOTLink: viewing the assembly graph

Figure 5 illustrates the main view of SPOTLink. At the left, the user may select the relevant machine from the list, but it is also possible to add, rename, copy or remove machines. Once a specific machine has been selected by the assembly worker, the assembly graph is displayed in the center of the window. It can be moved around by clicking on an empty spot and dragging. The size of the assembly graph can be changed via the slider placed at the bottom right of the editor. To be able to edit the graph, the editing mode may be toggled by clicking on the button at the top right corner of the editor. As can be seen by the yellow border, the editor has already been switched into the editing mode, in which each node of the graph can be re-positioned by dragging them with the mouse. A new assembly step may be added by clicking and dragging on the small circle of an already existing step. Similarly a new connection between two steps may be added by clicking and dragging the small circle of an existing step and dropping on the circle of another step.

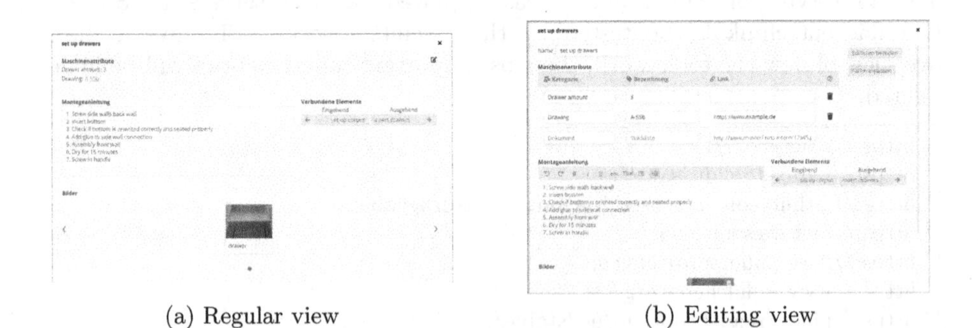

(a) Regular view (b) Editing view

Fig. 6. SPOTLink: Pop-ups

Figure 6 shows the information pop-up in both the regular view and the editing mode. As can be seen, nearly all data fields simple change to their respective input variants.

After the first test phase, the assembly workers requested the option to quickly switch to other assembly steps from within the information pop-up. As can be seen in Fig. 6a this was implemented with the help of a list of incoming and outgoing connections which the user may click to jump to another assembly step.

5 Evaluation

The SPOTLink Tool has been evaluated over multiple iterations with the target user group being extensively involved into the different test phases.

5.1 Creating a Mock-Up with Real Data

As already mentioned in Sect. 4.3 at the beginning of the development, a mock-up was created in cooperation with the employees of Spaleck. In this context, the assembly workers were asked to exemplary design a graph for an existing machine with real assembly steps. For a number of these steps, additional information was gathered in a mocked pop-up to better explain the functionality of the tool. This mock-up has then been converted to a clickable version which the employees reviewed internally.

One of the main results of this first mock-up based test was that the assembly workers already started to identify with the knowledge management tool and actively discussed possible extensions.

5.2 Evaluating the First Prototype

Subsequently, a first testable version of the prototype was developed and hosted on a server of the Westfälische Hochschule. Before it was made available, the prototype was presented within a workshop with the same employees as during the first phase. To find out whether the users were able to intuitively use the software they were instructed to add a new machine and create the assembly graph without any explanation. Meanwhile, the screen was recorded and the reactions of the users were collected with the help of the think aloud method.

As a result it became apparent that the users were able to easily use the software without extensive introduction except for some minor exceptions. In particular, they experienced difficulties when adding content to the information table. This was mainly due to a different saving paradigm.

Usually, data is saved automatically, once the user has finished editing. To add a new table row, however, the users needed to hit a plus sign to the right of the table. In the tests, the users tried to add rows multiple time, but never clicked the add button. This was fixed in a later revision of the prototype where the add button was removed and a auto save function has been introduced. After the workshop the prototype was made available for the first test phase.

5.3 Active In-House Tests of SPOTLink

After multiple optimizations, bug fixes and some functional extensions, the software was deployed on the company's in-house servers. Currently, SPOTLink is evaluated for a longer period by the assembly workers, after which the feedback will be incorporated into the prototype.

6 Discussion and Outlook

This case study describes the development of a knowledge management tool based on a graph-based visualization within the assembly department of one medium-sized company in the domain of mechanical engineering. The employees where strongly incorporated into the design as well as the implementation process, which highlight one key feature of this study. Due to the participatory design approach the tool was highly accepted by the assembly workers and could be intuitively used without extensive training.

The development process was described from the first idea that originates from production planning to a running application. As usability and user experience play a vital role, multiple evaluation steps were performed, but it became apparent that through the strong involvement of the target user group only small and incremental changes were necessary. Following this, it can be stated that if a software requires high acceptance the development process may significantly benefit from involving the potential users already in an early stage.

It can not be concluded just yet, if the decision to implement a custom tool instead of using existing software has been correct. This will be shown by the tool usage of the assembly workers as they will need to actively maintain the software.

In the future, continuous improvements like adding a mobile friendly design, extensive user guidance and features for better archiving are planned to assure that the knowledge management system is sustainably used to document machines.

References

1. Canonico, P., de Nito, E., Esposito, V., Fattoruso, G., Pezzillo Iacono, M., Mangia, G.: Visualizing knowledge for decision-making in lean production development settings. insights from the automotive industry. Manage. Decis. (2021). (ahead-of-print), https://doi.org/10.1108/MD-01-2021-0144
2. Hirsch, C., Hosking, J., Grundy, J.: Interactive visualization tools for exploring the semantic graph of large knowledge spaces. In: Workshop on Visual Interfaces to the Social and the Semantic Web (VISSW 2009), vol. 443, pp. 11–16 (2009)
3. Eppler, J.M., Andreas Pfister, R.: Best of both worlds: hybrid knowledge visualization in police crime fighting and military operations. J. Knowl. Manage. **18**(4), 824–840 (2014). https://doi.org/10.1108/JKM-11-2013-0462
4. O'Leary, D.E.: Is knowledge management dead (or dying)? J. Decis. Syst. **25**(sup1), 512–526 (2016). https://doi.org/10.1080/12460125.2016.1193930

5. Probst, G.J.B.: Wissen managen: Wie Unternehmen ihre wertvollste Ressource optimal nutzen. SpringerLink Bücher, Gabler Verlag, Wiesbaden, 7. aufl. 2012korr. nachdruck 2013 edn. (2012). https://doi.org/10.1007/978-3-8349-4563-1
6. Rambau, A. (ed.): Fit für den Wissenswettbewerb: Vorgehensweise und Fallbeispiele für die Praxis. Bundesministerium für Wirtschaft und Technologie, Berlin, stand November 2013 edn., November 2013
7. Renaud, K., van Biljon, J.: Charting the path towards effective knowledge visualisations. In: Masinde, M. (ed.) Proceedings of the South African Institute of Computer Scientists and Information Technologists on - SAICSIT 2017, pp. 1–10. ACM Press, New York, USA (2017). https://doi.org/10.1145/3129416.3129421
8. Riffelmacher, P.: Konzeption einer Lernfabrik für die variantenreiche Montage (2013)
9. Smuts, H., Scholtz, I.-I.: A conceptual knowledge visualisation framework for transfer of knowledge: an organisational context. In: Hattingh, M., Matthee, M., Smuts, H., Pappas, I., Dwivedi, Y.K., Mäntymäki, M. (eds.) I3E 2020. LNCS, vol. 12067, pp. 287–298. Springer, Cham (2020). https://doi.org/10.1007/978-3-030-45002-1_24
10. Spreckelsen, C., Liem, S., Winter, C., Spitzer, K.: Cognitive tools for medical knowledge management (Kognitive Werkzeuge für das Medizinische Wissensmanagement). IT-Inf. Technol. **48**(1), 33–43 (2006)
11. Wiesbeck, M.: Struktur zur Repräsentation von Montagesequenzen für die situationsorientierte Werkerführung, Forschungsberichte/IWB, vol. Bd. 285. Utz, München (2014)

An Ensemble Learning Method for Constructing Prediction Model of Cardiovascular Diseases Recurrence

Yen-Hsien Lee[1], Tin-Kwang Lin[2], Yu-Yang Huang[1], and Tsai-Hsin Chu[1](✉)

[1] National Chiayi University, Chiayi City, Taiwan
{yhlee,thchu}@mail.ncyu.edu.tw, ian801117@gmail.com
[2] Buddhist Dalin Tzu Chi General Hospital, Chiayi County, Taiwan
shockly@tzuchi.com.tw

Abstract. Cardiovascular diseases (CVDs) have been reported as one of the leading causes of death worldwide by World Health Organization (WHO). Although CVDs can be treated, it has high risk of recurrence. In this study, we intended to construct the predictive model of cardiovascular disease recurrence by machine learning approach. We used the 18-month prognosis tracing data to construct and evaluate the recurrence predictive model. We collected 36 physiological factors associated to the cardiovascular disease recurrence identified from literature from 1274 cardiovascular disease inpatients as they discharged from the hospital and their follow-up prognoses after six months. To address the imbalance data problems that are prevalent in medical dataset, we revised the ensemble learning method by performing multiple undersampling to construct a committee of SVM classifiers. The evaluation results show that our proposed approach outperforms all benchmarks in term of F1 measure and Area under ROC curve (AUROC). Our study has demonstrated an approach to address to construct an effective prediction model for cardiovascular diseases recurrence. It might also support physicians in assessing patients who are at the high risk of recurrence.

Keywords: Cardiovascular diseases · Recurrence prediction · Committee machine · Medical mining · Machine learning

1 Introduction

Cardiovascular diseases (CVDs) refer to the disorders of the heart and blood vessels, including coronary heart disease, cerebrovascular disease, peripheral arterial disease rheumatic heart disease, congenital heart disease, and deep vein thrombosis and pulmonary embolism. CVDs have been reported as one of the leading causes of death worldwide by World Health Organization (WHO) in 2016. The statistics presents 31% of global deaths caused by CVDs, and of which, 85% are due to two acute cardiovascular events, heart disease and stroke.[1] Nowadays, cardiovascular diseases can be treated

[1] https://www.who.int/news-room/fact-sheets/detail/cardiovascular-diseases-(cvds).

F. Fui-Hoon Nah and K. Siau (Eds.): HCII 2022, LNCS 13327, pp. 208–220, 2022.
https://doi.org/10.1007/978-3-031-05544-7_16

by percutaneous transluminal coronary angioplasty (PTCA) surgeries, such as balloon angioplasty, stenting, or bypass surgery, these treatments are costly and with a high possibility of recurrence within six months (about 30–40% with balloon angioplasty; about 20%–30% with stenting) [1–4]. The high recurrence rate is also confirmed by our 18-month prognosis tracing data of 1274 cardiovascular disease inpatients, where it shows the recurrence rate of 17.7% within six months, 24.4% within 12 months, and 28.7% within 18 months. Our dataset collectively highlights the importance of managing cardiovascular disease recurrence in a short period of time.

A few studies have investigated the risk factors of cardiovascular disease recurrence in patients with established cardiovascular disease. These studies examined the relationships between prognosis data and recurrence of cardiovascular disease to conclude the important influential factors of secondary cardiovascular event or death. They collectively proposed the potential important factors including age, gender, diabetes, stroke, coronary artery disease, chronic kidney disease, systolic dysfunction, heme decline index, albumin decline index, diastolic blood pressure, left ventricular ejection rate, myocardial infarction, heart failure, unstable angina, and cardiovascular reconstruction surgery [5–7]. For estimation purpose, Wilson, D'Agostino, Bhatt, Eagle, Pencina, Smith, Alberts, Dallongeville, Goto, Hirsch, Liau, Ohman, Röther, Reid, Mas and Steg [8] constructed a 20-month risk assessment score for the secondary cardiovascular events and cardiovascular death in outpatients with established atherothrombotic vascular disease (AVD). The risk assessment score facilitates efforts by patients, clinicians, and public health officials to prevent these morbid and mortal events [8].

Although the 20-month risk model for cardiovascular death and next cardiovascular event of outpatients has been developed, the high risk of cardiovascular diseases relapsing in a short period of time, that indicates the needs for a short-term recurrence prediction. In addition, the cardiovascular disease recurrence, for example, may involve with a complicated and twist interaction among physiological index, disease, vascular status, inpatient treatments, disease, and medicines. As a results, our study seeks to construct a prediction model of short-term recurrence of the cardiovascular disease by the medical mining method. Medical mining is to discover hidden information and knowledge from a large amount of medical clinical data that uses machine learning approaches to construct prediction models on major diseases or healthcare risks to improve the quality and efficiency of medical treatment decision [9–11].

We aim to predict the cardiovascular disease patients who have the higher risk of recurrence to provide a satisfactory suggestion for both patients and cardiologists when making clinical decision. The remainder of this paper is organized as follows. In Sect. 2, we review literatures relevant to our study including cardiovascular disease occurrence and recurrence to establish the domain knowledge on current understanding and approaches to cardiovascular disease prevention. In Sect. 3, we detail the development of the proposed cardiovascular disease recurrence prediction model followed by the empirical evaluations and results in Sect. 4. We finally conclude our research contributions and limitations in Sect. 5.

2 Literature Review and Motivation

Cardiovascular diseases usually take long time to develop, but may recur and threaten life in a short period of time after treatment. In addition, the high possibility of recurrence complicates the clinical treatment on cardiovascular diseases. This study aims to apply machine learning to construct a prediction model for predicting the recurrence risk of cardiovascular diseases. We first review the previous studies on the risk factors of the cardiovascular diseases followed by the risk assessment scores for the cardiovascular diseases recurrence. The risk factors might in turn become the attributes of the proposed prediction model and the existent risk assessment scores would be adopted as the performance benchmarks. Finally, the machine learning approaches we adopt for constructing the prediction model will also be discussed.

2.1 Risk Factors of Cardiovascular Diseases

Previous studies explored the factors that relevant to the cardiovascular diseases in order to better manage the risk of first attack or recurrence events. Some studies highlighted the effects of diseases. For example, Weiner, Tighiouart, Amin, Stark, MacLeod, Griffith, Salem, Levey and Sarnak [7] investigated the correlation between chronic kidney diseases and the recurrence of cardiovascular diseases. With 4,278 samples in the United States, they confirmed that patients with cardiovascular diseases have an increased risk of cardiovascular diseases recurrence due to chronic kidney diseases. Zöller, Ji, Sundquist and Sundquist [12] analyzed 34,666 cases of Sweden cancer patients, and concluded that cancer is positively associated with the risk of coronary artery disease and the effect is the most significant within 6 months of cancer confirmed diagnosis. Eisen, Bhatt, Steg, Eagle, Goto, Guo, Smith, Ohman and Scirica [13] suggested that patients with coronary artery disease and angina pectoris may have a higher risk of cardiovascular events in the future. This study used Cox regression to examine the 26159 samples participating in the REduction of Atherothrombosis for Continued Health (REACH) program. The findings showed that patients with angina pectoris have a greater risk of cardiovascular events than those without angina pectoris. Further analysis showed that angina pectoris is significantly related to heart failure, hospitalization due to cardiovascular disease, coronary artery reconstruction (PCI and CABG). De Hert, Detraux and Vancampfort [14] also examined the correlation between mental illness and coronary artery disease.

Some other studies concerned the effects of drugs. For example, Naderi, Bestwick and Wald [15] investigated the effects of drug compliance on preventing cardiovascular disease. This study applies meta-analysis to aggregate findings from 20 studies, and it concludes that drug compliance of aspirin, angiotensin type I convertase inhibitors, angiotensin type II receptor antagonists, beta-blockers, calcium channel blockers, thiazide diuretics, and statins correlated to the prevention of cardiovascular disease. It indicates that patient's poor drug compliance will decrease the effectiveness of cardiovascular diseases prevention. Karmali, Lloyd-Jones, Berendsen, Goff, Sanghavi, Brown, Korenovska and Huffman [16] examined the correlation between drugs and the prevention of cardiovascular diseases by reviewing the relevant studies from 2005 to 2015, and found that aspirin, antihypertensive drugs, statins, and smoking cessation drugs can reduce the risk of cardiovascular diseases.

Some studies included wider physiological indicators to explore the important factors. For example, analyzing 432 patient samples from three Canadian hospitals with multivariate Cox regression analysis, Harnett, Foley, Kent, Barre, Murray and Parfrey [5] propose the significant effects of age, diabetes, coronary artery disease, systolic dysfunction, heme decline index, albumin decline index, blood pressure and diastolic blood pressure, left ventricular injection rate to hear failure prevalence, morbidity, and prognosis. Among them, coronary artery disease, systolic dysfunction, decreased heme, and decreased albumin play the key factors to the heart failure recurrence. Yusuf, Hawken, Ounpuu, Dans, Avezum, Lanas, McQueen, Budaj, Pais, Varigos and Lisheng [17] also conducted a large-scale cardiac study on subjects with myocardial infarction in 52 countries. They used multivariate logistic regression models to analyze the factors for the risk of acute myocardial infarction, and found that smoking, various cholesterol detailed indicators, history of hypertension, diabetes, obesity, diet, physiological exercise, alcohol consumption, and psychological factors have a significant impact on acute myocardial infarction. Jernberg, Hasvold, Henriksson, Hjelm, Thuresson and Janzon [6] analyzed 97,254 Sweden patient samples who are initially diagnosed with myocardial infarction by Kaplan-Meier analysis and Cox regression and indicated that age, gender, stroke, diabetes, myocardial infarction, heart failure, unstable angina, and cardiovascular reconstruction surgery are significant correlated to the recurrence of cardiovascular disease. Their findings highlight that the patients with myocardial infarction have a high risk of recurrent within one year after the in-patient service and recommend monitoring for a long time to prevent the recurrence of cardiovascular events. Finally, Perret-Guillaume, Joly and Benetos [18] proposed that heart rate may be a harmful factor for cardiovascular disease. Their study reviewed the relevant literature on heart rate and the occurrence of cardiovascular disease and then concluded that heart rate is a risk factor for the patients who have coronary artery disease or heart failure to have cardiovascular disease or to die.

2.2 Risk Assessment for Cardiovascular Disease Events

Research also develops risk assessments for the cardiovascular disease events. Wilson, D'Agostino, Levy, Belanger, Silbershatz and Kannel [19] conducts a Framingham heart study to predict the risk of coronary artery disease in next ten years by tracing patients without coronary artery disease for twelve years. This study develops the Framingham Risk Score to identify the corresponding risk scores for important physiological factors, including age, gender, diabetes, smoking, systolic blood pressure, and cholesterol. Then it aggregates the scores to calculate the total risk of the first coronary artery event within ten years. Following the Framingham study, Jackson [20] continuously comprehend the cardiovascular disease risk assessment by analyzing New Zealanders samples to estimate the 5-year risk of New Zealanders suffering from cardiovascular diseases. Similar physiological risk factors are adopted in this study and are further converted into the risk probability by the comparison with color charts where each cell represents the risk level under different conditions, such as gender, age, blood pressure, diabetes and smoker. In addition, Assmann, Schulte and von Eckardstein [21] develop a PROCAM (Prospective Cardiovascular Münster) scoring scheme to evaluate the risk of acute coronary events. The study establishes a Cox proportional hazards model by examining a variety of physiological risk factors, such as age, cholesterol, blood pressure, and family history of myocardial infarction, and identify each risk factors into risk score based on the coefficients in the Cox proportional hazard model. The total PROCAM score is calculated and transformed into the probability of incidence of acute coronary events occurring within10 years. Similarly, Conroy, Pyörälä, Fitzgerald, Sans, Menotti, De Backer, De Bacquer, Ducimetière, Jousilahti, Keil, Njølstad, Oganov, Thomsen, Tunstall-Pedoe, Tverdal, Wedel, Whincup, Wilhelmsen, Graham and group [22] published the SCORE chart to evaluate Europeans' risk on fatal cardiovascular disease occurring within 10 years based on physiological risk factors, including blood pressure, smoking, gender, and cholesterol.

Besides the risk assessment of primary event of cardiovascular diseases, some studies focus on the secondary event of cardiovascular diseases (i.e., recurrence) [8]. Harnett, Foley, Kent, Barre, Murray and Parfrey [5] examined the relationship between prognosis and recurrence of cardiovascular disease and physical examination data. Their findings show the significant relevant factors include age, diabetes history, coronary artery disease, systolic dysfunction, heme decline index, albumin decline index, diastolic blood pressure, and left ventricular ejection rate. Weiner, Tighiouart, Amin, Stark, MacLeod, Griffith, Salem, Levey and Sarnak [7] repost that chronic kidney disease increases the chance of recurrence of cardiovascular disease. Jernberg, Hasvold, Henriksson, Hjelm, Thuresson and Janzon [6] proposed that age, gender, stroke, diabetes, myocardial infarction, heart failure, unstable angina, and cardiovascular reconstruction surgery have significant impacts on the recurrence of cardiovascular diseases. Except the significant factors, Wilson, D'Agostino, Bhatt, Eagle, Pencina, Smith, Alberts, Dallongeville, Goto, Hirsch, Liau, Ohman, Röther, Reid, Mas and Steg [8] constructed an assessment tool to predict 20-month risk assessment score for next cardiovascular event and cardiovascular death based on 49,689 patient samples in the REACH program by Cox proportional-hazard regression. Gender, age, smoking, diabetes mellitus, BMI, number of vascular beds, cardiovascular event in past year, congestive heart failure, atrial fibrillation, statins therapy, aspirin use, and geographical regions are identified as important factors, and risk score of each factor is assigned according to the result of multivariable analysis on the risk of the secondary cardiovascular event.

2.3 Research Gaps

Previous studies examine the risk factors and assessment instruments of cardiovascular diseases for identifying individuals at higher risk for more intensive investigation, follow-up, and treatment. These studies target the people who have never suffered from cardiovascular disease to assess their risks of the first cardiovascular event. Few prediction models focus on patients with cardiovascular diseases [8]. As cardiovascular diseases threaten to life with high recurrence risk, it is important to develop a method that can be easily applied to build a prediction model of cardiovascular disease recurrence. In addition, the previous assessments are developed based on statistic methods, such as regressions, to estimate the coefficient of the factors. However, this estimation may be limited by the independence assumption and linear estimation of the statistical methods. These limitations may be regrettable when it is applied to medical context where the factors caused the particular disease may be complicate and with high interactions among factors. To bridge these gaps, we seek the data-driven approach to explore the hidden relationship among variables to extend the existing understanding on cardiovascular disease prediction. Specifically, we apply the ensemble learning which is a kind of machine learning approach to construct the prediction model of cardiovascular disease recurrence.

3 Development of Cardiovascular Disease Recurrence Prediction Model

This study intends to construct a cardiovascular disease recurrence prediction model by using the machine learning approach to analyze real-world prognostic tracking data of cardiovascular disease patients. The development of the prediction model can be divided into three stage, including the prognostic data collection, data-preprocessing, and model construction.

3.1 Prognostic Data Collection

To collect the prognostic data of cardiovascular disease patients for constructing the prediction model, we cooperated with a regional teaching hospital in Taiwan. We collect the data of inpatients who received the treatments for first cardiovascular events and track their prognoses on receiving secondary cardiovascular disease treatment or death within six months. Table 1 lists the inpatients' attributes that we collect and adopt to construct the recurrence prediction model. These attributes are suggested by both literatures and cardiologists, and we further categorize them into five categories, including physiological index (11 features), vascular index (5 features), inpatient diagnosis (6 features), disease (9 features), and medicine (5 features).

Table 1. Inpatients' attributes for constructing recurrence prediction model

Category	Feature	Reference
Physiological index	Gender, Age, Smoke, BMI, TCH, LDL, HDL, TG, GLU, CABG, Cardiac catheterization	[5, 6, 8, 13, 17]
Vascular index	LVEF, SBP, DBP, NYHA, Hear rate	[5, 18]
Inpatient diagnosis	AF, PSVT, Other dyarhymia, CHF, CAD, AMI, STEMI, NSTEMI, UA	[5, 6, 8, 12, 13]
Disease	Diabetes, Kidney disease, Hypertension, Coronary heart disease, Myocardial infarction, Angina, Arrhythmia, Tumor, Mental illness	[5–8, 12–14, 17]
Medicine	Diuretics, Beta-blockers (BB), Angiotensin type I conversion inhibitors (ACEI), Angiotensin receptor blockers (ARB), Statins, Fibrates (Fibrate)	[8, 15, 16]

We totally collected 1426 cases from 2012 to 2014. Among them, 152 cases are discarded because of patient's rejection and the incomplete tracking process, and the remaining 1274 cases were used for following analysis in which consisted of 226 relapse cases and 1048 non-relapse cases. In addition, the missing data problem is prevalent in the dataset. There are 725 cases with incomplete records distributed among 13 attributes, including BMI (48 records), LVEF (182 records), SBP (15 records), DBP (15 records), NYHA (2 records), heart rate (54 records), TCH (317 records), LDL (355 records), HDL (391 records), TG (216 records), GLU (129 records), BB (1 records), ACEI (1 records).

3.2 Data Pre-processing

The main task of the data preprocessing stage is to handle the missing data problem and to categorize numeric attributes into meaningful categories. Missing data are prevalent in the medical records. As described above, there are nearly 60% of data records we collected have two missing values averagely. However, use of machine learning approaches to construct a prediction model requires the training examples having no missing values [23]. Because there are too many incomplete examples in our collected dataset, the exclusion of incomplete examples will greatly reduce the training examples and may lower the performance of the constructed prediction model. We therefore employ data imputation approach to impute the missing data values [24, 25]. In this study, we will explore four common imputation approaches, including k-nearest neighbor, median, mode, and mean value to determine the better approach to our experimental dataset.

Furthermore, we categorize some numerical attributes into categorical ones to fit the clinical practices to provide more comprehensive meaning on clinical diagnosis. For example, the LDL value over 130 represents abnormal, otherwise normal. In such case, the categorized attributes provide more meaningful information in clinical practice than the numerical ones. The numerical attributes in our dataset include age, BMI, LVEF (LV ejection fraction), SBP (systolic blood pressure), DBP (diastolic blood pressure), NYHA (New York Heart Association classification), heart rate, TCH (total cholesterol), LDL (low-density lipoprotein), HDL (High-density lipoprotein), TG (triglyceride), GLU (blood glucose). As shown in Table 2, we consulted a senior cardiologist with his clinical expertise to identify the scheme of data discretization.

3.3 Model Construction

The ratio of relapse cases to non-relapse cases among the 1274 valid cases we collected is about 1:4.5, presenting an imbalanced distribution of positive and negative examples. The class-imbalance problem may lead the constructed prediction model to favor the majority class when making predictive decisions. In other words, such prediction model will be less effective in identifying the examples belonging to the minority class, i.e., the recurrence cases in our study. Oversampling and undersampling are two common approaches to address the class-imbalance problem. To balance the class distribution, the prior uses synthetic sampling or resampling from the minority class to increase the number of minority examples; while the later resamples from the majority class

Table 2. Scheme of data discretization

Features	Data discretization
Age	0: ≦45; 1: >45 (male) 0: ≦55; 1: >55 (female)
BMI	0: ≦18.5; 1: 18.5–27.9; 2: >27.9
LVEF (LV ejection fraction)	0: ≦50; 1: >50
SBP (Systolic blood pressure)	0: ≦100; 1: 100–140; 2: >140
DBP (Diastolic blood pressure)	0: ≦90; 1: >90
NYHA (New York heart association classification)	0: ≦1; 1: >1
Heart rate	0: 50–120; 1: others
TCH (Total cholesterol)	0: ≦200; 1: >200
LDL (Low-density lipoprotein)	0: ≦130; 1 >130
HDL (High-density lipoprotein)	0: >45; 1: ≦45
TG (Triglyceride)	0: ≦150; 1: >150
GLU (Blood glucose)	0: 70–100; 1: others

to decrease the number of majority examples. However, both oversampling and undersampling approaches may degrade the performance of prediction model because of the increase of synthetic data or the discard of useful samples.

To address the class-imbalance problem, we propose a revised ensemble learning method which combines bootstrap aggregating and undersampling to construct the prediction model. Bootstrap aggregating, also called bagging, is one of the ensemble learning methods in machine learning. Ensemble learning generates multiple classifiers in learning process and make prediction decisions collectively to improve the accuracy and stability of prediction [26]. The use of ensemble classifier can achieve better prediction effectiveness than single classifier [27, 28]. Ensemble learning applies divide-and-conquer strategy by dividing the training sample into several subsets and constructing multiple classifiers in parallel. Specifically bagging uses random sampling with replacement to generate multiple subsets of training sample, uses each of training sample to construct an individual classifier, and then forms a "committee" to make joint decisions [29]. Instead of sampling with replacement, our study applies undersampling to generate a set of class-balanced training samples, each of which contains all the examples in the minority class (i.e., relapse cases) and the identical number of examples randomly sampling from the majority class (i.e., non-relapse cases). We generate the class-balanced training samples to avoid the problem that constructed classifier favoring the majority class. Furthermore, each example in the majority class will have the chance to be included in different training samples to reduce the possibility of discarding useful cases. Finally, we initially implemented support vector machine [30] as the basic classifier of the ensemble classifier.

4 Empirical Evaluations and Results

We conduct an empirical evaluation to assess the effectiveness of the proposed prediction model of cardiovascular disease recurrence. In the following, we describe the experimental design, parameter-tuning experiments, and experimental results.

4.1 Experimental Design

As described in Sect. 3.1, we collect 36 physiological factors from inpatients who received the treatments for the first cardiovascular events and their follow-up prognostic data on receiving secondary cardiovascular disease treatment or death within six months. Finally, a total of 1274 valid cases is collected, consisted of 226 relapse cases and 1048 non-relapse cases. We employ the 10-fold cross validation to evaluate the performance of the constructed prediction models. Furthermore, we adopt prediction accuracy, sensitivity, specificity and AUC ROC as the evaluation criteria in this study. Prediction accuracy indicates the number of examples assigned to correct class over the number of all testing examples; sensitivity refers to the proportion of the relapse cases that are correctly identified; specificity refers to the proportion of the non-relapse cases that are correctly identified; and AUC ROC, the area under the ROC curve, is used to show the discriminative power of the prediction model.

In addition, we employed the Framingham Risk Score developed by Wilson, D'Agostino, Levy, Belanger, Silbershatz and Kannel [19] as performance benchmark for evaluating the constructed prediction model. According to the suggestion of WHO Cardiovascular Risk Assessment and Management Code [31] which divides cardiovascular risk probability into different severity. In this study, we conservatively set the threshold probability to 10%; that is, the patient whose risk probability lower than 10% will be classified as non-relapse case; otherwise relapse.

4.2 Parameter-Tuning Experiments

The proposed prediction model has two parameters need to be determined, including data imputation and the number of classifiers before for the subsequent comparative evaluation. We first fix the number of classifiers to 101 and evaluate the effectiveness of prediction model under different data imputation methods, including the k-nearest neighbor, median, mode and mean values. As shown in Table 3, the performance achieved by the four imputation methods are comparable. The k-nearest neighbor method arrived at the relatively better sensitivity and AUC ROC. As the result, we adopted k-nearest neighbor method as the data imputation method.

Then, we fix the data imputation method to k-nearest neighbor and examine the effects of number of classifiers on the effectiveness of the constructed prediction model, ranging from 51 to 251 at an increase of 50. Table 4 shows that the prediction model can achieve better performance when setting the number of classifiers to 151.

Table 3. Parameter-tuning on data imputation methods

Method	Accuracy	Sensitivity	Specificity	AUC ROC
k-nearest neighbor	0.589	0.642	0.568	0.605
Median	0.590	0.625	0.576	0.601
Mode	0.591	0.623	0.578	0.600
Mean	0.585	0.615	0.574	0.594

Table 4. Parameter-tuning on number of classifiers

Number of classifier	Accuracy	Sensitivity	Specificity	AUC ROC
51	0.590	0.617	0.579	0.598
101	0.596	0.626	0.584	0.605
151	0.611	0.667	0.588	0.627
201	0.591	0.648	0.568	0.608
251	0.590	0.628	0.575	0.602

4.3 Experimental Results

We evaluation the performance of our constructed recurrence prediction model using the tuned parameter values and adopt the Framingham risk score as the performance benchmark. As shown in Table 5, both prediction accuracy and AUC ROC achieved by our proposed method are over 0.6 and obviously higher than that achieved by the performance benchmark. Furthermore, our proposed method attains acceptable sensitivity and specificity. Though the Framingham risk score is advantageous over our proposed method in sensitivity, it has a worse performance at specificity. Overall, our proposed method outperforms the Framingham risk score.

Table 5. Comparative evaluation results

	Accuracy	Sensitivity	Specificity	AUC ROC
Proposed method	0.633	0.607	0.638	0.623
Framingham risk score	0.483	0.673	0.442	0.557

We further analyze the experimental results to identify the important attributes that have greater impacts on the prediction model constructed using the support vector machine algorithm. According to the coefficient values obtained by particular attributes, we list the top 5 attributes that have the positive or the negative impacts on the prediction model. The top 5 attributes with positive effect are mental illness, angiotensin I type Conversion inhibitors (ACEI), diuretics, history of disease with coronary heart disease (CHD), history of disease with myocardial infarction (MI), and that with negative effect are coronary artery bypass graft surgery (CABG), tumors, inpatient diagnosis with arrhythmia (AF), hospitalized diagnosis of coronary artery disease (CAD), β-blocker (BB).

5 Conclusion

Cardiovascular disease prevention is one of the most important causes of death in the world. Past studies investigate the important risk factors and develop risk assessment instruments to assist preventing the risk of cardiovascular disease. At present, although there are some successful treatments of cardiovascular diseases, patients suffering from the secondary cardiovascular diseases events or death are high in months. Therefore, the prevention of cardiovascular diseases recurrence has become an important issue. This study applied machine learning approach to develop an ensemble classifier for the prediction of cardiovascular disease recurrence. The experimental results show that the proposed prediction model outperforms the benchmark which conventionally used in clinical literature and practice. Our findings show that the rates of prediction accuracy, sensitivity, specificity, and AUC ROC achieved by the proposed method are all above 0.6, indicating a satisfying performance. In addition, we also indicate the top 5 attributes that have positive or negative impacts on the recurrence prediction model.

The contributions of this study are both academic and practical. For academy, this study contributes to e-healthcare research by extending the machine learning approach on improving the prediction effectiveness of cardiovascular disease recurrence. For practices, this study contributes to the prevention of cardiovascular disease recurrence by revealing the important attributes for cardiologist reference on clinical diagnosis and patient cares. This study proposes a method to address the missing data and class imbalance problems that commonly exist in the medical dataset.

There are some limitations of this study. First, the physiological features we adopted to constructed the prediction model are limited. Those features are suggested by a cardiologist and by the literature we reviewed. Therefore, more potential features shall be included and evaluated in the future study. In addition, the medical data contains lots of missing value as its nature. Although our method has shown its robustness on such dataset, the effectiveness of the prediction model shall be degraded by data distortion to some degree. The future study could pay more attention to avoiding missing value or to developing an effective approach to reducing the impacts of missing data. Moreover, our findings on cardiovascular disease recurrence may be suitable to explain patients in Taiwan because the data is from one regional teaching hospital in particular. Since people's living habit different across regions and countries, the future study can extend the scope of dataset to improve the generalizability of the prediction model.

References

1. Grines, C.L., et al.: Coronary angioplasty with or without stent implantation for acute myocardial infarction. N. Engl. J. Med. **341**, 1949–1956 (1999)

2. Brodie, B.R., et al.: Six-month clinical and angiographic follow-up after direct angioplasty for acute myocardial infarction. Final results from the primary angioplasty registry. Circulation **90**, 156–162 (1994)
3. Salwan, R.: Choice of stent in STEMI interventions. In: Mehta, S. (ed.) Manual of STEMI Interventions, pp. 187–206. Wiley, Hoboken (2017)
4. Sharma, G.N.: Short term and long term complications after Percutaneous Transluminal Coronary Angioplasty (PTCA). Int. J. Cardiol. Heart Health **1**, 36–39 (2017)
5. Harnett, J.D., Foley, R.N., Kent, G.M., Barre, P.E., Murray, D., Parfrey, P.S.: Congestive heart failure in dialysis patients: prevalence, incidence, prognosis and risk factors. Kidney Int. **47**, 884–890 (1995)
6. Jernberg, T., Hasvold, P., Henriksson, M., Hjelm, H., Thuresson, M., Janzon, M.: Cardiovascular risk in post-myocardial infarction patients: nationwide real world data demonstrate the importance of a long-term perspective. Eur. Heart J. **36**, 1163–1170 (2015)
7. Weiner, D.E., et al.: Chronic kidney disease as a risk factor for cardiovascular disease and all-cause mortality: a pooled analysis of community-based studies. J. Am. Soc. Nephrol. **15**, 1307–1315 (2004)
8. Wilson, P.W., et al.: An international model to predict recurrent cardiovascular disease. Am. J. Med. **125**, 695-703.e691 (2012)
9. Manogaran, G., Lopez, D.: A survey of big data architectures and machine learning algorithms in healthcare. Int. J. Biomed. Eng. Technol. **25**, 182 (2017)
10. Murdoch, T.B., Detsky, A.S.: The inevitable application of big data to health care. JAMA **309**, 1351–1352 (2013)
11. Raghupathi, W., Raghupathi, V.: Big data analytics in healthcare: promise and potential. Health Inf. Sci. Syst. **2**, 3 (2014)
12. Zöller, B., Ji, J., Sundquist, J., Sundquist, K.: Risk of coronary heart disease in patients with cancer: a nationwide follow-up study from Sweden. Eur. J. Cancer **48**, 121–128 (2012)
13. Eisen, A., et al.: Angina and future cardiovascular events in stable patients with coronary artery disease: insights from the Reduction of Atherothrombosis for Continued Health (REACH) registry. J. Am. Heart Assoc. **5**, e004080 (2016)
14. De Hert, M., Detraux, J., Vancampfort, D.: The intriguing relationship between coronary heart disease and mental disorders. Dialogues Clin. Neurosci. **20**, 31–40 (2018)
15. Naderi, S.H., Bestwick, J.P., Wald, D.S.: Adherence to drugs that prevent cardiovascular disease: meta-analysis on 376,162 patients. Am. J. Med. **125**, 882-887.e881 (2012)
16. Karmali, K.N., et al.: Drugs for primary prevention of atherosclerotic cardiovascular disease: an overview of systematic reviews. JAMA Cardiol. **1**, 341–349 (2016)
17. Yusuf, S., et al.: Effect of potentially modifiable risk factors associated with myocardial infarction in 52 countries (the INTERHEART study): case-control study. Lancet **364**, 937–952 (2004)
18. Perret-Guillaume, C., Joly, L., Benetos, A.: Heart rate as a risk factor for cardiovascular disease. Prog. Cardiovasc. Dis. **52**, 6–10 (2009)
19. Wilson, P.W., D'Agostino, R.B., Levy, D., Belanger, A.M., Silbershatz, H., Kannel, W.B.: Prediction of coronary heart disease using risk factor categories. Circulation **97**, 1837–1847 (1998)
20. Jackson, R.: Updated New Zealand cardiovascular disease risk-benefit prediction guide. BMJ **320**, 709–710 (2000)
21. Assmann, G., Schulte, H., von Eckardstein, A.: Hypertriglyceridemia and elevated levels of lipoprotein (a) are risk factors for major coronary events in middle-aged men. Am. J. Cardiol. **77**, 1179–1184 (1996)
22. Conroy, R.M., et al.: Estimation of ten-year risk of fatal cardiovascular disease in Europe: the SCORE project. Eur. Heart J. **24**, 987–1003 (2003)

23. He, H., Garcia, E.A.: Learning from imbalanced data. IEEE Trans. Knowl. Data Eng. **21**, 1263–1284 (2009)
24. Ayilara, O.F., Zhang, L., Sajobi, T.T., Sawatzky, R., Bohm, E., Lix, L.M.: Impact of missing data on bias and precision when estimating change in patient-reported outcomes from a clinical registry. Health Qual. Life Outcomes **17**, 106 (2019)
25. Gomes, M., Gutacker, N., Bojke, C., Street, A.: Addressing missing data in Patient-Reported Outcome Measures (PROMS): implications for the use of PROMS for comparing provider performance. Health Econ. **25**, 515–528 (2016)
26. Polikar, R.: Ensemble based systems in decision making. IEEE Circuits Syst. Mag. **6**, 21–45 (2006)
27. Dietterich, T.G.: Ensemble methods in machine learning. In: Kittler, J., Roli, F. (eds.) MCS 2000. LNCS, vol. 1857, pp. 1–15. Springer, Heidelberg (2000). https://doi.org/10.1007/3-540-45014-9_1
28. Tresp, V.: Committee machines. In: Hu, Y.H., Hwang, J.-N. (eds.) Handbook of Neural Network Signal Processing. CRC Press, Boca Raton (2001)
29. Breiman, L.: Bagging predictors. Mach. Learn. **24**, 123–140 (1996)
30. Cortes, C., Vapnik, V.: Support-vector networks. Mach. Learn. **20**, 273–297 (1995)
31. World Health Organization: Prevention of Cardiovascular Disease: Guidelines for Assessment and Management of Total Cardiovascular Risk. World Health Organization, Geneva (2007)

Assessing the Effectiveness of Digital Advertising for Green Products: A Facial Expression Evaluation Approach

Chang Yueh Wang(✉) and Fang Suey Lin

Graduate School of Design, National Yunlin University of Science and Technology, Yunlin, Taiwan (R.O.C.)
d10930015@gemail.yuntech.edu.tw, linfs@yntech.edu.tw

Abstract. Effectiveness of advertisement can be measured in terms of a person's emotional response to the advertisement media. Besides traditional survey method, the emotional feedback of the consumers is valuable to understand the purchasing intention. Artificial intelligence and machine learning have provided researchers and practitioners of marketing and advertisement with new tools, such as facial expression recognition analysis, to explore the context of advertisement effectiveness. In this study, a facial expression recognition survey was carried out for analyzing the advertising impact and consumer feedback on types of advertisement. Participants were separated into two groups, where the first group was asked question related to one randomly chosen digital poster advertisements of green products and the second group was surveyed for an advertisement without sustainable properties. The facial expressions of participants were recorded and later classified into three states as positive, neutral, and negative using a machine learning model. The results reveal people show relatively higher positive emotion for green products and their purchasing intentions are driven by the willingness to save environment and perceived value. Whereas, for normal products, purchasing intentions are mainly driven by the brand image. This study also shows that sustainable cues in product advertisement leads to positive consumer feedback.

Keywords: Advertisement · Emotional feedback · Facial expression · Green product

1 Introduction

Green products are gaining popularity among the consumers, due to their environment friendly properties. In recent years, because of increasing levels of atmospheric carbon dioxide and global warming [1], many measures have been taken by government, including raising awareness of green products. Therefore, positive attitude towards the green products is on rise among consumers. Several manufacturers have shifted their attention towards making green and environment friendly products in recent years. These efforts are also visible in the way products are advertised. To attract customers towards

F. Fui-Hoon Nah and K. Siau (Eds.): HCII 2022, LNCS 13327, pp. 221–232, 2022.
https://doi.org/10.1007/978-3-031-05544-7_17

products, manufacturers and advertisement firms are focusing on advertisement techniques that convey the message that the product is indeed environment friendly, green and sustainable.

The key intermediate advertising effects can be studied with factors like affect, cognition and experience [2]. The studies related to green products suggest that the consumer intention to purchase the product is affected by consumer attitude towards the product [3]. The knowledge of green brand and attitude towards green brand increases the green/sustainable product purchase intention [4]. The awareness about green product is one of the driving factors for the consumers to be attracted towards the green and sustainable products. The health and safety values of the sustainable products also have a positive effect on consumers' green product purchase behavior [5]. Also, the role of government and media is significantly important for developing the understanding and knowledge about sustainable and green products among consumers. Determinants for green/sustainable purchase include individual factors, products attributes and marketing together with social influence [6]. Besides the availability of several researches and studies on green products and consumer behavior, most of the studies are concentrated in big countries like Mainland China, India, Australia and Europe but very few papers focus on countries with relatively smaller population with similar quality of life.

In this paper, a study is conducted to analyze the effectiveness of digital advertisement of green products using traditional Likert scale survey method combined with facial expression feedback. A sample size of 53 participants from different age group and professional background is used to analyze the effect of visual appearance, perceived value, product information and brand image on consumers purchasing intention. The participants are divided into two groups and given two separate surveys consisting of questions related to one green product and one general product. The results of the study reveal key factors that have a major impact on consumer's purchasing intention and digital advertisement effectiveness in Taiwan's demography.

2 Literature Review

In terms of advertising, the presentation of the product is critical for making an impression among consumers. Studies have suggested that attractive things make a long-lasting impact on a person's emotional and behavioral feedback towards it [7, 8]. People perceive attractive things to work better regardless of the reality. However, the consumer attitude towards the product is not only related to the attractiveness, brand image also plays an important part. A study in Taiwan suggests brand image has a positive influence on the purchasing intention of the consumers [9]. Similarly, brand communication, brand personality and brand identity affect the purchasing intention of the consumer [10]. However, for green products; quality, sense of caring and uniqueness are also key influential factors in purchasing intention [11]. The purchasing intention of green or sustainable products varies according to the consumer's awareness of environmental issues. As consumers involved with environment have higher purchasing intention of green products regardless of green appeal of the products while a green appeal of the product is important for the consumers from low-involvement group [12]. A more recent study

[13] suggests that consumers emotions also have a moderating effect on green product advertising. Also, the consumer's attitude and values towards green products acts as a moderating variable as revealed in [14]. Therefore, consumer awareness, consumer's emotions, consumer's attitude, product's appeal, sense of caring, quality and uniqueness are key advertising factors.

While measuring the effectiveness of advertisement, many methodologies are applied, which includes traditional methods such as survey, in-depth interviews, and focus groups. Many recent studies also use facial expressions, besides surveys and in-depth interviews to analyze the consumer behavior. By evaluating the facial expressions, the unconscious behavior [15] of the consumers towards the product, design or simulant can be observed. [16] applies emotion recognition for observing the emotional communication impact on consumers. In [17], emotional responses are analyzed to find the compatibility between quantitative and figurative responses. Similarly, in [18], facial expressions are analyzed to recognize the emotional feedback of consumers for different graphic styles. As suggested in [19], facial feedback has a major role in emotional processing and people are more likely to use facial expressions as a means of nonverbal communication, which may reflect their unconscious and conscious emotions. Facial expressions influence the positive, negative and other discrete emotions, but not the intense emotions like fear and surprise [20]. Although, the reason for such emotional effects is not clear, but it has great practical implication in the area of advertising. A study conducted in [21] for evaluating the emotional impact and influence of advertisements by evaluating the facial expression in a controlled study at laboratory and living room settings, suggests that explicit evaluations using self-reports are not always clear for determining the effectiveness of an advertisement, however using facial expression analysis, this process can be made easier. In [22], another evaluation for effect of facial expression in print advertising shows that model's facial expression has an influence on the product evaluation. In [23], emotional effectiveness of advertisement is also evaluated using facial expression. A framework for facial expression analysis is presented in [24] and its use for evaluating the advertisement effectiveness. For practical implementations, [25] suggests that signal processing based facial expression recognitions are suitable for applications using statistical methods and the top 10% peak values can be taken as the emotional reaction of participants.

3 Research Hypothesis

Several studies related to green product and consumer behavior suggest that consumer's attitude towards the product has a high influence on purchasing intention. This is usually related to the knowledge of green products and desire to be more environment friendly. Therefore, drawing from the evidences reported in [3–6], the strong relationship between product knowledge and purchasing intention for sustainable and green products can be hypothesized. Product knowledge and environmental-friendly values together can have an effect on the likeability of the green product advertisement. Therefore, first hypothesis for this study is suggested as follows:

H1: Environmental-friendly values increase the effectiveness of advertisement for green products.

While environmental-friendly values play an important role in purchasing intention for green products. Visual appearance and brand image are also key factors for effectiveness of an advertisement. Visual appearance and aesthetics of product [7, 8] and its presentation in advertisement media usually increase the effectiveness of overall advertising. As people perceive that the attractive things are more functional and they work better. Therefore, second hypothesis for advertisement media can be stated as follows:

H2: Visual appearance and brand image have strong contribution towards effectiveness of green product advertisement.

Using H1 and H2 as our base for the advertisement effectiveness model for sustainable products, we can present our model as shown in Fig. 1. Purchasing intention is used as the measure of advertisement effectiveness in this case. Where, the visual appearance, perceived value and brand image has a direct impact on advertisement effectiveness and information (product knowledge) and facial expression or emotional reaction has indirect impact.

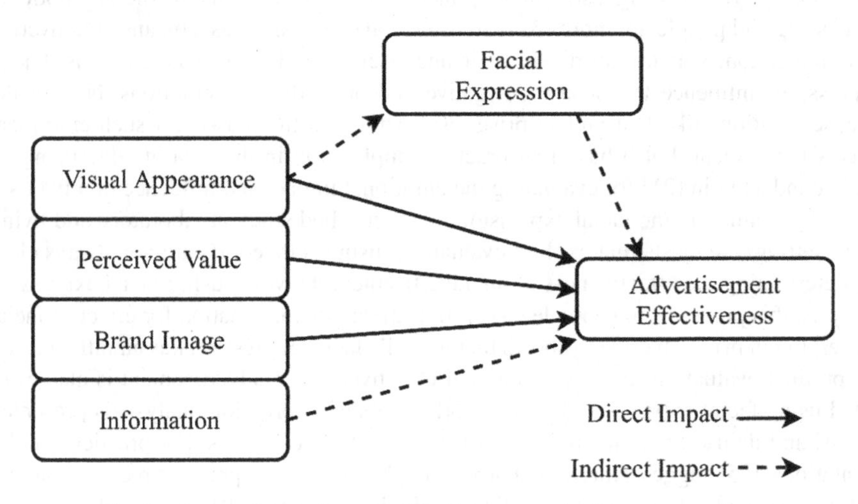

Fig. 1. Proposed advertisement effectiveness model for green products.

As many studies suggest that the facial expressions can evaluate the unconscious emotional feedback of consumers towards a product and facial expression recognition has been widely used in evaluating the effectiveness of advertisement. By evaluating the results and findings of [18, 19, 23], it can be assumed that effectiveness of an advertisement is directly correlated to its ability to invoke positive emotions in the consumers. In order to influence the purchasing intention of consumers for green products, the product advertisement must be able to positively communicate with consumers on an emotional level. This can be hypothesized as follows:

H3: Positive emotion for an advertisement corresponds to higher purchasing intention.

4 Methodology

4.1 Study Design

An open-source implementation of "frame-attention networks for facial expression recognition in videos" [26] was used for evaluating the facial expressions of respondents and top 10% peak emotions were considered as the emotional feedback for the advertisement poster. Top 10% peak emotions in this case are calculated with respect to the total interaction time of the participants with the interview images and questions. In this study, we categorized the emotions into three categories as positive (happiness and surprise), neutral (absence of facial expression) and negative (disgust, contempt, anger, fear and sadness). These negative, neutral and positive emotions feedbacks are encoded as 0, 1 and 2 numerical values for statistical analysis.

4.2 Participants

Participants for this study were recruited via online survey through an online survey platform (https://www.google.com/forms) and the personal identities were kept anonymous in the questionnaire.[1,2] The participants of this study comprise 22 male and 31 female respondents from Taiwan. 40% of the respondents in first group (Group 1) were between age of 20 to 30 years old while in second group (Group 2) 32.1% were in 21 to

Table 1. Description of characteristics of Group 1 & Group 2.

Characteristic		Group 1 N	(%)	Group 2 N	(%)
Gender	Male	9	36	13	46.4
	Female	16	64	15	53.6
Age	Under 20	0	0	1	3.6
	21–30	10	40	9	32.1
	31–40	3	12	9	32.1
	41–50	8	32	2	7.1
	51–60	4	16	6	21.4
	60 up	0	0	1	3.6
Occupational category	Sales	0	0	5	17.9
	Student	12	48	4	14.3
	Manufacturing engineering	3	12	5	17.9
	Management	3	12	8	28.6
	Education	6	24	4	14.3
	Design & arts	1	4	4	14.3

[1] Online survey for group 1 can be found at https://forms.gle/GTFe6WPHuqaNbK5m6.

[2] Online survey for group 2 can be found at https://forms.gle/3sMSeFFbuRLEevCG7.

30 years age group and other 32.1% in 31–40 years age group, the demographic of the participants is listed in Table 1.

4.3 Procedures

Respondents were made aware of the nature of survey and that their facial expression will be recorded during the session they fill in the survey information. For an unbiased study, we divided the respondents into two groups, first group (N = 25) was shown two sustainable product advertisements and second control group (N = 28) was shown the advertisement for a general product with no sustainable attribute in the advertisement media.

In this study, both groups were asked questions related to visual appearance, perceived value, purchasing intention, information and brand image. The questions were answered in Likert scales ranging from 1 to 5 where 1 corresponds to strongly disagree and 5 corresponds to strongly agree. Each correspondent's facial expression was recorded using a digital camera while filling the survey and when the advertisement poster was shown to them.

5 Results and Discussion

For the analysis of survey results, the Likert scale is encoded and transferred to IBM's SPSS statistical analysis software suite. The emotional feedback was also encoded numerically and arranged together with the factors included in the survey for analysis. We used the aforementioned software for all the statistical and regression analyses presented in this study. The descriptive and reliability statistics for attributes in questionnaire for the Group 1 and Group 2 are is shown in Table 3 and Table 4. Due to multivariate nature of our data, Cronbach's Alpha was used to test the reliability of the data as described in [27]. The standard deviation of each construct was over recommended value of 0.5 and therefore convergent validity can be confirmed. Cronbach's Alpha for each factor was higher than 0.7 for both groups with overall Cronbach Alpha of 0.797 for Group 1 and 0.876 for Group 2, therefore we can confirm the reliability of factor measurement. For both group's dataset, all item-total correlations were over 0.4, showing good internal consistency (Table 2).

Table 2. Descriptive statistics and reliability test of items for Group 1

	Mean	Std. deviation	Item-total correlation	Cronbach's alpha
Visual appearance	3.600	.8660	.743	.722
Perceived value	3.640	.9074	.738	.721
Purchasing intention	3.640	.9074	.570	.761
Brand image	3.160	1.1431	.728	.718
Information	3.360	.9522	.187	.847
Facial expression	1.360	.7000	.426	.792

Table 3. Descriptive statistics and reliability test of items for Group 2.

	Mean	Std. deviation	Item-total correlation	Cronbach's alpha
Visual appearance	3.714	1.1174	.594	.872
Perceived value	3.571	.9974	.725	.848
Purchasing intention	3.357	1.0261	.681	.855
Brand value	3.536	1.0357	.715	.849
Information	3.071	1.1198	.757	.841
Facial expression	1.286	.7629	.649	.864

Correlation Analysis between variables for experiments conducted for Group 1 and Group 2 are listed in Table 5 and Table 6. There is no correlation ($r = 0.6$, $p < 0.776$) of information/product knowledge with purchasing intention in results of Group 1, however, for Group 2 this correlation is relatively higher ($r = 0.557$, $p < 0.05$). For both groups, perceived value is correlated with visual appearances. For Group 1 ($r = 0.604$, $p < 0.01$) and Group 2 ($r = 0.517$, $p < 0.05$), which is compatible with the study in [7, 8]. Both perceived value and facial expression shows significant correlation with purchasing intention for Group 1, which validates hypotheses H1 and H3. Brand image however shows higher correlation with purchasing intention for Group 1 than Group 2. H2 is only partially verified as visual appearance shows significant correlation ($r = 0.445$, $p < 0.05$) with purchasing intention in Group 1, however brand image has no significant correlation. Therefore, for both sustainable and general advertising, visual appearance, perceived value and positive facial expression (emotion) show significant correlations with purchasing intention. It is worth noting that in both groups the brand image has significant correlation with perceived value.

Table 4. Correlation between different variables for Group 1.

	Visual appearance	Perceived value	Purchasing intention	Brand image	Information	Facial expression
Visual appearance	1	$.604^{**}$	$.445^{*}$	$.783^{**}$	0.333	0.316
Perceived value	$.604^{**}$	1	$.696^{**}$	$.741^{**}$	−0.037	$.541^{**}$
Purchasing intention	$.445^{*}$	$.696^{**}$	1	0.379	0.06	$.541^{**}$
Brand image	$.783^{**}$	$.741^{**}$	0.379	1	0.328	0.237
Information	0.333	−0.037	0.06	0.328	1	−0.015
Facial expression	0.316	$.541^{**}$	$.541^{**}$	0.237	−0.015	1

** Correlation is significant at the 0.01 level
* Correlation is significant at the 0.05 level

Table 5. Correlations between different variables for Group 2

	Visual appearance	Perceived value	Purchasing intention	Brand image	Information	Facial expression
Visual appearance	1	.517**	.383*	.393*	.727**	.360
Perceived value	.517**	1	.734**	.589**	.526**	.508**
Purchasing intention	.383*	.734**	1	.580**	.557**	.480**
Brand image	.393*	.589**	.580**	1	.604**	.737**
Information	.727**	.526**	.557**	.604**	1	.539**
Facial expression	.360	.508**	.480**	.737**	.539**	1

** Correlation is significant at the 0.01 level
* Correlation is significant at the 0.05 level

Factor analysis using principal component analysis (PCA) of Group 1 reveals two main factors affecting the advertisement effectiveness, as shown in Table 7. It was important to perform factor analysis in order to further evaluate the proposed hypotheses, since the correlation matrix results were not conflicting. First factor comprises visual appearance, perceived value and brand image, while the second factor consisting the information/product knowledge. PCA of group 2 doesn't reveal over one dominant factor, therefore not included in this study.

Table 6. Factor analysis using principal component analysis for Group 1.

Variables	Component	
	1	2
Visual appearance	.899	.060
Perceived value	.813	−.463
Brand image	.947	−.024
Information	.383	.902

In Table 7 and Table 8, a linear regression model fit is presented for Group 1 and Group 2. The model considers the advertisement effectiveness model (purchasing intention) shown in Fig. 1. From Table 7 and Table 8, while considering visual appearance, perceived value, brand image, information and facial expressions as independent variables and purchasing intention (advertisement effectiveness) as dependent variable, models for both groups show a R-Squared fitness value of 0.61. However, for Group 1, perceived value of the product shows a high standardized coefficient of 0.971, followed

Table 7. Linear regression model for variables related to purchasing intention for Group 1

Model	Unstandardized coefficients		Standardized coefficients	R-squared
	B	*Std. Error*	*Beta*	
(Constant)	−.117	.880		0.610
Visual appearance	.260	.250	.248	
Perceived value	.971	.290	.971	
Brand image	−.498	.252	−.627	
Information	.209	.166	.219	
Facial expression	.116	.240	.090	

Dependent Variable: Purchasing Intention

Table 8. Linear regression model for variables related to purchasing intention for Group 2.

Model	Unstandardized coefficients		Standardized coefficients	R-squared
	B	*Std. Error*	*Beta*	
(Constant)	.529	.636		0.610
Visual appearance	−.194	.187	−.212	
Perceived value	.628	.185	.610	
Brand image	.117	.220	.118	
Information	.301	.208	.328	
Facial expression	−.024	.270	−.018	

Dependent Variable: Purchasing Intention

by visual appearance (0.248) and information (0.219). This suggests the dependency of purchase intention for sustainable product advertisement is highest on perceived value, followed by visual appearance and information. Brand image has a negative standardized co-efficient of −0.627, which shows the brand value doesn't positively affect the purchasing intention for a sustainable product. Facial expression, however low, but has a positive co-efficient of 0.09, therefore can be concluded as a driving factor for sustainable product advertisement. Hence, we can partially validate our advertisement effectiveness model for sustainable products shown in Fig. 1. For Group 2, perceived value, information and brand image are the positively affecting variables, as shown in Table 8. However, visual appearance and facial expression do not show any positive effects. Therefore, for general product advertisement, two key factors can be concluded as perceived value, and brand image. From Group 1 regression result, perceived value variable has higher standardized coefficient, which supports our hypothesis H1. Also, earlier from the coefficient results in Table 8, the evidence to support H1 is clear. Therefore, conclusion for hypothesis H1 can be drawn definitively. Using correlation and factor analysis results we modified our initially proposed model as shown in Fig. 2.

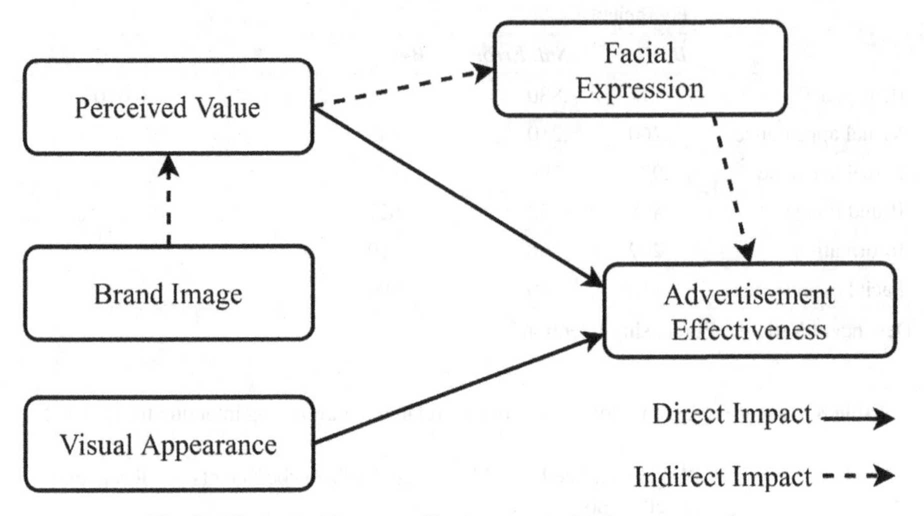

Fig. 2. Final advertisement effectiveness model for green products

After analysis, visual appearance and perceived value indeed has direct impact on advertisement effective and purchasing intention. We also observed that perceived value has an indirect effect on facial expression of the respondents, as a high correlation between two was observed. Brand image appears to moderately affect the perceived value of the green products. However, brand image and information of the green product advertisement overall have no effect on advertisement effectiveness. The total effect of facial expression is not clear however, the significant correlation can be classified as indirect effect. Therefore, the final model includes two factors with direct impact and one factor as indirect impact.

6 Conclusion

The findings of this study suggest that perceived value of the sustainable product strongly affects the consumers' purchasing intention. While visual appearance of green product advertisement is a significant factor, the same does not seem to positively affect the general product advertisement. Through our study, we can conclude that the variables directly affecting the advertisement effectiveness of green products are perceived value and visual appearance and facial expression also contributes indirectly. However, for general product advertisement, we only observed perceived value, information and brand image to be significant factors. This study proves the hypothesis H1 and H3 formulated in Sect. 2 and an advertisement effectiveness model for green products is proposed. Hypothesis H2 is only partially verified as brand image has no direct or indirect effect on purchasing intention of green products. Overall, study reveals that Taiwan's consumers are highly motivated by the visual appearance and perceived value when considering a green product. Facial expression is concluded to affect the advertisement efficiency indirectly, hence using facial expression during survey is advised for getting more in-depth understanding of consumer behavior.

The limitation of this study includes the relatively smaller sample size used for survey and facial analysis. The future work of this study includes increasing the sample size and developing a predictive model of effectiveness of green product advertisement based on facial expression analysis and other variables discussed in this study.

References

1. World Meteorological Organization: WMO statement on the status of the global climate in 2019 (2019)
2. Vakratsas, D., Ambler, T.: How advertising works: what do we really know? J. Mark. **63**, 26 (1999). https://doi.org/10.2307/1251999
3. Ayoun, S., Ben Cheikh, A., Abdellatif, T., Ghallab, N.: Purchase intention of green product: an approach based customer orientation. SSRN Electron. J. (2015). https://doi.org/10.2139/ssrn.2583517
4. Mohd Suki, N.: Green product purchase intention: impact of green brands, attitude, and knowledge. Br. Food J. **118**, 2893–2910 (2016). https://doi.org/10.1108/BFJ-06-2016-0295
5. Jan, I.U., Ji, S., Yeo, C.: Values and green product purchase behavior: the moderating effects of the role of government and media exposure. Sustainability **11**, 6642 (2019). https://doi.org/10.3390/su11236642
6. Zhang, X., Dong, F.: Why do consumers make green purchase decisions? Insights from a systematic review. Int. J. Environ. Res. Public Health. **17**, 1–25 (2020). https://doi.org/10.3390/ijerph17186607
7. Norman, D.: Emotion & design: attractive things work better. Interactions **9**, 36–42 (2002). https://doi.org/10.1145/543434.543435
8. Chawda, B., Craft, B., Cairns, P., Heesch, D., Rüger, S.: Do "attractive things work better"? An exploration of search tool visualisations. In: 19th British HCI Group Annual Conference: The Bigger Picture, Edinburgh, UK, pp. 46–51 (2005)
9. Lin, C.T.L., Chuang, S.S.: The importance of brand image on consumer purchase attitude: a case study of e-commerce in Taiwan. Stud. Bus. Econ. **13**, 91–104 (2018). https://doi.org/10.2478/sbe-2018-0037
10. Mao, Y., et al.: Apple or Huawei: understanding flow, brand image, brand identity, brand personality and purchase intention of smartphone. Sustainability **12**, 1–22 (2020). https://doi.org/10.3390/SU12083391
11. Song, S.Y., Kim, Y.K.: A human-centered approach to green apparel advertising: decision tree predictive modeling of consumer choice. Sustainability **10** (2018). https://doi.org/10.3390/su10103688
12. Schuhwerk, M.E., Lefkoff-Hagius, R.: Green or non-green? Does type of appeal matter when advertising a green product? J. Advert. **24**, 45–54 (1995). https://doi.org/10.1080/00913367.1995.10673475
13. Kao, T.F., Du, Y.Z.: A study on the influence of green advertising design and environmental emotion on advertising effect. J. Clean. Prod. **242**, 118294 (2020). https://doi.org/10.1016/j.jclepro.2019.118294
14. Liao, Y.K., Wu, W.Y., Pham, T.T.: Examining the moderating effects of green marketing and green psychological benefits on customers' green attitude, value and purchase intention. Sustainability **12** (2020). https://doi.org/10.3390/SU12187461

15. Kaiser, J., Davey, G.C.L., Parkhouse, T., Meeres, J., Scott, R.B.: Emotional facial activation induced by unconsciously perceived dynamic facial expressions. Int. J. Psychophysiol. **110**, 207–211 (2016). https://doi.org/10.1016/j.ijpsycho.2016.07.504
16. Yu, C.-Y., Ko, C.-H.: Applying emotion recognition to graphic design research. In: Fukuda, S. (ed.) Emotional Engineering, Vol. 7, pp. 77–95. Springer, Cham (2019). https://doi.org/10.1007/978-3-030-02209-9_6
17. Bao, Q., Burnell, E., Hughes, A.M., Yang, M.C.: Investigating user emotional responses to eco-feedback designs. J. Mech. Des. Trans. ASME **141** (2019). https://doi.org/10.1115/1.4042007
18. Yu, C.Y., Ko, C.H.: Applying FaceReader to recognize consumer emotions in graphic styles. Procedia CIRP. **60**, 104–109 (2017). https://doi.org/10.1016/j.procir.2017.01.014
19. Buck, R.: Nonverbal behavior and the theory of emotion: the facial feedback hypothesis. J. Pers. Soc. Psychol. **38**, 811–824 (1980). https://doi.org/10.1037/0022-3514.38.5.811
20. Coles, N.A., Larsen, J.T., Lench, H.C.: A meta-analysis of the facial feedback literature: effects of facial feedback on emotional experience are small and variable. Psychol. Bull. **145**, 610–651 (2019). https://doi.org/10.1037/bul0000194
21. Ausin, J.M., Guixeres, J., Bigné, E., Alcañiz, M.: Facial expressions to evaluate advertising: a laboratory versus living room study. In: Zabkar, V., Eisend, M. (eds.) Advances in Advertising Research VIII. European Advertising Academy. Springer, Cham (2017). https://doi.org/10.1007/978-3-658-18731-6_9
22. Isabella, G., Vieira, V.A.: The effect of facial expression on emotional contagion and product evaluation in print advertising. RAUSP Manag. J. **55**, 375–391 (2020). https://doi.org/10.1108/RAUSP-03-2019-0038
23. Otamendi, F.J., Sutil Martín, D.L.: The emotional effectiveness of advertisement. Front. Psychol. **11** (2020). https://doi.org/10.3389/fpsyg.2020.02088
24. Hamelin, N., Moujahid, O.E., Thaichon, P.: Emotion and advertising effectiveness: a novel facial expression analysis approach. J. Retail. Consum. Serv. **36**, 103–111 (2017). https://doi.org/10.1016/j.jretconser.2017.01.001
25. Lewinski, P., Fransen, M.L., Tan, E.S.H.: Predicting advertising effectiveness by facial expressions in response to amusing persuasive stimuli. J. Neurosci. Psychol. Econ. **7**, 1–14 (2014). https://doi.org/10.1037/npe0000012
26. Meng, D., Peng, X., Wang, K., Qiao, Y.: Frame attention networks for facial expression recognition in videos. In: 2019 IEEE International Conference on Image Processing (ICIP), Taipei, Taiwan, pp. 3866–3870. IEEE (2019)
27. Hair, J.F., Black, W.C., Babin, B.J., Anderson, R.E.: Multivariate Data Analysis. Pearson Education, New York (2009)

Predicting Hospital Admission by Adding Chief Complaints Using Machine Learning Approach

I-Chin Wu[1(✉)], Chu-En Chen[2], Zhi-Rou Lin[1], Tzu-Li Chen[3], and Yen-Yi Feng[4]

[1] Graduate Institute of Library and Information Studies, School of Learning Informatics, National Taiwan Normal University, Taipei, Taiwan
icwu@ntnu.edu.tw

[2] Fu-Jen Catholic University, New Taipei City, Taiwan

[3] National Taipei University of Technology, Taipei, Taiwan

[4] Emergency Medicine, Mackay Memorial Hospital, Taipei, Taiwan

Abstract. Overcrowded conditions in emergency departments (EDs) have increased patients' waiting time, while the variety of patient afflictions has caused difficulties in the allocation of medical resources. Therefore, the ability to predict a patient's hospital admission at the time of triage could allocate medical resources to patients who go to EDs in urgent need of immediate care. Using a dataset from the MacKay Memorial Hospital in Taipei (Taiwan), which contains 177,038 valid records collected from 2009 to 2010 in this research, we aim to have on hand chief complaints (CCs), demographic data, administration information and clinical information at the triage stage to predict the probability of a patient's hospital admission. Firstly, we select terms from the CCs to predict which patients may require eventual hospitalization. We then integrate the selected terms with several algorithms to predict the probability of patient admissions. Accordingly, this research includes a series of machine learning processes, such as data preprocessing for structure data and CC data, imbalanced data processing, models construction by logic regression, neural networks, random forest, XGBoost, and model evaluation. The research results show that the ensemble learning approach, XGBoost, can achieve 0.88, and 0.76 in terms of accuracy and AUC respectively. The results show that triage, fever status, age, and terms extracted from the CCs are important attributes to predict if patients should be hospitalized. The results of this study will provide a reference approach in the field of emergency hospital admissions prediction and help hospitals improve resource allocation in emergency rooms.

Keywords: Chief complaint · Prediction of hospital admission · Triage · XGBoost

F. Fui-Hoon Nah and K. Siau (Eds.): HCII 2022, LNCS 13327, pp. 233–244, 2022.
https://doi.org/10.1007/978-3-031-05544-7_18

1 Introduction

The emergency department has become the most important and busiest unit in most hospitals. Since the initiation of National Health Insurance (NHI) in Taiwan in 1995, outpatient numbers and medical expenses have grown quickly. In recent years, the demand for emergency medical care has increased year by year. According to statistics from the Health Department of the Executive Yuan, the number of emergency medical cases per year in Taiwan increased from 6,349,211 in 2009 to 6,783,280 in 2019. That is, over six million people have gone to EDs per year for the past ten years. In comparison, there were only 4,955,165 ED visits in 1999. The increasing numbers of patients are caused by ED overcrowding, which makes it impossible to properly allocate medical resources and results in staff being unable to provide perfect medical service for patients in a timely manner. Instead, patients must wait for a long time, which leads to a decline in medical outcome quality. Therefore, to solve the problem of emergency department overcrowding, this study will establish a demand prediction framework for EDs through data mining and machine learning techniques.

ED overcrowding is associated with increased length of stay and costs for admitted patients and decreased patient satisfaction (Carter et al. 2014; Fernandes et al. 2020; Guttmann et al. 2011; Parker et al. 2019). The growing demand for ED resources results in prolonged waiting times for the medical treatment of patients. The priority of ED treatment is based on patients' acuity, which is evaluated at the triage stage in the ED. Triage is basically determined by physicians and ED nurses based on the severity of the patient injury or illness. Thus, triage methods include overtriage and undertriage and require more number of reliable indicator to help nurses' make judgement (Ashour and Okudan Kremer 2016; Chonde et al. 2013; Moll 2010). As such, the triage level is can be only one of the initial indicators to analyze the relationship of patient behavior with ED overcrowding, and we do not use triage information to subjectively investigate patient behavior in the ED. Basically, attending physicians and nurses judge the severity of patient conditions and then determine the priority of patient treatments. Some indicators, for instance the subjective pain score for determining triage levels, are based on the experience of the attending physicians and nurses. Thus, the patient acuity state can not solely have determined based on the triage system. Although several models and applications have been presented based on the triage system developed by each country, there is still growing demand for timely and precise emergency services to assist in prioritizing patients, predicting hospital admission at discharge from the ED, predicting mortality, etc. (Araz et al. 2019; Fernandes et al. 2020; Raita et al. 2019).

In this research, besides information of triage level of patients based on the Taiwan Triage and Acuity Scale (TTAS), we aim to collect demographic and administrative data, chief complaints (CCs), and limited clinical data available at the triage stage to predict the possibility of hospital admission at the time the patient's arrival to the ED. The pilot study of Parker et al. (2019) adopted a logic regression model on the dataset of 1,382,896 visits over ten years, collected from the ED of Singapore General Hospital (SGH) to predict the possibility of hospital admission. The research finally used eight variables: age group, triage category, race, postal code, day of the week, time of day, fever status, and mode of arrival to achieve 0.825 AUC (area under the curve). Among eight variables, increasing triage category, increasing age, and fever symptoms are the most significant.

The research addressed notable findings to help us select suitable variables which can be collected at the triage stage. We found that few researchers used textual data of CCs with the structured data to construct the hospital admission prediction model. Accordingly, we aim to integrate CCs with demographic and administrative data and clinical data available at the triage stage to predict the possibility of hospital admission when the patient arrives at the ED.

In summary, this study proposes and constructs a data science framework for predicting emergency hospital admissions based on a series of data mining processes, such as data preprocessing, imbalanced data processing (Fernández Hilario et al. 2018), learning algorithms, hyperparameter optimization, model evaluation and feature ranking. Accordingly, we seek to explore the following main research questions.

(1) *What are important variables at the triage stage to predict whether the patient will be hospitalized?*
(2) *Which model with associated parameters can achieve better prediction results for future research reference?*

2 The Research Process

This study obtained 177,038 valid medical records related to the emergency room from the health information system of Mackay Memorial Hospital from 2009 to 2010. This study was approved by Mackay Memorial Hospital's Institutional Review Board (IRB) for protection of patients' privacy and data. The demographic and administrative attributes include age, mode of arrival, day of the week, and arrival time. The clinical data includes triage level based on TTAS and temperature (fever status). The statistical data of the target database are shown in Table 1. Since patients are admitted to the hospital (admission, in short) and later their discharges are unevenly distributed, we need to conduct an imbalanced data processing step for avoiding predicting in favor of the target label that has the greater proportion, i.e., discharge from the hospital. In our dataset the admitted-to-discharged ratio is 15% to 85%. We adopted random undersampling (RUS), random oversampling (ROS), the synthetic minority oversampling technique (SMOTE), and the adaptive synthetic (ADASYN) method to evaluate the effectiveness of the various methods.

The prediction models are established by using Logistic Regression, Deep Neural Networks, Random Forest, and eXtreme Gradient Boosting (XGBoost). Because no single metric can reflect the results of our evaluation models – especially for the imbalanced data condition – the models in this research are evaluated by multiple evaluation metrics. Several research studies in this area adopted the Accuracy, G-mean, AUC, true negative rate (TNR), true positive rate (TPR), and positive prediction value (PPV) metrics to measure the prediction performance of the approaches (Artetxe et al. 2018; Fernandes et al. 2020). In our research context, the admitted-to-discharged ratio is 15% to 85%, which is an imbalanced dataset. Thus, the metric Accuracy will have a higher probability of predicting the patients who will be discharged. Accordingly, we adopted Accuracy, area under the curve (AUC), G-mean, true positive rate (TPR, also called sensitivity) and true negative rate (TNR, also called specificity) to measure the effectiveness of the hospital admission results. The AUC value is the area under the ROC curve and is a common metric to show the prediction results in a medical context. The higher the AUC, the better the performance of the model at distinguishing between the positive and negative samples. An excellent model has an AUC near 1, which means it is a good measure of distinguishing between two extreme target attributes.

Figure 1 illustrates our overall evaluation process. Basically, we separated the data into training and testing data, which include structured attributes and unstructured features extracted from the chief complaints (CCs) in our research. We conducted imbalanced data preprocessing and then tested four prediction models to find the model with the best imbalanced data processing handling ways to predict if the ED visitors should be hospitalized. The features will be ranked by each model. Then, we carried out the same process on the testing data to evaluate the models mainly in terms of Accuracy, AUC, TPR, and TNR.

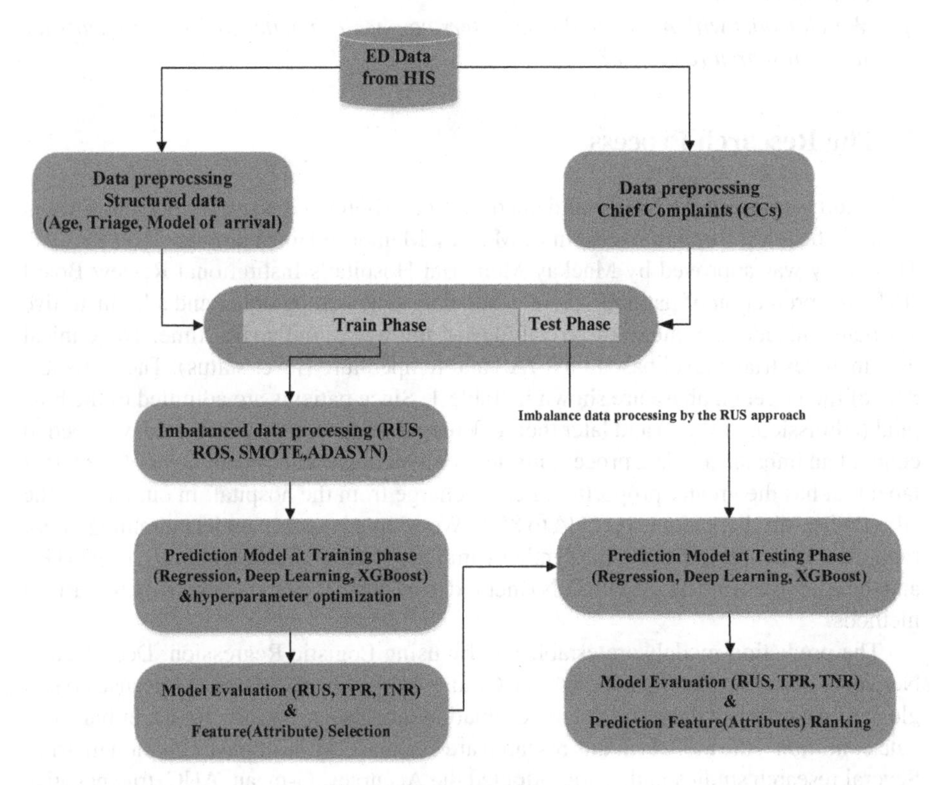

Fig. 1. The evaluation process

Table 1. Basic individual and treatment attributes with descriptive statistical data of ED patients (2009 and 2010 statistic data of Mackay Memorial Hospital)

Attributes	Values	2009		2010	
		# of records	Percentage	# of records	Percentage
Age	0–10	3,096	3.55%	3060	3.41%
	10–20	4,862	5.57%	5183	5.78%
	20–30	**19,979**	**22.88%**	**20106**	**22.41%**
	30–40	15,334	17.56%	15737	17.54%
	40–50	11,498	13.17%	11436	12.74%
	50–60	11,924	13.66%	12244	13.65%
	60–70	7,706	8.83%	8238	9.18%
	70–80	7,678	8.79%	7956	8.87%
	80–90	4,518	5.17%	4963	5.53%
	90–100	690	0.79%	791	0.88%
	Above 100	21	0.02%	18	0.02%
Mode of arrival	Referral from nursing homes	125	0.14%	96	0.11%
	Walk in/Come	18,291	20.95%	19763	22.02%
	Referral from out-patient clinic	1,110	1.27%	1115	1.24%
	Referral from other hospital	2,166	2.48%	2214	2.47%
	Medical referral 119	8,338	9.55%	9818	10.94%
	Escorted	**57,263**	**65.59%**	**56713**	**63.20%**
	Others	13	0.01%	13	0.01%
Arrival day	Weekday	**58,891**	**67.45%**	**61028**	**68.01%**
	Weekend	28,415	32.55%	28704	31.99%
Arrival time	Morning (0–8)	19,318	22.13%	20171	22.48%
	Afternoon (8–16)	30,985	35.49%	32106	35.78%
	Night (16–24)	**37,003**	**42.38%**	**37455**	**41.74%**
Triage	1 level (T1)	2,870	3.29%	1685	1.88%
	2 level (T2)	30,384	34.80%	12568	14.01%
	3 level (T3)	**52,872**	**60.56%**	**50889**	**56.71%**
	4 level (T4)	1,180	1.35%	21407	23.86%
	5 level (T5)	0	0.00%	3183	3.55%
Temperature	Non-fever	77,706	89.00%	79429	88.52%
	Fever	9,600	11.00%	10303	11.48%

(continued)

Table 1. (*continued*)

Attributes	Values	2009		2010	
		# of records	Percentage	# of records	Percentage
Disposition	**Discharge**	**66,681**	**76.38%**	**68133**	**75.93%**
	Death	190	0.22%	198	0.22%
	Non-admission	1,241	1.42%	1558	1.74%
	Against medical advice	5,019	5.75%	5260	5.86%
	Inpatient	13,203	15.12%	13209	14.72%
	Refund and discharged	40	0.05%	405	0.45%
	Surgical operation	301	0.34%	345	0.38%
	Transfer to another hospital	33	0.04%	12	0.01%
	Transfer to the general branch	598	0.68%	612	0.68%

3 The Evaluation Metrics and Experimental Results

3.1 Experiments in the Training Phase

Herein, we first evaluate the effectiveness of considering CCs in the prediction models and then evaluate the effectiveness of imbalanced data preprocessing methods. We show the evaluation results during the training phase and provide the discussions.

1. Figure 2 shows that if the CCs are to be considered in each model, it will increase the AUC value. The XGBoost will increase 6.44% in AUC value compared to the model without including the CCs. The results show that the CCs of ED visitors are helpful for the early prediction task at the stage of the triage and may streamline the following resource allocation process in the hospital.
2. We then adopted the XGBoost to conduct further experiments. We evaluated different types of data-level imbalanced data preprocessing methods – RUS, ROS, SMOTE, and ADASYN – under different ratios of data sampling conditions. Twenty sets of experimental combinations were tested and evaluated in terms of AUC. Table 2 shows that the RUS method of using an equal sampling ratio of hospitalized and discharged from the hospital can achieve the best AUC value, i.e., 0.762 on average. The sampling ratio is calculated by minority cases of the category (admitted inpatient) divided by majority cases of the category (discharged).
3. Figure 3 shows the AUC of the number of removing some of the patients' demographic and administrative attributes or keywords in the CCs by running the XGBoost model. Apparently, if more attributes are removed, there will be a lower AUC. In addition, the value of AUC will be similar if the number of removed attributes are

within the range of 1 to 31. The top 15 important attributes selected by the XGBoost model are *age*, *temperature*, *arrival time*, *arrival day*, *heartbeat*, *triage*, CC = "January visit," CC = "seek treatment," CC = "Just," mode of arrival = "Referral from," CC = "powerless," CC = "pain," CC = "several days," *fever status*, CC = "tarry stool."

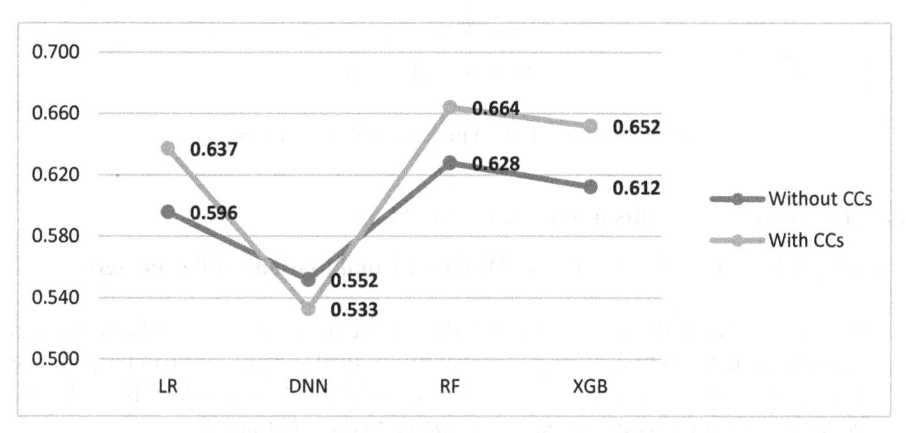

Fig. 2. Prediction models considering the features (terms) in the CCs or not considering the CCs (AUC)

Table 2. AUC values of the model under different combinations of imbalanced data processing methods with different sampling ratios

Method	Sampling ratio	AUC	Sdv	Method	Sampling ratio	AUC	Sdv
RUS	**1**	**0.7623**	**0.0017**	ADASYN	0.8	0.6742	0.0013
ROS	1	0.7590	0.0013	SMOTE	0.8	0.6701	0.0252
RUS	0.8	0.7563	0.0025	SMOTE	0.67	0.6648	0.0021
ROS	0.8	0.7514	0.0018	SMOTE	1	0.6638	0.0032
RUS	0.67	0.7504	0.0042	SMOTE	0.5	0.6630	0.0020
ROS	0.67	0.7448	0.0009	ADASYN	0.67	0.6628	0.0022
RUS	0.5	0.7378	0.0035	ADASYN	1	0.6625	0.0027
ROS	0.5	0.7300	0.0025	ADASYN	0.5	0.6617	0.0021
RUS	0.3	0.7053	0.0013	SMOTE	0.3	0.6614	0.0031
ROS	0.3	0.6982	0.0030	ADASYN	0.3	0.6604	0.0023

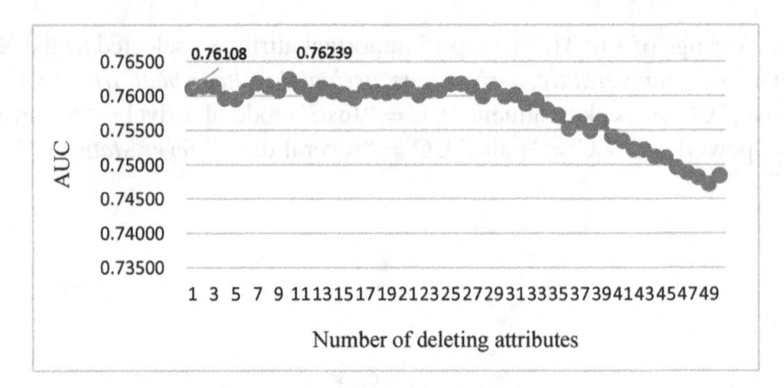

Fig. 3. Feature selection iterations (XGBoost model)

3.2 Experimental Results in the Evaluation Phase

Prediction Results of Models Under Different Combinations of Parameters

1. Overall, similar to the results of the XGBoost, the best prediction performance can be achieved in terms of AUC and G-means by an equal sampling ratio of hospitalized and discharged from the hospital. For the Accuracy, the DNN and XGBoost have a better predication task performance compared to the other models.
2. For the sensitivity (TPR) metric, most of the prediction models achieve the best results under an equal sampling ratio. This shows that if the data are to be balanced, the distribution of the target categories can improve the prediction results. For the specificity (TNR), the data without conducting imbalanced data preprocessing can achieve the best prediction results. It can infer that discharge from the hospital is the majority category of the dataset; thus, the model will favor the majority category (Table 3).

Table 3. Prediction results of models under different combinations of parameters

Metrics	Models	Imbalanced data	Sampling ration	Prediction result
Accuracy	LR	None	0.3	0.8684
	DNN	None	0.3	**0.8715**
	RF	SMOTE	0.3	0.8694
	XGB	ADASYN	0.8	0.8707
AUC	LR	RUS	1	0.7551
	DNN	ROS	0.8	0.7565
	RF	RUS	1	0.7627
	XGB	RUS	1	**0.7630**

(continued)

Table 3. (*continued*)

Metrics	Models	Imbalanced data	Sampling ration	Prediction result
G-mean	LR	RUS	1	0.7546
	DNN	ROS	1	0.7599
	RF	RUS	1	0.7627
	XGB	RUS	1	**0.7629**
Sensitivity (TPR)	LR	ADASYN	1	0.7362
	DNN	ROS	0.8	**0.8338**
	RF	RUS	1	0.7537
	XGB	RUS	1	0.7497
Specificity (TNR)	LR	None	0.3	0.9727
	DNN	None	0.3	0.9627
	RF	None	0.3	**0.9784**
	XGB	None	0.3	0.9679

Prediction Results of Models Under Different Sampling Ratio of the RUS Method. We list all the results of each model in terms of AUC, G-mean, Sensitivity, and Specificity in Table 4. Because the RUS imbalanced data processing method can achieve the best performance for most of the models (as shown in Table 4), we show the results of adjusting the data sampling ratio applying the RUS method. The results show that all of the prediction models can achieve the best AUC, G-mean and Sensitivity values while the sampling ratio is set to 1 to 1. Among the models, XGBoost can achieve the best predication results. When we increase the proportion of majority cases of the category, i.e., discharged category, the value of AUC, G-mean, and Sensitivity will decrease for each prediction model. If we include more non-admission patients' records in the dataset, the model will favor the indicator of specificity. Thus, if derived the model will have a higher specificity at the expense of sensitivity, as sensitivity and specificity are basically inversely proportional to each other.

Table 4. AUC values of the models under different sampling ratio by the RUS data processing method

AUC	**1**	**0.8**	**0.67**	**0.5**	**0.3**
LR	0.7551	0.7479	0.7412	0.7279	0.6838
DNN	0.7556	0.7444	0.7561	0.7259	0.6948
RF	0.7628	0.7577	0.7515	0.7306	0.6847
XGB	0.7630	0.7573	0.7551	0.7354	0.6987

(*continued*)

Table 4. *(continued)*

G-mean	**1**	**0.8**	**0.67**	**0.5**	**0.3**
LR	0.7546	0.7434	0.7315	0.7064	0.6299
DNN	0.7503	0.7558	0.7174	0.7425	0.6459
RF	0.7627	0.7548	0.7425	0.7105	0.6312
XGB	0.7629	0.7544	0.7481	0.7162	0.6549
Sensitivity	**1**	**0.8**	**0.67**	**0.5**	**0.3**
LR	0.7258	0.6664	0.6218	0.5520	0.4176
DNN	0.7494	0.6818	0.6539	0.5667	0.4405
RF	0.7537	0.6907	0.6357	0.5606	0.4194
XGB	0.7497	0.6915	0.6525	0.5685	0.4551
Specificity	**1**	**0.8**	**0.67**	**0.5**	**0.3**
LR	0.7845	0.8293	0.8607	0.9038	0.9501
DNN	0.7480	0.8554	0.8332	0.8751	0.9540
RF	0.7718	0.8247	0.8673	0.9006	0.9501
XGB	0.7762	0.8230	0.8578	0.9024	0.9422

Overall Results and Discussions. The contributions of this research are as follows.

1. The *XGBoost* approach achieved the highest performance in terms of AUC with a rate of 0.763 and specificity metric, equal to a true negative rate (TNR) of 0.968 compared to the other models. It also can achieve 0.750 of the sensitivity metric, i.e., a true positive rate. The high TNR value indicates that the hospital will not make incorrect judgments for which patients are discharged using the *XGBoost* approach.
2. We list the top six attributes ranked by the four models, as shown in Table 5. Apparently, the age, triage, and fever status are the most important variables to predict which patients will be inpatients. The result is similar to research findings by Parker et al. (2019). Furthermore, the heartbeat attribute and model of arrival hospital mentioned "transferred from other places/hospital" are also important factors. For the CCs, three of the four models have included the phrase "tarry stool." At least two key terms of the CCs have been selected as important factors by each model, as shown in Table 5. Handly et al.'s (2015) study has demonstrated that constructing the hospital admission model by adding CCs can improve the prediction results by a neural network model; however, the research did not list the keywords related to inpatient for reference.

Table 5. Top-six important attributes of the four models

	LR	DL	RF	XGBoost
1	Age	Age	Age	Triage
2	Fever	Triage	Fever status	Fever
3	Triage	CC = "powerless"	Triage	Age
4	*Heartbeat*	CC = "just"	*mode of arrival* = "Referral from"	CC = "tarry stool."
5	CC = "diagnosis"	Fever	*Heartbeat*	CC = "Inpatient"
6	CC = "tarry stool."	*Heartbeat*	CC = "tarry stool."	*mode of arrival* = "Referral from"

4 Conclusion and Future Works

This study aims to provide a better scientific data framework in the field of emergency hospital admissions prediction and help hospitals to reduce congestion in emergency rooms. The major contribution of this work is that we integrated the structured data and textual data of CCs to derive hospital admission prediction. The results of this study suggest that XGBoost can derive the best prediction results in terms of AUC and suggest the important attributes that could be considered for inclusion in the prediction model. Moreover, the imbalanced data processing can increase the Accuracy, G-mean, and AUC compared to the model without conducting it. However, decreasing the data of discharged cases by the random undersampling (RUS) method will increase sensitivity at the expense of specificity. The sensitivity and specificity metrics are inversely proportional to each other. The application of the ED visitors' admission prediction model should consider the real ED context to evaluate the relative importance of the two metrics. For alleviating the ED overcrowding condition, specificity will be more important than the sensitivity metric. That is, if the patient is not likely to be admitted, the ED can quickly assess what kind of medical resources are needed by the patient thus expediting inpatient discharges. Accordingly, the medical resources can be allocated to the patient who needs them most urgently. In line with the research context, we will apply the modern word-embedding approach in natural language processing, e.g., Word2vec, to select terms with a semantic relationship from the CCs for increasing the prediction results. Finally, we will continue to enlarge the dataset to validate the evaluation results. Implementation of the models as a decision support tool can help hospital manage resources effectively and may improve patient flow.

References

Carter, E.J., Pouch, S.M., Larson, E.L.: The relationship between emergency department crowing and patient outcomes: a system review. J. Nurs. Scholarsh. **46**(2), 106–115 (2014)

Fernandes, M., Vieira, S.M., Leite, F., Palos, C., Finkelstein, S., Sousa, J.M.C.: Clinical decision support systems for triage in the emergency department using intelligent systems: a review. Artif. Intell. Med. **102**, 101762 (2020)

Guttmann, A., Schull, M.J., Vermeulen, M.J., Stulel, T.A.: Association between waiting and short term mortality and hospital admission after departure from emergency department: population based cohort study from Ontario, Canada. BMJ **342**, d2983 (2011). https://doi.org/10.1136/bmj.d2983

Parker, C.A., Liu, N., Wu, S.X., Shen, Y., Lam, S.S.W., Ong, M.E.H.: Predicting hospital admission at the emergency department triage: a novel prediction model. Am. J. Emerg. Med. **37**(8), 1498–1504 (2019)

Ashour, O.M., Okudan Kremer, G.E.: Dynamic patient grouping and prioritization: a new approach to emergency department flow improvement. Health Care Manag. Sci. **19**(2), 192–205 (2016). https://doi.org/10.1007/s10729-014-9311-1

Chonde, S.J., Ashour, O.M., Nembhard, D.A., Okudan Kremer, G.E.: Model comparison in emergency severity index level prediction. Expert Syst. Appl. **40**(17), 6901–6909 (2013)

Moll, H.A.: Challenges in the validation of triage system. J. Clin. Epidemiol. **63**(4), 384–388 (2010). https://doi.org/10.1016/j.jclinepi.2009.07.009

Araz, O., Olson, D., Ramirez-Nafarrate, A.: Predictive analytics for hospital admissions from the emergency department using triage information. Int. J. Prod. Econ. **209**, 199–207 (2019)

Fernández Hilario, A., et al.: Learning from Imbalanced Data Sets. Springer, Heidelberg (2018). https://doi.org/10.1007/978-3-319-98074-4

Raita, Y., Goto, T., Faridi, M.K., Brown, D.F.M., Camargo, C.A., Jr., Hasegawa, K.: Emergency department triage prediction of clinical outcomes using machine learning methods. Crit. Care **23**, 1–23 (2019)

Artetxe, A., Beristain, A.,Grana, M.: Predictive models for hospital readmission risk: a systematic review of methods. Comput. Methods Programs Biomed. **164**, 49–64 (2018)

Handly, N., Thompson, D.A., Li, J., Chuirazzi, D.M., Venkat, A.: Evaluation of a hospital admission prediction model adding coded chief complaint data using neural network methodology. Eur. J. Emerg. Med. **22**(2), 87–91 (2015)

User Experience and Innovation Design

Easy Hand Gesture Control of a ROS-Car Using Google MediaPipe for Surveillance Use

Christian Diego Allena[1], Ryan Collin De Leon[1,2], and Yung-Hao Wong[1(✉)]

[1] Minghsin University of Science and Technology, 30401 Xinfeng, Hsinchu County, Taiwan
yvonwong@must.edu.tw

[2] Adamson University, 900 Ermita, Manila, Philippines

Abstract. Hand gestures are a relatively new way for humans to communicate with computers. The goal of gesture recognition is to bridge the physical and digital worlds. Hand gestures make it much easier to communicate our intentions and ideas to the computer. There are numerous methods for a computer to recognize a hand gesture, one of which is image recognition. The use of a Convolutional Neural Network (CNN) allows for the detection of human gestures. However, training a CNN necessitates a massive dataset of human gesture images. In this paper, we employ Google MediaPipe, a Machine Learning (ML) pipeline that combines Palm Detection and Hand Landmark Models, to develop a simple hand tracking method to control a Robot Operating System (ROS) based surveillance car with socket programming. The study demonstrates control of a ROS car's steering direction and speed. Hand-gesture-controlled surveillance vehicles could aid in the improvement of security systems.

Keywords: ROS · Hand tracking · Google MediaPipe · Socket programming · Hand gesture recognition

1 Introduction

At present, one of the goals of the researchers is to bridge the gap between computers and humans. They aim to improve communication between humans and computers more naturally [1, 2]. Some of the ways to achieve it are using hand gesture recognition [3–5], voice recognition [6], and eye motion [7]. Sensing and recognizing movements are two critical aspects of gestural user interaction recognition [8]. Some application of gesture recognition is handling presentations [9–11], controlling drones [12–15], controlling wheeled robots [16–18], controlling robot arms [19–21]. One recent advancement in hand gesture recognition is the use of a convolutional neural network (CNN) trained over images of hand gestures to allow the computer to recognize the hand gesture in the image [22–24].

Another aspect of gesture-controlled robots is how to communicate with them wirelessly. It is now possible to establish fast and effective wireless communication between a human and a robot thanks to advancements in wireless communication technology.

F. Fui-Hoon Nah and K. Siau (Eds.): HCII 2022, LNCS 13327, pp. 247–260, 2022.
https://doi.org/10.1007/978-3-031-05544-7_19

Currently, using hand gestures to control a robot is commonly used. Some of its application is for surveillance and military use. Humans, usually, communicate with a robot using external hardware or through image recognition. But using image recognition for hand gestures is a very challenging problem. To control the robot using images recognition, you need to have a lot of images and a high computational cost. In addition, controlling a robot wirelessly often requires external devices such as a receiver and transceiver and it is often limited to some range. The study will utilize the newly developed Google Media Pipe Hands, a recent Google's media processing library, for hand gesture recognition and socket programming for establishing communication wirelessly.

The rest of the study is organized as follows. Section 2 discussed the related concepts involved in this research. Section 3 is all about the methods used to achieve the project. Section 4 shows the results and demonstration of the project, and Sect. 5 is the conclusion and recommendation for future work.

2 Review of Literature

2.1 Socket Programming

The internet has become a necessity in our daily lives as technology advances. The web can not only be seen applied to our computers but also our smartphones, tablets, and other intelligent digital devices [25]. One application of the web is for server-client relationships between devices as shown in Fig. 1. The client is the one who sends requests while the server answers the request sent by the client [26]. In this project, the use of socket programming will be utilized to achieve wireless communication between a human and a robot.

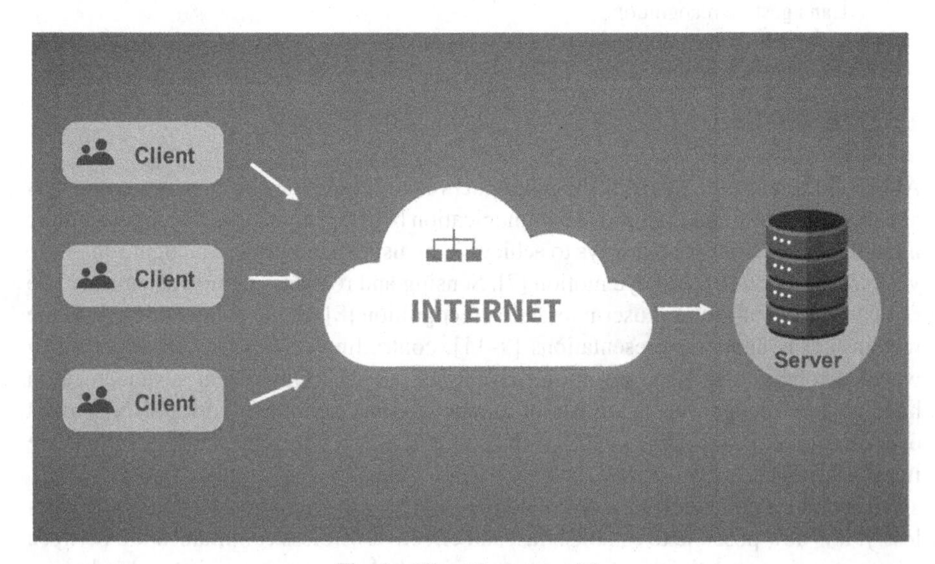

Fig. 1. Client-Server model.

With the help of the internet, the client and server can establish communication wirelessly. Every client establishes a connection with each other through the server [27].

According to Oluwatosin [25], some of the application of Server-Client architecture is as follows:

- File Transfer: files such as movies, images, documents can be sent from client to server and vice versa.
- Mail Transfer: using Mail Transfer Protocol, a server can pass messages to the client
- Hypertext Transfer Protocol (HTTP): using HTTP to create better client-server communication for transferring multimedia files.

Another method for the client and server to establish a connection is to use a socket. A socket is a type of abstract data structure produced by an operating system that acts as an access in the absence of a connection between the process of sending to receiving information [28]. The client and server can transfer data with each other by writing and reading from these sockets.

2.2 Google MediaPipe

Recently Google Researchers created a hand skeleton predictor of a human hand using a palm detector and hand landmark model [29].

Palm Detector. As the name suggests, this detector aims to find the palm of the human hand. It used a convolutional neural network. The model used for their detector can be seen in Fig. 2.

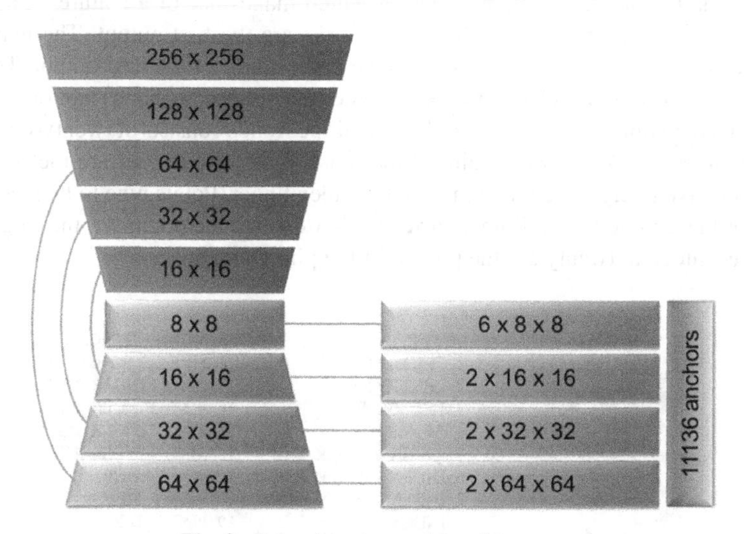

Fig. 2. Palm detector model architecture

Starting from 256 × 256 size, they reduced it to 128 × 128. Then they reduce it again by two folds until the size gets to 8 × 8. Furthermore, by using an alternative anchor scheme they were able to reduce the number of necessary layers. In their experiment, they used a total of 11136 anchors as described in Fig. 2.

Hand Landmark Model. To detect the hand, they created a hand landmark model as shown in Fig. 3.

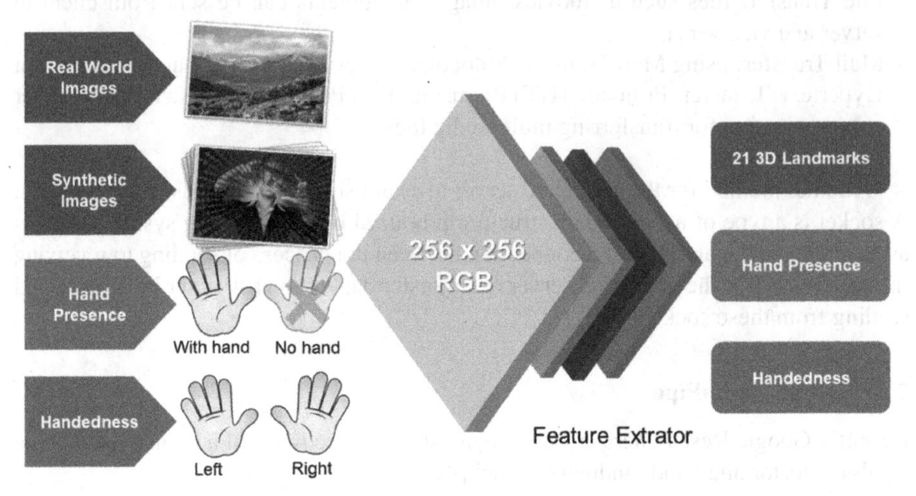

Fig. 3. Hand landmark model architecture

Their model has three outputs, each of which makes use of a feature extractor as shown in above. The twenty-one hand landmarks are the first output. The hand flag, which determines whether a hand exists in the input image, is the next output. The final output is handedness, which determines whether the detected hand is right or left [29].

The hand landmark model of Google MediaPipe Hands consists of twenty-one points as shown in Fig. 4. Zero is the point of the wrist. From one to four are the points for the thumb. Five to eight are the points for the index finger. Points nine to twelve are the points for the middle finger. Points thirteen to sixteen are the points for the ring finger. Lastly, seventeen to twenty are the points for the pinky finger.

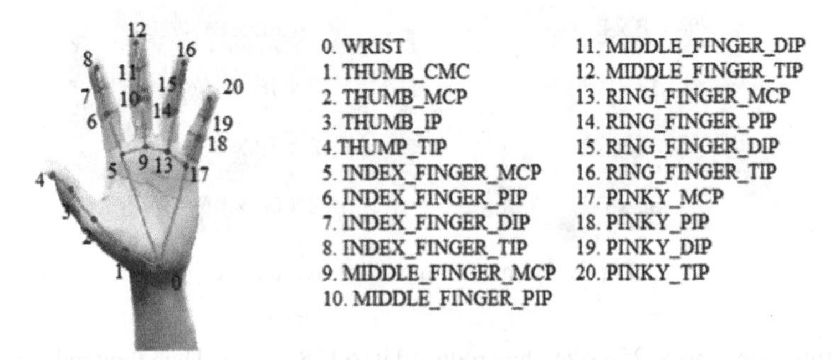

Fig. 4. Hand landmark model

Datasets and Annotation. In their project, they gathered data from various sources to create a very reliable model and address the various issues that could arise when detecting a human hand [29].

- In-the-wild Dataset – this dataset aims to address the problem of having geographical diversity, hand characteristics, and different lighting conditions
- In-house collected gesture Dataset: – this dataset aims to solve the problem of detecting various angles of all physically plausible hand gestures.
- Synthetic Dataset – to further improve their model they provide a high-quality synthetic hand model that is placed in different background

Google MediaPipe Hands recognizes the landmarks of a captured hand in an image It can detect the skeleton of the hand based on this sample image with a good estimate as shown in Fig. 5. The Google MediaPipe Hands will be used in this project to simplify the project's image recognition.

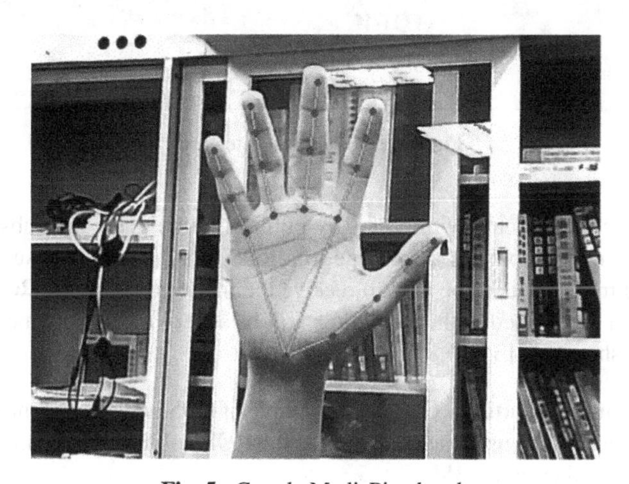

Fig. 5. Google MediaPipe hands

2.3 Robot Operating System (ROS)

ROS is a robot programming platform that allows for the creation of flexible robot software. It's a set of tools, libraries, and conventions designed at making the development of complex and dependable robot behavior on a variety of robotic systems easier [30].

ROS Node Operation Architecture. ROS node architecture consists of a node, a topic, and a message as described in Fig. 6.

- Nodes – is a program that can be executed to communicate with other nodes [31].
- Messages – this is the ROS data type utilized upon subscribing or publishing to a topic [32].

- Topics – this is where the publisher sends the message and where the subscriber gets the message [33].

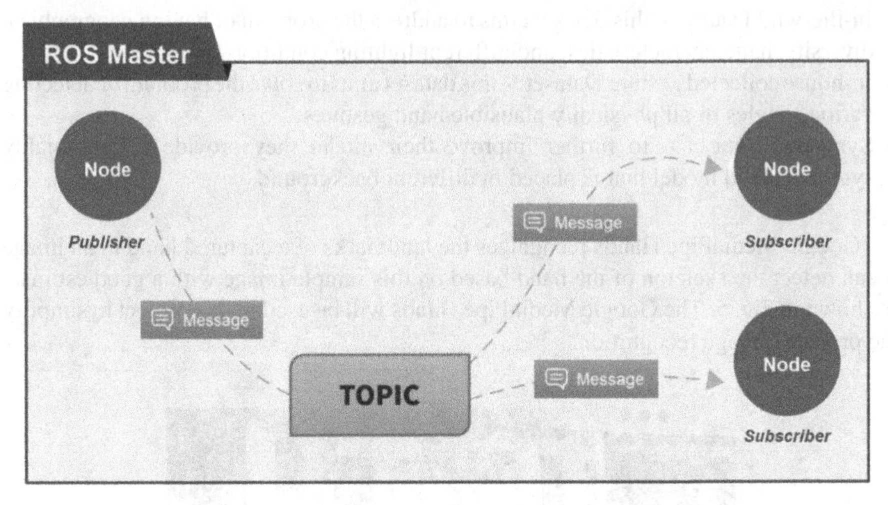

Fig. 6. ROS node architecture.

The nodes communicate with others by publishing the topic and subscribing to it. ROS publisher node publishes a message to a ROS topic. Then ROS subscriber node will subscribe to the topic to be able to receive the messages from the ROS topic. Any ROS subscriber subscribed to the ROS topic will be able to receive the message sent by the ROS publisher node [34].

ROS Serial Communication. One good example of ROS node communication is ROS serial communication. It uses the computer as the ROS publisher node and the Arduino as the ROS subscriber node [35].

The computer, being the publisher, will publish to a ROS topic then send a message to the ROS topic as described in Fig. 7. Subsequently, the Arduino will subscribe to the

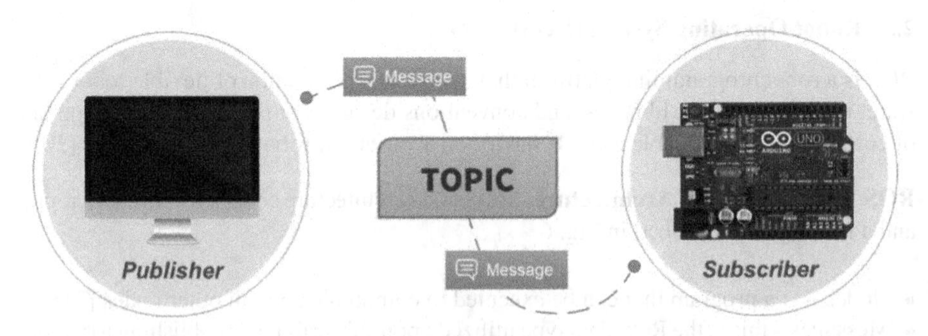

Fig. 7. ROS serial communication

ROS topic to be able to receive the message sent by the computer to the ROS topic. In this project, ROS serial communication will be utilized to control the motors attached to the ROS-Car.

3 Methodology

3.1 Structure of the Experiment

The components and materials used in this project are purchasable commercially. The ROS-Car used in this project is purchased at RealPlus Technology Ltd.

ROS-Car

In this project, the researchers used RealPlus Technology Ltd's AI Car as shown in Fig. 8. It is made up of Nvidia's Jetson Nano, 2 Cameras, AT-Mega 1284p, DC motor, and servo motor. Moreover, it uses Robot Operating System to communicate with the AT-Mega 1284p.

Fig. 8. ROS-Car

3.2 Design of Network

To successfully establish communication between the ROS-Car and the computer, the researchers needs to establish the network by using a modem and a router to communicate wirelessly. The network architecture of the project is illustrated in Fig. 9. For this project, the researchers limits the study to using a local using area network by using a router. In this experiment, the assigned private IP address for the ROS-Car is 192.168.1.5 whereas the designated private IP address for the computer is 192.168.1.2. To check that it is using a private IP address the beginning of the IP address must start with 192.168.*.*.

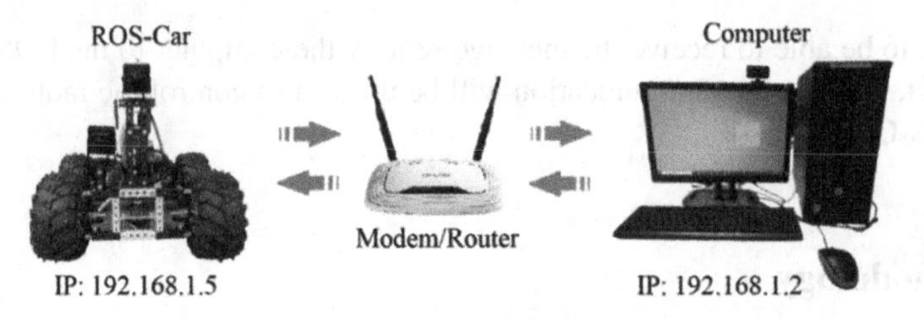

Fig. 9. Network architecture

3.3 Car Modules Interconnection

The ROS-Car has a camera module for the ROS-car to have a vision, a Wi-Fi module to be able to connect to a network, input devices like a keyboard and mouse to access the operating system of the Jetson Nano, output device like the monitor to view the operating system, and a power supply to power the Jetson Nano as described in Fig. 10. Furthermore, it has an AT-Mega 1284p module that will control the motors attached to it namely the DC motor, to control the movement speed of the ROS-car, and the servo motor, to control the steering.

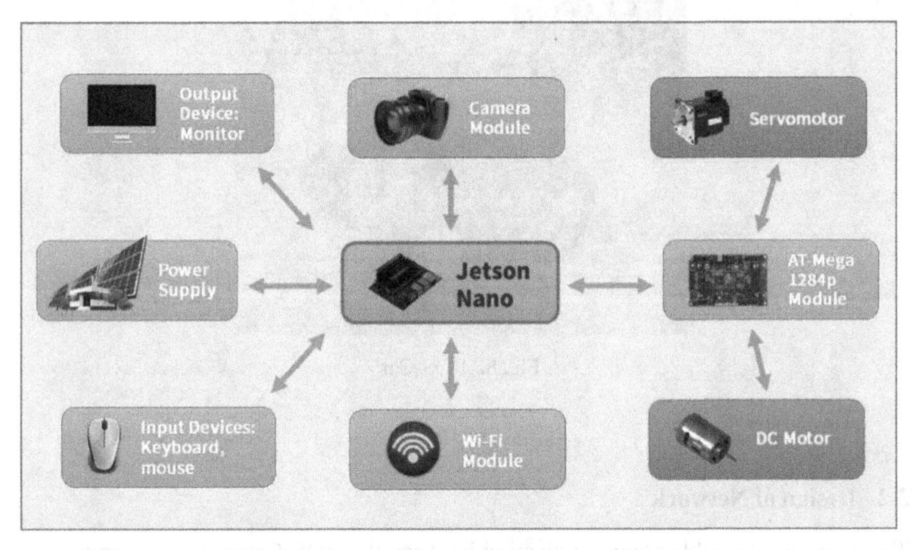

Fig. 10. Diagram of car modules interconnection

4 Results and Discussion

The general description of the project is summarized in Fig. 11. The user will form some hand gesture then the camera of the computer will catch it. Then it will be processed by Google MediaPipe to determine the hand gesture. Every hand gesture has specific command. After determining the command of the hand gesture, it will be sent to the ROS car. Then the ROS-Car will decide its action depending on the command received. Then the robot operating system will communicate to the Arduino. Then the Arduino will instruct the motors attached to the ROS-car namely the DC motor to control the speed and the servo motor to steer the robot to the left or right. After that, the images captured by the camera attached to the ROS car will send it back to the personal computer.

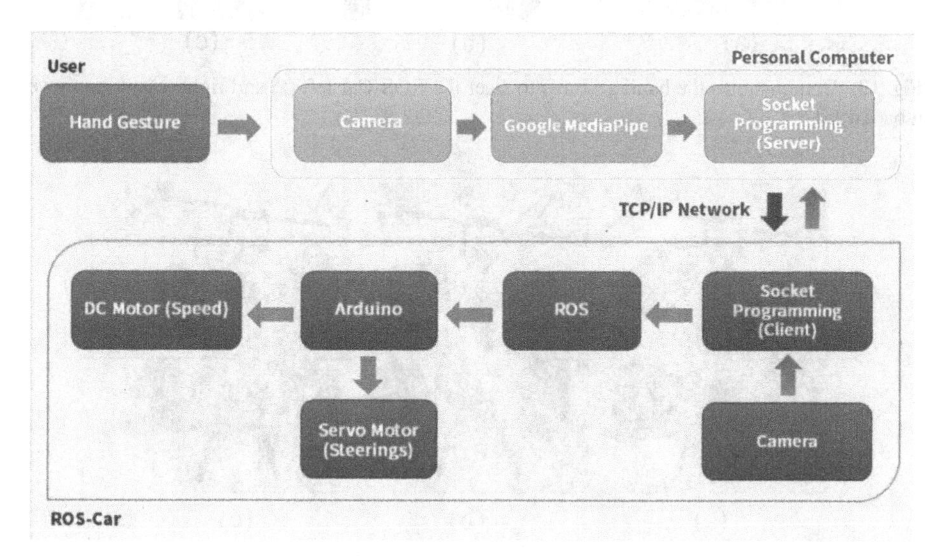

Fig. 11. Description of the system

Figure 12 and Fig. 13 demonstrate how to change the speed and steering respectively. Meanwhile Fig. 14 illustrate the ROS-Car's each steering positions.

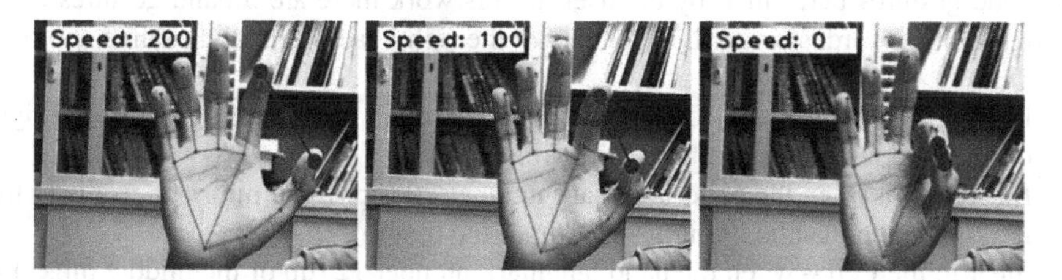

Fig. 12. Demonstrates how to change the speed of the ROS-Car using the hand gestures

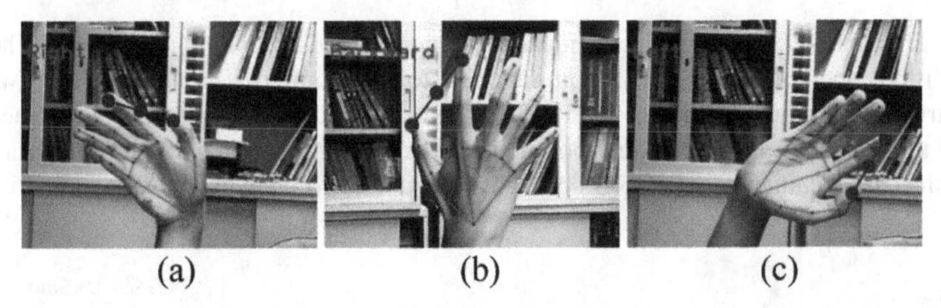

Fig. 13. Demonstrates the hand gestures to steer the ROS-Car left (a) and right (c) and to move it backward (b).

Fig. 14. ROS-Car positions; (a) steer to the left (b) center position (c) steer to the right

4.1 Specification of Messages Through Gestures

Using the hand landmarks described in Fig. 15, the wheeled robot can recognize different hand gestures performed by the user. In this work there are 5 hand gestures to be recognized: forward, steer to the left, steer to the right, backward, and change speed.

1. Forward – when the hand landmark number 20 (tip of the pinky finger) is to the right of the hand landmark number 4 (tip of the thumb)
2. Backward – when the hand landmark number 20 (tip of the pinky finger) is to the left of the hand landmark number 4 (tip of the thumb)
3. Steer to the Left – when the hand landmark number 12 (tip of the middle finger) is to the left of the hand landmark 0 (wrist)
4. Steer to the Right – when the hand landmark number 12 (tip of the middle finger) is to the right of the hand landmark 0 (wrist)
5. Change Speed – the distance between the hand landmark number 8 (tip of the index finger) and the hand landmark number 4 (tip of the thumb) determines the speed of the car. There are 5 possible speeds. The maximum speed is 200 while the minimum speed is 0. While the 3 remaining speeds are 150, 100, 50.

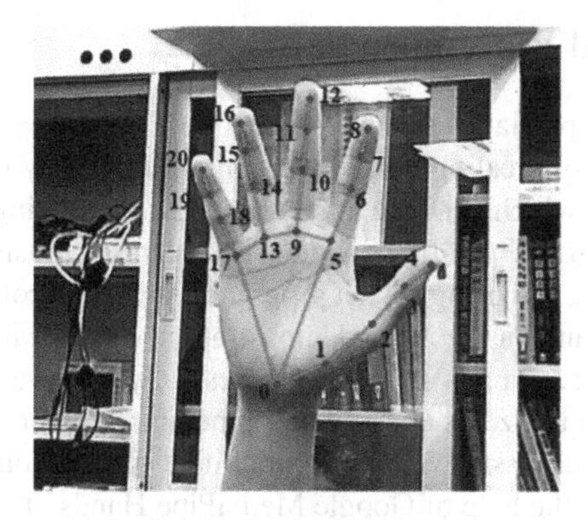

Fig. 15. Hand landmark

4.2 Experimental Result

Figure 16 (a) shows the user doing the hand gesture. The camera captured the hand gesture of the user then it will be processed by the program created. The Python program, with the help of Google MediaPipe, aids the computer to recognize the hand gesture. (b) shows the location of the ROS-Car where it is doing surveillance inside the laboratory of the researchers.

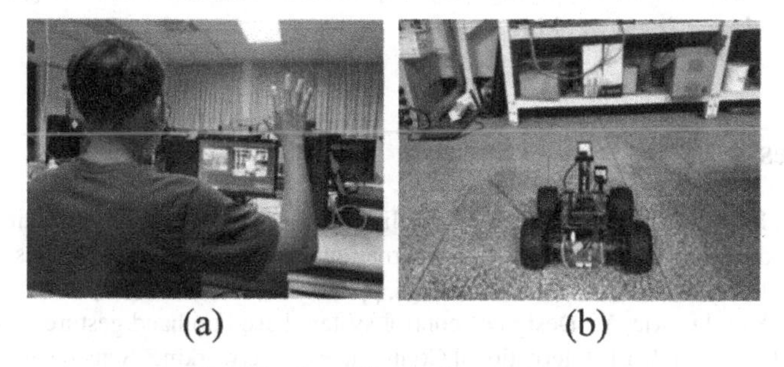

(a) (b)

Fig. 16. Surveillance demonstration

The remote computer screen is shown in Fig. 17. On the left side of the screen is the received video surveillance of the ROS-Car while doing its surveillance while on the right side is the image of the captured hand gesture of the user.

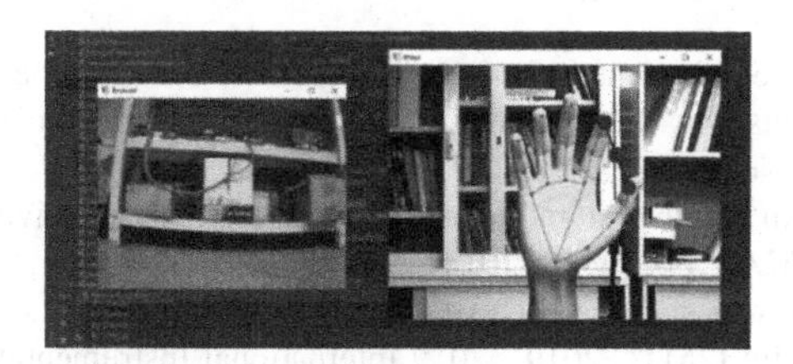

Fig. 17. Remote computer screen

5 Conclusion and Recommendation

Using Google MediaPipe Hands for hand recognition is much simpler and faster than the traditional method of creating a hand recognition model by gathering datasets and training a good model, which can take a long time just by training alone.

The system to control a ROS-based video surveillance car using Google MediaPipe Hands for hand tracking and using socket programming to control it wirelessly has been successfully designed and has shown a demonstration of its great performance.

In conclusion, the research's goal was successfully achieved with an outstanding outcome. This research utilizes the newly developed module for hand tracking, Google MediaPipe Hands, to successfully create a simplified way of controlling a robot using hand recognition. With the help of Google MediaPipe Hands, it was able to control the robot with the following command: backward, forward, steering left, steering right, and control of speed. In addition, it takes advantage of the use of socket programming to be able to create communication between the user and the robot wirelessly.

Succeeding the result of the research, the researcher strongly suggests the following recommendations:

a. This study suggests adding more gesture commands to control the robot car
b. This study suggests making use of the video captured by the robot and applying a convolutional neural network
c. This study recommends applying a recurrent neural network to the Google MediaPipe hand for a better hand tracking.

References

1. Sun, J.H., Ji, T.T., Bin Zhang, S., Yang, J.K., Ji, G.R.: Research on the hand gesture recognition based on deep learning. In: 2018 12th International Symposium on Antennas, Propagation and EM Theory, ISAPE 2018 - Proceedings, pp. 10–13 (2019)
2. Song, S., Yan, D., Xie, Y.: Design of control system based on hand gesture recognition. In: ICNSC 2018 - 15th IEEE International Conference on Networking, Sensing and Control, no. 16, pp. 1–4 (2018)
3. Kabir, R., Ahmed, N., Roy, N., Islam, M.R.: A novel dynamic hand gesture and movement trajectory recognition model for non-touch HRI interface. In: 2019 IEEE Eurasia Conference on IOT, Communication and Engineering, ECICE 2019, pp. 505–508 (2019)
4. Meng, Z.Y., Pan, J.S., Tseng, K.K., Zheng, W.M.: Dominant points based hand finger counting for recognition under skin color extraction in hand gesture control system. In: Proceedings - 2012 6th International Conference on Genetic and Evolutionary Computing, ICGEC 2012, pp. 364–367 (2012)
5. Lamb, K., Madhe, S.: Automatic bed position control based on hand gesture recognition for disabled patients. In: International Conference on Automatic Control and Dynamic Optimization Techniques, ICACDOT 2016, pp. 148–153 (2017)
6. Koo, Y., et al.: An intelligent motion control of two wheel driving robot based voice recognition. In: ICCAS, pp. 13–15 (2014)
7. Chew, M.T., Penver, K.: Low-cost eye gesture communication system for people with motor disabilities. In: I2MTC 2019 - 2019 International Instrumentation and Measurement Technology Conference Proceedings, vol. 2019-May, pp. 8–12 (2019)

8. Mankar, S.M., Chhabria, S.A.: Review on hand gesture based mobile control application. In: 2015 International Conference on Pervasive Computing, ICPC 2015, vol. 00, no. c, pp. 58–59 (2015)
9. Wang, K., Zhao, R., Ji, Q.: Human computer interaction with head pose, eye gaze and body gestures. In: Proceedings - 13th IEEE International Conference on Automatic Face and Gesture Recognition, FG 2018, p. 789 (2018)
10. Wardhany, V.A., Kurnia, M.H., Sukaridhoto, S., Sudarsono, A., Pramadihanto, D.: Smart presentation system using hand gestures and Indonesian speech command. In: Proceedings - 2015 International Electronics Symposium, IES 2015, pp. 68–72 (2016)
11. Salunke, T.P.: Recognition based on hog feature extraction and K-NN classification. In: ICCMC, pp. 1151–1155 (2017)
12. Tsai, C.C., Kuo, C.C., Chen, Y.L.: 3D hand gesture recognition for drone control in unity. In: International Conference on Automation Science and Engineering, vol. 2020-August, pp. 985–988 (2020)
13. Yu, Y., Wang, X., Zhong, Z., Zhang, Y.: ROS-based UAV control using hand gesture recognition. In: Proceedings of the 29th Chinese Control And Decision Conference, CCDC 2017, vol. 410072, pp. 6795–6799 (2017)
14. Natarajan, K., Nguyen, T.H.D., Mete, M.: Hand gesture controlled drones: an open source library. In: Proceedings - 2018 1st International Conference on Data Intelligence and Security, ICDIS 2018, pp. 168–175 (2018)
15. Ghasemi, H., Mirfakhar, A., Masouleh, M.T., Kalhor, A.: Control a drone using hand movement in ROS based on single shot detector approach. In: 2020 28th Iranian Conference on Electrical Engineering, ICEE 2020 (2020)
16. Jain, M., et al.: Object detection and gesture control of four-wheel mobile robot. In: Proceedings of the 4th International Conference on Communication and Electronics Systems, ICCES 2019, pp. 303–308 (2019)
17. Wang, Y., Song, G., Qiao, G., Zhang, Y., Zhang, J., Wang, W.: Wheeled robot control based on gesture recognition using the Kinect sensor. In: 2013 International Conference on Robotics and Biomimetics, ROBIO 2013, no. December, pp. 378–383 (2013)
18. Sriram, K.N.V., Palaniswamy, S.: Mobile robot assistance for disabled and senior citizens using hand gestures. In: 1st International Conference on Power Electronics Applications and Technology in Present Energy Scenario, PETPES (2019)
19. Raheja, J.L., Shyam, R., Kumar, U., Prasad, P.B.: Real-time robotic hand control using hand gestures. In: ICMLC 2010 - 2nd International Conference on Machine Learning and Computing, pp. 12–16 (2010)
20. Bularka, S., Szabo, R., Otesteanu, M., Babaita, M.: Robotic arm control with hand movement gestures, pp. 543–546 (2018)
21. Choudhary, G.B., Chethan, R.B.V.: Real time robotic arm control using hand gestures. In: 2014 International Conference on High Performance Computing and Applications, ICHPCA 2014, pp. 5–7 (2015)
22. Islam, M.R., Mitu, U.K., Bhuiyan, R.A., Shin, J.: Hand gesture feature extraction using deep convolutional neural network for recognizing American sign language. In: 2018 4th International Conference on Frontiers of Signal Processing, ICFSP 2018, pp. 115–119 (2018)
23. Dhall, I., Vashisth, S., Aggarwal, G.: Automated hand gesture recognition using a deep convolutional neural network model. In: 2020 10th International Conference on Cloud Computing, Data Science & Engineering (Confluence), pp. 811–816 (2020). https://doi.org/10.1109/Confluence47617.2020.9057853
24. Zhan, F.: Hand gesture recognition with convolution neural networks. In: Proceedings - 2019 IEEE 20th International Conference on Information Reuse and Integration for Data Science, IRI 2019, pp. 295–298 (2019)

25. Oluwatosin, H.S.: Client-server model. IOSR J. Comput. Eng. **16**(1), 57–71 (2014)
26. Kratky, S., Reichenberger, C.: Client/server development based on the apple event object model, Atlanta (2013
27. Yuqing, D.L.Z., Wu, W.: Efficient client assignment for ClientServer Systems. IEEE Trans. Netw. Serv. Manag. **13**(4), 835–847 (2016)
28. Xue, M., Zhu, C.: The socket programming and software design for communication based on client/server. In: Proceedings of the 2009 Pacific-Asia Conference on Circuits, Communications and Systems, PACCS 2009, pp. 775–777 (2009)
29. Zhang, F., et al.: MediaPipe hands: on-device real-time hand tracking. arXiv (2020)
30. About ROS. https://www.ros.org/about-ros/
31. Meier, L.: Understanding ROS Nodes (2019). http://wiki.ros.org/ROS/Tutorials/UnderstandingNodes
32. Kurzaj, D.: Messages (2016). http://wiki.ros.org/Messages
33. Topics (2019). http://wiki.ros.org/Topics
34. Kutluca, H.: Robot Operating System 2 (ROS 2) Architecture. https://medium.com/software-architecture-foundations/robot-operating-system-2-ros-2-architecture-731ef1867776
35. Jazba, M.: Introduction to rosserial_arduino (2018). https://atadiat.com/en/e-rosserial-arduino-introduction/

A Survey-Based Study to Identify User Annoyances of German Voice Assistant Users

Annebeth Demaeght[1(✉)], Josef Nerb[2], and Andrea Müller[1]

[1] Hochschule Offenburg, Badstrasse 24, 77652 Offenburg, Germany
annebeth.demaeght@hs-offenburg.de
[2] Pädagogische Hochschule Freiburg, Kunzenweg 21, 79117 Freiburg, Germany

Abstract. Voice user interfaces (VUIs) offer an intuitive, fast and convenient way for humans to interact with machines and computers. Yet, whether they'll be truly successful and find widespread uptake in the near future depends on the user experience (UX) they offer. With this survey-based study (n = 108), we aim to identify the major annoyances German voice assistant users are facing in voice-driven human-computer interactions. The results of our questionnaire show that irritations appear in six categories: privacy issues, unwanted activation, comprehensibility, response quality, conversational design and voice characteristics. Our findings can help identify key areas of work to optimize voice user experience in order to achieve greater adaptation of the technology. In addition, they can provide valuable information for the further development and standardization of voice user experience (VUX) research.

Keywords: Voice assistants · Voice user experience · Voice user interfaces · Conversational user experience

1 Introduction

Voice user interfaces enable natural, convenient and intuitive forms of human-computer interaction inspired by human-human communication [1]. The idea of enabling machines to understand and respond adequately to natural language has a long history in science [2]. Thanks to developments in natural language recognition, machine learning and cloud computing, voice-based applications are now broadly available in everyday life [3], the most well-known examples being Amazon's Alexa, Apple's Siri, Google's Assistant, Microsoft's Cortana and Samsung's Bixby.

However, as Seaborn and Urakami [1] point out, the technical quality of voice recognition and voice expression is just one aspect of voice-based interaction. Research has shown that user perception of voice plays a significant role [1]. Measuring not only the usability but also the quality of user experience (UX) is therefore of great interest to evaluate and optimize voice assistants [4].

With this paper, we want to contribute to the research on voice user experience and present the results of two surveys done amongst German voice assistant users, all digital natives.

F. Fui-Hoon Nah and K. Siau (Eds.): HCII 2022, LNCS 13327, pp. 261–271, 2022.
https://doi.org/10.1007/978-3-031-05544-7_20

2 Related Work

2.1 Voice Assistants

Voice assistants enable interaction between a human user and a machine via natural language input and output [5]. The software components of voice-controlled assistance systems are based on five core technologies [3]:

1. Automated speech recognition (ASR): speech input is converted into text form (speech-to-text STT).
2. Natural language processing (NLP): this component is mainly responsible for recognizing user intent and assigns meaning to individual text fragments.
3. Dialog manager (DM): this component is responsible for the flow of the conversation and provides a response to human utterance in the form of code.
4. Natural language generation (NLG): provides the response to the user request in human language.
5. Text-to-speech (TTS): provides the acoustic speech output of the response.

Natural language processing (NLP) is at the heart of the system, because contextual interpretation of the search words and recognition of the user's intention is a fundamental prerequisite for a successful dialog [3].

User annoyances may occur in all of the above described processing steps.

2.2 Voice User Experience

One of the challenges of digital voice assistance systems is to enable natural interactions. While people learn the social conventions for a regulated flow of conversation in daily interactions from early childhood, this is a complex task for digital systems [6].

In linguistics, the sub-discipline of pragmatics deals with the analysis of conversational structures. The cooperation principle, introduced by Paul Grice in 1975 [7], occupies an important place in the research field. The principle states that a conversational contribution should be designed to serve the accepted purpose or direction of the conversation [7].

This principle and further insights from pragmatics, for example on the topic of speaker change or politeness, offer fundamental insights for the design of voice user interfaces between humans and computers.

However, dialog principles have not only been postulated in pragmalinguistics, but also in the research field of human-computer interaction. For example, the ISO 9241-110:2020 [8] standard provides generally applicable design recommendations for interactive systems. The document describes standardized interaction principles between user and system.

If a conversational user interface does not follow the interaction principles, this can lead to user confusion or frustration [6]. For the design and optimization of a dialog model, it is therefore important to be able to identify and analyze such user irritation-triggering elements.

2.3 Measuring Voice User Experience

Klein and colleagues [4] correctly state that measuring voice user experience will be of great interest to evaluate and optimize voice assistants. Yet, according to Seaborn and Urakami [1] it remains an open question how to measure voice user experience. They point out that UX-testing methods are typically developed for graphical user interfaces (GUIs). Voice user interfaces (VUIs) however, tend to be non-visual, intangible and are considered to be disembodied [1]. Therefore, we briefly explore how classic UX-testing methods, as described in Miclau et al. [9], can be applied to voice.

Eye-Tracking. Eye Tracking is considered a central measuring instrument for determining user behavior, since it delivers valuable insights into the users' gaze course [9]. In the context of measuring voice user experience however, this method can only be useful when testing devices with a display. For example, graphical content is displayed in the Alexa App or the Google Assistant smartphone-application. In addition, visual output can be displayed to screen devices, like the Echo Show smart speakers or Google Nest hub devices [3]. When testing voice-only devices, this method drops out.

Think Aloud. Asking users to actively comment their experience while interacting with the system is a frequently used UX-testing method [9, 10]. In the case of voice user interfaces however, the Concurrent Think Aloud (CTA) method [9, 10], meaning the verbalization happens simultaneously while performing a task, cannot be used since the user cannot verbally express his thoughts while talking to the assistant. This only leaves the Retrospective Think Aloud (RTA) method, where the questioning and the comments take place after the interaction with the system [9, 10].

Emotion Recognition. It is recognized that affect and emotion need to be considered when measuring user experience (e.g. [11]). The method most frequently used to do so is facial expression analysis using the Facial Action Coding System (FACS) developed by Ekman & Friesen [12]. It is possible to use this method even if the participant is talking, yet it makes it more difficult to detect the muscle movements that indicate emotion. Thus, the method can be applied only in a limited way. A more reliable method might be speech emotion recognition, using specialized software which detects emotion from the voice signal (e.g. [13]).

Questionnaire and Interview. Klein and colleagues [4] discuss that there already exist several questionnaires to measure the usability of voice-based systems, but that their concentration on task-related properties is not enough. They point out that the complete UX should be considered and thus also non-task-related UX aspects. Therefore, they developed a new questionnaire concept, considering three scales for measuring the UX aspects of voice assistance systems.

This short review has shown that researchers are currently working on developing a suitable VUX-testing method. In order to support this work, we conducted a survey to explore which VUX-aspects are particularly important to voice assistant users.

3 Study

The aim of our study was to identify current irritations in the interaction between users and voice assistants. In a pre-study, we asked 20 digital natives to actively name three to five situations that annoy them when using voice user interfaces. The answers obtained were expanded with results of various studies on voice user experience, resulting in a list of 26 items.

In the second step, 108 participants were presented these 26 items in a questionnaire. To answer the individual items, the participants were given a six-point Likert scale with the poles "doesn't annoy me at all – 1" and "annoys me a lot – 6". This method has been previously developed and applied by Gast [11] in a study on user experience in e-commerce. In both studies, we used the online survey application LimeSurvey [14].

3.1 Pre-study

The pre-study was conducted with 20 German participants, all digital natives. In a short online-survey they were shown a list of voice assistants with the question which ones of these they had used before ("*Welchen dieser Sprachassistenten haben Sie schon einmal benutzt?*"). The list included Alexa, Siri, Google Assistant, Cortana, Bixby, Telekom Magenta and the voice assistants of BMW, Mercedes, Škoda and Volkswagen. Figure 1 shows the distribution of the answers: As expected, most participants had experience with Alexa, Siri and/or Google Assistant. Cortana, Mercedes, Bixby and Magenta were rarely ticked. None of the participants had experience with the voice assistants of BMW, Skoda or Volkswagen.

In a second question, participants were asked to actively name three to five situations which annoy them when using voice user interfaces ("*Welche Dinge oder Probleme ärgern Sie bei der Nutzung von Sprachassistenzsystemen?*"). The participants' answers were complemented with results of various studies and literature on voice user experience (e.g. [3, 6, 8]), resulting in a list of 26 items (cf. Table 1) which can be structured in 6 thematic clusters:

- Comprehensibility (in terms of content or acoustically)
- Unwanted activation
- Response quality
- Privacy issues
- Conversational design issues
- Voice characteristics

3.2 Main Study

Population. The study was conducted with 108 German participants (57 male, 49 female, 1 diverse and 1 not specified) of which 91 were between 18 and 25 years old, 16 between 26 and 35 years old and only one person was older than 35 years. Figure 2 shows which VA systems our respondents had experience with.

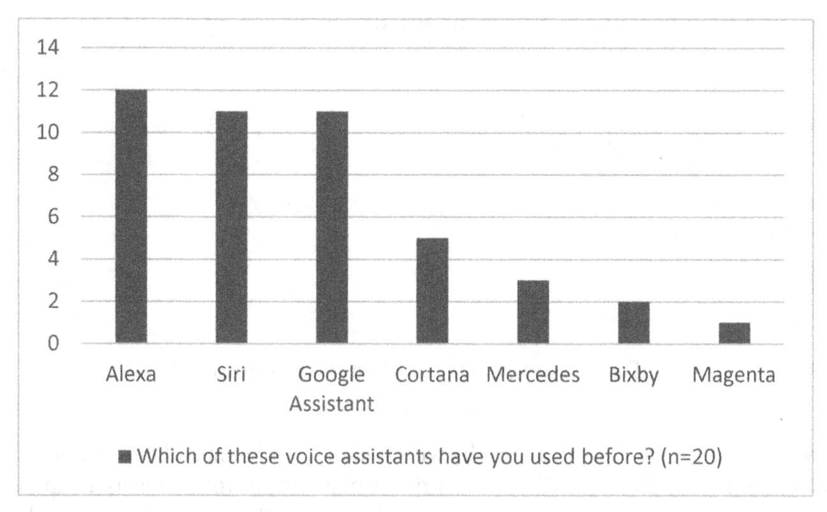

Fig. 1. The VA systems participants of the pre-study had experience with (multiple choice).

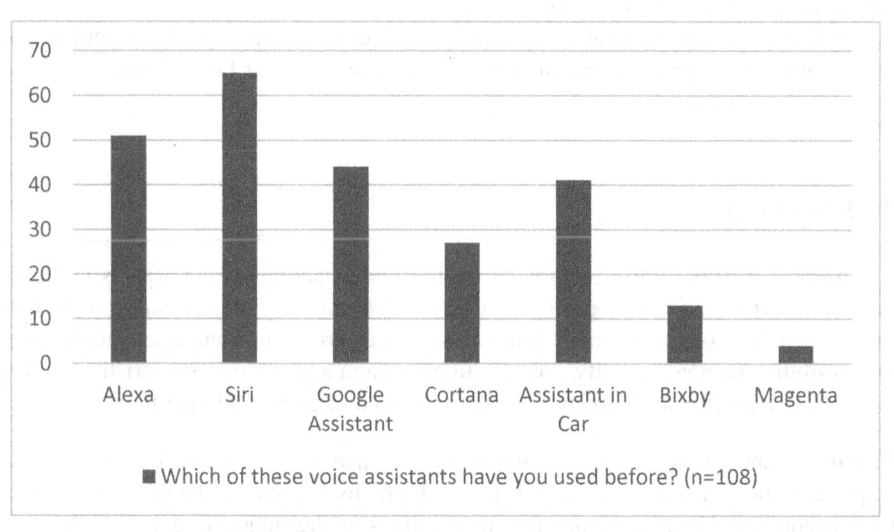

Fig. 2. The VA systems participants of the main study had experience with (multiple choice).

Questionnaire. Out of the 26 items of the pre-study a questionnaire was built with the question "You get annoyed when using digital voice assistants…" ("*Sie ärgern sich bei der Nutzung digitaler Sprachassistenten …*").

To answer the individual items, the participants were given a six-point Likert scale with the poles "doesn't annoy me at all – 1" and "annoys me a lot – 6". Figure 3 shows an excerpt of the original questionnaire in German.

Sie ärgern sich bei der Nutzung digitaler Sprachassistenten ...

	Ärgert mich gar nicht 1	2	3	4	5	Ärgert mich sehr 6
... wenn die Antwort des Sprachassistenten nicht genau genug ist.	○	○	○	○	○	○
... wenn die Antwort nicht zufriedenstellend ist.	○	○	○	○	○	○
... wenn die Ergebnisse auf Ihrer Frage schlecht dargestellt werden.	○	○	○	○	○	○
... wenn die Reaktionszeit zu lange dauert.	○	○	○	○	○	○
... wenn der Sprachassistent sich oft wiederholt.	○	○	○	○	○	○

Fig. 3. Excerpt of the questionnaire in German.

Results. Table 1 shows the mean results with standard deviation and number of respondents for each of the 26 items. We see that privacy plays a huge role in the user experience with voice assistants. Having a feeling of surveillance was the item which annoyed users most, followed by items concerning the system's response quality and comprehensibility, unwanted activation and conversational design.

On the other hand, the voice of the assistant does not seem to play a major role for users, since our responders indicated that a monotone voice or lack of voice selection does not annoy them.

4 Discussion

The purpose of this study was to gain a better understanding of the aspects influencing voice user experience, focusing on items which may trigger user annoyances. Our findings can be grouped in 6 main categories: privacy issues, unwanted activation, comprehensibility, response quality, conversational design and voice characteristics. In this discussion, we will mainly focus on the items with a mean result larger than 4.

Privacy Issues. Experiencing a feeling of surveillance was the item which annoyed our responders most. This result is consistent with previous research, the relevance of this topic is confirmed by a vast amount of publications on the subject (e.g. [15–19]).

A key point which has a negative impact on the current adaptation and acceptance of voice assistants is the lack of transparency on the part of platform operators with regard to data processing and data protection [3]. The main issue is where exactly users' voice input is processed and whether it is stored. According to Kahle and Meißner [3], the large corporations are somewhat too light-handed with the legitimate questions of current or potential users on this regard.

There are only little independent initiatives, especially since Snips, a company with the goal to design the very first "Privacy By Design Voice Assistant", was acquired by Sonos [20]. Some examples of open-source software are Mycroft AI [21] and Rhasspy [22].

Table 1. Survey "You get annoyed when using digital assistants…", including M, SD and n

You get annoyed when using digital voice assistants …	M	SD	n
… when the voice assistant listens permanently	4,96	1,39	108
… when the voice assistant does not understand you acoustically and you have to repeat yourself several times until the system recognizes your voice input	4,93	1,03	107
… when the voice assistant activates itself unintentionally	4,83	1,26	107
… when the voice assistant misunderstands you acoustically	4,62	1,03	106
… when the voice assistant does not respond when you say something	4,61	1,06	106
… when the voice assistant aborts	4,46	1,22	107
… when the voice assistant responds to the radio or television	4,42	1,37	105
… when unnecessary info is given before you can use a function	4,41	1,25	104
… when there is no possibility of correction and you have to start all over again when a mistake is made	4,39	1,20	104
… when the voice assistant misunderstands you in terms of content	4,34	1,16	107
… when you have to confirm your entries several times	4,31	1,14	105
… when you have to download a skill first before you can use a certain function	4,20	1,52	107
… when the voice assistant does not remember information already given and you have to repeat yourself	3,97	1,43	104
… when the statement or question of the voice assistant is not understandable	3,97	1,24	103
… when the voice assistant cannot be interrupted and you can only continue after the voice assistant has spoken	3,94	1,36	106
… when you have to adjust your pronunciation (e.g. speak louder)	3,87	1,33	107
… when the voice assistant often repeats itself	3,87	1,23	105
… when the response time is too long	3,59	1,22	105
… when the answer is not satisfactory	3,56	1,29	107
… when the voice assistant cannot be used while music is playing	3,56	1,72	102
… when the results on your question are poorly presented	3,43	0,99	105
… when the voice assistant's response is not accurate enough	3,25	1,20	106
… when the voice assistant does not understand your dialect	3,09	1,51	104
… when the voice assistant responds to the voices of others	3,08	1,52	106
… when the voice assistant has too monotone a voice	2,16	1,08	106
… when you cannot choose the voice of the voice assistant	2,02	1,21	106

Unwanted Activation. The issue of unwanted activation is closely related to the feeling of surveillance and was rated as the third most annoying feature by our respondents. Furthermore, the item "when the voice assistant responds to the radio or television." had a mean result larger than 4.

This corresponds with previous research by Pins and Alizadeh [23]. They conducted 17 in-depth interviews with VA users and all of their participants reported situations in which their VA suddenly began to speak, even if it was not directly activated. While in some situations it was quite easy for their participants to explain the unwanted situation, there were also situation were users were confused and tried to find a logical reason for what was happening. If the user could not find an explanation for the activation, this would lead to irritations. Pins and Alizadeh [23] found that users typically did nothing to clarify or change the situation. Only one of their participants was annoyed enough by repeated unintended activations to be on the verge of turning the microphone off.

Comprehensibility. If the voice user interface offers a good comprehensibility, the users have the impression that the VA correctly understands their voice input, they feel that their intention is recognized [4]. Three items with a mean result larger than 4 had to do with unsatisfying comprehensibility:

- when the voice assistant does not understand you acoustically and you have to repeat yourself several times until the system recognizes your voice input (M = 4,93).
- when the voice assistant misunderstands you acoustically (M = 4,62).
- when the voice assistant misunderstands you in terms of content (M = 4,34).

The results show that users are particularly annoyed when they're misunderstood acoustically. Furthermore, a false interpretation of the user's intent leads to irritations. Our responders were more forgiving when the assistant demands an adjustment of their pronunciation (M = 3,87) or does not understand their dialect (M = 3,09).

Response Quality. A satisfying response would implicate that the user has the impression that the VA answers clearly & distinctly and thereby fulfills the user's intent [4]. In our survey, only one item concerning response quality had a rating larger than 4: "when the voice assistant does not respond when you say something" (M = 4,61). Other items concerning response quality had ratings between 3 and 4. For example: "when the voice assistant often repeats itself", "when the response is too long" and "when the answer is not satisfactory".

Conversational Design. As functionalities like comprehensibility and response quality are continuously getting better, the topic of conversational design becomes more apparent. Conversational designers map out what users can do in a space, considering both the technological constraints and the user's needs. They define the flow and the underlying logic of a conversation [24]. In our survey, three items with a mean result larger than 4 had to do with design issues:

- when unnecessary info is given before you can use a function.
- when there is no possibility of correction and you have to start all over again when a mistake is made.
- when you have to confirm your entries several times.

VUI-designer Cathy Pearl addresses these issues in her book on the principles of conversational experiences [6]. She states that it's important to include different strategies in the design if the users will be using a system regularly. Novice and expert users have different information needs and the system should consider this. Ideally, the application should be designed to fulfill the user's needs in his current state. If, for example, the user already knows how to use the application, superfluous instructions should be avoided, whereas these might be helpful for novice users.

Furthermore, a stable and well thought-out error management is necessary to avoid user annoyances. Conversational designers should not only design for things to work, they need to design for when things go wrong as well [6]. If something is misunderstood, the user should always have the possibility to go back and correct the voice input.

Finally, confirmation strategies are an important part of conversational design. Asking users to confirm their voice input might ensure accuracy, but - as our survey shows - it also leads to annoyances. Therefor VUI-designers have to consider the necessity of a confirmation. If, for example, a user wants to transfers money via the VA, confirmation would be of great importance, whereas it would not be necessary if the users plays a quiz for fun [6].

Voice Characteristics. Finally, we would like to highlight the two items that had a score smaller than 2,50 and were therefore not considered disturbing by our participants. These items had to do with the assistant's voice characteristics: "when the voice assistant has too monotone a voice." (M = 2,16) and "when you cannot choose the voice of the voice assistant." (M = 2,02). Both of them had a mean result close to the pole "doesn't annoy me at all". This result is somehow surprising, since it is not consistent with previous literature on the topic. For example, a 2019 research collaboration by voicebot.ai, Voices.com and pulselabs [25] found that users have clear preferences between a human and a synthetic voice. According to the study, users prefer a human voice and described the synthetic voices as "monotonous, robotic and computerized". Furthermore, they found a small preference for female voices.

Limitations. Our findings can help identifying the most important working areas to optimize voice user experience in order to increase the acceptance of the technology. In addition, they can provide valuable information for the further development and standardization of voice user experience (VUX) research.

However, the results of this study have to be seen in the light of some limitations, as the sample is relatively small and all respondents are German native digitals. A survey amongst a different or a larger target group might bring other results. Furthermore, the participants' self-assessment may differ from their behavior. According to the survey, the assistant's voice characteristics do not evoke negative feelings, but previous research has shown that users have preferences regarding the assistant's voice personality.

5 Conclusion

Voice user interfaces are predicted a bright future, since they offer an intuitive, fast and convenient way to interact with machines and computers. Yet, whether they'll be truly successful and find widespread uptake depends on the user experience (UX) they offer.

With this paper, we've contributed to the research on voice user experience and presented the results of two surveys done amongst German voice assistant users, all digital natives. We especially focused on user annoyances and questioned which UX-aspects annoy users most. We found that the issue of data privacy plays a major role and therefore will have an effect on the adaptation and acceptance of voice-based interfaces. Further UX-relevant aspects are unwanted activation, response quality, comprehensibility and conversational design issues. The participants rated the assistant's voice characteristics as rather unimportant.

References

1. Seaborn, K., Urakami, J.: Measuring voice UX quantitatively. In: Kitamura, Y., Quigley, A., Isbister, K., Igarashi, T. (eds.) Extended Abstracts of the 2021 CHI Conference on Human Factors in Computing Systems, Article 416, pp. 1–8. Association for Computing Machinery, New York (2021). https://doi.org/10.1145/3411763.3451712
2. Juang, B.H., Rabiner, L.R.: Automatic speech recognition - a brief history of the technology development. In: Georgia Institute of Technology, Atlanta Rutgers University and the University of California, Santa Barbara (2004). https://web.ece.ucsb.edu/Faculty/Rabiner/ece259/Reprints/354_LALI-ASRHistory-final-10-8.pdf. Accessed 8 Feb 2022
3. Kahle, T., Meißner, D: All about voice. Konzeption, Design und Vermarktung von Anwendungen für digitale Sprachassistenten. Haufe Group, Freiburg (2020)
4. Klein, A.M., Hinderks, A., Schrepp, M., Thomaschewski, J.: Measuring user experience quality of voice assistants. In: 15th Iberian Conference on Information Systems and Technologies (CISTI), Sevilla, Spain, pp. 1–4. IEEE (2020). https://doi.org/10.23919/CISTI49556.2020.9140966
5. Thar, E.: Ich habe Sie leider nicht verstanden. Linguistische Optimierungsprinzipien fuer die muendliche Mensch-Maschine-Interaktion. Peter Lang, CH (2015)
6. Pearl, C.: Designing Voice User Interfaces. Principles of Conversational Experiences. O'Reilly, Beijing (2017)
7. Grice, H.P.: Logic and conversation. In: Cole, P., Morgan, J.L. (eds.) Syntax and Semantics, vol. 3, Speech Acts, pp. 41–58. Academic Press, New York (1975)
8. DIN EN ISO 9241-110:2020-10: Ergonomie der Mensch-System-Interaktion - Teil 110: Interaktionsprinzipien. Beuth Verlag, Berlin (2020)
9. Miclau, C., Demaeght, A., Müller, A.: Empirical research as a challenge in day-to-day teaching during the pandemic of 2020/21 - practical solutions. In: Nah, F.-H., Siau, K. (eds.) HCII 2021. LNCS, vol. 12783, pp. 608–618. Springer, Cham (2021). https://doi.org/10.1007/978-3-030-77750-0_40
10. Jo, M.Y., Stautmeister, A.: Don't make me think aloud! – Lautes Denken mit Eye-Tracking auf dem Prüfstand. In: Brau, H., Lehmann, A., Petrovic, K., Schroeder, M.C. (eds.) Tagungsband UP11, pp. 172–177. German UPA e.V., Stuttgart (2011)
11. Gast, O.: User experience im e-commerce. Messung von Emotionen bei der Nutzung interaktiver Anwendungen. Springer, Cham (2018). https://doi.org/10.1007/978-3-658-22484-4

12. Ekman, P., Friesen, W.: Manual for the Facial Action Coding System. Consulting Psychologists Press, Palo Alto (1978)
13. Audeering Homepage. https://www.audeering.com/de/. Accessed 6 Feb 2022
14. LimeSurvey Homepage: LimeSurvey: An Open Source Survey Tool; LimeSurvey GmbH, Hamburg, Germany. https://www.limesurvey.org/de/. Accessed 6 Feb 2022
15. Dallmer-Zerbe, S., Haase, J.: Adapting smart home voice assistants to users' privacy needs using a Raspberry-Pi based and self-adapting system. In: 2021 IEEE 30th International Symposium on Industrial Electronics (ISIE), pp. 1–6 (2021). https://doi.org/10.1109/ISIE45552.2021.9576469
16. Anniappa, D., Kim, Y.: Security and privacy issues with virtual private voice assistants. In: 2021 IEEE 11th Annual Computing and Communication Workshop and Conference (CCWC), pp. 0702–0708 (2021). https://doi.org/10.1109/CCWC51732.2021.9375964
17. Germanos, G., Kavallieros, D., Kolokotronis, N., Georgiou, N.: Privacy issues in voice assistant ecosystems. In: 2020 IEEE World Congress on Services (SERVICES), pp. 205–212 (2020). https://doi.org/10.1109/SERVICES48979.2020.00050
18. Sweeney, M., Davis, E.: Alexa, are you listening? In: Information Technology and Libraries **39**(4) (2020). https://doi.org/10.6017/ital.v39i4.12363
19. Pal, D., Arpnikanondt, C., Razzaque, M.A., Funilkul, S.: To trust or not-trust: privacy issues with voice assistants. IT Prof. **22**(5), 46–53. (2020). https://doi.org/10.1109/MITP.2019.2958914
20. Snips Homepage. https://snips.ai/. Accessed 6 Feb 2022
21. Mycroft Homepage. https://mycroft.ai/. Accessed 6 Feb 2022
22. Rhasspy Homepage. https://rhasspy.readthedocs.io/en/latest/. Accessed 6 Feb 2022
23. Pins, D., Alizadeh, F.: Without being asked: identifying use-cases for explanations in interaction with voice assistants (2021). https://doi.org/10.13140/RG.2.2.18764.33923
24. Google Assistant Developers Homepage. https://developers.google.com/assistant/conversation-design/what-is-conversation-design. Accessed 6 Feb 2022
25. Voicebot.ai, Voices.com, pulselabs: What consumers want in voice app design. https://voicebot.ai/wp-content/uploads/2019/11/what_consumers_want_in_voice_app_design_voicebot.pdf. Accessed 6 Feb 2022

Factors that Influence Cookie Acceptance

Characteristics of Cookie Notices that Users Perceive to Affect Their Decisions

Julia Giese and Martin Stabauer(✉)

Johannes Kepler University, Linz, Austria
martin.stabauer@jku.at

Abstract. Especially in e-commerce and associated online marketing, web cookies play an essential role as they provide information that is key, for instance, to improving website functionality and customization. With the 2019 ruling of the Court of Justice of the European Union, cookie notices became mandatory in the EU. Companies seek to measure and improve cookie opt-in rates to avoid large data losses relevant for online marketing. We tested in an experiment the most common cookie variants – the binary-choice cookie notice and the category-choice cookie notice – for their acceptance rates. The results showed that the former achieved a slightly, but statistically significantly, higher opt-in rate, and the highest opt-in rate was found among users browsing on mobile devices. The decision to accept or reject cookies when presented with a cookie notice is made within seconds and can be influenced by various external factors, which we sought to identify and examine in this study with the use of a survey following the experiment. None of the external influencing factors examined were perceived as influential by more than half of the participants. Simplicity of use, the speed with which the cookie notice is dismissed and time pressure when browsing were the most frequently mentioned external influencing factors. However, all factors examined had some effect on users' attitudes to cookie notices.

Keywords: Cookies · Cookie notices · GDPR

1 Introduction

Cookie notices are no longer new to web users in Europe. As soon as they visit a website – especially when accessing it for the first time or when changing device while browsing – users are confronted with a cookie notice and the request to confirm which cookies they want to allow for functionality and tracking. Although cookie notices were initially implemented on most websites in recent years, users were often not given a meaningful choice regarding the collection of their data. Cookie notices are mandatory under EU law whenever personal data is collected on the internet. Thus, website operators are forced to transparently disclose what data about visitors to their site is tracked and what is done with it [14].

F. Fui-Hoon Nah and K. Siau (Eds.): HCII 2022, LNCS 13327, pp. 272–285, 2022.
https://doi.org/10.1007/978-3-031-05544-7_21

Collection and use of data for personalized advertising has recently become increasingly important. In e-commerce in particular, it is essential for web shop operators to store specific data in cookies in order to improve the usability of the site and to assess numerous design and process changes. Cookies help to collect data for conversion optimization and various marketing activities [3].

In seeking to collect as much data as possible about the user behavior, operators of web shops and websites depend on the maximizing the proportion of visitors agreeing to the collection of all data and thus to all cookies in the cookie notices. Without acceptance of certain optional cookies, for example, A/B testing of a new design or check-out process on the website is not possible [5].

Studies have shown that even small changes to the cookie notice itself can have an impact on the opt-in-rate, and corresponding influencing factors have already been investigated in previous work (see next section). However, most of these surveys were conducted before the 2019 ruling of the Court of Justice of the European Union (CJEU) against planet49[1], so it was possible for users to close a cookie notice without selecting a specific setting such as only agreeing on necessary cookies and not allowing any tracking for online marketing purposes at all [5].

The goal of this study was to determine and assess holistically factors that influence users' cookie notice decisions. Situational and personal perceived factors were studied in addition to those concerning the cookie notice itself. Perceived factors are defined as aspects that consist of a temporal and a situational component or relate to the user themself [14].

The decision is defined as that which relates to permitting only functional or all cookies, including all optional cookies, in a cookie notice. In addition, we examined in detail how strongly the two most frequently used variants of cookie notices – category-choice and binary-choice – differ in terms of consent to the options offered. The category-choice cookie notice displays all the categories that a user can choose from when giving consent on one screen, whereas the binary-choice cookie notice only offers two choices, one to accept all cookies or to update preferences on the second screen, which requires an additional click from the user. We focused on the following research questions: (a) *In the context of a web shop, are users more likely to accept all cookies when presented with a category-choice cookie notice or a binary-choice cookie notice?*, and (b) *Which factors do users perceive as influencing their decision to accept cookies?* To answer these questions, we conducted (i) an experiment with 46,512 participants using an Austrian web shop and (ii) a follow-up survey with 627 participants.

[1] Ruling C-673/17 made clear that pre-ticked check boxes and similar technologies authorizing the use of cookies do not constitute valid consent under the e-Privacy Directive.

2 Related Work

Cookie notices have increasingly become the subject of research, especially due to the enforcement of the European Union's General Data Protection Regulation (GDPR) and privacy directive, and the latest rulings by the CJEU.

Kulyk et al. [8] analyzed the users' perception of cookie notices, their reactions and whether these led to users feeling more informed about use of their data. Their study was concerned primarily with testing the cookie notices, which were relatively new at the time, and how they were perceived. Variants were examined that became obsolete under EU law in October 2019, since they did not elicit a decision, but rather informed users of each website or web shop of data collection and processing. However, some of the findings stayed relevant and were confirmed in later studies (e.g., [11]).

Utz et al. [14] studied the effects of position, number of consent options and language of cookie notices, and their impact on visitors. It was shown that the position of a cookie notice has a considerable effect on user decisions. They tested a range of cookie notices and found that binary-choice cookie notices achieved significantly greater consent. Since this data was collected before the 2019 CJEU ruling, some variants were included that no longer comply with EU law. In addition, users were not nearly as familiar with cookie notices as they are today, and could avoid making a decision, as most of the implemented variants could be clicked away and browsing could be continued without giving any consent. In the meantime, cookie notices have become a key part of today's browsing experience. A more recent study by Bauer et al. [1] showed that small changes in design such as color adjustments, can now increase the consent rate by up to 85% .

Nouwens et al. [10] examined the impact of the most commonly used consent management platform (CMPs) designs in the UK, which contain deployable cookie notice solutions for webshops or websites. Only 11.8% of the CMP variants examined met the minimum legal requirements, and they used dark patterns to trick users into consenting. The data was collected in September 2019 and thus immediately before the CJEU ruling in October 2019.

The effects of technical and legal knowledge on the rejection of cookie tracking were examined in an experiment by Strycharz et al. [13] They showed that being informed of cookie tracking made users feel less threatened and made them more likely to allow cookies. Furthermore, they identified a high relevance of consumer informedness, especially in matters concerning online privacy. The presence of cookie notices can have a negative impact on consumers' perception of threat to digital privacy. However, when enough choices are available, the consumers' perceived control increases, as Bornschein et al. [2] found in their study.

A study by Schmidt et al. [12] showed that displaying cookie notices can have another side effect: If users consciously agree to tracking, they accept frequent price changes based on their individual browsing behavior. According to the study, this is due to consumers seeing themselves as being responsible after making the cookie decision and thus having a stronger perception of price fairness, which can lead to an increased purchase intention.

In addition, the legal compliance of cookie notice implementation was also investigated in terms of whether they actually stored and implemented the choices made by visitors. Matte et al. [9] showed that of the websites examined in their study in 2019, over 54% violated the desired settings or the legal requirements.

3 Implementation of Cookie Notices

Cookie notices must be implemented to comply with legal requirements such as the European Union's Directive 2009/136/EC. The main purpose of a cookie notice is to enable a visitor to make the best possible consent decision about the processing of their personal data when browsing. [4]

Visitor consent to cookies that process personal data must be obtained before they are set. It is therefore not sufficient to provide information and an optionally clickable link to legal information or data protection page: the disclosure must form part of the cookie notice itself. However, cookie notice implementations must allow compliant consent to be given. Article 4 (11) of the EU GDPR defines compliant consent as follows: *'consent of the data subject means any freely given, specific, informed and unambiguous indication of the data subject's wishes by which he or she, by a statement or by a clear affirmative action, signifies agreement to the processing of personal data relating to him or her'*. This consent must also be obtained when processing personal data through cookies. The ruling by the CJEU in 2019 triggered new developments in cookie notice implementation and design. Approximate design guidelines can be derived from this ruling, because variants in which the cookie notice can be closed or is shown purely informatively were declared legally invalid. [14]

The results of this study can also contribute to the legal discussion around implementation of cookie notice variants, by showing that cookie notice variants that make it difficult for users to give compliant consent lead to higher opt-in rates. In addition, the question arises of whether cookie notice variants that do not make rejection as easy as giving consent have any legal validity at all and whether they should be permitted.

4 Methodology

Our first step in answering the research questions, was to conduct an online field experiment in the form of an A/B test on the website of an Austrian online shop. The second step consisted of an online questionnaire that was presented to website visitors via a pop-up during the experiment.

4.1 Study Setup

The study took place from 5th to 6th May 2021 as a field experiment in the online shop of a manufacturer of cosmetics, skincare and dietary supplements with a follow-up survey. The experiment and the survey were displayed and rolled out only in German-speaking countries, and therefore the cookie notices and the survey were in German.

At the beginning of the experiment, any previous cookie selection settings of the visitors were reset, such that one of two cookie notice variants was displayed randomly. As soon as consent of any kind was given, a pop-up linking to the online survey was displayed. Like the cookie notices, it was shown on all browsing devices during this period. The placement was chosen at the bottom left on desktop devices to avoid an overlap with the icon of the live chat on the right-hand side. On mobile devices, the pop-up was positioned in the center of the screen. Both variants and the pop-up were created in the company's corporate design (CD) and the mobile design of the variants are shown in Figs. 1 and 2, respectively. For the purpose of this paper, both variants were translated to English.

In order to investigate the first research question regarding the consent choice of the cookie notice variants, *In the context of a web shop, are users more likely to accept all cookies when presented with a category-choice cookie notice or a binary-choice cookie notice?*, two hypotheses were formulated. These were derived from previous findings which also indicated that choices are made differently concerning the cookie notice variants and are also made differently on mobile devices differ from those on desktop.. The following hypotheses were formulated: (i) Binary-choice cookie notices achieve higher approval of all cookies than category-choice cookie notices. (ii) Binary-choice cookie notices are more likely to achieve consent to all cookies on mobile devices than on desktops.

By clicking on the 'participate now' button, participants were redirected to the online survey by opening a new tab. This pop-up could also be closed, which meant that it was no longer displayed to that user. Its text was written in German and invited visitors to participate in the survey. In order to avoid complaints to the company, no incentive was offered for participation, as only visitors from German-speaking countries would have been able to benefit.

4.2 Online Experiment

In order to conduct the field experiment and test our hypotheses, both cookie notice variants had to be created in both desktop and mobile form, as none of the variants to be tested had previously been implemented in the online shop. Both variants contained the same text and the same cookies, which were approved by the company's legal department. In addition, both cookie notices were created in the corporate design of the company, not only because the company required it, but also because visitors would be familiar with it. The three further links – 'imprint', 'privacy policy' and 'more information' – specified in the cookie notice were also displayed in the bottom of each notice.

Variant 1: Binary-Choice Cookie Notice. The first cookie notice variant was placed in the center of the screen of all browsing devices as a wall, with the content of the online shop slightly hidden under a gray overlay. Browsing further

THIS WEBSITE USES COOKIES

This website uses cookies to personalise content and advertisements and to analyse visits to the website. Information on your usage of our website is passed on to our partners for analysis, advertising and for social media. They may combine your information with other data that you have already provided them or that they have collected while you were using their services. By clicking on 'Accept all', you are agreeing to the use of these services. By clicking the button 'Select cookies', you can manage your cookie settings and the use of cookies separately. By clicking the button 'Only use necessary cookies', you are only agreeing to the use of those cookies that are required for the provision of our website.

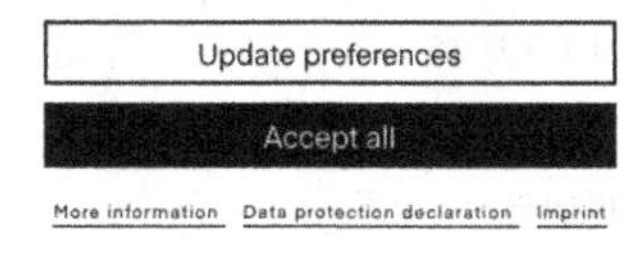

Fig. 1. Binary-choice mobile

THIS WEBSITE USES COOKIES

This website uses cookies to personalise content and advertisements and to analyse visits to the website. Information on your usage of our website is passed on to our partners for analysis, advertising and for social media. They may combine your

Essential

More Details

Functional

More Details

Statistics & Analysis

More Details

Save Selection | Accept all

More Information Imprint Data protection declaration

Fig. 2. Category-choice mobile

was therefore only possible after clicking one of the two buttons. By clicking the button with the text 'accept all', all cookies were accepted. The buttons in the first layer were shown in the same color as those in the second layer, while the buttons for accepting all cookies were highlighted in black with white text. By clicking on the button above, labeled 'Update preferences', the second layer of the cookie notice appeared, where more detailed settings could be selected. Here, by clicking 'Save selection', users could agree to accept essential cookies only.

Variant 2: Category-Choice Cookie Notice. The second variant tested was also placed in the center as a wall on top of a gray overlay such that the framework conditions for testing both variants were as similar as possible. In this variant, visitors also had to make a decision about cookies before they could access the online shop. Its text was taken from the first notice, and it also featured the company's CD. By clicking the button 'Save selection', only essential cookies were accepted. However, if users clicked the right button labeled 'accept all', they accepted all cookies. This variant had only one layer in which all categories were displayed. Here, detailed settings could be chosen and by clicking 'more details' individual categories could be approved. The options were displayed in the same layer.

4.3 Online Questionnaire

To answer the second research question, in the same period an online questionnaire was displayed after the cookie notice via a pop-up link in the web shop. In order to ensure that the respondents understood the concept of a cookie notice and could answer the questions in the best possible way, especially with regard to influencing factors, a brief explanation with an accompanying illustration of a cookie notice was provided at the beginning of the survey. The factors were then displayed to the participants. All participants selected the factors that were relevant to them, that is, those by which they felt influenced. Subsequently, participants were asked which kind of effect (positive to negative on a four-point scale) these aspects had on their attitudes towards cookie notices; this included factors they had selected as irrelevant to them in the previous question. In addition, participants were given the opportunity to add further external influencing factors. The other sections of the survey focused on general statements regarding cookie notices, specific questions about the cookie notice variants shown in the experiment and demographic questions. The external influencing factors studied are listed in Table 1. Some of these aspects were taken from previous studies, and others were added based on our own assumptions based on our own experiences. Further, an open text field was provided for the respondents to list additional factors.

Table 1. Influencing factors.

Influencing factor	Adopted or adapted from
Personalized design of the cookie notice	Kettner et al. (2020) [7]
Humorous presentation of the cookie notice	Authors' assumption
Time pressure when surfing	Authors' assumption
Simplicity of use of the cookie notice	Bornschein et al. (2020) [12]
Smartphone as a browsing device	Authors' assumption
Laptop as surfing device	Authors' assumption
Type of website: Online shop	Bornschein et al. (2020) [12]
Type of website: Newspaper/news site	Bornschein et al. (2020) [12]
Brand/online shop awareness	Authors' assumption
Language of the website/cookie notice	Bornschein et al. [2]
High number of choices in cookie notice	Bornschein et al. [2]
Low number of choices in cookie notice	Bornschein et al. (2020) [2]
Dismissal speed of the cookie notice	Gradow/Greiner (2021) [6]
Large amount of text in the cookie notice	Kulyk et al. (2021) [8]
Position of the cookie notice: in the middle	Utz et al. (2020) [14]
Position of the cookie notice: left or right	Utz et al. (2020) [14]

5 Results

A total of 46,512 valid entries were recorded in the database. Of these, 38.3% (17,792 accesses) were made via desktop devices, and 61.7% (28,720 accesses) via mobile devices. This accords with the usual distribution of devices accessing the company's online shop in 'normal' operation. A total of 627 people who participated in the experiment also completed the survey fully. Since the web shop sells skincare, cosmetics and dietary supplements, the vast majority of respondents were female (95.1%) and only 4.9% were male. The respondents had a mean age of 42.48 years with a standard deviation of 10.74 years. Most participants logged in from Germany (45.9%), followed by Austria (37.5%), Switzerland (9.6%) and other countries (7%).

5.1 Opt-In Rates for the Variants Tested

The binary-choice cookie notice led to consent to all cookies in 93.1% of users with mobile devices and 90.5% of users on desktop devices, while the category-choice notice achieved only 81.5% on mobile devices and 80.5% on desktop devices.

For the first hypothesis, we found that a correlation exists between the two cookie notices regarding consent to all cookies, with a $\mathcal{X}^2$ of 1,222.90. This correlation is weak according to a Cramér's V of 0.16, but statistically significant with $p < 0.001$. This corroborates that, overall, binary-choice cookie notices achieved slightly higher opt-in rates than category-choice cookie notices, although the statistically significant correlation between the two cookie notice variants and consent was weak.

For the second hypothesis we tested whether the binary-choice notice resulted in higher opt-in rates on mobile devices than on desktop devices. With a $\mathcal{X}^2$ of 51.791, a weak correlation was found between the two browsing devices with regard to accepting all cookies in a binary-choice cookie notice. This correlation is statistically significant with $p < 0.001$, but weak according to a Cramér's V of 0.047. In summary, binary-choice cookie notices resulted in a slightly higher opt-in rate on mobile devices than on desktop devices.

5.2 Perceived Influencing Factors and Their Effects on Personal Attitude Towards Cookie Notices

The second part of the empirical study consisted of an online survey in which all respondents of the previously conducted A/B test could participate. A total of 1914 people started the survey, 627 of whom completed it fully and submitted their answers. In the evaluation of the questionnaire, only entirely filled out forms were taken into account.

Below we give an overview (also see Fig. 3) of the influencing factors and their effects on people who felt influenced by them.

Presentation of the Cookie Notice in the Style of the Online Shop. This was identified as influential by 130 respondents (20.7%), who were asked whether the influence was positive or negative. The majority (88.7%) of all participants found that a customized design of the cookie notice had a somewhat positive to very positive effect. **Humorous presentation of the cookie notice.** This was indicated as having an influence by only 85 people (13.6%), and was described as having a somewhat positive to very positive effect by the majority of all respondents (68.5 %). **Time pressure when surfing.** Around a third (34.6%) of the respondents stated that time pressure when surfing influenced their decision. This factor had a somewhat negative to very negative effect on the majority (76.9%) of all respondents' personal attitudes towards cookie notices.

Ease of Use of the Cookie Notice. This was a factor that influenced the decisions of about half of the respondents (46.3%), and was thus the aspect that had the greatest impact according to the most responses. Most people (91.1%) answered that ease of use had a somewhat positive to very positive effect. **Smartphone as a surfing device.** This factor was rated as influential by 20.7% of the participants and had a somewhat positive to very positive effect on the majority of respondents (72.7%). **Type of website: Online shop.** This was selected as an influencing factor by 98 people (15.6%), many (87.1%) rated online shops as somewhat positive to very positive in their attitude towards cookie notices.

Type of Website: Newspaper/News Site. In contrast to online shop as type of website, type of website: Newspaper/news site was described by only 36 people (5.7%) as having an influence on their decisions. Almost half of the respondents (49.1%) said that the type of website being a newspaper had a somewhat negative to very negative effect on their attitude. **Awareness of the brand or the online shop.** 122 of the respondents (19.5%) answered that awareness of the brand or the online shop influences their decision. Being familiar with the brand or the online shop has a somewhat positive to very positive effect on the attitude towards cookie notices for the majority of respondents (90.5%). **Laptop as a surfing device.** Only 29 (4.6%) of the respondents thought that laptop as a surfing device influenced their decision. Although this factor was rated as insignificant by most, 73.4% said that the device tended to have a somewhat positive to very positive impact on their attitude. **The language of the cookie notice.** Only 73 respondents (11.6%) considered the language of the cookie notice to influence their decision, and 88.9% said that it had a somewhat positive to very positive effect.

High Number of Choices in the Cookie Notice. Although more than half of the respondents said they preferred the category-choice variant, which offered a high number of choices in the cookie notice, only 40 respondents (6.04%) said that this had an impact on their decision, and most respondents (71.5%) reporting a somewhat negative to very negative effect. **Low number of choices.** In contrast, 125 people (19.9%) thought that a low number of choices in the cookie notice influenced their decision, and 76.3% felt it had a somewhat positive to very positive effect on their attitudes towards cookie notices. **Speed with which the cookie notice could be dismissed**. The speed with which the cookie notice could be dismissed influenced the decision of 227 people (36.2%),

90.3% stated that speed had a somewhat positive to very positive effect on their personal attitudes towards cookie notices.

Large Amount of Text in the Cookie Notice. This was a factor that affected the decisions of only 130 people (20.7%), 51.7% stated that this factor results in a somewhat negative to very negative effect on their the attitudes towards cookie notices. **Position of the cookie notice: in the center.** Although it is clear from previous research that the position of a cookie notice has an effect on the opt-in rate, only 51 people (8.1%) stated that position of the cookie notice: in the center influenced their decisions. Most of the respondents (61.3%) said that a central position of the cookie notice on the screen had a somewhat positive to very positive effect on their attitudes. **Position of the cookie notice: left or right** This was mentioned as an influencing factor by only 19 people (3%). In addition, 53.6% answered that this had a somewhat positive to very positive effect on their attitudes towards cookie notices.

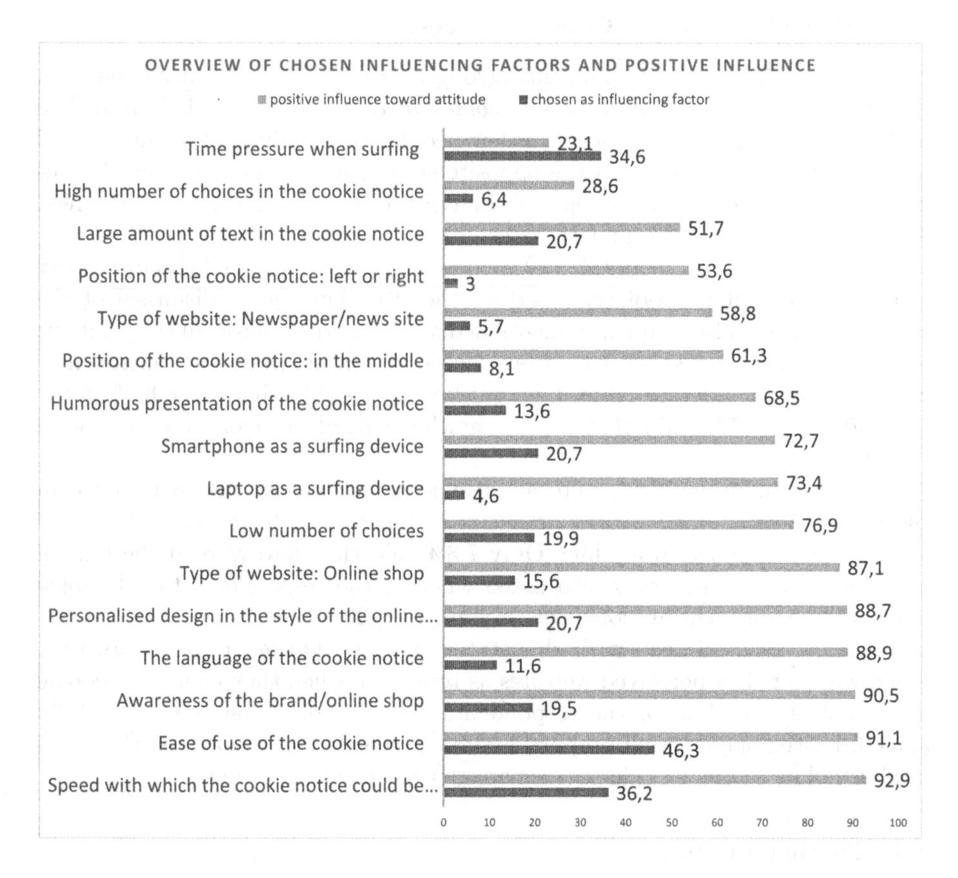

Fig. 3. Overview of the influencing factors examined and positive influence

All participants had the option of listing further factors they considered to be influential in an open text field. 35 of the respondents used this option. The majority of additions related to their general attitudes towards cookies. Twelve respondents indicated that they only ever agreed to essential cookies, regardless of how they are designed, and thus did not allow any of the factors mentioned to influence their decisions. Seven respondents mentioned the issue of trust: They stated that, if they trusted the company or the website, they were more likely to agree to all cookies. In addition, the frequency of use of a website or online shop was mentioned in relation to trust: The more frequently a website or online shop was used, the greater their trust in the company/website. Four respondents mentioned that they thought cookies and the associated cookie notices were annoying or disturbing. Some respondents mentioned that they used cookie blockers and, in order for them to accept cookies on a website or web shop, they had to be frequent visitors. Further, they mentioned that openness and honesty played a central role.

5.3 Attitudes Towards Cookie Notices

23.6% and 37.3% of the respondents strongly agreed or somewhat agreed with the statement *I always agree to all cookies in a cookie notice.* This is in line with the results of the experiment, as few people chose individual settings or did not agree to all cookies. With the opposite statement, namely whether all non-essential cookies were always rejected, half of the participants somewhat agreed or strongly agreed (32.3% and 18.3%).

Most of the respondents (63.2%) answered that they decided spontaneously whether to accept all cookies, based on the situation. Quick dismissal of the cookie notice was also frequently mentioned as an external factor. In this context, respondents were asked whether they clicked the most prominent button in a cookie notice in order to close it as quickly as possible. Just over half of the respondents (56.7%) said that they somewhat agreed or strongly agreed with this statement.

In addition, respondents were asked whether they read the text shown in the cookie notice. A small proportion of respondents (29.3%) said they read the text at least once in a while. Only 7.3% said they always read the text of a cookie notice. The survey also asked whether the respondents found cookie notices annoying. The majority of respondents (30.8% and 57.1%) somewhat agreed or strongly agreed with this statement. In addition, respondents were asked whether they perceived websites as reputable when they contained cookie notices. Just over half of the respondents (58.5%) somewhat agreed or fully agreed. Lastly, only a small proportion of respondents (12.7%) agreed that they used a tool or a browser add-on to manage their cookie decisions.

5.4 Further Findings

In the course of further investigations in the survey, users were shown the two cookie notice variants, and then questions were displayed.

The first additional finding concerns the respondents' opinions on the **Number of choices** given in the two variants. The majority felt that the number of options was appropriate for both the binary-choice cookie notice (77.4%) and the category-choice notice (76.2%) was appropriate. Only among the remaining respondents did the distribution differ: 20.6% stated that the number of options in the category-choice notice was too high, while only 12.1% stated that this was the case for the binary-choice notice. Furthermore, we were interested in respondents' **Perception of control** over the choices in the two cookie notices. Respondents were able to rate their feelings in this regard on a four-point scale. The majority of participants strongly agreed that they had control over the choices both for the binary-choice cookie notice (53.1%) and the category-choice (53.4%). Almost one third of the respondents (28.6%) somewhat disagreed with this statement for the choices in the binary-choice variant, while fewer people (16.4%) stated the same for the category-choice variant.

Participants were shown pictures of both variants to ask them about their **Preferred cookie notice variant**. More than half of the respondents (53.5%) answered that they preferred the category-choice cookie notice and 37% preferred the binary-choice variant; a small percentage (9.5%) expressed no preference. Respondents could state in an open text field why they preferred a particular variant. Those who chose the *category-choice* notice as their preferred variant mentioned the following reasons:

- Category-choice provided a better overview of the cookies and therefore better transparency.
- All decisions could be made on the first level of the notice and it was not necessary to click several times to accept only the essential cookies.
- Trust was won by presenting all cookies and communicating transparently, since this was perceived as the company not being afraid to address this topic openly and not seeking a particular outcome. In general, the category-choice variant was considered more trustworthy the binary-choice.
- Some of the responses referred to the design and the amount of text. For example, the design of the category-choice notice was perceived as more attractive and the amount of text as more pleasant, although both notices followed the same design guidelines and displayed the same text.

The respondents who preferred the *binary-choice variant* stated the following reasons:

- Some said that it was easier to navigate than the category-choice variant.
- They mentioned that selection was faster, as the notice was perceived to be simple and straightforward.
- The amount of text was also mentioned as a factor: Respondents had no time or desire to read all the text and they considered the text on this variant to be adequate.
- As there were only two choices, participants stated that this variant seemed clearer.
- Existing trust in the brand was also cited by some respondents as a reason for their preferences.

These results indicate that cookie notice variants that make users feel like they do not have control over their choices undermine user trust, which makes these variants poor choices not only from a legal point of view.

6 Conclusion

The experiment showed that binary-choice cookie notice resulted in a 10.7% higher consent rate than the category-choice variant; an even higher consent rate on mobile devices was found. This can be attributed to the fact that factors such as speed with which the cookie notice could be dismissed were mentioned often and in the case of mobile browsing, the entire screen is blocked by the cookie notice, so people tend to tend to close it as quickly as possible.

Although none of the factors examined stood out particularly, this study provides some insights of practical relevance. As expected, users felt predominantly annoyed by cookie notices (87.9%). Nevertheless, these are usually displayed upon first contact with a website or online shop. Agreement with the statements that the most prominent button is clicked in order to dismiss cookie notices as quickly as possible or that all cookies are usually accepted also confirms that people want to deal with cookie notices as little as possible. However, 58.5% of respondents associated the presence of a cookie notice with the credibility or trustworthiness of the site.

When designing cookie notices, website operators should therefore consider that showing their users a complicated cookie notice with digital nudges risks annoying them further. Due to the dynamics associated with cookie notices both in the digital sphere and within the legal framework, switching to consent management platforms to significantly reduce effort might make sense. Nevertheless, changes to the cookie notice, such as adjustments to the design or even the positioning of the buttons, can also lead to better opt-in rates. These options should be tested in any case, as they are not very risky. Moreover, all possible choices for the users should be made visible on the first layer of the cookie notice.

The research work presented here also has limitations. The company serving as the practical example belongs to the cosmetics manufacturing sector, which is probably the reason for the significant gender bias among the participants. Tracking in the experiment was restricted: It was only recorded whether all or a selection of individual cookies were accepted, but not which selection was made in detail. This could offer a starting point for further research, as it would be interesting to determine whether further selections are made or whether people opt to allow essential cookies only. Furthermore, the empirical study was limited to visitors from German-speaking countries. A comparison of countries and sectors in this context seems a promising avenue for further research.

References

1. Bauer, J.M., Bergstrøm, R., Foss-Madsen, R.: Are you sure, you want a cookie? - the effects of choice architecture on users' decisions about sharing private online data. Comput. Hum. Behav. **120**, 106729 (2021). https://doi.org/10.1016/J.CHB.2021.106729
2. Bornschein, R., Schmidt, L., Maier, E.: The effect of consumers' perceived power and risk in digital information privacy: the example of cookie notices. J. Public Policy Mark. **39**(2), 135–154 (2020). https://doi.org/10.1177/0743915620902143
3. Boßow-Thies, S., Hofmann-Stölting, C., Jochims, H. (eds.): Data-driven Marketing. Springer, Wiesbaden (2020). https://doi.org/10.1007/978-3-658-29995-8
4. Dürager, S.: Der EuGH zur Zulässigkeit des Setzens von Cookies - eine endlose Geschichte. jusIT **6**, 241 (2019)
5. Friedmann, V.: Privacy UX: better cookie consent experiences (2019). https://www.smashingmagazine.com/2019/04/privacy-ux-better-cookie-consent-experiences/
6. Gradow, L., Greiner, R.: Quick Guide Consent-Management. QG, Springer, Wiesbaden (2021). https://doi.org/10.1007/978-3-658-33021-7
7. Kettner, S., Thorun, C., Spindler, G.: Innovatives Datenschutz-Einwilligungsmanagement. Tech. rep., German Federal Ministry of Justice (2020). https://www.bmjv.de/SharedDocs/Downloads/DE/Service/Fachpublikationen/090620_Datenschutz_Einwilligung.pdf
8. Kulyk, O., Hilt, A., Gerber, N., Volkamer, M.: "This website uses cookies": users' perceptions and reactions to the cookie disclaimer. In: Proceedings of the European Workshop on Usable Security, EuroUSEC. Internet Society (2018). https://doi.org/10.14722/eurousec.2018.23012
9. Matte, C., Bielova, N., Santos, C.: Do cookie banners respect my choice?: measuring legal compliance of banners from IAB Europe's transparency and consent framework. In: Proceedings - IEEE Symposium on Security and Privacy, vol. 2020, pp. 791–809, November 2020. https://doi.org/10.1109/SP40000.2020.00076
10. Nouwens, M., Liccardi, I., Veale, M., Karger, D., Kagal, L.: Dark patterns after the GDPR: scraping consent pop-ups and demonstrating their influence. In: Conference on Human Factors in Computing Systems - Proceedings. Association for Computing Machinery (2020). https://doi.org/10.1145/3313831.3376321
11. Schiefermair, J., Stabauer, M.: The effects of cookie notices on perceived privacy and trust in E-Commerce. In: Nah, F.F.-H., Siau, K. (eds.) HCII 2020. LNCS, vol. 12204, pp. 535–549. Springer, Cham (2020). https://doi.org/10.1007/978-3-030-50341-3_40
12. Schmidt, L., Bornschein, R., Maier, E.: The effect of privacy choice in cookie notices on consumers' perceived fairness of frequent price changes. Psychol. Mark. **37**(9), 1263–1276 (2020). https://doi.org/10.1002/mar.21356
13. Strycharz, J., Smit, E., Helberger, N., van Noort, G.: No to cookies: empowering impact of technical and legal knowledge on rejecting tracking cookies. Comput. Hum. Behav. **120**, 106750 (2021). https://doi.org/10.1016/j.chb.2021.106750
14. Utz, C., Degeling, M., Fahl, S., Schaub, F., Holz, T.: (Un)informed consent: studying GDPR consent notices in the field. In: Proceedings of the ACM Conference on Computer and Communications Security, pp. 973–990. Association for Computing Machinery (2019). https://doi.org/10.1145/3319535.3354212

The Factors Influencing the Willingness of Investors to Use Robo-Advisors

Yi-Cheng Ku(✉) and Hai-Xuan Wang

Fu Jen Catholic University, New Taipei City 242062, Taiwan
ycku@mail.fju.edu.tw

Abstract. This study conducted an empirical survey to investigate the factors influencing investors' willingness to use Robo-advisors. The research results show that both perceived ease of use and perceived control have a positive and significant impact on perceived usefulness, which, in turn, increases the willingness to use Robo-advisors. Social presence also positively affects users' willingness to use Robo-advisors. Moreover, users' trust in vendors increases users' trust in Robo-advisors, while perceived risk decreases trust in Robo-advisors. Additionally, users' anxiety enhances users' perception of risk. The findings show that users' trust both in vendors and in Robo-advisors positively affects users' willingness to use Robo-advisors. This study provides useful insights into the marketing strategy of Robo-advisor services.

Keywords: Robo-Advisor · IT adoption · Usage Intention

1 Introduction

The rise of the digital wave and the impact of the global financial turmoil on the financial industry in 2008 have created good opportunities for the development of financial technology (Fin-tech). As a product of the combination of financial services and technology, Fin-tech relies on the ICT (information and communication technology) devices to provide financial-related services, which not only simplifies the otherwise complicated transaction process, but also frees financial services from the time and place restrictions caused by cash transactions. As a result, current transactions have become more flexible, efficient, and convenient [19].

Robo-advisors, as a typical application of financial technology, can provide investors with personalized asset management advice through customer profile analysis. Compared with traditional financial advisors, Robo-advisors have significant advantages such as a low investment threshold, low management costs, instant availability, and a transparent operation and management process [4]. As the main purchasers of financial products, bank users mostly have small savings, lack investment experience, and are risk-averse. Robo-advisors provide these investors with more diversified choices, which can help them obtain the opportunity to use professional asset management services at lower costs [13]. In addition, for banks and financial companies, the use of artificial

F. Fui-Hoon Nah and K. Siau (Eds.): HCII 2022, LNCS 13327, pp. 286–299, 2022.
https://doi.org/10.1007/978-3-031-05544-7_22

intelligence technology not only effectively reduces labor costs but also enhances customer interaction. Therefore, Robo-advisors have received more and more attention in the financial market as an emerging investment channel.

In recent years, financial institutions have launched a variety of Robo-advisors to capture market opportunities. According to the data of Statista, in the global Robo-advisor market, there are nearly 300 million users and the total assets under management are more than $1.36 trillion. And the market will maintain an annual growth rate of 20.1% (2021–2025), which shows a promising market prospect[1]. However, despite the trend of rapid growth, compared with other investment and finance tools, Robo-advisors are significantly lower than expected in terms of both growth rate and user acceptance [11]. Taking the global penetration rate of Robo-advisors for example, by 2020, the penetration rate on the global scale was only 3%. In addition, the growth rate has been declining since 2018, indicating that in the current market, there are still many people who are not users of Robo-advisors. Therefore, identifying the factors that influence investors' willingness to use Robo-advisors has become an important issue. This helps financial institutions design intelligent investment instruments that are more acceptable to users, and realize the value created by both financial institutions and users. Meanwhile, the popularization and promotion of Robo-advisors in the world can also be furthered.

However, most of the previous studies discussed the influence mechanism of the willingness to use Robo-advisors from the perspective of system function optimization. Considering that Robo-advisors are both an emerging technology and a service, they need to focus on delivering core values, as well as social emotions and relationship elements. While researchers attach importance to the functional design of the system, they also need to pay attention to the influence of social-emotional factors and the relationship between users and Robo-advisors on users' adoption intention. But so far, there is still insufficient research to explore and consider these aspects.

The motivation of this study is to explore and consider the influencing factors of different aspects of Robo-advisors from a user-centered perspective, and establish the corresponding influencing mechanism. Therefore, the main purposes of this study are:

(1) To explore the influence of users' cognition of system function on users' willingness to use Robo-advisors.
(2) To explore the influence of users' cognition of platform emotion on users' willingness to use Robo-advisors.

2 Literature Review

Robo-advisors are an emerging automated asset management service, which can quickly analyze customer information and provide customized asset management recommendations through the application of interactive intelligent assistance systems, big data analysis, machine learning, and cloud computing technologies [13, 24]. The main functions of Robo-advisors include user assessment, portfolio advice, portfolio monitoring, and asset rebalancing. The service process can be roughly summarized into three stages:

[1] https://www.statista.com/outlook/dmo/fintech/personal-finance/robo-advisors/worldwide.

Configuration, Matching & Customization, and Maintenance. It also includes six steps, namely Initiation, Profiling, Assessment, Offer, Implementation, and Maintenance [13].

Compared with traditional financial advisers, Robo-advisors have the advantages of a low investment threshold, low costs, high efficiency, and instant availability because the suggestions are completely based on algorithms. Robo-advisors can also help users effectively reduce the impact of behavioral biases (e.g., trend-chasing and ranking effects) [7]. In addition, because the operation management is completely based on the system, the cost structure of Robo-advisors is relatively more transparent, which strengthens security to a certain extent [4]. Therefore, in recent years, Robo-advisors have become popular with many young people and users who fear financial fraud [4]. It is estimated that by 2025, the global Robo-advisor market will have 470 million users and $2.9 trillion in assets under management.

However, Robo-advisors also have many limitations. First of all, the current Robo-advisors are still unable to achieve a high degree of interaction with customers, so it is difficult to establish a close relationship with customers and provide customized financial advice for them. At the same time, because the current Robo-advisors usually only use a general questionnaire to collect and measure the status of users, and the questionnaire is usually too simple to fully summarize their financial status and needs, failing to provide the most suitable suggestions [15]. In addition, there are also some conflicts of interest between Robo-advisors and customers. In theory, though, Robo-advisors are expected to offer fully personalized investment advice to users as algorithms continue to improve. Therefore, Robo-advisors still have great potential for development [19].

2.1 Relevant Research on Robo-Advisors

Since the application of intelligent Robo-advisors involves interdisciplinary knowledge, in previous studies, many scholars have studied the relevant issues of Robo-advisors from the perspectives of information management, financial management, and legal supervision. In addition to the discussion of the basic concepts of Robo-advisors, the research topic also involves the market development of Robo-advisors. In the field of information management, relevant studies focus on the optimization of their systems, models, algorithms, and interface design [1, 14]. In addition, from the perspective of users, some scholars focus on the factors that influence users' adoption of Robo-advisors [2, 5, 12]. In the financial field, some studies discuss the development status and prospects of Robo-advisors from the perspective of wealth management, as well as the substitution effect of Robo-advisors on traditional financial advisers [4]. In the field of law, relevant studies focus on the controversial issues and challenges in the legal aspect of Robo-advisors and the impact on the relevant laws [25]. And in terms of supervision, research discussed the important part of Robo-advisors and the key points of supervision, and put forward corresponding improvement suggestions for related security issues [27].

2.2 Factors Influencing the Adoption of Robo-Advisors

Previous studies have shown that the design of Robo-advisors needs to integrate the perspectives of financial service providers and users and explore the value jointly created by both parties. In addition to focusing on the optimization of interface design, risk assessment, and financial advice, research on Robo-advisors also needs to observe the behavior of users as an important source of information to improve user experience [14]. Among them, the observation of users' behavior is to analyze their behavior patterns from the perspective of users, including their trust and acceptance of Robo-advisors. In addition, Wirtz et al. [26] believe that the acceptance of service robots depends not only on their core functions but also on the emotions of users in the process of using the system and their relationship with the robots. Accordingly, they proposed the service robot acceptance model (sRAM) and argued that functional elements, social-emotional elements, and relational elements will significantly affect the willingness to use service robots. Therefore, based on the Service robot acceptance model (SRAM), this study will summarize the factors affecting the adoption of Robo-advisors mentioned in the previous literature from three dimensions: system function, social emotion, and relationship.

Functional Dimensions. System function-related research mainly explores users' perception of Robo-advisor function. From the perspective of behavioral intentions, many scholars discuss the factors that affect users' willingness to use Robo-advisors based on the technology acceptance model (TAM) and its development model, the unified theory model (UTAUT). Belanche et al. [2] found that users' willingness to use Robo-advisors would be positively influenced by perceived ease of use, attitude, and subjective norms. Although perceived usefulness positively affects attitude, it has no significant effect on adoption intention. In addition, the decrease of technology familiarity will enhance the influence of subjective norms on users' adoption intentions.

Sa et al. [24] verified cost, reliability, convenience, personal innovation, self-efficiency, social impact and perceived ease of use can enhance users' perception of the usefulness of Robo-advisors. And the strengthening of both perceived ease of use and perceived usefulness increases users' willingness to use the platform. Moreover, research also found that the reduction in prior knowledge not only weakens the positive effects of cost and social influence on perceived ease of use, but also weakens the positive effects of cost, convenience, and self-efficiency on usefulness. Some studies have also used integrated technology acceptance model 2(UTAUT2) to identify potential determinants of fintech adoption [10]. The research results show that performance expectancy, social influence, and technical conditions also have a significant positive impact on the willingness to use Robo-advisors.

Other studies have demonstrated that perceived user control is also an important factor in users' adoption of Robo-advisors. Because a high degree of system automation may enhance users' perception of risk, leading to users' distrust of the system. This negative impact can be effectively alleviated by improving users' control level over the system. In delegating tasks, losing control of the task also increases the client's perception of risk [21]. Therefore, Ruhr et al. [22] proved that perceived automation can improve adoption intention by influencing performance expectations. While perceived user control can also improve adoption intention by reducing users' risk perception of the system. In addition, Ruhr [23] also found that system transparency and user control have a significant positive impact on the improvement of adoption intention. The degree of automation has been shown to have an "inverted U" effect on adoption intention.

Social-Emotional Dimensions. In the research on social emotion, social presence is considered to be an important factor affecting users' feelings about Robo-advisors.

Because social presence can well explain the positive influence of the anthropomorphic design on system affinity and users' trust in Robo-advisors. Adam et al. [1], based on anchoring-and-adjustment effect, anthropomorphism theory, and social response theory, explored whether personalized anchor and the Anthropomorphism of Robo-advisors can enhance investors' willingness to invest by enhancing investors' social presence. The results confirm that personalized anchor and users' awareness of the social presence of Robo-advisors have greatly increased users' investment. Other studies further explored the relationship between anthropomorphism, social presenteeism, trusting beliefs, and users' likeliness to follow advice [18]. The results show that the enhancement of the degree of anthropomorphism can significantly enhance investors' perception of social presence. Then social presence increases users' willingness to use the advice by enhancing users' trust in Robo-advisors.

Relational Dimensions. In terms of relational elements, trust, as a key factor affecting adoption intention, has always been the focus of research. Many studies have shown that the trust relationship between users and Robo-advisors significantly increases users' adoption willingness [18]. Users' trust in vendors and technologies as well as users' perceived risk are considered to be important factors affecting users' trust in Robo-advisors. From the perspective of users' trust in vendors and technologies of Robo-advisors, some scholars explored the trust influence mechanism of Robo-advisors [5], and found that users' trust in the vendor and technology would significantly improve users' trust in the platform. Other studies have proved that the type of vendor is also an important factor in affecting the trust of users [17]. In addition, some users are worried about Robo-advisor's failure to achieve expected performance and financial information leakage, and thus feel that they are taking more risks. Furthermore, perceived risk has been shown to significantly weaken users' trust in Robo-advisors. Gerlach and Lutz [10] found that users' perception of operational risk would reduce users' willingness to use Robo-advisors. Moreover, affective reactions are also considered to be an important factor in affecting the relationship between users and Robo-advisors. Joy and anxiety have been proved to be closely related to adoption intention [12].

3 Research Method

3.1 Research Framework and Hypotheses

The framework of this study is based on the service robot acceptance model. As discussed in the previous literature, in the dimension of system function, perceived ease of use and perceived usefulness usually have a significant impact on the willingness to use the system, and perceived control has also been proved to have a certain connection with adoption intention. Therefore, the variables of "perceived ease of use", "perceived usefulness" and "perceived control" were included in this study. In the dimension of social emotion, social presence was proved to be an important factor in affecting adoption intention. Therefore, the "social presence" variable was taken into consideration in this study. In relation dimension, trust turns out to be an important relationship between users and Robo-advisors and can significantly affect adoption intention. Furthermore, users' trust in vendors, perceived risk of cognition, and anxiety are closely related to users' trust in the platform. Therefore, "trust in vendors", "perceived risk", "anxiety" and "trust in Robo" were also taken into account in this study.

In the field of fintech, users' perception of the ease of use and usefulness of fintech also significantly increases adoption willingness. Perceived ease of use and perceived usefulness significantly affect users' willingness to use Robo-advisors, while perceived ease of use has an impact on the role of perceived usefulness [2, 24]. In addition, when the perceived control is improved, users will believe that the system gives them more permission and the possibility to change system decisions, thus reducing the risk perception of task failure. This way, users will believe that the system is practical and reliable [23]. Therefore, this study inferred the following hypotheses:

H1: Perceived usefulness will positively influence users' willingness to use Robo-advisors.

H2: Perceived ease of use will positively influence users' willingness to use Robo-advisors.

H3: Perceived ease of use will positively influence the perceived usefulness of users.

H4: Perceived control will positively influence the perceived usefulness of users.

In terms of social emotions, if Robo-advisors can make users feel as if interacting with real people, it means that this interaction process has a high degree of social interaction. It will increase their willingness to use Robo-advisors [26]. Similarly, when users perceive that Robo-Advisors have the presence of a human-like service, their willingness to use them will be increased. Therefore, the following hypotheses are inferred in this study:

H5: Social presence will positively influence users' willingness to use Robo-advisors.

As for the relationship dimension, it has been proved that when users have higher trust in Robo-advisors, they will be more likely to accept financial advice [11, 17]. However, previous studies have also confirmed that users' trust in Robo-advisor vendors will be transferred to users' trust in the platform [5]. Other studies have also confirmed that the higher the perceived risk of users to Robo-advisors, the lower their trust and acceptance [10]. In addition, some studies have also proposed that anxiety about new technology will reduce users' psychological acceptance, which will have a negative impact on users' intention to use Robo-advisors [8]. Therefore, the following hypotheses are inferred in this study:

H6: The trust in Robo-advisor vendors will positively influence users' willingness to use Robo-advisors.

H7: The trust in Robo-advisor vendors will positively influence users' trust in Robo-advisors.

H8: Users' trust in Robo-advisors positively influences users' willingness to use Robo-advisors.

H9: Perceived risk will negatively influence users' trust in Robo-advisors.

H10: Perceived risk will negatively influence users' willingness to use Robo-advisors.

H11: Anxiety positively affects users' perceived risk.

3.2 Empirical Survey

This study conducted a survey to collect empirical data through a questionnaire. The development of the questionnaire is based on the measurements proposed by previous studies. And the content is adjusted according to the actual situation of Robo-advisors. The questionnaire consists of three parts. The first part introduces the concept of Robo-advisors. The second part is the measurements of research constructs, and the third part is the collection of demographic data of the subjects. The questionnaire was measured by a Likert seven-point scale. The operational definitions and measurement items of the variables in this study are shown in Table 1.

Table 1. Measurement instruments

Constructs	Operational definition	Measurement items	Reference
Perceived ease of use	The extent to which users perceive Robo-advisors to be easy to use	Overall, I found that: 1. The RA is easy to use 2. Learning to use the RA was easy 3. My interaction with the RA is clear and understandable	[3, 9]
Perceived usefulness	The extent to which users' perception of Robo-advisors can improve their investment performance	Overall, I found this RA: 1. Useful in finding suitable investment portfolios 2. Makes it easier to search for investment portfolios 3. Enhances my effectiveness in finding suitable investment portfolios	[3, 9]
Perceived control	The extent to which the user perceives himself to be able to control Robo-advisors	When using RA, I think: 1. I felt in control with the RA 2. The RA lets me be in charge 3. When searching for investment portfolios, RA gives me more control over the process	[6]

(continued)

Table 1. *(continued)*

Constructs	Operational definition	Measurement items	Reference
Social presence	The extent to which users feel they are talking to a real person while using the Robo-advisor	When I was interacting with RA, I felt: 1. A sense of human contact in the RA 2. A sense of human warmth in the RA 3. A sense of human sensitivity in the RA	[20]
Trust in vendor	The extent to which consumers trust the Robo-advisor vendors	Based on my experience with the vendor (Banks/fund companies). I know: 1. It is honest 2. It is not opportunistic 3. It is trustworthy	[9]
Perceived risk	The degree of uncertainty risk perceived by users in the process of using the Robo-advisor	If I use this RA for investment, I think: 1. There is a high potential for money wasted 2. There is a significant risk 3. This is a risky choice	[16]
Anxiety	Users' anxiety during the process of using a Robo-advisor	When using an RA, I would be afraid that: 1. Mistakes that cannot be corrected will occur in the process (e.g., investment plan, registration) 2. I could cause a technical error 3. I could make errors in my investment	[12]
Trust	The degree to which the user trusts the Robo-advisor	I think this RA: 1. can understand my needs and preferences about investment portfolios 2. can provide an unbiased recommendation 3. is overall honest	[3]
Willing to use	User's intention to use the Robo-advisor	I am willing to use this RA as a tool that: 1. suggests to me a number of investment portfolios from which I can choose 2. assists me in deciding which investment portfolio to buy 3. helps with my decisions about which product to buy	[3]

This study focuses on investors who can do financial practice independently and have certain financial knowledge and experience. Samples come from two regions, mainland China and Taiwan. Considering the limitation of the physical questionnaire, this study collected data using an online questionnaire, which was sent to each participant by email. According to users' experience of Robo-advisors, the research samples were divided into "novice" and "experienced". Therefore, for the first category of samples, which is "those without experience", the research process was divided into two stages. In the first stage, samples watched a video introducing Robo-advisors provided by this study and entered the trial platform for trial. In the second stage, the samples entered the website of the online questionnaire to fill in the answers. For the second category of samples, which is "those with experience", they skipped the first stage and directly completed the online questionnaire in the second stage.

4 Findings and Discussions

In this study, data were collected employing an online questionnaire. A total of 136 questionnaires were finally recovered, and 16 invalid questionnaires were deleted after inspection. A total of 120 valid questionnaires were collected, and the valid rate was 88.2%. Among the effective samples, the gender ratio was 55% female and 45% male. Age distribution is 18 to 25 (56.7%), followed by 26 to 35 (31.7%) and 36 to 45 (10.8%). Students (59.2%) were in the majority of occupations, followed by the service, wholesale and retail trades (all 8%). The majority of educational attainment was graduate school (53.3%) and college (41.7%). And among the respondents, 28.3% had experience in smart investment platforms.

4.1 Reliability and Validity Analysis

Before the model validation, the reliability and validity of each construct were tested in this study. In terms of reliability, as shown in Table 2, the composite reliability (CR) and Cronbach's alpha of each construct in this study both exceeded 0.8, indicating that the measurement questions of each construct had credibility. In terms of construction validity, as shown in Table 3, the factor loading of all items in this study was higher than 0.8, indicating that the construction validity of this questionnaire was good. In terms of discriminant validity, according to previous studies, the square root of the average variance extracted (AVE) of self-constructs should be greater than the correlation coefficient between self-constructs and different constructs. As shown in Table 4, all constructs in this study meet the requirements, indicating that each construct in this study has certain discriminant validity.

Table 2. Construct mean and reliability

Constructs	CR	α	Mean	SD	AVE
Perceived ease of use	0.917	0.864	5.892	0.851	0.786
Perceived usefulness	0.900	0.834	5.781	0.889	0.751
Perceived control	0.923	0.875	5.119	1.161	0.800
Social presence	0.954	0.928	4.239	1.376	0.873
Trust in vendor	0.911	0.854	5.167	0.987	0.773
Perceived risk	0.915	0.860	4.267	1.317	0.783
Anxiety	0.956	0.931	4.589	1.557	0.879
Trust	0.896	0.827	5.317	0.906	0.743
Willing to use	0.904	0.840	5.183	1.061	0.759

Table 3. Factor loading of constructs

Constructs	Item	Factor loading	Constructs	Item	Factor loading	Constructs	Item	Factor loading
Perceived ease of use	PE1	0.888	Social presence	SP1	0.927	Anxiety	AN1	0.935
	PE2	0.894		SP2	0.941		AN2	0.961
	PE3	0.878		SP3	0.934		AN3	0.916
Perceived usefulness	PU1	0.835	Trust in vendor	TIV1	0.878	Trust	TR1	0.755
	PU2	0.862		TIV2	0.852		TR2	0.920
	PU3	0.901		TIV3	0.907		TR3	0.902
Perceived control	PC1	0.929	Perceived risk	PR1	0.835	Willing to use	WTU1	0.909
	PC2	0.883		PR2	0.950		WTU2	0.861
	PC3	0.872		PR3	0.864		WTU3	0.842

4.2 Data Analysis

This study test research model by Smart PLS, and the results are shown in Fig. 1. On the functional dimension, perceived usefulness had a significant positive effect on willingness to use RA ($\beta = 0.213, p < 0.05$), so hypothesis H1 is significantly supported. However, although perceived ease of use positively affects willingness to use RA, its path coefficient does not reach a significant level, so hypothesis H2 is not supported. In addition, perceived ease of use and perceived control had a significant positive effect on perceived usefulness, and the path coefficients reached a significant level ($\beta = 0.514, p < 0.001$; $\beta = 0.369, p < 0.001$), thus supporting hypotheses H3 and H4. For the social-emotional dimension, the path coefficient of social presence to willingness to use RA also reached a significant level ($\beta = 0.192, p < 0.01$), thus supporting hypothesis H5. On the relational dimension, users' trust in vendors has a positive effect on willingness

Table 4. Discriminant validity of constructs

Constructs	PE	PU	PC	SP	TIV	PR	AN	TR	WTU
PE	**0.887**								
PU	0.645	**0.867**							
PC	0.353	0.551	**0.895**						
SP	0.094	0.151	0.410	**0.934**					
TIV	0.382	0.477	0.677	0.503	**0.879**				
PR	−0.178	−0.173	−0.109	−0.002	−0.146	**0.885**			
AN	−0.111	−0.089	−0.167	0.097	−0.121	0.513	**0.938**		
TR	0.433	0.554	0.676	0.409	0.662	−0.215	−0.192	**0.862**	
WTU	0.401	0.537	0.581	0.460	0.628	−0.174	−0.044	0.668	**0.871**

Note: Elements in the diagonal represent the square root of AVE for the corresponding construct

to use RA and trust in Robo-advisors, and the path coefficient also reached a significant level ($\beta = 0.205$, $p < 0.05$; $\beta = 0.644$, $p < 0.001$). Therefore, hypotheses H6 and H7 are supported. However, users' trust in Robo-advisors has a positive influence on adoption intention, and its path coefficient reached a significant level ($\beta = 0.319$, $p < 0.01$), so hypothesis H8 is significantly supported. In addition, the path coefficient of perceived risk on users' trust in Robo-advisors was significant ($\beta = -0.121$, $p < 0.05$), but the path coefficient of its negative influence on willingness to use RA did not reach a significant level, so hypothesis H9 is supported while hypothesis H10 is not supported. Finally, perceived risk was also positively affected by anxiety, and its path coefficient was significant ($\beta = 0.513$, $p < 0.001$), so hypothesis H11 is supported.

4.3 Discussions

The purpose of this study was to investigate the influence of user perception on the willingness to adopt Robo-advisors. In the framework of this study, in terms of system function, perceived usefulness had a significant positive effect on adoption intention, while perceived ease of use and perceived control improved users' perception of platform usefulness. In terms of social emotions, social presence will enhance users' willingness to use Robo-advisors. In terms of relationship, users' trust in vendors of Robo-advisors and in Robo-advisors will enhance users' adoption willingness. However, the trust in vendors will enhance users' trust in Robo-advisors. On the contrary, the perceived risk will weaken this trust. In addition, users' anxiety during the use of Robo-advisors has been shown to significantly increase perceived risk.

For the perceived ease of use of the functional dimension in the framework of this study, hypothesis H2: "Perceived ease of use will positively influence users' willingness to use Robo-advisors" is not supported. According to the results of the study, although perceived ease of use can indirectly affect adoption intention by affecting perceived usefulness, its positive effect on adoption intention is not significant. This study concludes

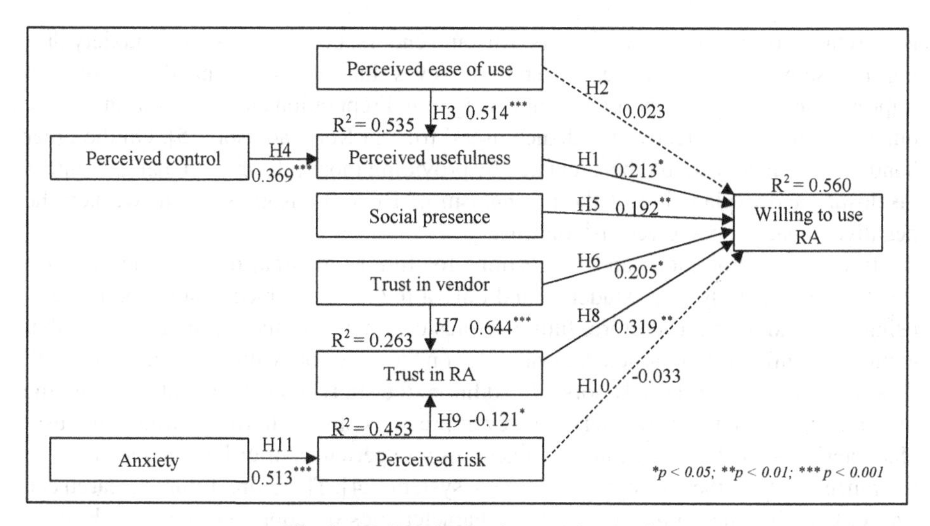

Fig. 1. PLS analysis of research model

that the main function of Robo-advisors is to guide investors to complete automatic investment and financial services. Because this is a fully automated service and involves financial investment, the focus of users' attention may be more inclined to ensure the improvement of investment performance and the guarantee of investment safety, while the impact of ease of use on adoption intention may be relatively weak, so it cannot have a significant positive impact on adoption intention.

For the perceived risk of the relational dimension in the framework of this study, hypothesis H10: "Perceived risk will negatively influence users' willingness to use Robo-advisors" is not supported. This study inferred that because the perceived risk variable in the study measured "the degree of uncertainty risk perceived by users in the process of using Robo-advisors", such perceived risk could weaken the adoption intention only by reducing users' trust in the platform. Therefore, the negative effect of perceived risk on adoption intention is not significant.

5 Conclusion

According to the research results, this study provides the following suggestions for Robo-advisor vendors to improve users' adoption intention. First, vendors can improve the ease of use of Robo-advisors by optimizing the design of the system and interface, or by increasing the selectivity of the system (for example, by adding expected investment years, investment target options, or more investment and financing options to choose from). In this way, users' perception of the system's usability can be enhanced, which may effectively promote users' willingness to use Robo-advisors. Secondly, Robo-advisor vendors can also improve the anthropomorphic degree of Robo-advisors by designing anthropomorphic images and names or adding chat boxes to them [1], which makes it easier for users to generate a sense of social presence in their interaction with Robo-advisors, thus effectively promoting adoption intention. In addition, trust in Robo-advisors significantly affects users' adoption intention, and users' trust in Robo-advisors is correlated to some extent with users' trust in vendors, perceived risk, and anxiety during the use process. Therefore, this study believes that, on the one hand, vendors can improve users' trust in vendors by improving brand reputation and information service quality, which may effectively enhance users' trust in Robo-advisors [5]. On the other hand, users' anxiety should be alleviated by providing more information that guarantees disclosure and explanation of the mechanism of Robo-advisors, so as to weaken the negative impact of perceived risk on trust.

In terms of limitations and suggestions for future research, most of the effective samples in this study were students, and only a few of the subjects had experience in using Robo-advisors. Therefore, future studies can increase study samples and further explore the influence of the selected variables on the adoption willingness of "those with experience in using Robo-advisors". In addition, the study did not take into account the individual characteristics of users. However, previous studies have confirmed that user characteristics, such as user innovativeness, user experience, prior knowledge, and risk tolerance all affect the willingness to use the system [24]. Therefore, future research can further examine the influence of users' characteristics on adoption intention. Finally, this study explores the factors that influence the willingness to use Robo-advisors, and the design of the system function is often influenced by user perception. Therefore, future research can explore the effects of interface design on these factors based on the conclusions of this study, so as to improve users' feelings through the optimization of interface design and promote their adoption intention.

Acknowledgement. This work was supported by the Ministry of Science and Technology (MOST) under the grants MOST 110-2410-H-030-030.

References

1. Adam, M., Toutaoui, J., Pfeuffer, N., Hinz, O.: Investment decisions with robo-advisors: the role of anthropomorphism and personalized anchors in recommendations. In: 27th European Conference on Information Systems (ECIS2019), Stockholm & Uppsala, Sweden (2019)
2. Belanche, D., Casaló, L.V., Flavián, C.: Artificial Intelligence in FinTech: understanding robo-advisors adoption among customers. Ind. Manag. Data Syst. **119**(7), 1411–1430 (2019)
3. Benbasat, I., Wang, W.: Trust in and adoption of online recommendation agents. J. Assoc. Inf. Syst. **6**(3), 72–101 (2005)
4. Brenner, L., Meyll, T.: Robo-advisors: a substitute for human financial advice? J. Behav. Exp. Financ. **25** (2020). Article 100275
5. Cheng, X., Guo, F., Chen, J., Li, K., Zhang, Y., Gao, P.: Exploring the trust influencing mechanism of robo-advisor service: a mixed method approach. Sustainability **11**(18), 4917 (2019)
6. Collier, J.E., Sherrell, D.L.: Examining the influence of control and convenience in a self-service setting. J. Acad. Mark. Sci. **38**(4), 490–509 (2010)
7. D'Acunto, F., Prabhala, N., Rossi, A.G.: The promises and pitfalls of robo-advising. Rev. Financ. Stud. **32**(5), 1983–2020 (2019)
8. Gelbrich, K., Sattler, B.: Anxiety, crowding, and time pressure in public self-service technology acceptance. J. Serv. Market. **28**(1), 82–94 (2014)
9. Gefen, D., Karahanna, E., Straub, D.W.: Trust and TAM in online shopping: an integrated model. MIS Q. **27**(1), 51–90 (2003)

10. Gerlach, J.M., Lutz, J.K.: Digital financial advice solutions–evidence on factors affecting the future usage intention and the moderating effect of experience. J. Econ. Bus. **117** (2021)
11. Hildebrand, C., Bergner, A.: Conversational robo advisors as surrogates of trust: onboarding experience, firm perception, and consumer financial decision making. J. Acad. Mark. Sci. **49**(4), 659–676 (2020). https://doi.org/10.1007/s11747-020-00753-z
12. Hohenberger, C., Lee, C., Coughlin, J.F.: Acceptance of robo-advisors: effects of financial experience, affective reactions, and self-enhancement motives. Financial Planning Review **2**(2), e1047 (2019)
13. Jung, D., Dorner, V., Glaser, F., Morana, S.: Robo-advisory. Bus. Inf. Syst. Eng. **60**(1), 81–86 (2018)
14. Jung, D., Dorner, V., Weinhardt, C., Pusmaz, H.: Designing a robo-advisor for risk-averse, low-budget consumers. Electron. Mark. **28**(3), 367–380 (2017). https://doi.org/10.1007/s12525-017-0279-9
15. Kaya, O., Schildbach, J., AG, D.B., Schneider, S.: Robo-advice–a true innovation in asset management. Deutsche Bank Res. (2017)
16. Liébana-Cabanillas, F., Sánchez-Fernández, J., Muñoz-Leiva, F.: Antecedents of the adoption of the new mobile payment systems: the moderating effect of age. Comput. Hum. Behav. **35**, 464–478 (2014)
17. Lourenco, C.J., Dellaert, B.G., Donkers, B.: Whose algorithm says so: the relationships between type of firm, perceptions of trust and expertise, and the acceptance of financial Robo-advice. J. Interact. Mark. **49**, 107–124 (2020)
18. Morana, S., Gnewuch, U., Jung, D., Granig, C.: The effect of anthropomorphism on investment decision-making with robo-advisor chatbots. In: Proceedings of the 28th European Conference on Information Systems (ECIS2020) (2020)
19. Phoon, K., Koh, F.: Robo-advisors and wealth management. J. Alternative Invest. **20**(3), 79–94 (2017)
20. Qiu, L., Benbasat, I.: Evaluating anthropomorphic product recommendation agents: a social relationship perspective to designing information systems. J. Manag. Inf. Syst. **25**(4), 145–182 (2009)
21. Rijsdijk, S.A., Hultink, E.J.: "Honey, have you seen our hamster?" Consumer evaluations of autonomous domestic products. J. Product Innovat. Manage. **20**(3), 204–216 (2003)
22. Rühr, A., Berger, B., & Hess, T.: Can I control my robo-advisor? Trade-offs in automation and user control in (digital) investment management. In: 25th Americas Conference on Information Systems, Cancun, Mexico (2019)
23. Rühr, A.: Robo-advisor configuration: an investigation of user preferences and the performance-control dilemma. In: 28th European Conference on Information Systems (ECIS2020), Marrakech, Morocco (2020)
24. Sa, J.H., Lee, K.B., Cho, S.I., Lee, S.H., Gim, G.Y.: A study on the influence of personality factors on intention to use of Robo-advisor. J. Eng. Appl. Sci. **13**(19), 7795–7802 (2018)
25. Strzelczyk, B.E.: Rise of the machines: the legal implications for investor protection with the rise of robo-advisors. DePaul Bus. Commercial Law J. **16**(1), 54–85 (2017)
26. Wirtz, J., et al.: Brave new world: service robots in the frontline. J. Serv. Manag. **29**(5), 907–931 (2018)
27. Yan, Z.W.: Key components and regulatory recommendations for robo-advisor in the financial services industry. Finance **9**(6), 557–563 (2019)

Holistic Approach to the Social Acceptance of Building Information Modelling Applications

Jari Laarni(✉) and Esa Nykänen

VTT Technical Research Centre of Finland Ltd., Espoo, Finland
jari.laarni@vtt.fi

Abstract. Building Information Modelling (BIM) applications are widely used in construction industry. Since there is some resistance to their adoption, we have to understand better the process of their acceptance and adoption. We introduce a three-legged approach to social acceptance assessment consisting of three main phases: The first leg/phase introduces a case-based approach for Adoption Impact Map construction, in which the aim is to model impacts of BIM adoption on end-user value. In the second leg/phase, BIM adoption among stakeholders is evaluated through the SPHERE BIM Digital Twin adoption survey, which will be distributed among professional stakeholders at the ending phase of the project. The survey is divided into two main parts: measurement of BIM acceptance readiness and measurement of individual and organizational intention. In the third leg/phase, post-occupancy evaluation is implemented through the SPHERE post-occupancy evaluation questionnaire. It gathers feedback on what occupants like in the building in which they live. The results of the assessment will be used to further support the development of SPHERE BIM Digital Twin Platform tool.

Keywords: Social acceptance · Building Information Modelling · Adoption Impact Map

1 Introduction

Building Information Modelling (BIM) applications are widely used in construction industry. Since there is some resistance to their adoption, we have to understand better the process of their acceptance and adoption. In an EU-funded project entitled SPHERE, in which the ultimate goal is to provide a BIM-based Digital Twin Platform to optimize the building lifecycle, reduce costs, and improve energy efficiency in residential or non-residential buildings, our aim is to develop an holistic approach to the acceptance and adoption of the SPHERE BIM platform and/or using selection of tools linked to it. According to the project's webpage [11], SPHERE integrates two planes of research, innovation and improvement: 1) development of a building-centered Digital Twin Environment, involving not only the design and construction of the building but including also the manufacturing and the operational phases; and 2) the seamless and efficient updating and synchronization of SPHERE's Digital Twin platform based on an Integrated Design and Delivery Solutions framework.

F. Fui-Hoon Nah and K. Siau (Eds.): HCII 2022, LNCS 13327, pp. 300–310, 2022.
https://doi.org/10.1007/978-3-031-05544-7_23

We introduce a three-legged approach to social acceptance assessment in SPHERE consisting of three main phases. Figure 1 describes the methodology for SPHERE social acceptance assessment. The first leg/phase introduces a case-based approach for Adoption Impact Map (AIM) construction, in which the aim is to model impacts of BIM adoption on end-user value. AIM specifies a set of relationships between the actions taken by main stakeholders and the effects of these actions on other stakeholders. The construction of Adoption Impact Maps is based on interview data collected at two different moments of the project. In the second leg/phase, BIM adoption among stakeholders is evaluated through the SPHERE BIM Digital Twin adoption survey, which will be distributed among professional stakeholders at the ending phase of the project. The survey is divided into two main parts: measurement of BIM acceptance readiness and measurement of individual and organizational intention. In the third leg/phase, post-occupancy evaluation is implemented through the SPHERE post-occupancy evaluation questionnaire. It gathers feedback on what occupants like the building in which they live. The results of social acceptance assessment will be used to further support the development of SPHERE BIM Digital Twin Platform tool.

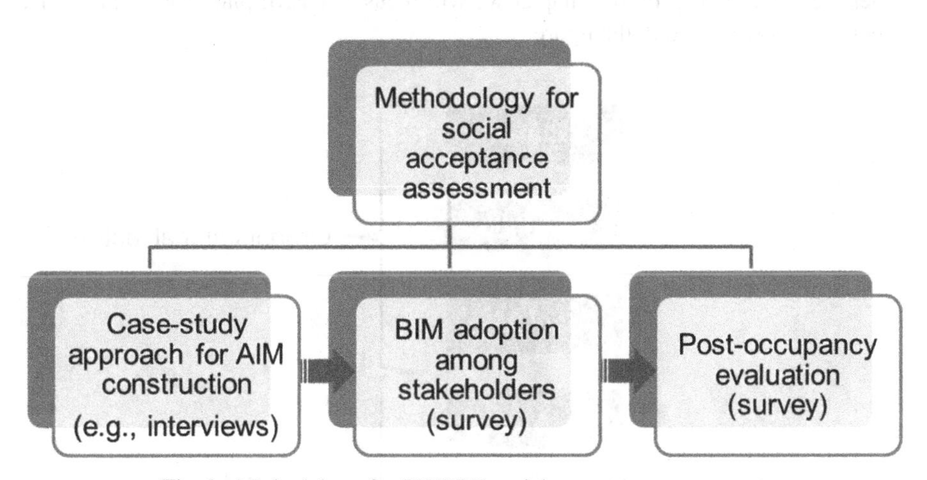

Fig. 1. Methodology for SPHERE social acceptance assessment.

2 Social Acceptance

2.1 Definition of Key Terms and Outline of Our Approach

In social psychology, acceptance means that other people wish to include a particular person in their group. Social acceptance of information technology means that people show their desire to include a particular technology in their life. According to [13], acceptance can be defined as the changes in attitudes, perceptions, and actions that steer humans to develop new practices and innovations different from what they normally use. It typically occurs on a continuum ranging from hardly tolerating a particular system to actively trying to acquire it. Adoption can be defined as the action of choosing to decide by a stakeholder to use a new system in the organization [2]. Implementation is the final phase in which the new system has been put into use. Social acceptance and adoption can be considered as different phases in the information system adoption process. If you look at the process from the organization's perspective, adoption precedes acceptance, because the organization has to adopt and use the information system before an individual can say whether it is acceptable or not. Technology acceptance models, however, traditionally look at the adoption process from the individual's perspective. According to them, the user can try and test the system to determine whether it is useful and easy to use, and based on these tests, he/she makes a decision to intend or not to intend to use the system.

Acceptance and adoption are desirable features: the aim is that a released information technology is accepted and adopted by users. Their opposite is resistance and rejection, which should be avoided. But since people may have good reasons for their reservations about new technologies, the designers and manufactures should also listen to these skeptical voices attentively.

Figure 2 shows the whole adoption process from initiation to diffusion based on the model presented by [4]. In this chapter we will focus SPHERE platform adoption and acceptance shown in red in the figure.

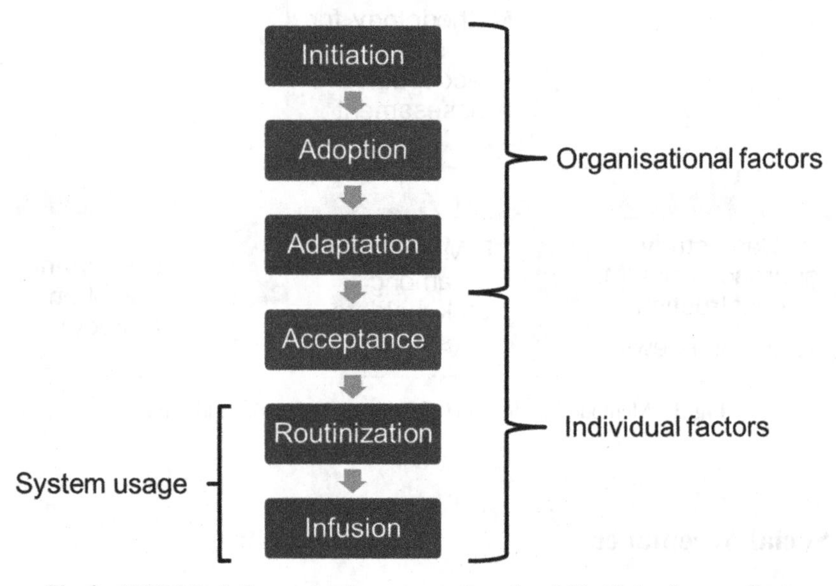

Fig. 2. SPHERE platform adoption process (based on [4]). (Color figure online)

Measuring Social Acceptance. A classical approach to social acceptance of technology is based on the technology acceptance model (TAM), according to which perceived usefulness and perceived ease of use are key elements to explain the use of a particular technology. According to the TAM, social adoption of technology is determined by a person's behavioral intention [5].

The original TAM has been further elaborated and revised by taking into account a larger set of external factors [18, 19]. Still, the TAM has been criticized to not taken into account a variety of factors related to the user task environment and constraints, and therefore it has been suggested that the TAM should be integrated into a larger framework covering a larger set of aspects [6]. In Fig. 3, the original TAM model has been supplemented by a characterization of motivational processes and phases of implementation to use a particular technology. Users' motivation to use a system consists of four phases, Attitude, Motivation, Intention and Behavior, whereas the phases of implementation are Adoption, Acceptance and Implementation.

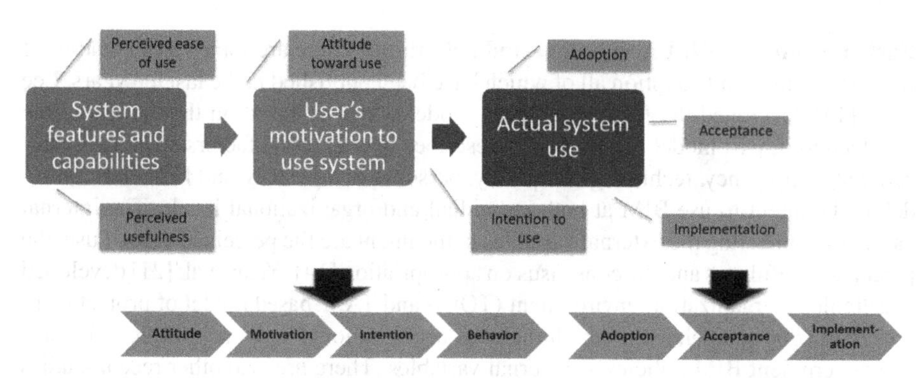

Fig. 3. An elaborated version of the technology acceptance model (TAM).

There are different characterizations of this larger framework. According to [15], technology adoption is jointly influenced by technological, organizational and environmental contexts so that organizational support, technical features and government policies have an impact on social acceptance and adoption of a technology. Another holistic framework has been proposed by [20] in which three dimensions of social acceptance have been identified, markets, socio-political and community. Based on Wustenhagen et al.'s distinction, Devine-Wright et al. presented an interdisciplinary approach to social acceptance integrating theoretical ideas from social representations theory, governance and human geography [6, 20]. According to their framework, market, political and community acceptance have to be analyzed at three different levels, international, national and local.

At the lowest level it could be investigated the process of SPHERE platform adoption and diffusion, drivers and barriers affecting the adoption and the relationships between organizational characteristics and willingness to adopt the SPHERE platform. In SPHERE we are interested in the acceptance and adoption of the BIM Digital Twin Platform both at the individual and organizational level, but the macro-level examination is mostly excluded.

Another classification has been proposed by [17], according to which adoption barriers can be clustered into four main dimensions, structure, people, technology and task dimensions. The structure dimension consists of factors such as legislation, awareness about BIM, government incentives and regulation, and spread of use; the people dimension includes factors such as acceptance vs. resistance to change, level of expertise, amount of training, information sharing, collaboration and trust, and management support; the technology factor includes standards and interoperability, infrastructure, requirements for adoption, applicability and practicability, availability, and quality of information; and finally the task factor includes investment costs, proven benefits, and investment capital needed [17]. In order to have a complete picture of SPHERE BIM Digital Twin Platform acceptance and adoption, we have to address social acceptance and adoption of the SPHERE System of Systems at all these different levels (i.e., structure, people, technology and task levels) and select relevant participants from all stakeholder groups.

Earlier Studies of BIM Acceptance and Adoption. There are some earlier studies of BIM acceptance and adoption all of which have been published in the last ten years. Lee et al. [14] developed the BIM acceptance model which is based on the TAM and the task technology fit model (TTF; [8]). It describes how external factors such as organizational competency, technological quality, personal competency and behavior control define the intent to use BIM at both individual and organizational levels. The internal variables connecting the external variables to the intent are the perceived ease of use, the perceived usefulness and the consensus on appropriation [14]. Yuan et al. [21] developed a technology-organization-environment (TOE)- and TAM-based model of project owners' BIM adoption behavior, which includes organizational support, technical features and government BIM policies as external variables. There are also other recent studies in which acceptance models for BIM have been developed, such as [3, 10].

There are several limitations in these TAM-based acceptance models [9]. In the first hand, they are too narrow, since they do not take into account a variety of factors related to the user's task environment and constraints. They neither consider the role of different actors and their interactions, nor a comprehensive list of beliefs about and responses to technological change, such as resistance or engagement. Individual and organizational maturity has an impact on BIM adoption, but it has not taken into account in TAM models very well. As 'black-box models' they do not describe the mechanisms through which BIM use can enhance the value users and occupants receive from construction projects, i.e., they do not predict the actual use of BIM tools. They focus on whether a technology is accepted or not, not on how people use it and what effects the technology has on people's life. It is also difficult to gather enough data for statistical analyses to evaluate causal relationships between BIM adoption actions and the effects of BIM use on users.

All in all, BIM acceptance and adoption should be seen from a holistic, systemic, socio-technical and life-cycle perspective. There is a continuum with different phases and stages of maturity, and different perspectives can/should be taken, since there is no single answer to BIM acceptance and adoption. In addition, more effort should be placed on the processes, impacts and implications related to BIM adoption, and there should be a shift from 'simple' survey-based methods to case-study approaches.

In order to completely understand social acceptance/adoption of the SPHERE BIM Digital Twin Platform, we have to consider all relevant stakeholders at both individual and organizational levels. Especially interesting is to study the benefits of SPHERE platform for the occupants of pilot facilities, because there is very little research on impacts of BIM adoption efforts on occupant value and experiences.

3 SPHERE Approach for Measurement of Social Acceptance and Adoption

3.1 Case Study Approach for AIM Construction

Impact mapping is a visual modelling technique for system development. An Adoption Impact Map specifies a set of relationships between the actions taken by main stakeholders and the outcomes of these actions on other stakeholders [1]. These interactions are mainly identified by interviews and focus group discussions. An impact map has four main components, Goal, Actor, Impact and Deliverable ([1]; Fig. 4): Goal defines what is desired to be achieved; Actor defines who will produce the desired effect; Impact defines what role would the actor play to help to achieve the goal; and Deliverable defines what makes it possible to create the desired impacts [1].

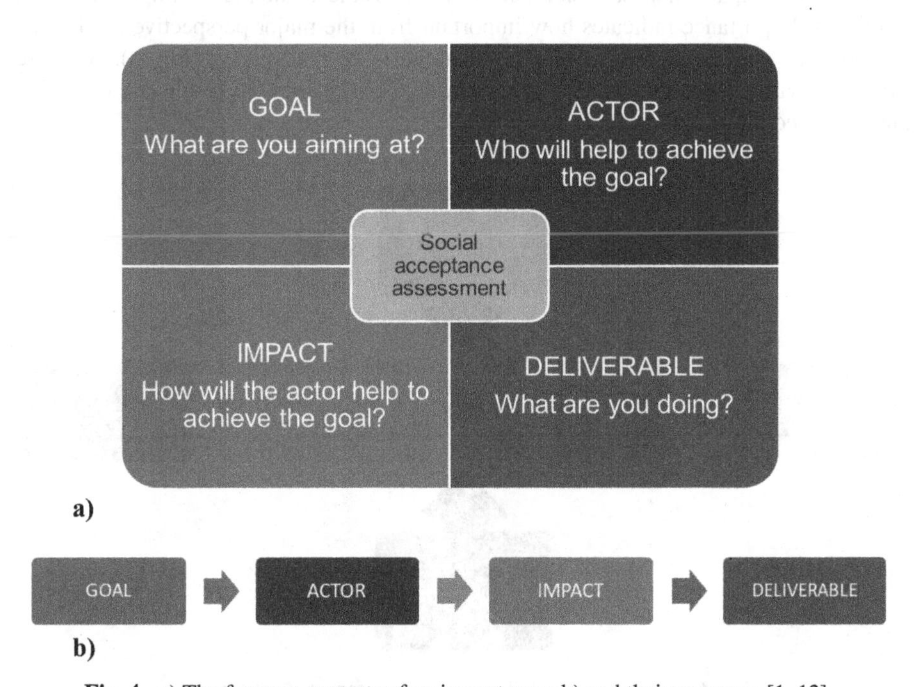

Fig. 4. a) The four components of an impact map, b) and their sequence [1, 12].

Interview transcripts are analysed by extracting above-mentioned main components from the transcripts and by constructing impact map statements using the identified components. Figure 5 illustrates an impact map statement for the promotion of social acceptance of BIM tools.

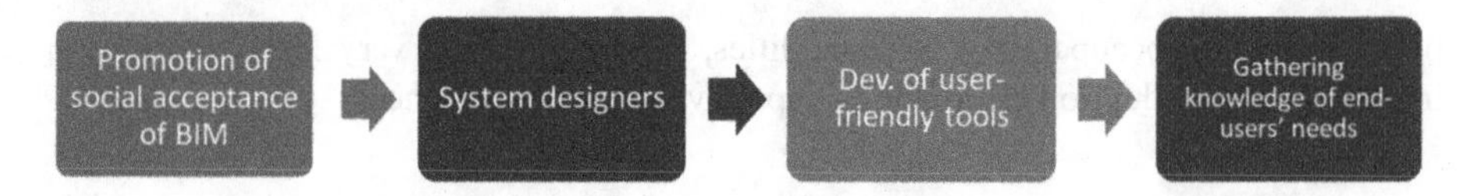

Fig. 5. An example of an impact map statement.

Each impact map statement can be decomposed into two simple statements by 'reading' the map backwards. As an example, the statement in Fig. 5 can be broken down into the following statements:

- Knowledge of end-users' needs enables development of user-friendly tools.
- Development of user-friendly tools by system designers promotes social acceptance of BIM tools.

These simple statements resemble those that Gurevich et al. [9] called 'adoption impact maps', but we prefer the description that is in accordance with how impact mapping is used in software development.

For each impact map statement, it is also possible to define its importance and likeliness. Importance indicates how important from the major perspective the impact is, and likeliness indicates how likely it is about to be realized [9] (Fig. 6). Both are evaluated with a three-item scale: low – medium – high, and the results are classified and tabulated along these two dimensions.

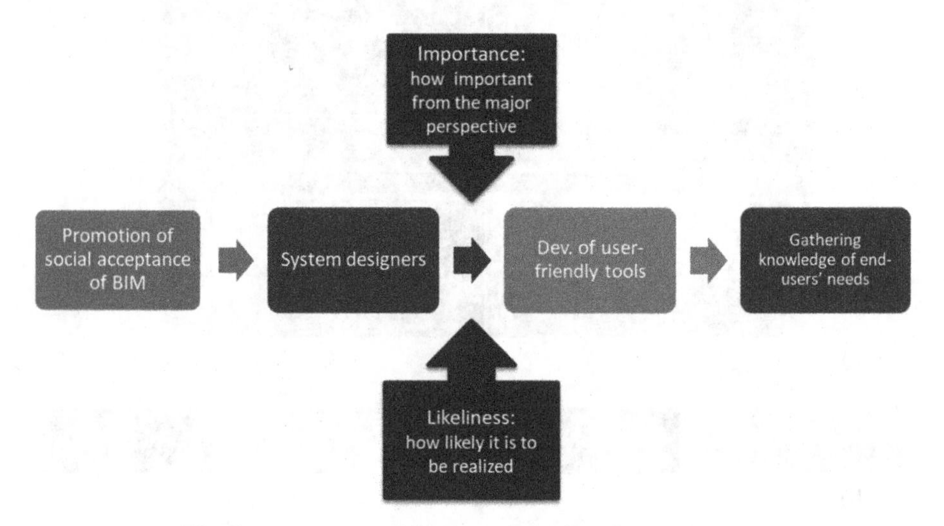

Fig. 6. Importance and likeliness for an impact statement [9].

3.2 SPHERE Platform Adoption Survey

A SPHERE BIM Digital Twin platform acceptance questionnaire will be distributed among professional stakeholders at the ending phase of the project. The questionnaire is divided into two main parts: measurement of BIM acceptance readiness and measurement of individual and organizational intention [14]. BIM acceptance readiness comprises factors such as Organizational Efficacy and Innovativeness, Personal Efficacy Innovativeness, Consensus on Appropriation, Collaborative Easiness, and Organizational Support and Pressure [14]. Individual and organizational intention comprises two factors, Individual and Organizational intention of SPHERE BIM DIGITAL TWIN acceptance [14].

3.3 Post-occupancy Evaluation Survey

A Post-occupancy Evaluation (POE) questionnaire gathers feedback on what occupants like or dislike in the building in which they live. The results of the survey will be used to further support the development of SPHERE BIM Digital Twin Platform tool. POE questionnaires were prepared based on existing POE survey tools both for residential and non-residential buildings (for a review, see, e.g., [16]).

4 Collection and Analysis of Interview Data

An inductive mixed method case-study approach is adopted, which makes it possible to approach the social acceptance of SPHERE platform with several relevant methods. A grounded theory approach suitable for case-study research is applied in the analysis of interview data. Based on continuous comparison between data and theory, the grounded theory approach enables, e.g., to identify opposing views, construct typologies and simple models and draft propositional statements.

Our case study approach consists of several phases that are shown in Fig. 7. The names of these phases are adopted from [7].

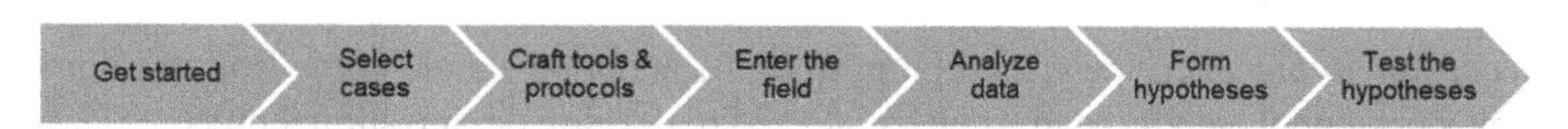

Fig. 7. Phases of the case study method applied in examination of social of acceptance of SPHERE BIM Digital Twin Platform (based on [7]).

A set of questions will be prepared for the interviews/focus group discussions. Pilot case descriptions and SPHERE BIM Digital Twin Platform information flow requirements will be used in preparation of the set of questions. The questions are aimed to generate a comprehensive picture of, e.g., each stakeholder's organization, adoption processes, the steps it will take to adopt the SPHERE platform and the results at each stage of the adoption process.

Especially, the interviews and focus groups provide answers to the questions such as (Fig. 8):

- what social benefits are aimed at;
- who will produce the desired effect;

- what is done to achieve the goal; and
- how it can be promoted.

The interviews and focus group discussions will be transcribed, and a summary of the interviews/discussions will be prepared. The findings are compared with existing research, results are discussed among SPHERE partners, and their feedback is received and processed.

5 Preliminary Results

Data collection and analysis is still ongoing, but some preliminary results can be presented. According to our first interviews, it is important that the goals of a development project support the company's strategic objectives, i.e., digitalization in construction projects. The project like SPHERE provides guidelines to what direction the company should develop their own tools. The project does not necessarily open new businesses: its main value lies in providing support for renewed businesses.

One of the main challenges is on how to integrate the new tools with existing solutions in terms of background information, system architecture, and operating system. It is typical that changing one system affects all other systems in a big company partly due to the fact that decision making is centralized. All in all, it is very difficult to develop a platform that is compatible with every other system. The evaluation of new tools may also be challenging, because it is very difficult to compare things that are not commensurable. For example, the older system might be quite slow, but the information was reliable - it might be the other way around with the new one. In any case, there is a trend towards more integrated processes and systems that support these processes.

5.1 User Acceptance and Adoption

According to our interviews, the first push to the adoption should come from the company's personnel, and there must be a real need to develop organizational processes and a permission to do so. It is feasible to start piloting in one site, and after that, extend the piloting to several other sites. In addition, systematic testing can be launched to select the best option among several alternatives: A promising tool is tested more thoroughly, and its fit with other systems is evaluated.

One of the main reasons for a possible failure of a procurement of a BIM solution is the lack of technical user support, since the need for technical support is often underestimated. Basic use of a BIM tool is quite easy to learn, but after that, more advanced support is often needed for a quite long period of time, since new kinds of problems may emerge at later phases.

According to the interviewees, new programs must be easy to use and implement. It is also important that user interfaces are stable and intuitive. Overall, developers should better know and understand end-users' needs. They often think that they can develop new tools in such a way that end-users are urged to try out the new artefact and say for what purpose it can be applied. It was considered as a totally wrong approach to developing user-friendly tools.

Figure 8 shows some examples of answers to the four AIM questions based on expert interviews. Key experts' proficiency, e.g., in project management and sales and meter installation and control promotes energy savings, carbon footprint reduction, thermal comfort and other benefits. The key experts' proficiency is, in turn, based on their skillful use of SPHERE BIM Digital Twin with associated tools such as energy simulation programs and Human Thermal Model.

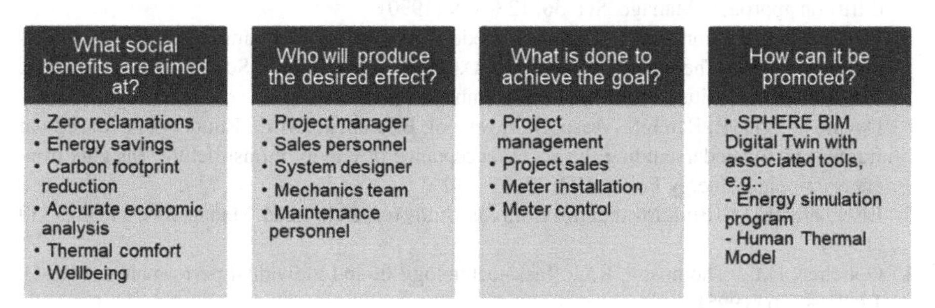

Fig. 8. Examples of answers to the AIM questions based on expert interviews.

6 Conclusions

BIM acceptance and adoption should be seen from a holistic, systemic, socio-technical and life-cycle perspective. Individual and organizational maturity has an impact on BIM adoption. There is a continuum with different phases and stages of maturity. Different perspectives can/should be taken: there is no single answer to BIM acceptance and adoption. More effort should be placed on the processes, impacts and implications related to BIM adoption, and there should be a shift from 'simple' survey-based methods to case-study approaches.

We advocate a three-legged approach to social acceptance assessment consisting of three main phases: an interview-based approach for Adoption Impact Map construction, evaluation of BIM adoption among stakeholders through a SPHERE BIM Digital Twin adoption survey, and post-occupancy evaluation implemented through a SPHERE post-occupancy evaluation questionnaire. According our first results, the first push to the adoption should come from the company's personnel, and there must be a real need to develop organizational processes and a permission to do so. It was also found that one of the main reasons for a possible failure of a procurement of a BIM solution is the lack of technical user support.

Acknowledgment. The SPHERE project has received funding from the EU's Horizon 2020 research and innovation program under grant agreement no. 820805.

References

1. Adzic, G.: Impact mapping. Making a Big Impact with Software Products and Projects. Leanpub (2014)

2. Bouwman, H., van den Hooff, B., van de Wijngaert, L., van Dijk, J.: Information & Communication Technology in Organizations; Adoption, Implementation, Use and Effects. Sage Publications, London (2005)
3. Cao, D., Li, H., Wang, G., Zhang, W.: Linking the motivations and practices of design organizations to implement building information modeling in construction projects: empirical study in China. J. Manag. Eng. **32**(6), 04016013 (2016)
4. Cooper, R.B., Zmud, R.W.: Information technology implementation research: a technological diffusion approach. Manage. Sci. **36**, 123–139 (1990)
5. Davis, F.D.: A technology acceptance model for empirically testing new end-user information systems: Theory and Results (Ph.D. Thesis). MIT Sloan School of Management, Massachusetts Institute of Technology, Cambridge (1986)
6. Devine-Wright, P., Batel, S., Aasc, O., Sovacool, B., Labelle, M.C., Ruud, A.: A conceptual framework for understanding the social acceptance of energy infrastructure: Insights from energy storage. Energy Policy **107**, 27–31 (2017)
7. Eisenhardt, K.M.: Building theories from case study research. Acad. Manag. Rev. **14**, 532–550 (1989)
8. Goodhue, D.L., Thompson, R.L.: Task-technology fit and individual performance. MIS Q. **19**, 213–236 (1995)
9. Gurevich, U., Sacks, R., Shrestha, P.: BIM adoption by public facility agencies: impacts on occupant value. Build. Res. Inf. **45**, 610–630 (2017)
10. Hong, S.-H., Lee, S.-K., Kim, I.-H., Yu, J.-H.: Acceptance model for mobile building information modeling (BIM). Appl. Sci. **9**, 3668 (2019)
11. https://sphere-project.eu/
12. https://www.plays-in-business.com/impact-mapping/
13. Khosrow-Pour, M.: Encyclopedia of Information Science and Technology. IGI Global, Hershey (2005)
14. Lee, S., Yu, J., Jeong, D.: BIM acceptance model in construction organizations. J. Manag. Eng. **31**, 04014048 (2015)
15. Legris, P., Ingham, J., Collerette, P.: Why do people use information technology? A critical review of the technology acceptance model. Inf. Manag. **40**, 191–204 (2003)
16. Li, P., Froese, T.M., Brager, G.: Post-occupancy evaluation: state-of-the-art analysis and state-of-the-practice review. Build. Environ. **133**, 187–202 (2018)
17. Oesterreich, T.D., Teuteberg, F.: Behind the scenes: Understanding the socio-technical barriers to BIM adoption through the theoretical lens of information systems research. Technol. Forecast. Soc. Chang. **146**, 413–431 (2019)
18. Venkatesh, V., Bala, H.: Technology acceptance model 3 and a research agenda on interventions. Decis. Sci. **39**, 273–315 (2008)
19. Venkatesh, V., Davis, F.D.: A theoretical extension of the technology acceptance model: four longitudinal field studies. Manage. Sci. **46**, 186–204 (2000)
20. Wüstenhagen, R., Wolsink, M., Bürer, M.J.: Social acceptance of renewable energy innovation: an introduction to the concept. Energy Policy **35**, 2683–2691 (2007)
21. Yuan, H., Yang, Y., Xue, X.: Promoting owners' BIM adoption behaviors to achieve sustainable project management. Sustainability **11**, 3905 (2019)

Attracting Future Students' Attention by an UX-Optimized Website

Christina Miclau(✉), Luisa Herzog, and Andrea Müller

Hochschule Offenburg – University of Applied Sciences, Badstrasse 24, 77652 Offenburg, Germany
christina.miclau@hs-offenburg.de

Abstract. As a university it is more and more difficult to reach all target groups equally. Common problems like information overload, numerous institutions with same focuses or multi-channel-communication make it hard to gain the attention of the target group. This paper is four-fold: we present an overview of the state of art and the importance of the study (I), based on which we highlight the approach to user experience analysis. First, we identified the irritations in the course of an expert evaluation (II) and verified them within the test, including the target groups (III). Finally, based on the results, we were able to pro-vide recommendations for action to improve the UX and to be used for the conception of an intranet (IV).

Keywords: Empirical research · User experience test · Intranet concept · University Marketing Communication

1 Relevance

As with companies, a website is the digital flagship of a public educational institution, which is why it is essential to review the online presence for its reasonability, design and interaction - to evaluate the user experience. The main challenge is to address the many different target groups a university has and to identify their various needs.

Interactive technologies are becoming increasingly important in the course of digitalization and globalization, as they enable communication regardless of time and place. This aspect also proved to be beneficial during the Covid 19 pandemic [1]. People seek to be connected and accessible, as demonstrated by the megatrends of the global society of 2020: Individuality, Knowledge Culture, Connectivity, New Work, Globalization and the resulting higher comparability [2]. The relevance of accessible communication is also increasing in-house. In many companies, the intranet is already the hub for a comprehensive exchange. And each interaction with users is characterized by emotions and the experience of using the interface [3].

The present study deals with the identification of recommendations for the conception and design of an intranet website for universities in Baden-Wuerttemberg. The necessary analysis of user-centered needs and requirements was carried out using the example of a university website in the form of an empirical study with different target groups such as pupils, students and employees. The needs of the users were measured and

F. Fui-Hoon Nah and K. Siau (Eds.): HCII 2022, LNCS 13327, pp. 311–324, 2022.
https://doi.org/10.1007/978-3-031-05544-7_24

examined on the basis of a university website. As it is currently the digital interface between the university and the users, it is very similar to an intranet in some respects. The new homepage thus represents the object of investigation in order to be able to derive recommendations for action for a user-centered intranet for the universities in Baden-Wuerttemberg on its basis.

The following questions for the evaluation of the contents and necessities of an intranet will be examined more closely: What expectations do the various target groups have of the university website and how is it evaluated after use? Which areas of the new university website show irritations and how can these be improved? How should an intranet be structured and designed, taking into account the test results of the website? In order to answer these questions, an understanding of the target groups, the procedure for investigating the user experience and the relevance of an intranet platform is essential.

2 The Status Quo – A Qualitative Study of a University's Website

The user-centric study starts by identifying the specific target groups and their needs. For this purpose, four different categories were determined as test subjects in the context of this work, which reflect the most relevant user groups (see Fig. 1).

Category 1	Category 2	Category 3	Category 4
Students at the University of Offenburg	Students from other universities	Seniors, High school graduates, Prospective students	Scientific workers at the University of Offenburg
Number of test subjects	Number of test objects	Number of test subjects	Number of test subjects

Fig. 1. Relevant User Groups of a University's Website

For an appropriate and representative sample of end users, it is also efficient to include people without prior knowledge of the specific platform and to identify any irritations and problems that would be bypassed specifically, if they were familiar with the platform and were used to interacting with it [4]. According to Rauterberg et al., in order to aim for heterogeneity in groups, individuals with different prior experience, age, gender, and user perspective should be selected. In addition, some individuals should be integrated who will continue to interact with the system in the future [5].

For the qualitative study of the university website, the Customer Experience Tracking procedure (short: CXT) was applied. It is a multi-stage, modular and scalable method for investigating user experiences in a marketing context. Here, not only the pure usability, i.e. functionality, of the application is relevant, but the entire user experience - expectations, interaction (usability) and experiences [6]. In this context, emotions have a significant impact [7]. "User experience includes all emotions, notions, preferences, perceptions, well-being or discomfort, behaviors, and performances that occur before, during, and after use" [8]. These emotions are becoming increasingly relevant in the context of interactive, networked systems and represent an important factor in the decisions made by users [9]. The user experience is therefore a key factor in the optimization of interactive systems, as it determines whether or not a service will be accepted and used again in the future [6].

In order to ensure a positive user experience, it is important to ascertain the needs of the target groups. Perceived impressions are highly individual, which is why all negatively influencing aspects of a service are first identified before users test it. This is why the CXT method can also be described as a combined analysis, which consists of two phases: 1. Expert-based analysis; 2. User-based analysis. In the expert-based analysis, all obvious problems are encountered first. These include, for example, poor usability of a product, as well as any frustrations that may be caused by its design or unappealing use (look and feel) [10].

User-based analysis is required to identify all other specific problems of the target group and to verify the irritations already previously uncovered in the expert-based analysis [10]. The information is obtained directly from the participants to whom the analysis is directed and is highly relevant for investigating the user experience of a service. Of particular interest are the individual motivations and opinions, which can be obtained with the help of various survey techniques [11]. Due to this combination of different perspectives and the creation of an interface of relevant irritations, combined analyses are a target-oriented method.

2.1 Relevant Instruments Used to Measure the UX

The CXT method of the Offenburg University uses different instruments for the acquisition of relevant information and the examination of the user experience, which are briefly described in the following.

AttrakDiff: AttrakDiff enables a qualitative survey of the subjectively perceived usability and appearance of a product [12]. Hassenzahl has succeeded in developing an evaluation procedure that can measure the hedonic and pragmatic quality of a service. These two factors are significantly responsible for a positive overall assessment of a product. The questionnaire uses the semantic differential method (28 opposing pairs of adjectives with seven degrees of gradation) and makes it possible to compare the expectation with the subsequent experience after use [4].

Eye Tracking: People usually spend very little time when visiting a website. For this reason, it is even more important to direct their gaze to relevant content. Eye tracking is a scientific method that uses infrared light to track the user's gaze during use, so that it is possible to check whether relevant functions of a platform are recognized. Eye tracking provides a valuable insight into the test person's initial impulses and answers the questions such as: Which elements are perceived when? Where do users expect to find information or elements? [4].

Facial Expression Analysis: The facial expression of a person during the use of a service is an essential characteristic for the experience of an emotion [6]. In 1977, Ekman determined not only the four basic emotions (joy, anger, sadness, and dis-gust), but also various mixed forms. In his model, the Facial Action Coding System (FACS), he defined 44 action units, i.e. 44 specific muscle movements in the face. With the help of a reduction to 19 action units, emotions can be reliably assigned. The advantages of facial expressions are their non-verbal comprehension as well as their cross-cultural uniformity and gender-independence [13].

Think Aloud: Think Aloud requires respondents to describe their subjective thoughts and actions. Two variants are possible: Concurrent Think Aloud (CTA) or Retrospective Think Aloud (RTA). The communicative exchange with the test per-son can thus take place either during the test phase or retrospectively afterwards [14].

Qualitative Interviews: Interviews provide a holistic insight into the thoughts and values of the test person. Emotions through expectations and experiences are incorporated into every rational decision [4]. In contrast to Think Aloud, users are actively questioned about the functions of the service, but also about their subjective experiences. It is essential to define in advance whether the research is qualitative or quantitative and whether the interview should be standardized, structured or non-structured [11]. A decision about this depends on the object of research and the aim of the investigation.

In the context of emotion measurement in the UX, there are a number of tried and tested instruments which, when considered independently, always have advantages as well as disadvantages [4]. For this reason, a combination of suitable methods will be chosen in the context of CXT for a fully comprehensive evaluation of the users.

2.2 The Need for Emotional Activation of Users

The constantly growing mass of information in our everyday life is a challenge nowadays. Research on human sensory perception has shown that only a small fraction of all information passes into long-term memory. Repetition, associations with different senses and especially emotions enhance information intake. In the worst case, frequent information overload in everyday life and thus a failure to take into account the limited human capacity for reception can lead to systems being avoided or even not used at all [15]. To counteract this problem, interactive systems should be simple (usability) and emotional (experience). If both characteristics are combined, we can refer to a holistic user experience. Every person has his or her own value system and individual needs [15]. It is therefore increasingly complex to design a product that is perceived as equally appealing by everyone. As a result, it is important to focus on a target group and find out where the greatest intersection exists in order to address as many people as possible [4]. Creating emotions in marketing is a great challenge as they cannot simply be evoked and transported to users. They originate in the respective minds and are based on one's own values and experiences. Emotions can therefore not be transferred, but only the requirements for them can be created in the product and are everywhere where users act [15].

3 Relevance and Design of Intranet Websites

The basic contents of the CXT and the knowledge of emotions in the UX context form the foundation of the conception of the testing and help to gain a better understanding regarding the relevance and design of intranet websites. In addition to other digital, in-house communication media, such as e-mails, internal social media, apps or blogs, the intranet in particular has become an indispensable part of corporate communication [1]. First, we will show how an intranet is defined and what requirements it must meet in order to be considered as such. At the same time, it is essential to clarify the significance of internal platforms in the world of digital networking and to demonstrate proven design elements and principles for digital communication platforms.

3.1 Definition and Relevance

Establishing a standard definition by which to define the intranet is becoming increasingly complex [16]. A final and limited definition should be avoided in any case, since an intranet is an individual and company-specific construct in its existence [17]. There is no binding definition of components, which is why there is a large degree of flexibility in the design. A definition according to Meier et al. states that an intranet is the basis for many other IT platforms. It is also a central instrument for internal communication and an important tool for supporting change and knowledge management as well as a lively feedback culture in a world that is changing even faster [16]. Dewitte defines content, communication, activity and collaboration as the central goals of the intranet [18]. In this context, the term social intranet is taking on greater significance. It involves a classic intranet enriched with social media functions [17]. Instead of centralized provision of information in the form of top-down communication, interactivity and two-way communication are promoted [16].

The Intranet – An All-Rounder: An intranet is a digital, networking platform with a wide range of applications and interfaces. It supports users in their everyday activities, especially in all aspects of knowledge work [19]. With the help of a user-centric, strategically designed platform, a fast and uniform information flow as well as optimal internal communication can be ensured [20]. A professional intranet should therefore contain all the information needed for the day-to-day work of stakeholders. A final checklist does not exist, which is why it allows a free definition of content and design elements. Thus, it is up to the respective institution how an intranet is structured and maintained. The experience and intensity of use, however, also depends on this. The COVID 19 pandemic has further increased the importance of digital communication platforms [21]. More and more institutions see the advantages of intranets with their characteristics as a digital workplace, i.e., a central provision of digital work tools. These can be applied to a high degree of compatibility by the users [16]. However, with its function as a jack-of-all-trades, the intranet faces a number of challenges.

The Intranet – A Challenge: (1) Diverse needs: A major challenge arises especially from the aspect of different needs and demands colliding with each other. In addition, intranets encounter a high degree of similarity with globally operating platforms and face difficulties in meeting rising expectations [16].

(2) Unused potential: In most cases, only a fraction of the possible functions is used and content is not maintained regularly. An intranet is often offered just to follow the current zeitgeist and rather than to take advantage of benefits [20].

(3) Free design: Hardly any two intranets are alike. In practice, there are considerable differences in implementation and also different assessments by users. Irritations can arise at both the content and structural level, for example, in the ineffectiveness of addressing target groups [17]. Intranets are thus confronted with increasing demands due to the rapid changes in communication in society [22].

4 Findings

To derive relevant recommendations for the conception of an intranet website for universities in Baden-Wuerttemberg, the expectations of the target groups, their evaluations after testing and the irritations found serve as a basis. The objective is to determine characteristics, designs and implementation possibilities of a user-centered intranet based on the research results of the homepage.

Adhering to the SMART attributes (Specific, Measurable, Attractive, Realistic, Timed), the following guiding questions were constituted for this investigation [23]:

a) What expectations are placed on the homepage of the Offenburg University of Applied Sciences by different target groups during testing and how is it evaluated after use?
b) Which areas of the new university homepage show irritations during the testing and how can these be remedied in order to fully satisfy the target groups?
c) How should a user-centered intranet, taking into account the results of the testing of the homepage, be created and designed for universities in Baden-Wuerttemberg?

The three objectives listed as guiding questions meet the requirements of the SMART method since they are specific to the object of study. In addition, they are measurable as they are achieved with the selected research methods and attractive due to the advantages of a user-centered platform.

The design of the UX testing was based on the results of the expert evaluation. Here, obvious irritations affecting usability as well as user experience were identified based on heuristics and the principles for standardizing the ergonomics of hu-man-system interaction [24]. The recorded irritations were then classified into 18 different categories for ease of understanding. These categories include, for example, legal spelling, layout, incorrect linking, choice of images, gender-appropriate language, and logic errors. In order to ensure the quality of the results, it was important for the test persons not to be familiar with the website in question. Thus, problems cannot be circumvented by the participants and already learned behavior, as a result of previous interactions with the object of investigation, does not lead to a falsification of the results [4].

During the expert evaluation, in addition to irritations associated with an inconsistent layout or design (a), functional irritations (b) could also be identified. Examples thereof: (a) If a left-justified display was chosen in some places, larger banners in a more prominent design existed in other places on the same page and covered the entire width. (b) Some windows in the header could only be closed by clicking directly on the corresponding icon.

Based on the expert evaluation, the next step in the process is to define the tasks for the UX testing. This way, the user-based analysis guarantees that the test persons inevitably encounter the corresponding irritations and the corresponding reaction and opinion can be obtained. Furthermore, a comparability of the results is ensured due to the same procedure.

4.1 Analyses of the Instruments Used

AttrakDiff: The online survey via AttrakDiff, which contrasts the test subjects' perceived hedonic and pragmatic quality and assigns the rating to either a redundant, neutral or desirable area, showed a positive result for all four test groups (see Fig. 2). The portfolio reflecting whether the test group perceived the hedonic and pragmatic quality as equally fulfilled or whether the homepage was perceived as too self-oriented (HQ) or too action-oriented (PQ) indicated that a classification of the examined university website to the "desirable" area was given. Hence, the two qualities were classified as balanced. Within the test groups, only slight deviations between the expectation before and the experience after the product use can be seen in the portfolio representation: Whereas the students of the university assess the homepage minimally better after use, this assessment is slightly worse for the employees.

In a comparison between the test group portfolios, on the other hand, the result of the students from other universities is noticeable, which has a significantly greater variance, which is why the evaluation also lies in the neutral range. The areas "self-oriented" and "action-oriented" are also tangentially affected. From this result it can be determined that the participants of this test group have a differing opinion about the homepage than the other test groups. Their assessments were more homogeneous, which is why there is less variation in the portfolios.

Based on the analysis with the help of the online tool, it can be stated, that there are no significant differences in the assessment between the test groups, despite different ages, genders and previous experience.

Emotion Measurement: The perception of the investigated website was positive by all user groups and was described as clear, modern and subtle in its presentation. Furthermore, in addition to the clarity of the categories, the quick achievement of purposes was also appreciated, characterized by performing only a few clicks, which enables simple as well as purposive navigation. The use of images and visualizations, such as large themes and clusters, contribute to a pleasant sense of well-being. However, this decent presentation was not rated positively by all test persons. For example, some people mentioned that the website was designed too subtle in some places and that individual headings did not noticeably highlight themselves. Others noted in the overall assessment that the website was "nothing special" and did not distinguish itself significantly from other college and university websites. Moreover, a uniform and authentic overall image of the pictures as well as a harmonious layout and dimensions were important.

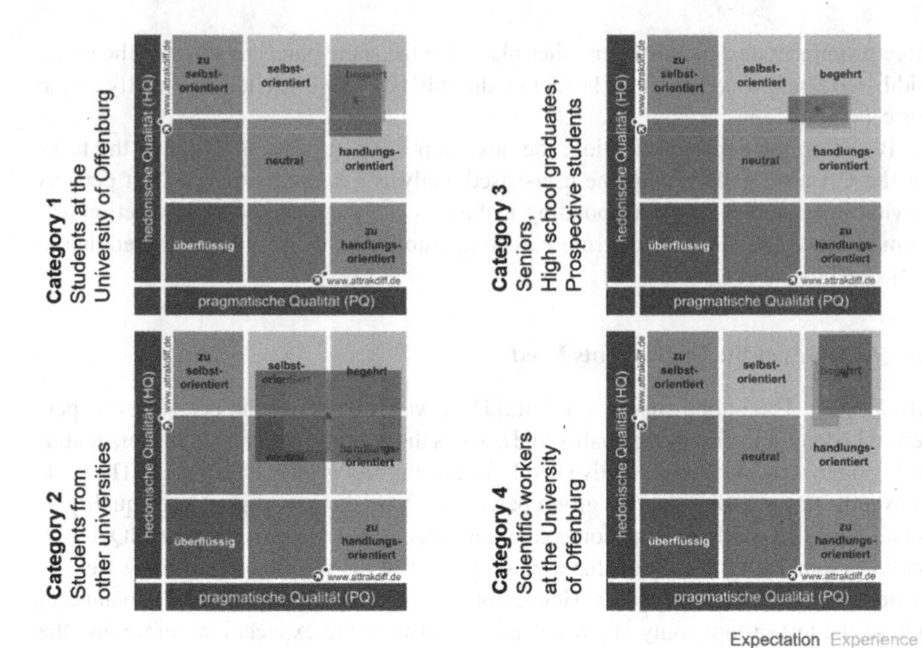

Fig. 2. Results of the AttrakDiff Analysis

Interviews: After running through the scenarios, all participants indicated having navigated the website well, with the exception of the aforementioned irritations, and would also recommend it to their personal contacts. In addition to questions about individual errors, the test persons were asked to describe the university in three words based on their experiences and impressions, with the most frequently mentioned terms being "familiar", "innovative", "international", "modern", "practice-oriented", "technical" and "diverse".

In order to obtain additional information about the impression of the website and possible transfer points to the conception of an intranet, the question about opinion and general conditions of an intranet was essential. Thereby a two-sided point of view could be ascertained. Some respondents consider a networking platform as a central place to handle all matters that can be communicated and accessed in one place to be useful - regular news, access to other platforms, uploaded documents by lecturers, jobs and student jobs. This would also counteract the existing flood of e-mails. However, some users mentioned the risk that an intranet would not be used intensively enough and that the advantages would not be taken into account.

This concern is consistent with the results of the Staffbase study, which supports the statements of the test subjects in this study. Staffbase, one of the largest providers of corporate communications software, conducted an intranet study with its customers in 2021. With the help of 140 intranet managers in German-speaking countries, the study determined what the main problems are with existing intranets in companies. The results show that a considerable proportion of existing intranets have potential for optimization. This means that companies are not taking full advantage of the benefits offered by such a platform [4]. Accordingly, it is important to communicate interesting and up-to-date content in a way that can be personalized, and to convey important information briefly and clearly. In addition, usage is guaranteed if all matters can only be handled via the network in the long term. Some respondents took a critical stance toward an intranet at the beginning, but changed their minds after advantages and disadvantages were reflected upon. This behavior reflects the classic reaction to change. During implementation, a university should be aware of this behavior and involve stakeholders at an early stage to allay their fears.

4.2 Recommendations for Action in Designing Intranet Platforms

The results indicate that a major challenge for the university is to satisfy different target groups. Primarily, the tested homepage is designed for prospective students or other interested members of the public. On the other hand, some target groups need the homepage for their everyday life (lecture schedules, Moodle, library, mailing). In order to ensure satisfaction for all addressees, a central, self-explanatory interface for all everyday applications is needed for university members - an intranet. This chapter will now derive recommendations for action to answer the research question and design a potential section of an intranet landing page for a university.

Based on the results from the expert evaluation and the user-centered analyses with the participants from phase 4, the recommendations for action can be classified into four categories. The following Fig. 3 summarizes these graphically:

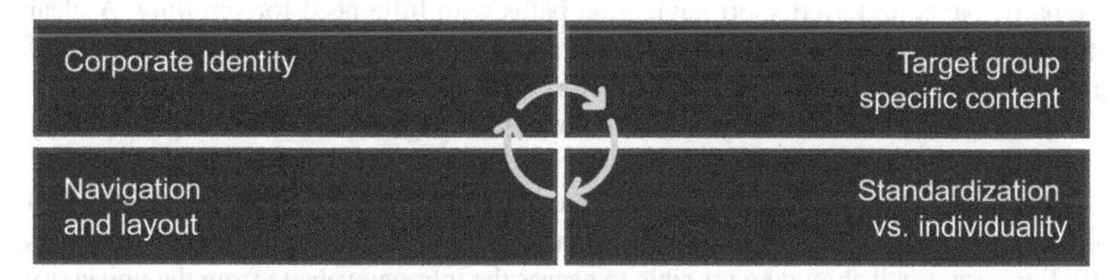

Fig. 3. Categories of Recommendation for Actions for Offenburg University's Intranet-Site

Corporate identity is the first category of recommended actions. It is defined as the company's personality, which is expressed in its behavior, communication and appearance (corporate design) [9]. For an institution, it is important that its own and external images match as closely as possible in order to achieve the desired goals, such as acquiring new students or informing existing ones.

When designing an intranet website, this specifically means that the published content must match the essence of the university and that standards communicated in the mission statement must be applied, such as the use of gender-appropriate language. In addition, a company or university should remain true to the design scheme used [9].

The second recommendation for action relates to a target group-specific design of content. During the interview, respondents stated that interesting and up-to-date content is important to ensure that an intranet is used on a long-term basis. Besides, the content should be customizable and clearly arranged.

Besides, the third recommendation for action describes a tension between standardization and individuality, which should be taken into account in design. Where a standardized presentation was preferred in some places, users wanted a high degree of individuality in others. A website should therefore not be exclusively standardized or individualized in design. Rather, it should be used in the right places.

For instance, the page design should be standardized, color concepts should be consistently enforced, and graphics should be appropriately displayed in the overall picture. At the same time, content ought to be customizable and thus adaptable to the individual needs of the user. Personal images and functions should also be used in addition to standardized features in order to create a unique selling point compared to other university websites.

Standardization arouses trust and openness in people towards a new system. At the same time, every individual has a natural desire for stimuli and individuality [25]. It is for this reason that an intranet must satisfy both standardization and individuality and reconcile these two factors.

Finally, navigation and layout make up the last category. Overall, it is important for an intranet website to be designed as simply and clearly as possible so that users can easily find their way through and no moments of annoyance arise. Based on the results, this can be ensured on the one hand by short navigation paths - users should be able to reach the desired goals on the platform with just a few clicks and without having to think about it too much. A simple and not overcrowded page structure with few, but large, tiles makes it easier to find one's way around. In addition, the results showed that the participants preferred short navigation paths with little need for scrolling. A clear structure of the content, a coherent layout and a correct and accurate presentation of images and content are also important factors for user satisfaction.

A design of a section of a potential intranet start page for the Offenburg University of Applied Sciences is presented in Fig. 4.

It is important to adhere to the corporate identity of the associated university, such as the use of the logo, colors and font, the basic design and the arrangement of graphics [9]. Furthermore, it should be possible to access the intranet website from the university homepage in a clearly visible manner in order to avoid long navigation paths. Test subjects also felt it was important that the design of the new homepage should not differ significantly from the previous one. A completely different presentation of the articles and content would not fit in with the character of the university and would therefore meet with aversion from the target groups because the expectations would not be fulfilled.

In addition, the pictured intranet offers an optimal combination of standardization and individuality. For example, the layout of the platform is standardized and simply structured. At the top right of the header are classic tools such as settings, language or login/logout. The search function should also be positioned at the top right to retain familiar placements and ensure positive usability. The eye tracking results during the testings showed that the test persons often looked at this page area to search for clues or help when they were feeling despondent.

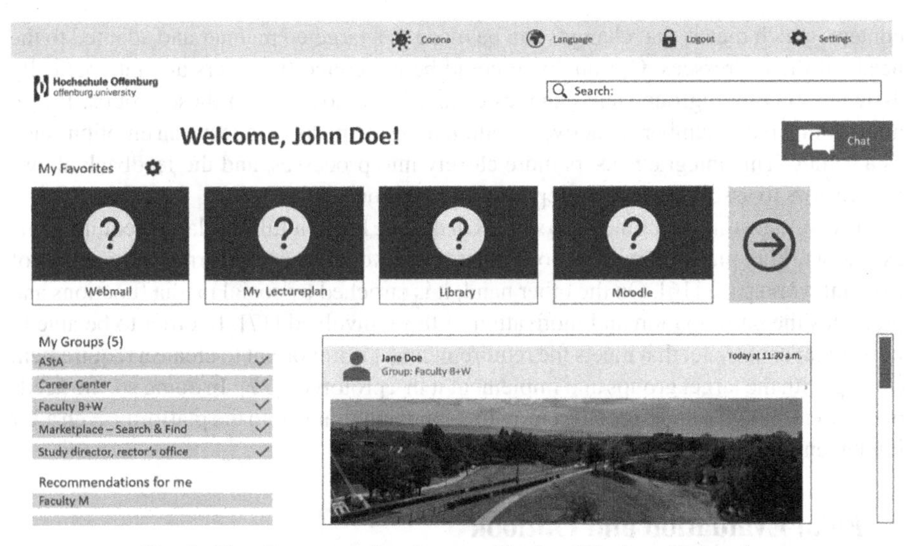

Fig. 4. First Prototype of an intranet-Site for Offenburg University

In the center are web tiles with a link to important platforms. The bar can be moved by clicking an arrow on the right side, as the majority of the test subjects wished. On the right side of the page, you will find the option of a chat function for an interactive exchange with other users. This is highlighted in a blurry orange. The corresponding button follows the course of the page when scrolling and thus always remains visible on the right side. This way, it can be used at any time.

Finally, the published intranet posts by university members can be accessed under the tiles. On the left side, users can use a filter function to select between the content and contributions of different groups. This was integrated due to the positive feedback during the testing to categorize the courses. When designing the postings, it is important to pay attention to the use of relevant and current topics. Depending on the group, editing and publishing rights can be granted to the users so that a review of the contributions is possible. In the case of a sports or meditation group, for example, the creation of a post would be possible for all participants; university-related posts only from central authorities.

In addition to some standardized elements, the page should also be customizable and personally appealing. At the beginning, the university members are greeted with their name. Personalization is made possible, for example, with the large tiles. With the help of the settings wheel, the users can determine their own favorites and the order in which they appear. The symbols with the question marks are placeholders for graphics that describe the interface and create a unified overall picture.

Moreover, personalization is made possible by the filter. Currently, all information at the university is communicated by e-mail, which is why relevant information can be overshadowed by less important messages [26]. With the help of filtering, everyone would receive just the content that would be of interest and individual relevance. In addition, the respective group receives real-time feedback on who is interested in the content, which means that changes can be made in a targeted manner and adapted to the needs of the addressees. Optionally, it could be integrated that users are automatically assigned to certain groups and changes cannot be removed from these groups. It also allows university members to actively comment on the posts or to leave an emotion, such as a smiley. This integrates users more closely into processes, and the feedback allows the creators to assess current perceptions and opinions.

The intranet thus combines two sides. On the one hand, it fulfills classic requirements as a networking platform with important interfaces to relevant platforms and features of a digital workspace [16]. On the other hand, it is enriched by social media functions and promotes the participation and motivation of those involved [17]. In order to be able to implement an intranet that meets the requirements, it is important to create a requirement catalog with the target groups and implement it in a prototype [27]. In doing so, the needs of the users should be continuously monitored and integrated into the platform, even after implementation [11].

5 Final Evaluation and Outlook

The present study highlighted the importance of a user-centered design of interactive communication platforms for universities in Baden-Wuerttemberg. Colleges and universities increasingly find themselves in a competitive market environment that requires entrepreneurial thinking as well as the successful execution of research and teaching [28]. These challenges are complemented by various target group needs that have to be met. For this reason, the design of an intranet should focus on these very needs. A user-oriented design is an essential success factor for a sustainable and intensive use of the platform [4].

The development towards an information society requires a reaction to the resulting challenges [22]. The provision of an intranet platform helps to cope with the challenges and to present content in a structured and personalized way [1]. In times of fast-moving and constant change, services and products should be placed in a dynamic process, which is why the intranet of a university must also be subject to regular review and a continuous improvement process in order to be able to realize the successes achieved through implementation on a long-term basis. The principle applies here as well: "The users always have the last word - anything they don't accept is not successful" [29].

References

1. Engelhardt, K.: Erfolgreiche Interne Kommunikation im Digital Workplace: Basics und Tools: Social Intranet, Mitarbeiter-App, Mitarbeitermagazin, pp. 1, 5, 39. Springer Fachmedien Wiesbaden, Wiesbaden (2020)
2. Deutsches Zukunftsinstitut. Die Megatrends. https://www.zukunftsinstitut.de/dossier/megatrends/. Accessed 18 Oct 2021
3. Thüring, M.: Nutzererleben – komponenten, phasen, phänomene. In: Boll, S., Maaß, S., Malaka, R. (eds.) Mensch & Computer 2013 – Workshopband, 13, fachübergreifende Konferenz für interaktive und kooperative Medien, München, pp. 113–120 (2013)
4. Sarodnick, F., Brau, H.: Methoden der Usability Evaluation. Wissenschaftliche Grundlagen und praktische Anwendungen, Bern, 3rd edn, pp. 119, 167–193 (2016)

5. Rauterberg, M., Sinas, P., Strohm, O., Ulich, E., Waeber, D.: Benutzerorientierte Softwareentwicklung. Konzepte, Methoden und Vorgehen zur Benutzerbeteiligung, Stuttgart, p. 89 (1994)
6. Müller, A., Gast, O.: Customer-experience-tracking – online-Kunden conversion-wirksame Erlebnisse bieten durch gezieltes emotionsmanagement. In: Keuper, F., Schmidt, D., Schomann, M. (eds.) Smart Big Data Management, Berlin, pp. 313–343 (2014)
7. Thüring, M., Mahlke, S.: Usability, aesthetics and emotions in human technology interaction. Int. J. Psychol. (London) **42**(4), 253–264 (2007)
8. DIN EN ISO 9241-210. Ergonomie der Mensch-System-Interaktion - Teil 210: Menschenzentrierte Gestaltung interaktiver Systeme (ISO 9241-210). Deutsche Fassung EN ISO 9241-210: 2020. DIN Deutsches Institut für Normung e.V., Berlin (2020)
9. Meffert, H., Burmann, C., Kirchgeorg, M., Eisenbeiß, M.: Marketing. Grundlagen marktorientierter Unternehmensführung, Wiesbaden, 13th edn, pp. 281, 739 (2019)
10. Rahn, J.: Usability und User Experience, Universität Ulm. https://docplayer.org/4972908-Johannes-rahn-29-07-2010-usability-und-user-experience.html. Accessed 13 Oct 2021
11. Thommen, J.-P., Achleitner, A.-K., Gilbert, D.U., Hachmeister, D., Jarchow, S., Kaiser, G.: Allgemeine Betriebswirtschaftslehre. Umfassende Einführung aus managementorientierter Sicht, 9th edn, Wiesbaden, p. 77 (2020)
12. User Interface Design GmbH. AttrakDiff. http://www.attrakdiff.de. Accessed 17 Oct 2021
13. Ekman, P., Friesen, W.V., Hager, J.C.: Facial Action Coding System. Investigator's Guide, Salt Lake City, p. 5 (2002)
14. Peute, L.W.P., De Keizer, N.F., Jaspers, M.W.M.: The value of retrospective and concurrent think aloud in formative usability testing of a physician data query tool. J. Biomed. Inf. **55**, 1–10 (2015)
15. Robier, J.: Das einfache und emotionale Kauferlebnis. Mit Usability, User Experience und Customer Experience anspruchsvolle Kunden gewinnen, Wiesbaden, pp. 4, 13 (2016)
16. Meier, S., Lütolf, D., Schillerwein, S.: Herausforderung Intranet. Zwischen Informationsvermittlung, Diskussionskultur und Wissensmanagement, Wiesbaden, pp. V, 12, 18 (2015)
17. Däbritz, V., Frömder, E., Anke, J.: Social Intranets als Grundlage für die interne Unternehmenskommunikation und Zusammenarbeit. In: HMD Praxis der Wirtschaftsinformatik, vol. 57, no. 1, pp. 133–149. Springer (2020)
18. Dewitte, D.: Intranet oder ein digitaler Arbeitsplatz? Eine kurze Geschichte. https://blog.amplexor.com/de/intranet-oder-ein-digitaler-arbeitsplatzkurze-geschichte. Accessed 27 Oct 2021
19. Pagel, P.: Social Intranet ist wie Gespräche an der Kaffeemaschine. Wirtschaftsinformat. Manag. **4**, 64–69 (2016)
20. NetFederation GmbH. Intranet-Studie 2016 – Informationsarchitekturen. https://www.digital-workplace-report.de/intranet-studie/studien/intranet-studie-2016/ergebnisse. Accessed 24 Oct 2021
21. Brechlin, F.: Strukturierte cloud transformation in unternehmen – Veränderungen durch Covid-19? In: HMD Praxis der Wirtschaftsinformatik, vol. 58, no. 4, pp. 739–753. Springer (2021)
22. Kollmann, T.: E-Business. Grundlagen elektronischer Geschäftsprozesse in der Digitalen Wirtschaft, Wiesbaden, 7th edn, p. 38, 97 (2019)
23. Eremit, B., Weber, K.F.: Individuelle Persönlichkeitsentwicklung: Growing by Transformation, Wiesbaden, pp. 95–98 (2016)
24. DIN EN ISO 9241-110. Ergonomie der Mensch-System-Interaktion - Teil 110: Interaktionsprinzipien (ISO 9241-110). Deutsche Fassung EN ISO 9241-110: 2020, p. 1. DIN Deutsches Institut für Normung e.V., Berlin (2020)

25. Schramm-Klein, H.: Multi Channel Retailing – Erscheinungsformen und Erfolgsfaktoren. In: Zentes, J., Swoboda, B., Morschett, D., Schramm-Klein, H. (eds.) Handbuch Handel, Wiesbaden, 2nd edn, pp. 419–437 (2012)
26. Köhler, T.R.: Informations- und Kommunikationstechnologie als Treiber und Leitplanken der neuen Arbeitskultur. In: Widuckel, W., De Molina, K., Ringlstetter, M.J., Frey, D. (eds.) Arbeitskultur 2020, pp. 89–97. Springer, Wiesbaden (2015). https://doi.org/10.1007/978-3-658-06092-3_6
27. Engelhardt, K.: Interne Kommunikation mit digitalen Medien. Learnings aus der Covid 19-Krise zu Prozess-Steuerung Mitarbeiterführung und Kommunikation, Wiesbaden, p. 40 (2020)
28. Schreiterer, U.: Hochschulen im Wettbewerb. Mehr Markt, mehr Freiheit, mehr Unübersichtlichkeit, Bundeszentrale für politische Bildung (BpB). https://www.bpb.de/gesellschaft/bildung/zukunft-bildung/185865/hochschulen-im-wettbewerb. Accessed 21 Oct 2021
29. Jacobsen, J., Meyer, L.: Praxisbuch Usability und UX: Bewährte Usability und UX-Methoden praxisnah erklärt, Bonn, 2nd edn, p. 219 (2019)

Developing Personas for Designing Health Interventions

Gaayathri Sankar[1(✉)], Soussan Djamasbi[1], Yunus Dogan Telliel[1], Adarsha S. Bajracharya[2], Daniel J. Amante[2], and Qiming Shi[2]

[1] User Experience and Decision Making (UXDM) Laboratory, Worcester Polytechnic Institute, Worcester, MA 01609, USA
{gsankar,djamasbi,ydtelliel}@wpi.edu
[2] University of Massachusetts Medical School (UMMS), 55 Lake Avenue North, Worcester, MA 01655, USA
adarsha.bajracharya@umassmemorial.org, {daniel.amante, qiming.shi}@umassmed.edu

Abstract. As more human needs are addressed with technology, designing positive user experiences becomes increasingly important in developing effective health interventions. Designing successful user experiences for digital health interventions requires a deep understanding of patient challenges. In this paper, we attempted to identify challenges that diabetic patients face adhering to guideline-recommended care through persona development. Previous user experience research suggests that such an approach can be particularly beneficial in designing digital health interventions. We explain how we developed data personas from Electronic Health Records (EHR) and combined them with proto personas that were generated by a group of medical experts for the same patient population. Our results support previous research that suggests combining data and proto personas is beneficial for intervention design. Additionally, our results reveal that combining data and proto personas is likely to improve intervention design by addressing fairness issues that may result from the underrepresentation of certain populations in EHR datasets.

Keywords: Persona development · Data personas · Proto personas · UX design · Intervention design · Cluster analysis · Electronic Health Records (EHR) · Type 2 diabetes

1 Introduction

The proliferation of digital goods has created crowded markets for developing products and services that address human needs. Because of this trend, user experience (UX) has become the central focus of product/service innovation [1]. Identifying opportunities for user-centered innovations requires a deep understanding of user needs, goals, and challenges [1]. One way to achieve this objective is through persona development [1–3]. Persona development refers to a methodology that categorizes user information that is relevant to product design into representative characters. Personas can be created from

F. Fui-Hoon Nah and K. Siau (Eds.): HCII 2022, LNCS 13327, pp. 325–336, 2022.
https://doi.org/10.1007/978-3-031-05544-7_25

existing user information (data personas) such as data in medical records. Personas can also be created by collecting and synthesizing information from a group of experts with knowledge about users (proto personas). They can also be created through conducting user interviews (user personas). Combining more than one type of persona development method can enhance knowledge about user needs [2].

The objective of this study is to create a set of personas that can be used to design digital interventions for type 2 diabetes patients. We achieved this objective by developing a set of data personas from electronic health records for type 2 diabetes patients in the UMass Memorial Accountable Care Organizations (ACOs). We then combined these data personas with proto personas that were generated in a previous study for the same population of patients to create a more comprehensive set of personas [4]. The final personas developed in our study will be used in subsequent projects that aim to design innovative self-supporting digital interventions for type 2 diabetes patients in the ACOs that provide care for Worcester and Central Massachusetts. Under Medicare, ACOs are formed by a group of doctors, hospitals, and other health care providers who voluntarily come together with the aim of avoiding unnecessary duplication of services, preventing medical errors, and spending health care dollars more wisely [5, 6].

2 Background

In HCI research, developing personas provide an effective method for summarizing and communicating user needs that are essential for designing interventions. By creating a shared understanding about user groups, personas help decision-makers to prioritize intervention goals and strategies. Because user needs change over time, personas must be updated before any major design decision [2]. In the following section, we discuss three major types of personas in UX research: proto, data, and user personas [1, 2]. Each persona development type described in this paper alone can provide a powerful tool for investigating user needs for technology design. Naturally, by combining any of these types of persona development together, researchers can provide a more comprehensive set of insights for design [1, 2].

2.1 Proto Personas

Developing proto personas is the most cost-effective of the three persona development methods because it can be achieved by gathering assumptions about user needs from experts or project stakeholders (e.g., managers, administrators). Despite being assumption-based, proto personas provide a great deal of value by helping project stakeholders to create a shared understanding of their users' needs. By doing so, proto personas often facilitate the cross-pollination of ideas. They also help project stakeholders to prioritize design and/or intervention strategies [2]. Proto personas typically provide information about user demographics, needs, challenges, and goals. Because proto personas are assumption-based, they benefit from validation through other types of persona development methods discussed in the following sections.

2.2 Data Personas

Data personas are typically generated from user data that is available on organizational systems. For example, customer analytics commonly leverages social media data to help develop customer profiles [7]. With the availability of big data, developing data personas is becoming increasingly relevant in designing UX-driven innovations [2]. For example, using EHR data to create data personas can facilitate the design of user-centered solutions customized for the diverse needs of various patient groups. Designing such customized solutions is particularly effective in addressing the needs of patients with chronic diseases, such as diabetes, which typically require long-term interventions. While data personas represent an abstraction of users' behaviors generated from data over a given period of time, they lack information about users' perceptions and attitudes. Such information is typically provided through assumption-based proto personas or by developing user personas.

2.3 User Personas

User personas are developed based on generative UX research that is typically conducted via one-to-one interviews with target users [1].

Hence, developing user personas is typically more resource-intensive than developing proto and/or data personas. However, this persona development method provides firsthand information about users. For example, user personas typically capture information about user perceptions and emotions that is generally not available in data personas. While proto personas typically include affective and attitudinal aspects of user needs, this information is not gathered directly from users.

3 Methodology

In this study, we investigated type 2 diabetes patients' needs in the UMass ACOs by combining data and proto personas. To achieve this goal, we first developed a set of data personas from electronic health records (EHR) records for this population. Next, we mapped the generated data personas to proto personas that were developed in a prior study for the same patient population [4]. In the following sections, we explain how we used cluster analysis to generate data personas from the EHR. Then we explain the methodology that we used to combine data and proto personas.

3.1 Developing Data Personas

Data Extraction and Preparation. A total of 6348 deidentified electronic health records of UMass ACO patients above 18 years of age were extracted from the Epic Clarity Database. Epic is a cloud-based EHR software solution that focuses on patient engagement and facilitates remote care. On the front end, it achieves this through a patient portal called MyChart which is an app available on both Android and iOS platforms [8]. At the backend, it uses the Clarity database which is an Oracle/Microsoft SQL server database that aids in exporting, transforming, and loading data from the MyChart environment [9]. The datasets extracted from Clarity are therefore static and do not provide real-time information such as those supplied by IoT devices. This means that the data used in this study is historical in nature. User personas are developed based on generative UX research that is typically conducted via one-to-one interviews with target users [1].

While the database used in our study is home to a large number of variables, only those variables that were identified as relevant and important in defining and analyzing data personas for our study were extracted and included in the analysis. To achieve this goal, we first looked at patient attributes that were deemed as considered relevant and important by a group of medical experts who developed the proto personas for the same population such as physical and mental health status, socio-economic status, culture, language barriers, etc. [4]. We then identified a set of variables in the EHR dataset that could represent these patient attributes and organized them into geographic, demographic, technological, diabetic health risk, and healthcare variables (see Fig. 1).

The dataset in our study contained both continuous and categorical variables. NULL values (or missing data) were embedded in all of the variables in the dataset. Because we wanted to preserve the presence of missing information in the continuous variables, we decided to convert the continuous variables (e.g., HbA1C) in the dataset to categorical variables. The continuous variables in our dataset that were converted to categorical variables and their respective categorical groupings are displayed in Fig. 2.

Cluster Analysis. The objective of this study was to create data personas that could aid in designing effective interventions for type 2 diabetes patients. To achieve this goal, consistent with a recent study, we used the two-step cluster analysis to group desired patterns in the EHR dataset [12]. A two-step cluster analysis is an exploratory tool that identifies groupings by first running a pre-clustering followed by a hierarchical method clustering [13]. Cluster analysis is an unsupervised technique that does not test any specific hypothesis. Hence, the solutions at best reflect the needs for which it is run, and the definition of a satisfactory solution is best defined by experts [13], in our case, by two co-authors who are medical professionals.

We created hierarchical clusters based on the 3 major health risk variables: HbA1C, LACE+, and BMI. Hemoglobin A1C (HbA1C) levels refer to blood glucose levels. HbA1C is considered a direct measure of risk for developing type 2 diabetes. LACE+, the second health risk variable we used in our study, is a predictor of mortality rates. LACE+ scores are calculated by considering factors such as age, sex, comorbidities, whether the patient was admitted through the emergency department (ED), number of ED visits in the six months prior to admission, and the number of days the patient was in an alternative level of care during admission [10]. Obesity, the third health risk variable in our study, is considered a global pandemic. Studies show that a majority of people with type 2 diabetes are obese [14]. Obesity is measured by a score representing a person's body mass index (BMI) [15].

Geographic variables:

- Zip code

Demographic variables:

- Income
- Age
- Gender
- Race
- Ethnicity
- Marital status
- Primary language

Technological variables:

- Email address
- MyChart status

Health risk variables:

- LACE+ score
- Body Mass Index (BMI)
- HbA1C levels

Health care variables:

- Care duration
- Appointment distance
- Insurance type

Fig. 1. List of variables considered to create data personas

It is important to note that clustering algorithms, such as the one used in our analysis, use the mode of variables to identify disjunct groups of data records. Hence, the characteristics identified by our analysis in each cluster represent attributes of the largest groups of users. That is, characteristics of a cluster reveal attributes of a persona that represents the majority of patients in that cluster. Conversely, a cluster can have some patients with some attributes that are different from those considered as characteristics of the cluster.

Organizing the data into health risk hierarchical clusters first based on HbA1C categories (a direct measure of diabetes), then by LACE+ (an indicator of mortality rate), and then by BMI (another major diabetes risk factor) allows us to identify and organize data based on the severity of diabetes risk variables.

HbA1C levels:

- NULL – missing data
- Below 6.5 – low
- 6.5-8 – moderate
- Above 8 – high

LACE+ score [10]:

- NULL – missing data
- 0-28 – low risk
- 29-58 – moderate risk
- 59-90 – high risk

BMI:

- NULL – missing data
- Below 18.5 – underweight
- 18.5-24.99 – normal
- 25.0–29.99 – overweight
- 30.0 and above – obese

Income [11]:

- NULL – missing data
- Up to $32048 – Poor or near-poor
- $32048 up to $53413 – Lower-middle class
- $53413 up to $106827 – Middle class
- $106827 up to $373894 – Upper middle class
- above $373894 – Rich

Fig. 2. Continuous variables and their respective categories

4 Results

The analysis of 6348 ACO patient records revealed the data set in our study included almost an equal number of male and female patients (52% and 48% respectively). Patients' age groups in the dataset ranged from below 30 to above 100 with the majority of patients belonging to 60–70 and 70–80 age groups (25% and 39% respectively). About 94% of patients were English speaking and 5% were Spanish speaking. 80% of the patients were white, not Hispanic or Latino, and a little more than 50% were married. About 45% of the patients had an email and an active MyChart status. Nearly 54% belonged to the middle class and about 90% were enrolled in Medicare. Over 95% have been receiving care for more than 5 years and about 70% have made between 1–5 appointments. Overall, a larger number of patients were found in the zip codes belonging to Shrewsbury, Worcester, and Fitchburg.

We used IBM SPSS Statistics v26 to conduct the cluster analysis based on the diabetes risk variables HbA1c, LACE+, and BMI. For cluster quality, as recommended, we set the silhouette measure to be above 0.0 and set the ratio of cluster sizes to be below 3 [16]. Because our dataset consisted of categorical variables, we used the log-likelihood distance measure to conduct our analysis [17].

To explore the data, an automatic determination of the number of clusters was performed which resulted in 6 clusters with an average silhouette score of 0.4 and a ratio of sizes at 2.97. This was followed by trials involving the manual determination of cluster sizes ranging from size = 6 to size = 9. This analysis showed that 9 clusters yielded the highest silhouette measure of 0.5 with the ratio of sizes <3 (1.89). Therefore, the number of clusters was manually determined to be 9.

As stated previously, the formation of these 9 clusters was defined by the variables HbA1C, LACE+, and BMI, which proved to be equally and highly important in defining the clusters (each had predictor importance of 1). The largest cluster in the dataset had 15.40% of the records and the smallest cluster, 8.20%.

Next, we visualized the clusters based on their health variables in a hierarchical tree structure, first based on HbA1C categories (high, moderate, low), then by LACE+ (high, moderate, low), and then by BMI (obese, overweight, normal, underweight). This organization, which implies a health-risk factor hierarchy, allowed us to visually organize clusters based on the severity of their categories. For example, clusters B, C, D, E, and F in Fig. 3, represent a relatively healthy population because they all have HbA1C less than 6.5. Based on their LACE+ and BMI, however, these clusters can be ordered as E > B > C > D.

Cluster F is not included in this sequence because we do not have any information about its LACE+ score (that is the value for LACE+ for the majority of records in this cluster was NULL).

In a similar manner, clusters G, H, and I have an HbA1C level between 6.5 and 8, which indicates that these groups are tending more towards having the disease. The health risk for these clusters can be ordered as H > I > G based on their LACE+ and BMI data.

There is only one cluster with a fairly higher risk of the disease, cluster A (high HbA1C, moderate LACE+, and high BMI).

4.1 Preparing Data Personas for Matching

After creating clusters based on factual major health risk factors, we used the demographic, technological, and health and healthcare variables to provide more nuanced data personas that could be mapped to proto personas.

For example, we created a more nuanced data persona for type 2 diabetes patients by incorporating information about insurance type and income in their cluster. Information about a patients' insurance type can support or enhance information about the patient's socioeconomic status, an attribute that was considered important in developing proto personas for the UMass ACO patient population [4]. Consider Medicaid, a state and federal program for very low-income populations [18]. While the EHR dataset in our study provided information about patients' household income, it did not include the number of people in a household, which is needed to accurately categorize socioeconomic status [11]. Hence, the two EHR variables *insurance* and *income* together provided a more accurate picture of the data persona's socio-economic status. Similarly, information about insurance type in the dataset provided insight about a patient's disability status, which was yet another attribute that was considered important for intervention design when proto personas for UMass ACO patients were being developed [4].

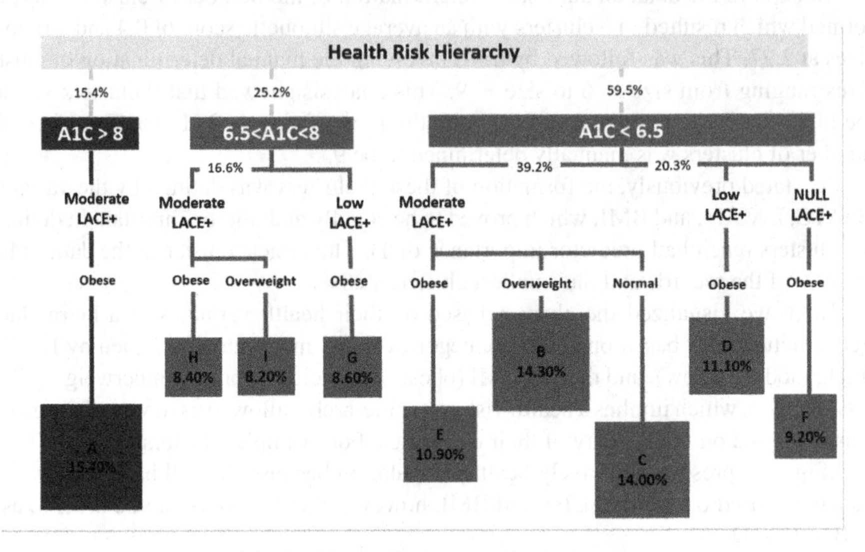

Fig. 3. Hierarchical tree of the clusters

For example, Medicare is a federal program for people above 65 years of age or those under 65 who have a disability [18]. Hence, having Medicare when patients are not yet 65 reveals that they suffer from some type of disability.

Likewise, we used having an email address and an active MyChart status in the EHR dataset as a sign that a patient had access to technology and felt somewhat comfortable using email and web technologies. These attitudinal attributes were considered in proto personas that were developed to represent UMass ACO type 2 diabetes patients [4].

Finally, we used the location of patients, represented by the geographic variables in Fig. 1, to identify the areas with the most patient population within each cluster. Including such information in data personas can help to design customized interventions and services that need to be provided locally.

4.2 Mapping Data and Proto Personas

The next step in our study was to manually map the 9 data personas (clusters) developed in this study to 8 proto personas that were developed by a group of ACO healthcare experts for the same patient population in a different study provides an example of a proto persona that was used in mapping in our study [4]. As displayed in Fig. 4, proto personas used in our study included qualitative data about patients' goals and frustrations as well as observational data such as disease knowledge, social and family support, cognitive and physical wellbeing, access to care, intention to get treatment and follow through the treatment program. The mapping of 9 data personas developed in this study and 8 proto personas developed in a prior study resulted in a set of 5 triangulated personas that combined factual EHR data such as HbA1C with qualitative and observational information that was lacking in the EHR dataset.

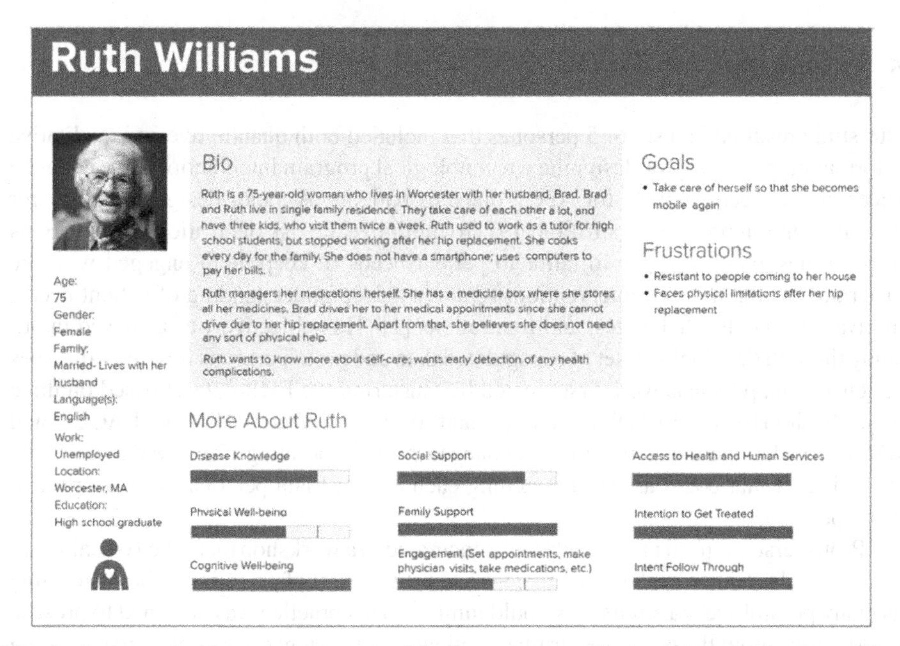

Fig. 4. An example of a proto persona used in the study created using IKE software (https://ike.wpi.edu/)

As discussed earlier, data personas identified in our study represented the largest groups of patients in each cluster. In contrast to data persona which typically represents the majority, proto personas are typically developed through a process that focuses on diversity. That is, proto personas are typically developed by considering all possible user groups, not only those that make up the majority of the population.

Given the difference in the focus of representation between proto and data personas, it is natural that some proto personas are not represented by data personas. For example, proto personas that portrayed female Spanish-speaking ACO patient populations were not sufficiently represented in the data persona clusters. This is not surprising because in our EHR dataset only 6% of records belonged to non-English speaking patients.

The remaining 5 proto personas were well represented by 94% of records that belonged to English-speaking patients. In particular, clusters A, B, C, F, H, and I were mapped to 3 proto personas that represented older English-speaking male patients. Moreover, qualitative information revealed that these patients might be single and may nor may not need assistance with tech usage and in making appointments. Those with family support relied either on their spouse or children for tech usage and in making appointments. Clusters E, D, and G were mapped to 2 proto personas that represented older English-speaking female patients. Qualitative data for this mapping revealed that they could either be single and tech-averse while those with family support relied on them for tech access and in making appointments.

Thus, the mapping made it richer by providing more nuanced personas that depicted both quantitative factual as well as qualitative assumption-based information.

5 Discussion

Our study resulted in a set of 5 personas that included both quantitative and qualitative information necessary for designing a technological program intervention to provide the necessary self-care support for type 2 diabetes patients. A major design challenge for any intervention program is attrition [19]. To minimize the risk of attrition, interventions must be designed in a way to tailor to patient needs to keep them engaged with the program. Designing such interventions requires a deep understanding of patient needs, motivations, goals, challenges, and frustrations [1]. This objective was achieved in our study through developing a set of triangulated data and proto personas for type 2 diabetes patients. Data personas were first created by clustering the EHR dataset based on three major health risk variables that are important to type 2 diabetes (HbA1C, LACE+, and BMI). The data personas were then enhanced by incorporating proxy variables in the EHR dataset that could assist in matching each cluster (data persona) to its respective proto persona.

Proto personas used in our study were generated in a workshop that asked 6 healthcare professionals and medical experts to generate a diverse set of personas by brainstorming as many possible scenarios as they could think of. This practice was designed to provide information about the entire spectrum of patient groups. Hence, the proto personas used in our study were representative of the range rather than the frequency of cases. The data personas in our study, using the SPSS two-step clustering algorithm, provided representation for the majority of patients. Because of their different focus (range vs. frequency of cases) combining the proto and data personas in our study provided the opportunity to develop a fairer representation of the patient population that was considered in our study. Additionally, supporting prior research, our results show that mapping data and proto personas resulted in a more nuanced picture of patient groups, which has been shown to be instrumental in designing successful products and services [2]. The information about patient location in our developed personas can help intervention design by providing access to needed services locally and/or by informing users about the nearest local services that are relevant or important to them.

Our study contributes to HCI literature by providing a methodology for developing data personas for type 2 diabetes patients from electronic health records. It also provides information for enhancing data personas with proxy information so that they can be matched with their respective proto personas to create a more comprehensive patient representation, which is needed for creating successful interventions [1]. The results of our study also show that in addition to creating more nuanced patient representation, combining data and proto personas can help to reveal some biases that may be inherent in EHR data clustering.

6 Limitations and Future Research

Despite the richness of qualitative information provided by proto personas in our study, this information was not gathered directly from patients. These limitations in our study can be addressed in future studies that focus on the development of user personas which is achieved by conducting unstructured or semi-structured interviews with patients.

Deidentified census tract IDs had been extracted for neighborhoods along with median household income. However, these IDs require GIS software for further analysis of socioeconomic status within a particular zip code. Future work can undertake analysis of these details to aid health officials and the local government in ensuring better access to care for socioeconomically disadvantaged neighborhoods. Furthermore, these IDs can also be used to analyze access to care and subsequent problems in transportation (an issue that was identified by proto personas). Transportation has not been touched on in the current study for two main reasons. One, the covid-19 pandemic at the time this study was conducted has made telehealth the only option, especially in cases such as long-term care of chronic diseases, hence, the current study focused on access to technology. Two, with time, people's addresses change, and this may or may not be reflected in the database. Therefore, assessing their distance from the hospital, healthcare, and service providers may or may not produce fruitful results. Nevertheless, it is important to note that challenges to access have been identified as a major issue by a number of proto personas used in this study.

On a similar note, qualitative aspects of proto personas such as patients' challenges in managing their health are difficult to capture in data personas. The Epic database does not have variables that monitor a patient's management of his/her medicines on a daily basis. However, advances in data science approaches such as text mining may help to enhance patient records by including insight from patient comments and feedback to understand patient emotional needs and challenges.

In this study, cluster analysis was conducted using historical data. Future projects can create software that can conduct this analysis automatically in real-time to create data personas, thus enabling easy access to the most up-to-date data personas which is necessary before any major design decision [1].

References

1. Djamasbi, S., Strong, D.: User experience-driven innovation in smart and connected worlds. AIS Trans. Hum.-Comput. Interact. **11**, 215–231 (2019)
2. Jain, P., Djamasbi, S., Wyatt, J.: Creating value with proto-research persona development. In: Nah, F.-H., Siau, K. (eds.) HCII 2019. LNCS, vol. 11589, pp. 72–82. Springer, Cham (2019). https://doi.org/10.1007/978-3-030-22338-0_6
3. Persons, B., Jain, P., Chagnon, C., Djamasbi, S.: Designing the empathetic research IOT network (ERIN) chatbot for mental health resources. In: International Conference on HCI in Business, Government and Organizations (HCIBGO) (2021)
4. Bajracharya, A., McDonald, A., Girardi, C., Ahumada-Zonin, G., Djamasbi, S., Amante, D.: Proto-research persona development: a user experience design approach. In: Third Annual Diabetes Center of Excellence, UMass Medical School, Diabetes Day (2019)

5. Accountable Care Organizations (ACOs): General Information | CMS Innovation Center: Innovation.Cms.Gov. (n.d.). https://innovation.cms.gov/innovation-models/aco. Accessed 6 Sept 2021
6. UMass Memorial Health – Accountable Care Organization (ACO): UMass Memorial Health (n.d.). https://www.ummhealth.org/none/umass-memorial-medicare-accountable-care-organization-aco. Accessed 6 Sept 2021
7. An, J., Cho, H., Kwak, H., Hassen, M.Z., Jansen, B.J.: Towards automatic persona generation using social media. In: 2016 IEEE 4th International Conference on Future Internet of Things and Cloud Workshops (FiCloudW), pp. 206–211. IEEE (2016)
8. Epic EHR Software - Pricing, Demo & Comparison Tool: EHR in Practice (n.d.). https://www.ehrinpractice.com/epic-ehrsoftware-profile-119.html. Accessed 6 Sept 2021
9. Epic Clarity | Data Analytics Center | Perelman School of Medicine at the University of Pennsylvania (n.d.). Med.Upenn.Edu. https://www.med.upenn.edu/dac/epic-claritydata-warehousing.html. Accessed 6 Sept 2021
10. Weiss, A.P.: Morning CMO Report. Upstate University Hospital (2017)
11. Snider, S.: Where Do I Fall in the American Economic Class System? U.S. News & World Report, December 2020. https://money.usnews.com/money/personalfinance/family-finance/articles/where-do-i-fall-in-theamerican-economic-class-system
12. Lee, C.Y.: A two-step clustering approach for measuring socioeconomic factors associated with cardiovascular health among older adults in South Korea. Korean J. Adult Nurs. **32**(6), 551 (2020). https://doi.org/10.7475/kjan.2020.32.6.551
13. Norušis, M.J.: IBM SPSS Statistics 19 Guide to Data Analysis. Prentice Hall, Upper Saddle River (2011)
14. Eckel, R.H., et al.: Obesity and type 2 diabetes: what can be unified and what needs to be individualized? Diabetes Care **34**(6), 1424–1430 (2011). https://doi.org/10.2337/dc11-0447
15. Body Mass Index (BMI): Centers for Disease Control and Prevention, 7 June 2021. https://www.cdc.gov/healthyweight/assessing/bmi/index.html
16. Norušis, M.J.: SPSS 16.0 Guide to Data Analysis. Prentice Hall, Upper Saddle River (2008)
17. IBM: Support: IBM (n.d.). IBM https://www.ibm.com/support/pages/howloglikelihooddistance-method-appliedtwostepclusteranalysis
18. Differences between Medicare and Medicaid:. Medicare Interactive, 5 August 2021. https://www.medicareinteractive.org/getanswers/medicarebasics/medicarecoverageoverview/differences-between-medicareand-medicaid
19. Huh, J., et al.: Personas in online health communities. J. Biomed. Inform. **63**, 212–225 (2016). https://doi.org/10.1016/j.jbi.2016.08.019

An Analysis of Gender Differences in the Innovative Function Design of Supermarket Self-service Checkout Kiosk

Sheng-Ming Wang[1] and Chen Han[2](✉)

[1] Department of Interaction Design, National Taipei University of Technology, 1, Sec. 3, Zhongxiao E. Rd., Taipei 10608, Taiwan
ryan5885@mail.ntut.edu.tw

[2] Doctoral Program in Design, College of Design, National Taipei University of Technology, 1, Sec. 3, Zhongxiao E. Rd., Taipei 10608, Taiwan
han.ceng987@gmail.com

Abstract. This research provides insight into applying self-service checkout kiosks in supermarkets to create innovative service experiences based on gender. A self-service checkout kiosk is essentially a device that allows consumers to interact directly with a supermarket, receive service at their convenience, increase revenues and streamline all purchasing processes. This innovative technology allows customers to have a better user experience without queues. To know the gender difference of self-service checkout kiosks' innovative service, this study began with the user-centered process that integrates emotional design and user experience design to analyze the supermarket's application scenarios of self-service checkout kiosks. Then we proposed eighteen innovative functions to create better service experiences for customers. We use Kano's Model method to design a questionnaire and invite candidate users to evaluate the proposed innovative functions by converting each respondent's answer to a score of "satisfaction potential." The Kano's Model evaluation results are then classified the proposed innovative functions into must-be, one-dimensional, attractive, and indifference categories and show gender differences. The results of this research show that gender does affect system performance and users' satisfaction classification, priority, and user experience to the proposed self-service checkout kiosk innovative functions. The results also show a comprehensive analysis of the gender difference in the innovative function design of supermarket self-service checkout kiosks. The contributions of this research can not only be used by the stakeholders of supermarkets to draw their strategies for deploying self-service checkout kiosks but also provide a better user experience design to the supermarket customers. Thus, in future designs, designers can refer to the results of this study to design self-checkout kiosks for offline retail shops based on brands with different positioning.

Keywords: Self-service checkout kiosk · Gender difference · Innovative function design · Kano's model · User experience design

F. Fui-Hoon Nah and K. Siau (Eds.): HCII 2022, LNCS 13327, pp. 337–349, 2022.
https://doi.org/10.1007/978-3-031-05544-7_26

1 Introduction

Self-checkout systems provide users with a new way to check out, an alternative to manual checkout. Several scholars have analyzed the current solutions to the check-out queue problem and their advantages and drawbacks in various areas and suggested the concept of 'self-checkout, fast payment' based on the perspective of supermarkets and consumers. Self-developed software has been applied to address the existing queue problem and boost the settlement efficiency of supermarkets and the flow rate of customers [1]. The global self-checkout system market is rapidly growing [2]. Still, the COVID-19 outbreak has hurt the market size in 2020, during which the use of self-service technology (SST) in retail environments has significantly increased [3–6]. SST can attend to customers without direct contact with employees [7, 8] and is mutually beneficial for customers and retailers. For customers, self-service offers convenience and can save queue time [9]. It can reduce labor costs and increase productivity [10, 11]. In recent years, fashion brands have deployed many self-checkout kiosks. Rebecca Minkoff, Nike, and Zara have implemented the self-checkout technology in their shops to provide customers with ease in purchasing products [12] and avoid lengthy and laborious manual checkout queues. SST can enhance the competition by providing customers with a seamless, bespoke, personalized shopping experience. This study explores the core requirements of self-checkout kiosks and assesses the priority development and improvement order of various functions based on gender. The technology adoption life-cycle is used to identify the primary users of self-service kiosks. Kano's model is used to analyze the development priority of various functions of self-checkout kiosks for the gender difference, which can be used as a reference for designers and developers when designing self-checkout kiosks for future supermarkets.

The main objectives of this study are as follows:

1. To summarize the functions of self-checkout kiosks in the current case study and propose the innovative service functions.
2. To evaluate the ranking of proposed functional satisfaction and importance based on genders using the Kano model.
3. To analyze the differences in the ranking of proposed functional satisfaction and importance based on genders.
4. To analyze and sort out different proposed functions of self-checkout to prioritize the development of future system functions based on genders.

2 Literature Review and Related Works

With the widespread use of self-checkout kiosks, several developers have improved their kiosks to improve customer perception and experience. If customers are satisfied with the 'usability' experience or feel that their expectations have been exceeded, the frequency of customer visits would increase and, thus, the spending of the customers [13, 14]. Almajid [15] suggests that the biggest motivation for users to choose a self-checkout kiosk is to save time and improve efficiency, while reasons for not choosing a self-checkout kiosk include complicated systems and time-consuming operations. Therefore, self-checkout systems must provide users with an efficient, convenient, and clear shopping experience. A majority of customers prefer self-checkout, enhanced convenience, accuracy, and checkout speed. Self-checkout kiosks are not limited to supermarkets as these can be used in all offline branded retail shops. Brands and shopping malls have diverse positioning and key consumer groups (different genders, ages, occupations, and hobbies). Similar to human personality, brand personality is multidimensional and comprises masculinity and femininity, which demonstrates the self-perception and temperament of men and women while purchasing products [16, 17].

The choice of products and brands differs based on gender. Women are more likely to buy beauty products, and men are more likely to buy alcohol and tobacco. Consumers often associate personal traits with brands as an extension of themselves [18], or as a result of consumer demand, brands produce gender-specific traits related to fashion, maturity, cuddly nature, and dependability. Therefore, the gender difference of consumers is regarded as the category to process purchase information for brands. Consumer gender is an important variable in market segmentation and is one of the key indicators to distinguish target markets based on gender-specific audience segmentation [19, 20]. Moreover, there are differences in the preferences of male or female about the style, color, and shape of the product. Often, products are designed to meet the needs of men or women based on their characteristics [21]. Therefore, the design of self-checkout kiosks used in supermarkets should also take gender differences into account. Differences between men and women in terms of characteristics, attitudes, and/or activities result in different emphasis or considerations in purchasing processes, gender differences, and consumption patterns [22]. There are differences between men and women in terms of purchases, amounts spent, risk perceptions, and attitudes towards advertising. Even when buying or consuming the same product, consumption patterns vary between genders.

Men and women engage in shopping, where time-saving, efficiency and rational calculations are important. However, women tend to engage in leisure-oriented shopping, where spending is not always planned or compulsory and is impulsive. Thus, the efficiency of the spending process is not important and may only be for leisure and enjoyment or for socializing [23]. Barletta argues that the consumption of women is rational and emotional, and that, in contrast to men's linear consumption decision-making process, women use a 'spiral route' to repeatedly collect a great deal of information and compare. Women pay more attention to details and the quality of the content to find a suitable product [22]. Therefore, women carefully choose the products they want to buy and enjoy happy shopping hours, finding it fun to buy novel products. Men prefer to shop at regular shops for low prices and to save time, but they can be confused about which shops or brands to choose from to make shopping quicker and more convenient [24]. When confronted with too much information, men tend to simplify their decision-making and prefer repetitive and planned purchases [25]. Men partly care about brand awareness because they can use brands to reduce search costs, boost their self-esteem and save time [26, 27].

As summarized from the above discussion, self-checkout kiosks have become more widespread and popular, and several brands and shops are experimenting with self-checkout technology to offer users a more diverse, comfortable, and fast shopping experience. As men and women differ in their shopping behavior, with men preferring to shop purposefully and directly and women preferring to shop casually, we assume that the order of features required differs between male and female while using self-checkout kiosks. Therefore, we consider it necessary to evaluate the functions of self-checkout kiosks based on gender.

3 Research Plan and Methodology

3.1 Emotional Design Analysis of Self-checkout Kiosks

Emotional design proposed by Don Norman is the concept of how to create designs that evoke emotions which result in positive user experiences. Designers aim to reach users on three cognitive levels—visceral, behavioral and reflective—so users develop only positive associations with products, brands, etc. [28]. The visceral is what you feel at first sight and is concerned with appearance, color, sound, material, and smell; behavioral is related to use and includes function, comprehensibility, ease of use, and feeling; reflective is at the heart level and relates to the meaning of an object.

The visceral, behavioral, and reflective of self-checkout kiosks are analyzed and listed (Table 1).

Table 1. Emotional design analysis self-checkout kiosks

Visceral	Behavioral	Reflective
New technology	Ease of use of interfaces and systems	Produce pleasure
New checkout method	Operation time of the self-checkout process	Reduce time and improve efficiency
The appearance of self-checkout kiosks	Time and efficiency of use	We will use self-checkout kiosks again
The interface of self-checkout kiosks	Reduce social interaction	
	Rethink and confirm purchased items	
	Check out according to the user's habit	

3.2 Kano's Model Analysis and Evaluation

The Kano's model, developed by Professor Kano in the 1980s, is related to product development and customer satisfaction. It can be used to rank services or functions, measure the depth of pain points, help reveal existing pain points, facilitate in-depth research to understand pain points, and understand why these features affect user satisfaction [29]. Each function can be separately listed in the questionnaire to evaluate satisfaction and importance. The Kano model collects data from large user groups to exhibit market trends and expectations, facilitates researchers to understand user perceptions of the product over time, and helps teams to understand when their products have reached a standstill and when innovative ideas should be recovered [30]. The five quality elements of the Kano model are: one-dimensional (O), also known as performance (P); must-be (M); attractive (A); indifferent (I), and reverse (R).

Based on the quality improvement coefficient proposed by Matzler (1998), the satisfying influence (also known as the better coefficient) and the dissatisfied influence (also known as the worse coefficient) were calculated to show the extent to which achieving this quality element affects increasing satisfaction or decreasing dissatisfaction [31].

The satisfaction and dissatisfaction coefficients were calculated based on the questionnaire results using the Kano's model.

Satisfied Influence:

$$SI = \frac{A + O}{A + O + M + I}$$

Dissatisfied Influence:

$$DI = \frac{(O + M)}{(A + P + M + I)}$$

Selection and Classification of Proposed Innovative Service Functions. Based on case study and analysis of preliminary user experience, we propose 18 innovative service functions of self-checkout kiosks in this research. The functions include: 1. The outdoor weather forecast, 2. Membership points, 3. Member login via mobile number, 4. Member login via face recognition, 5. On-screen touch, 6. Voice prompt, 7. Self-scan product barcode, 8. Weighing, 9. Display of product list in the shopping cart, 10. Display of discount details in the shopping cart, 11. Online payment, 12. Multi-platform payment, 13. Pay with face recognition, 14. Scan to pay, 15. Print receipts automatically, 16. Print receipts selectively, 17. Leave a comment, 18. Call for help.

Subjects Selection and Experiment Design. The technology adoption lifecycle is divided into five stages based on the characteristics of users adopting technology: innovators, early adopters, early majority, late majority and laggards, which accounted for 2.5%, 13.5%, 34%, 34% and 16% of the total number of users, respectively [32]. In order to follow the innovative service evaluation principles, this study recruited 80 subjects, 40 males and 40 females, according to the main user groups in the technology adoption lifecycle. They were early adopters, early majority or late majority in the technology adoption lifecycle, i.e. the main users of self-checkout kiosks. They were not laggards because they were willing to try out a self-checkout kiosk after it was launched and already had prior experience, but they were not innovators either, because they did not try out a self-checkout kiosk in the first place.

This study evaluated the priority development of innovative service functions of self-checkout kiosks; therefore, the Kano's model was used as the evaluation method. The

Kano's model quality satisfaction questionnaire was designed based on the 18 functions of self-checkout kiosks in supermarkets. Following the methodology of Kano's model, the questions in the questionnaire included positive and negative questions based on the satisfaction of each function and the importance of the function, with one set of questions for each function (18 sets in total) and three questions for each set (54 questions in total). The follow-up research was performed among the 80 subjects using a web-based questionnaire.

4 Experimental Results and Discussions

In this study, after the two-way questionnaire of the Kano's model was considered, the quality elements were categorized and evaluated based on the questionnaire results. Considering the differences among all users and between male and female users, as well as the user requirements of different genders, we evaluated the functions based on the order of importance, analyzed the necessity of each function, and proposed the development order of these functions.

4.1 Kano's Model Quality Elements Analysis

Based on the questionnaires collected in this research, the Kano'a quality element analysis of all subjects is shown in Table 2:

Table 2. Knao's model quality element analysis

A (Attractive quality)	9. Display of product list in the shopping cart; 13. Pay with face recognition; 14. Scan to pay; 17. Leave a comment
O (One dimensional)	2. Membership points; 15. Print receipts automatically
M (Must-be)	3. Member login via mobile number; 5. On-screen touch; 7. Self-scan product barcode; 10. Display of discount details in the shopping cart; 11. Online payment; 12. Multi-platform payment; 16. Print receipts selectively; 18. Call for help
I (Indifference)	1. Outdoor weather forecast; 4. Member login via face recognition; 6. Voice prompt; 8. Weighing

Based on the results, the better (SI)–worse (DI) values for each function were calculated using the satisfied influence and dissatisfied influence formulae, and the results of all subjects and the difference of male and female subjects are listed in Table 3.

When analyzing the satisfied influence and dissatisfied influence, the value of the former being closer to 0 indicates a low influence, whereas the value is closer to 1 indicates a greater possibility of increasing user satisfaction when this quality element is present. For the latter, the value being closer to 0 indicates a high influence, whereas the value being closer to 1 indicates greater dissatisfaction when the quality element is missing.

Table 3. Analysis of satisfied and dissatisfied influence with the Kano's model

Subject	All subjects		Male subjects		Female subjects	
	Better	Worse	Better	Worse	Better	Worse
1. Outdoor weather forecast	0.36	−0.055	0.47	−0.032	0.2	−0.08
2. Membership points	0.53	−0.277	0.65	−0.347	0.41	−0.182
3. Member login via mobile number	0.55	−0.426	0.62	−0.385	0.5	−0.455
4. Member login via face recognition	0.29	−0.31	0.29	−0.239	0.27	−0.364
5. On-screen touch	0.53	−0.575	0.56	−0.64	0.52	−0.522
6. Voice prompt	0.28	−0.246	0.31	−0.116	0.25	−0.358
7. Self-scan product barcode	0.54	−0.426	0.58	−0.484	0.5	−0.334
8. Weighing	0.33	−0.032	0.42	−0.031	0.24	−0.031
9. Display of product list in shopping cart	0.39	−0.24	0.41	−0.186	0.33	−0.334
10. Display of discount details in shopping cart	0.48	−0.492	0.58	−0.452	0.39	−0.549
11. Online payment	0.51	−0.489	0.5	−0.417	0.55	−0.6
12. Multi-platform payment	0.51	−0.359	0.62	−0.276	0.4	−0.44
13. Scan to pay	0.26	−0.283	0.21	−0.211	0.33	−0.381
14. Pay with face	0.24	−0.04	0.33	−0.084	0.14	0
15. Print receipts automatically	0.43	−0.233	0.5	−0.219	0.36	−0.24
16. Print receipts selectively	0.5	−0.417	0.43	−0.429	0.63	−0.375
17. Leave a comment	0.07	−0.104	0.06	−0.059	0.08	−0.154
18. Call for help	0.51	−0.568	0.55	−0.546	0.5	−0.625

Based on the quality element results, the properties of each function can be analyzed based on the better–worse values, which can be used to initially prioritize the functions for development and improvement. The development order is set up as follows: Must-be (M) > One-dimensional (O) > Attractive (A) > Indifferent (I). In this study, better–worse values were calculated based on the questionnaire results, and 18 functions were classified based on the four elements of the Kano model.

To present the properties of each function more clearly, the satisfied influence and dissatisfied influence of each function were calculated, and a matrix was plotted with the horizontal axis being the satisfied influence, the vertical axis being the dissatisfied influence and the boundary being their mean. The attribute distribution chart was divided into four quadrants, and the locations of the functions were marked on the matrix to investigate the importance that users attached to them. The importance represented by each quadrant: usually follows the order M (must) > O (one-dimensional) > A (attractive) > I (indifferent).

Figure 1 shows the distribution chart of Kano's quality elements based on all subjects and gender differences. We can see from the results that males and females are not quite the same for each of the elements. It can be concluded that males and females have commonalities in items A (attractive) and M (must-be), but not in items I (indifference) and O (one-dimensional). In A, males, and females are consistently identified: 1. Outdoor weather forecast; 8. Weighing; 9. Print receipts automatically is also an Attractive element for females; whereas for males, 4. Member login via face recognition; 6. Voice prompt; 13. Pay with face.

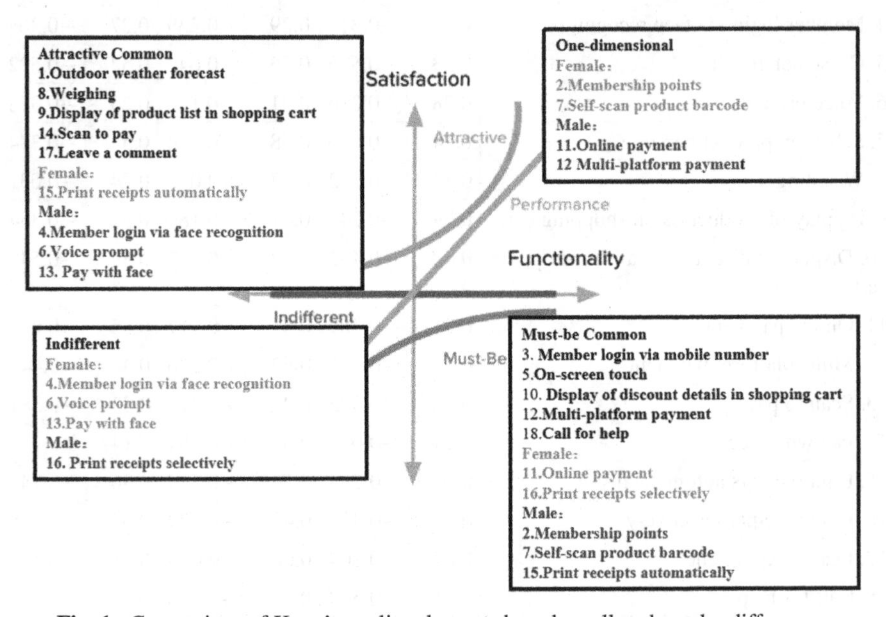

Fig. 1. Comparison of Kano's quality elements based on all and gender differences

In the M category: both males and females think that 3. Member login via mobile number; 5. On-screen touch; 10. Display of discount details in the shopping cart; 12. Multi-platform payment; 18. Call for help. females think 11. Online payment; 16. Print receipts selectively are also necessary conditions; while males think 2. Membership points; 7. Self-scan product barcode; 15. Print receipts automatically are necessary conditions.

Females consider 2. Membership points; 7. Self-scan product barcode as O items; while males consider 11. Online payment; 12. Multi-platform payments as O items. Females think 4. Member login via face recognition; 6. Voice prompt; 13. Pay with face recognition belongs to the I category; while males think 16. Print receipts selectively belong to the I category.

Summarized from an analysis of the distribution of Kano's quality elements. The post-test feedback shows that most users responded positively to the self-checkout kiosks and would be happy to use them again.

4.2 Ranking of Functions Based on Importance

The importance scores for each function differ across gender are shown in Table 4.

Table 4. Function importance by all and gender subjects

Function	All	Male	Female
1. Outdoor weather forecast	5.1(17)	4.5(17)	5.7(16)
2. Membership points	7.3(10)	7.2(10)	7.4(10)
3. Member login via mobile number	7.1(11)	6.7(13)	7.6(7)
4. Member login via face recognition	6.6(13)	6.9(11)	6.3(14)
5. On-screen touch	7.8(5)	7.6(7)	8.0(4)
6. Voice prompt	6.3(14)	6.0(16)	6.6(13)
7. Self-scan product barcode	7.9(4)	8.2(2)	7.5(8)
8. Weighing	5.9(16)	6.4(14)	5.5(17)
9. Display of product list in shopping cart	7.6(7)	7.7(6)	7.5(8)
10. Display of discount details in shopping cart	7.8(5)	7.9(5)	7.7(6)
11. Online payment	8.1(2)	8.0(4)	8.3(2)
12. Multi-platform payment	8.5(1)	8.5(1)	8.6(1)
13. Scan to pay	8.1(2)	8.1(3)	8.1(3)
14. Pay with face	6.2(15)	6.4(14)	6.1(15)
15. Print receipts automatically	7.6(7)	7.3(9)	7.9(5)
16. Print receipts selectively	7.0(12)	6.9(11)	7.1(12)
17. Leave a comment	4.3(18)	3.8(18)	4.7(18)
18. Call for help	7.4(9)	7.4(8)	7.3(11)

Results and Discussion. Considering design thinking as the basis, this study evaluated the core requirements of self-checkout kiosks through the design and analysis of user experience and evaluated the priority of developing and improving various functions for distinct gender groups.

Among the functions of self-checkout kiosks listed in this study, the sensation dimension included 'member login via mobile number', 'display of product list in shopping cart', 'display of discount details in shopping cart', 'outdoor weather forecast' and 'leave a comment'. For such functions, the emphasis lies on the visual display design, and the visual experience of the interface should be considered during development. The behavior of the dimension of the elements included 'membership points', 'member login via mobile number', 'member login via face recognition', 'on-screen touch', 'voice prompt', 'self-scan product barcode', 'weighing', 'online payment', 'multi-platform payment', 'scan to pay', 'pay with face recognition', 'print receipts automatically', 'print receipts selectively', 'leave a comment' and 'call for help'. Such functions focus on operational behavior and should be developed with an emphasis on the ease of use of the system and interface. The reflection dimension included 'outdoor weather forecast', 'membership points', 'member login via face recognition', 'member login via mobile number', 'voice prompt', 'self-scan product barcode', 'display of product list in shopping cart', 'display of discount details in shopping cart', 'online payment', 'multi-platform payment', 'scan to pay', 'pay with face recognition', 'print receipts automatically', 'print receipts selectively' and 'leave a comment'. These functions focus on services and on building relationships with users through user experience design. As shown in the distribution chart of Kano quality elements (Fig. 1), 'all users', 'member login via mobile number', 'On-screen touch', 'Self-scan product barcode', 'Display of discount details in shopping cart', 'Online payment', 'Multi-platform payment', 'Print receipts selectively', 'Call for help are mandatory elements'. Without these elements, the user experience would drop drastically if they are absent or poorly performed. One-dimensional elements include 'Membership points', 'Print receipts automatically'. User satisfaction would improve if such requirements are met. All the above functions are 'must-be elements' and 'one-dimensional elements', to which users have attached importance as a result of core system requirements. Failure to fulfill such functions would significantly surge user dissatisfaction. Therefore, it is recommended that they be the first functions to be considered for system development. Functions that are subsequently developed can be 'attractive elements', and their development can increase, but would not decrease, user satisfaction even if they are not developed. The functions in 'indifferent elements' can be considered at the last stage or do not require development.

Combining Kano results with the ranking of the importance of functions among all users, the prioritized order for the development of the functions of self-checkout kiosks among all users is set as follows: print receipts automatically > member login via mobile number > multi-platform payment > online payment > scan to pay > self-scan product barcode > display of discount details in shopping cart > on-screen touch > display of product list in shopping cart > call for help. Thus, the improved design of self-checkout kiosks in branded offline retail shops, regardless of gender, functions, and services, should be developed.

According to the distribution chart of Kano quality elements based on genders (Fig. 1), the Kano results of male users were consistent with those for all users. Combining Kano results with the ranking of the importance of functions among male users, the prioritized order for the development of the functions of self-checkout kiosks among male users is set as follows: print receipts automatically > member login via mobile number > multi-platform payment > self-scan product barcode > on-screen touch > call for help > display of discount details in shopping cart > scan to pay > online payment > display of product list in the shopping cart. Therefore, for the improved design of self-checkout kiosks in male-specific branded offline retail shops, functions and services should be developed based on the aforementioned order.

In the analysis of Kano's model results for female users, online payment, print receipts selectively are 'must-be elements'. Membership points, Self-scan product barcodes are 'one-dimensional elements'. Combining Kano results with the ranking of the importance of functions among female users, the prioritized order for the development of the functions of self-checkout kiosks among female users is set as follows: member login via mobile number > self-scan product barcode > membership points > multi-platform payment > online payment > on-screen touch > display of discount details in shopping cart > scan to pay > print receipts automatically > display of product list in the shopping cart.

5 Conclusion and Future Works

Previous studies found that men and women had different consumption habits and psychology, with men paying more attention to the shopping efficiency while women to the shopping process. This study revealed that the importance of and satisfaction with the functions of self-checkout kiosks varied across different genders, with emphasis on the functions of self-checkout kiosks. The results are presented as follows:

1. There were differences in the order of functional importance across different genders, including the overall, male and female groups.
2. According to the Kano model, the overall and male groups produced identical results but differed from the female group.
3. Combining the Kano results of these three groups with the importance-ranking results leads to the priority development order of the functions of self-checkout kiosks in Mainland China.

As a result, in the current stage of the improvement and design of self-checkout kiosks for different brands and shops, when brands and shops are positioned to target groups with no significant gender preference, it is recommended to follow the priority development order of the overall group. When brands and shops target male consumers, the recommendation would be to follow the priority development order of the male group. With female consumers, it is recommended to follow the priority development order of the female group.

As suggested by the data collected, during the physical purchasing process in shops, the salesperson would make certain adjustments or changes to the sales or service methods according to the gender of the customer, as they found certain differences in the purchasing behavior of men and women owing to their years of experience. In future applications, this could also be applied to self-checkout services by incorporating technologies such as face recognition, or customers can have the freedom to choose the mode of sale that best suits them when using self-checkout. The age of automation is approaching, and the self-service experience is becoming more important and needs more attention to enhance and popularize automated services. The present study is based on the current situation in Mainland China. Depending on the acceptance of the population and the level of technological development in different regions, there will be some differences in the priority development order of the functions of self-checkout kiosks, as well as some differences in the functional requirements of distinct genders. The user experience of a self-checkout kiosk is correlated to the promotion of the shop; thus, research and design on how to improve user satisfaction with self-checkout kiosks are important. In the future, extensive research needs to be conducted based on the following aspects:

1. As technological development, stage of automation, local habits, and acceptance of technology vary from region to region, for future research and use, the priority development order of self-checkout kiosks should be re-evaluated subject to the situation in each region to integrate into the local market.
2. The 18 functions in this study can be combined or reduced according to the needs when using self-checkout kiosks in shops or areas, as simplified self-checkout systems can improve efficiency and accuracy.
3. The results refer to the proposed innovative service functions of self-checkout kiosks The functional priorities of users over time need to be re-evaluated, which is particularly the case when new technologies are widely available or when automation and self-service are developing rapidly.

This study can be used as a reference for designers and developers in the functional development order of self-checkout kiosks. In future designs, designers can refer to the results of this study to design self-checkout kiosks for offline retail shops based on brands with different positioning. In future applications, to meet customized requirements, face recognition technology can be adapted to automatically switch the interface and operation modes for male or female users, providing the most suitable way for the users to interact with the kiosk. This study examined and integrated the existing self-checkout kiosks in Mainland China, revealing that the focus of self-checkout functions varied across different genders. The findings will boost the market growth of self-checkout kiosks and improve the shopping experience in offline retail shops of different brands, as well as the human-computer interaction experience between users of different genders and self-checkout kiosks.

Acknowledgements. We would like to show our gratitude to the funding support from the Ministry of Science and Technology with the project (108-2410-H-027-013-MY2). We also thank Bin-ying Zhong for sharing her pearls of wisdom with us during the course of this research.

References

1. Bulmer, S., Elms, J., Moore, S.: Exploring the adoption of self-service checkouts and the associated social obligations of shopping practices. J. Retail. Consum. Serv. **42**, 107–116 (2018)
2. Arnfield, R.L.: Supermarket self-checkout technology approaching point. Tipping (2014). Kioskmarketplace Retrieved from http://www.kioskmarketplace.com/articles/supermarket-self-checkout-technology-approaching-tipping-point/ArnfieldR
3. Yang, Y.C., Liu, S.W., Ding, M.C.: Determinants of self-service technology adoption. Afr. J. Bus. Manage. **6**(40), 10514–10523 (2012)
4. Jamal, A.: Retail banking and customer behaviour: a study of self concept, satisfaction and technology usage. Int. Rev. Retail Distrib. Consum. Res. **14**(3), 357–379 (2004)
5. Burke, R.R.: Technology and the customer interface: what consumers want in the physical and virtual store. J. Acad. Mark. Sci. **30**(4), 411–432 (2002)
6. Merrilees, B., Miller, D.: Superstore interactivity: a new self-service paradigm of retail service? Int. J. Retail Distrib. Manage. **29**(8), 379–389 (2001)
7. Oyedele, A., Simpson, P.M.: An empirical investigation of consumer control factors on intention to use selected self-service technologies. Int. J. Serv. Ind. Manage. **18**(3), 287–306 (2007)

8. Meuter, M.L., Ostrom, A.L., Roundtree, R.I., Bitner, M.J.: Self-service technologies: understanding customer satisfaction with technology-based service encounters. J. Mark. **64**(3), 50–64 (2000)
9. Collier, J.E., Kimes, S.E.: Only if it is convenient: understanding how convenience influences self-service technology evaluation. J. Serv. Res. **16**(1), 39–51 (2013)
10. Dabholkar, P.A., Bobbitt, L.M., Lee, E.J.: Understanding consumer motivation and behavior related to self-scanning in retailing: Implications for strategy and research on technology-based self-service. Int. J. Serv. Ind. Manage. **14**(1), 59–95 (2003)
11. Kleemann, F., Voß, G.G., Rieder, K.: Un (der) paid innovators: the commercial utilization of consumer work through crowdsourcing. Sci. Technol. Innov. Stud. **4**(1), 5–26 (2008)
12. Anitsal, I., Schumann, D.W.: Toward a conceptualization of customer productivity: the customer's perspective on transforming customer labor into customer outcomes using technology-based self-service options. J. Mark. Theory Pract. **15**(4), 349–363 (2007)
13. Park, J.S., Ha, S., Jeong, S.W.: Consumer acceptance of self-service technologies in fashion retail stores. J. Fash. Mark. Manage. Int. J. **25**(2), 371–388 (2020)
14. McNamara, N., Kirakowski, J.: Measuring the human element in complex technologies. Int. J. Technol. Hum. Interact. (IJTHI) **4**(1), 1–14 (2008)
15. De Ruyter, K., Bloemer, J.: Customer loyalty in extended service settings: the interaction between satisfaction, value attainment and positive mood. Int. J. Serv. Ind. Manage. **10**(3), 320–336 (1999)
16. Turner, J.J., Szymkowiak, A.: An analysis into early customer experiences of self-service checkouts: lessons for improved usability. Eng. Manage. Prod. Serv. **11**(1), 36–50 (2019)
17. Bem, S.L.: The measurement of psychological androgyny. J. Consult. Clin. Psychol. **42**(2), 155 (1974)
18. Freimuth, M.J., Hornstein, G.A.: A critical examination of the concept of gender. Sex Roles **8**(5), 515–532 (1982)
19. Belk, R.W.: Possessions and the extended self. J. Consum. Res. **15**(2), 139–168 (1988)
20. Meyers-Levy, J., Maheswaran, D.: Exploring differences in males' and females' processing strategies. J. Consum. Res. **18**(1), 63–70 (1991)
21. Meyers-Levy, J., Sternthal, B.: Gender differences in the use of message cues and judgments. J. Mark. Res. **28**(1), 84–96 (1991)
22. Chun-Hsien, L.: Measurement constancy in young men and women on the consumption pattern scale. J. Manage. Syst. **23**(2), 273–302 (2016)
23. Li, Y.: Shopping studies: a review essay. Taiwan J. Sociol. **37**, 207–236 (2006)
24. Bakewell, C., Mitchell, V.W.: Male versus female consumer decision making styles. J. Bus. Res. **59**(12), 1297–1300 (2006)
25. Shim, S.: Adolescent consumer decision-making styles: the consumer socialization perspective. Psychol. Mark. **13**(6), 547–569 (1996)
26. Mitchell, V.W., Walsh, G.: Gender differences in German consumer decision-making styles. J. Consum. Behav. Int. Res. Rev. **3**(4), 331–346 (2004)
27. Woodruffe-Burton, H.: Private desires, public display: consumption, postmodernism and fashion's "new man". Int. J. Retail Distrib. Manag. **26**(8), 301–310 (1998)
28. Norman, D.A.: Affordance, conventions, and design. Interactions **6**(3), 38–43 (1999)
29. Verhoeven, P., Sha, N.: A study based on Kano model to improve satisfaction of students at university libraries in Sweden (2017)
30. Berger, C., Blauth, R., Boger, D.: Kano's methods for understanding customer-defined quality, pp. 3–35 (1993)
31. Huang, D., Pan, H., Wu, J., Liang, R.: The application of Kano model in improving the service quality of college sports venue. Sports Sci. Technol. (Guangxi) **38**(1), 144–446 (2017)
32. Matzler, K., Hans, H.H.: How to make product development projects more successful by integrating Kano's model of customer satisfaction into quality function deployment. Technovation **18**(1), 25–38 (1998)

HCI in the Workplace

Evaluation of the Change in the Quality of Reports with the Application of Gamification in a Corporative Institution

Publio Pastrolin Cavalcante and Sergio Antonio Andrade Freitas(✉)

Programa de Computação Aplicada, Campus Universitário Darcy Ribeiro, Universidade de Brasília, Brasília, DF 70904-970, Brazil
publio@caval.com.br, sergiofreitas@unb.br

Abstract. In this paper we present a gamification process developed in a Brazilian public institution. We used gamification to engage users in the reports' quality improvement of the organization. Gamification is the application of game elements and game principles in non-game contexts [1]. The Octalysis framework proposed by Yu Kai-Chou [2] was used to model the users' profile and the gamification itself. Based on the identified profiles, a gamification was created using the techniques of Classification, Medals, Level Up, Attachment Monitor, Mentoring and Build from Scratch in addition to the techniques of Countdown, Grave Tombstone and Group Challenge. In new gamified process, the players interact collaboratively and develop skills and competences over the same shared space. The result shows that use of the gamified process both enable a better reports' production and creates a fun and attractive way to improve learning, affective, sociocultural aspects, collaboration, and improvement at the organizational model.

Keywords: Gamification · Reports' quality · Public organization

1 Introduction

This work presents a gamification process developed in sectors of a corporate institution after using gamification strategies to build a manual game based on the profile of users (players). Gamification is the application of game elements and game principles in non-game contexts [1]. The use of gamification in the corporate context is an innovative practice that brings the playfulness of the game world to the institution's organizations.

The idea of games being part of people's daily lives is not new, it is considered older than the culture itself [6] and demonstrates a very important aspect, even in its simplest forms, at the animal level, which is more than a physiological phenomenon or a psychological reflex [5]. According to one theory, the game is a preparation of the young for the serious tasks that later life will demand of him, according to another, it is an indispensable self-control exercise to the individual. Others see the game principle as an innate urge to exercise a certain faculty, or as a desire to dominate or compete [5].

People are motivated to play by four specific reasons that can be seen together or separately as individual motivators: mastering a certain subject; stress relieve; training;

F. Fui-Hoon Nah and K. Siau (Eds.): HCII 2022, LNCS 13327, pp. 353–367, 2022.
https://doi.org/10.1007/978-3-031-05544-7_27

and socialization [7]. Such aspects can be analyzed together or separately. The authors also cite four aspects of fun during the act of playing: competition and the pursuit of victory (hard fun), immersion in the exploration of the game universe (easy fun), alteration of the player's feelings during his experience (altered state fun), player engagement with other players (social fun).

We used as a basis the Octalysis framework proposed by Yu Kai-Chou [2] to model and design the user profile and the gamification itself. We observed and measured users' experiences as players in the game: the elements of the conquest game, their strategies for achieving goals and their involvement in the learning process.

2 Related Work

McGonigal [3] proposed four characteristics for games: objective, rules, feedback, and voluntary participation. The goal provides a sense of purpose. The rules impose limitations on how players can achieve the goal. The feedback system tells players how close they are to reaching the goal. Voluntary participation requires each player to accept the purpose, rules, and feedback consciously and voluntarily.

Fredericks et al. [4] present three types of engagement, emotional, behavioral, and cognitive engagements, each type of engagement has different indicators.

Gamification is not just the application of technology to old models of engagement, as, for example, in the case of awarding skiers with different insignia. Gamification creates completely new models of engagement. Its target is the new communities of people, and the objective is to motivate them to reach goals that they do not know yet [8].

It is a relatively new concept [9, 10], which consists of use of game design elements in non-game environments, such as social networks, healthcare applications, education systems and in companies, to motivate people. and changing behaviors [11], this statement corroborates several previous studies [12–14].

Gamification theory proposes that, by introducing game elements into an environment to satisfy some of the user's needs, it is possible to make activities more attractive, even if the user is not intrinsically motivated [15]. The motivations for gamification can be many, from making certain content more captivating, to reinforcing specific information [16].

Werbach and Hunter [14] emphasize that the purpose of gamification is to bring game elements and design to non-game contexts. The idea is not to make everything a game, it's just the opposite. It's not about playing games while working, the idea is to make it clear enough that the user is still in the real world, that they're still on the desktop, or that they're browsing a website to buy a product, but in a slightly different way. more playful. This technique seeks to make the experience better, learn from the games, seek to improve the user experience, find the focus of that experience, and make it more rewarding, create better motivation, but without taking it away from the real world.

Belinazo et al. [17] argues that gamification should develop meaningful and engaging game rules, integrated with business rules. To achieve this objective, the system interface must be improved, making it attractive to its user, also applying gameplay characteristics, making the "player" experience fun and enjoyable.

When talking about non-gaming context, it is related to the objective for which one is playing. It may be to learn something relevant to the job, to achieve some objective of the company where he works, or even playing a game to get a job, this is a non-gaming context [14]. Good gamification contains a "hook", something that initially grabs your audience's attention, and holds it, at least for a period, and strategically, from the organization's perspective, good gamification is successful gamification [18].

Some gamified activities [8, 19] are:

- The speed camera lottery (the fun theory) - was implemented in Stockholm, Sweden in 2010, by the Swedish national society for traffic safety and The Fun Theory, by an initiative of Volkswagen. The idea is that all drivers would be photographed by the speed camera, however, those who obeyed the speed limit competed in a lottery that was financed by the amount collected from the fines of violators. In the three days the experiment took place, the average speed in Stockholm dropped from 32 to 25 km/h.
- Google Recruitment - a billboard was installed in Silicon Valley with the following sentence: "first 10-digit prime found in consecutive digits of e.com". This was the first puzzle in a selection process implemented by the company in 2004. As a result of the last puzzle, the next step was to participate in a job interview at Google's headquarters.
- Nike+: exercising can be fun - applied by Nike, Nike + Nike Plus was an application that the user could install on their smartphone to monitor their location, pace, distance, laps, and calorie burn, being able to follow and share your progress, optimize your training, set personal goals, and challenge your friends. Nike's focus was on promoting its brand, selling running shoes and associated products.
- Stack overflow: auto moderation of the technical forum - The figure of the forum moderator was created to avoid answers out of context, insults, fights, etc. Using game elements on the forum, users accumulate points for their interactions, and as their reputation grows, new functionalities are made available to the user who knows the system and the rules better, users still receive medals that end up stimulating even more the system usage.
- Waze - users themselves feed the service with real-time traffic information. For this, the system works with points, user levels, ranking, avatar customization, missions, and others. Waze still achieves a strong social bond between participants.
- Pain Squad - SickKids hospital used a gamified application in which children play the role of police officers in a special force. Children, when performing activities such as patient reports, can be promoted within the "game" until reaching leadership positions. At Headquarters, children can see their badges (medals) and already have access to when they should fill out their next report. Characters from shows like Flashpoint and Rookie Blue encourage children to fill reports through videos that are broadcast during missions. The app gave the children a sense of control over their own pain management, according to one of the mothers.

2.1 The Gamification Framework

A gamification Framework (Fig. 1) is a support structure around which the gamification can be built, a system of rules, ideas or beliefs that is used to plan or decide something [20]. Several companies are studying and proposing their own technologies to implement gamification techniques, both from software companies and companies specifically focused on gamification [21].

We used the Yu-kai Chou Octalysis framework [22] due to our affinity with the methodology and a deeper understanding of how this framework works, which suited the needs of the present study.

According to Yu-kai Chou, gamification is the form of project focusing on human motivation, therefore, he highlights the emphasis on the human being to the detriment of projects focused on increasing productivity [23].

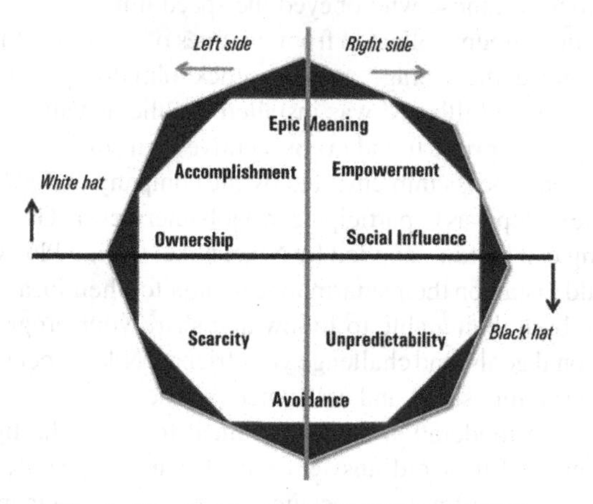

Fig. 1. The Octalysis visualization chart [22]

Octalysis [22] consists of eight main units of human motivation, named Core Drivers, as follows:

- Epic Meaning and Calling - User believes they are doing something greater than he, or he was "chosen" to do something.
- Development and Accomplishment - The feeling of making progress, developing skills and, eventually, overcoming challenges. The word "challenge" is extremely important, as a trophy or medal without a challenge means almost nothing.
- Empowerment of Creativity and Feedback - Expressed when users are involved in a creative process where they must repeatedly discover new things and try different combinations. It is very important that people receive feedback from activities to continue their process.
- Ownership and Possession - Users are motivated because they feel they control or own something. When the user feels ownership, he wants to improve and expand his possessions. In addition to being the main core of the desire to accumulate wealth, it handles many virtual goods or currencies between systems.

- Social Influence and Relatedness - Incorporates all the social elements that motivate people, including mentoring, social acceptance, social responses, companionship and even competition and envy.
- Scarcity and Impatience - It's about wanting things simply because they're extremely rare, exclusive, or just because you can't have them right now. A lot of games have Scheduling Dynamics or Torture Breaks (come back in 2 h to get your reward) - the fact that people can't have something right now can motivate them to think about it all day. As a result, they return to the product whenever they get a chance.
- Unpredictability and Curiosity - When something doesn't fit your regular pattern cycle, your brain pays more attention to the unexpected and you keep thinking about it. This drive is the main factor behind gambling addiction, but it is also present in all company-run sweepstakes or lotteries.
- Loss and Avoidance - Unsurprisingly, this drive is based on the motivation to avoid something negative happening. On a smaller scale, it might be to avoid losing a previous job or to change one's behavior. On a larger scale, it could be admitting that everything that has been done up to the present point has been unnecessary, because now you are willing to give up.

The Octalysis framework was constituted in this way so that the units that focus on creativity, personal expression and social dynamics were organized on the right side of the octagon which is called the Right Brain Core Units. The units that are most associated with logic, analytical thinking, and possession are imprinted on the left side of the octagon and are called the Left-Brain Core Units.

"Left side" units tend to rely on Extrinsic Motivation - you get motivated because you want to get something, whether it's a goal, a good, or something you can't get. On the other hand, Units on the "right side" are more associated with Intrinsic Motivations - you don't need a goal or reward to use your creativity, hang out with friends, or feel suspense or unpredictability - the activity itself is your reward.

Another factor to note in Octalysis is that the "top" Core Units are considered to be very positive motivations, while the "bottom" Core Units are considered more negative. Chou calls the techniques that use the Main Units "from above" White Hat Gamification, while techniques that use the Main Units "from below" are called Black Hat Gamification.

3 The Gamified Report Process

Based on the Octalysis framework [22], we develop and apply a gamification in the organizational environment called Gamified Report Process – GRP (Fig. 2). The gamification was built from the elements and mechanisms of the games which contains the dynamics, rules, stages, and awards with playfulness of the games and applied by a manager. This allows us to both produce a good game and evaluate how users are transforming their interaction in the production of reports [4].

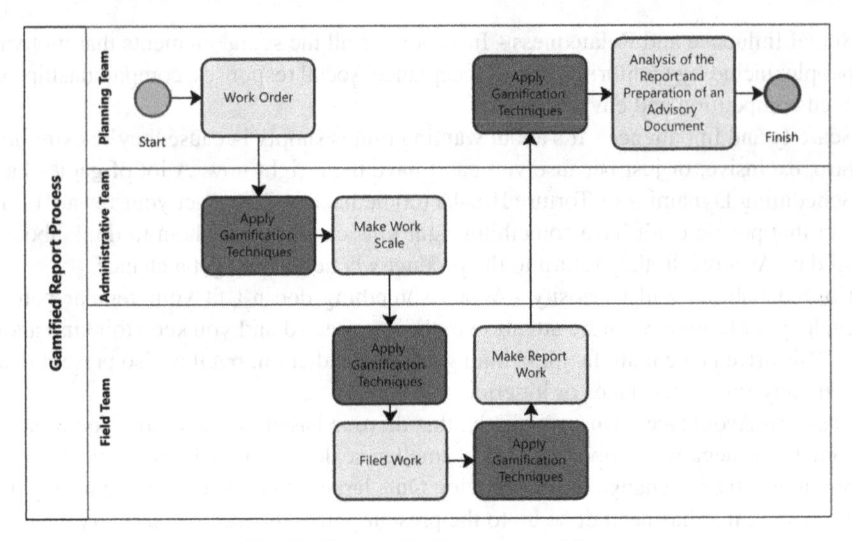

Fig. 2. The gamified process workflow

The original organization workflow has been modified to suit the gamification process. This change implied the introduction of gamified elements on the system screens, as well as the definition and implementation of new rules for the report production activity. In summary (or the main ones), the rules are:

- All activities that need to be evaluated by judges will have their score credited only after this step.
- Teams
  - They will be composed by the members of the participating units.
  - Every 180 days the team leader will be able to ask to change the name of his team.
  - Each month, the team with the most points will be able to change its name and badge.
- Points/Score
  - It is given to players as a form of stimulus and incentive to improve involvement and engagement.
  - Every activity will be rewarded with the number of points equivalent to the complexity, volume, and quality of his report.
  - Utilized points will be deducted from the total earned over time.
- Levels and Virtual Professions
  - Levels will be automatically assigned to players according to the number of accumulated points.

- The Virtual profession is one of the ways that the player can express his experience in a certain subject.

  - The title of the virtual profession must be acquired within the game using the accumulated score.
  - To acquire a title of a virtual profession, it is also necessary to fulfill the specific requirements for that title.

- Judges

  - The judges are the players who will evaluate the quality of the activities carried out within the game.
  - Judges must follow pre-defined criteria regarding:

    - Work orders:

      - Document clarity.
      - Order detailing.

    - Service reports:

      - Final report quality.
      - Quantity and adequacy to the context/need of photos attached to the report.
      - Text coherence, writing cohesion, readability.
      - Opportunity in the delivery of information.
      - Details of what was requested in the Work Order.

  - Will be selected as judges:

    - For service reports - The team leaders or the master leader if the task requires it.
    - For work orders – The players who will receive work orders for execution.

- Medals

  - Medals or badges will be awarded for personal or collective merits and will be divided into bronze, silver, gold, diamond, and platinum or unique/rare according to the importance or merit.

Levels presented in Table 1 are intended to demonstrate the player's experience in the institution in general. It is possible to establish a relationship of productivity with the time the player is working in the institution.

Table 1. Levels

Level	Description	Points
1	Newbie	20
2	Apprentice	60
3	Veteran	1500
4	Master	5000
5	Grand Master	15000

Table 2. Virtual professions

Title	Description	Criteria	Points needed
Junior photographer	Photography can be a hobby	15 photos with a score above 8	200
Full photographer	Becoming a professional photographer	100 photos with a score above 8	500
Senior photographer	Photography is like breathing	400 photos with a score above 8	1400
Mentor	Already guided other team members	100 guidelines or more	1000
Investigator	Points out hidden details in the reports	100 fixes or more in reports from other team members	900
Scribe	Write the reports directly and clearly	40 or more reports with maximum concept given by the judges	1100
Protector	Someone who when performing a given activity can see more security consequences beyond the original request	30 or more positive extra observations reported by the judges	1200

Virtual professions in Table 2 are used by team leaders to select players for training, as well as to identify players with the necessary qualifications to perform certain activities and possible multipliers or mentors, in addition to encouraging teamwork.

As a way of making the merits of the teams and players explicit to all players, the medals awarded can be viewed by all players (Table 3). Along with levels and virtual professions, medals make it possible to map the skills of players and teams, making planning easier in addition to daily activities.

Table 3. Medals

Badge	Name	Description	Criteria
	Photography Lover (bronze)	Recognition for dedication to sending quality images	Submission of 50 photos with note above 8 attached to reports
	Photography Lover (silver)	Recognition for dedication to sending better quality images	Submission of 100 photos with note above 8 attached to reports
	Photography Lover (gold)	Recognition for dedication to sending excellent quality images	Submission of 150 photos with note above 8 attached to reports
	In search of perfection	Recognition for focusing on quality when writing and submitting reports	Submission of 100 with maximum concept given by the judges

The game has a news board where the latest positive activities of the players appear, placing them in a prominent position before the group. Such a picture resembles face-to-face meetings in which employees receive praise in front of co-workers and demonstrates the appreciation that the institution places on positive and proactive attitudes of employees.

The ranking was implemented by teams as a way of encouraging social bonds and collective work. It was decided not to list the individual ranking, nor the complete ranking, so that there is no comparison between the first and last ones, always drawing attention to the teams that stand out.

The team leaders were responsible for visualizing the individual performance of their own team members to reorganize, train and manage their own team, seeking improvement, teamwork. They should encourage mentoring of those who stand out positively to help those who have not been able to obtain good results.

It is possible for players to view the medal board, title, and virtual profession of other players (Fig. 3). That way, if they need the help of an expert in a certain area, or advice from someone with more experience, the player will be able to know who to turn to. For the most competitive players, such visualization awakens the will to overcome.

In the individual profile (Fig. 4), the player sees his current level and can choose which of the virtual professions he wants to show among those acquired. The medal board is displayed on this screen as well.

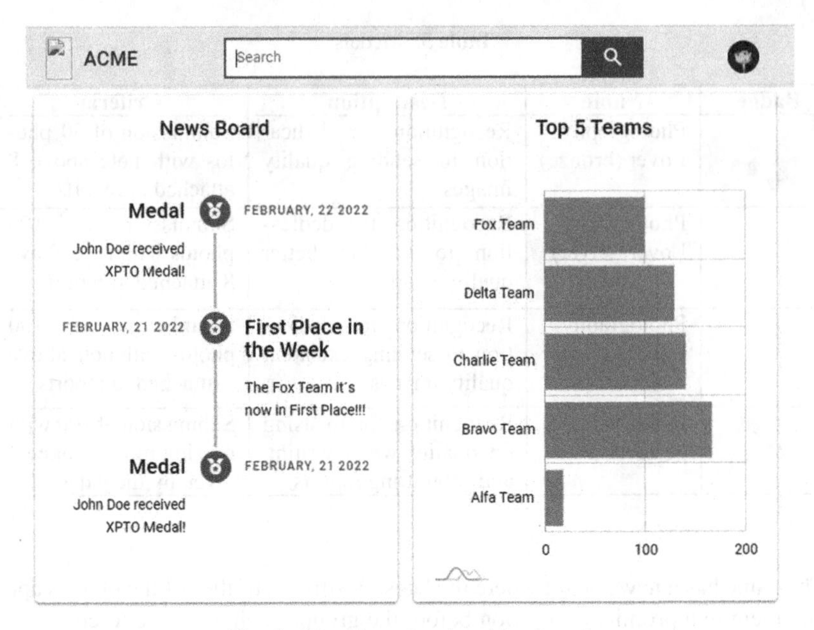

Fig. 3. News board and ranking board.

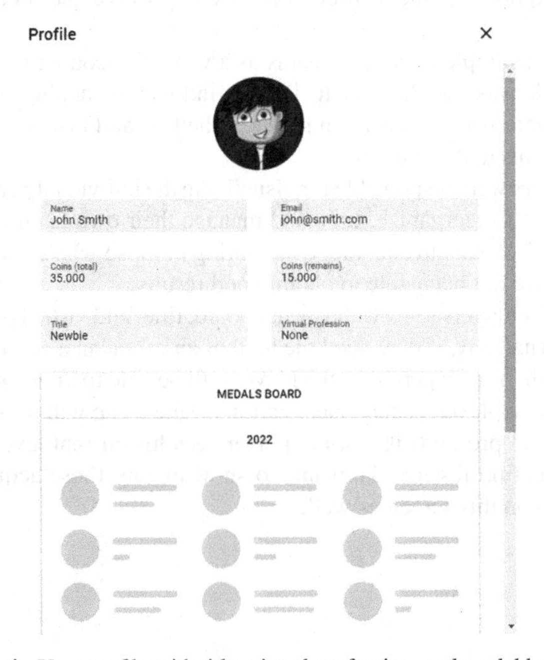

Fig. 4. User profile with title, virtual profession, and medal board

4 The GPR Construction

For the construction of GPR workflow, we follow the steps:

1. A survey built on the Intrinsic Motivation Inventory (IMI) [25].
2. Application of the survey to the audience.
3. Data collection from previous reports for evaluation and comparison.
4. Analysis of survey data.
5. Identification of respondent's profiles to build the gamification according to the report improvement proposal.

In addition to collecting and analyzing data from previously produced reports, it was necessary to know the profile of the target audience. The use of the questionnaire based on IMI was idealized, which allowed the subjective capture of information about intrinsic motivations and the motivation of the respondents. Demographic characterization questions were also used.

From the data collected in the questionnaire it was possible to identify (Table 4):

- How players value the activity they perform daily, its importance to the institution and how they believe they can learn new skills while working.

Table 4. Five IMI-related questions with the highest averages

Question	Average
I think my tasks are important	4.73
I think my tasks are important because they can provide quality information to the institution's leaders	4.64
I think my tasks are useful for the institution to achieve its goals	4.60
I enjoy doing my tasks	4.53
I believe that acting in the activities I perform can help me develop new skills	4.51

- Challenge, play with friends and pass time are the game features that most attract our target audience and are linked to the core drives of Development and Accomplishment and Social Influence and Relatedness (Fig. 5).

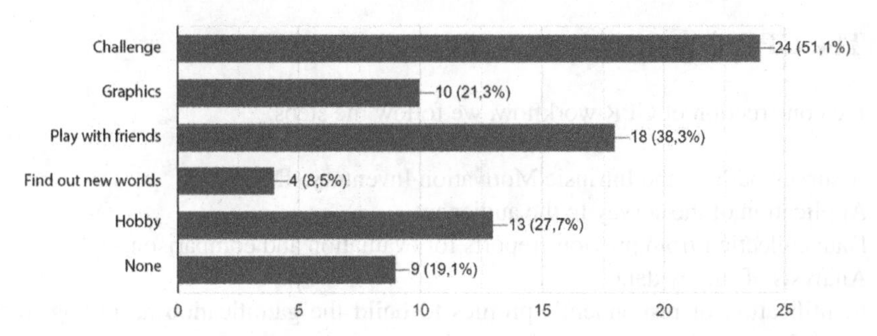

Fig. 5. Features that most appeal to the respondents

- More than 70% have already played digital games, which indicates that most respondents know the mechanics common to games such as leaderboards and trophies. This steered the gamification planning towards the Development and Accomplishment core drive (Fig. 6).

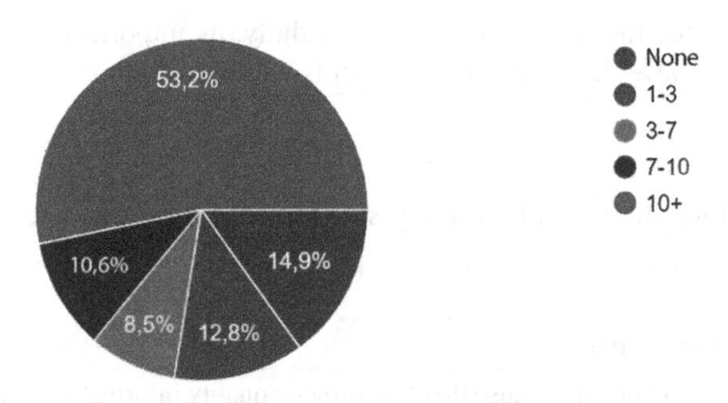

Fig. 6. Number of digital games played

Reynolds [24] states that a questionnaire based on the Intrinsic Motivation Inventory identifies the motivational profiles of the respondents. In her studies, she reports that the IMI is a multidimensional measurement instrument used in the subjective evaluation of the experience lived by the target audience of an activity. The theoretical framework for its development stems from studies of the Cognitive Assessment Theory [25, 26] and the Self-Determination Theory (SDT) [27]. It is an instrument used in several experiments related to the measurement of intrinsic motivation and can perceive/evaluate interest, perceived competence, effort, valuation, feeling of pressure and tension and the possibility of choice during the execution of some activity received [27].

Based on the literature and our experience, we first designed an Octalysis gamification [2] that comprises: the best "core drives" for 49 users, their gamification skills, and a built working environment. Then, we produced the game in collaboration with the users and using the strategies provided by the Octalysis framework; as well as the use of the game in some sectors to motivate and engage users in their activities.

5 Evaluation of the Results

We are applying the gamification to 49 users divided into two groups in the same organization. The experiment is conducted for four months. The gamification project allows us to easily propose assessment tests and observations.

Based on the identified profiles, the techniques Classification, Medals, Level Up, Crown (affects the basic motivations of Development and Accomplishment), Countdown (affects the basic motivation of Loss and Avoidance), Mentoring, Group Challenge (affects basic motivation of Social Influence and Relatedness), Attachment Monitor, Build from Scratch (affects basic motivation of Ownership and Possession), and Gravestone (affects basic motivation of Loss and Avoidance).

Using gamified play, users can discover, play, build and learn best practices. The working environment becomes more engaging and fun, users are more motivated in a playful and challenging environment.

Players began to view the profiles of other players, and within the teams, leaders began to work with their members to reduce the standard deviation of those who had lower productivity or produced reports with lower quality.

Collaboration between players also increased after game application. The most experienced players from each team helped the less experienced with a focus on increasing points gain and beating other teams.

The evaluation also takes place in the correction of each activity, before moving on to the level of the most difficult phases of the game. The participation of users in the game shows the individual evolution of each player when "navigating" through the learning objectives proposed by the manager in the phases of the game.

6 Conclusion

We observed that the labor and learning are developed in a pleasant and effective way, with pleasure in acquiring new knowledge. The gamified process GPR created an environment where users engaged in the language of games in a fun and attractive way, improving learning in the cognitive, affective, sociocultural aspects, among others; in a democratic, collaborative, and interactive organizational model.

At the end of the collection of the results of the work, interviews are being carried out with those responsible for each unit, as well as with the head of the institution to raise the perceptions regarding the motivation and engagement of those involved, apply a new questionnaire based on the IMI and analyze the new questionnaire focused on identifying changes in behavior patterns.

The results can be used for the evolution of the gamification already applied and can be applied in other areas of the organization.

References

1. Deterding, S., Dixon, D., Khaled, R., Nacke, L.: From game design elements to gamefulness: defining "gamification". In: Proceedings of the 15th International Academic MindTrek Conference (2011)

2. Chou, Y.: Actionable Gamification: Beyond Points, Badges, and Leaderboards, Leanpub (2015)
3. McGonigal, J.: Reality Is Broken: Why Games Make Us Better and How They Can Change the World, Penguin Books, New York (2011)
4. Fredricks, J.A., Blumenfeld, P.C., Paris, A.H.: School engagement: potential of the concept, state of the evidence. Rev. Educ. Res. **74**, 59–109 (2004)
5. Huizinga, J., Nachod, H., Flitner, A.: Homo ludens: vom Ursprung der Kultur im Spiel. Rowohlt Taschenbuch Verlag (2006)
6. Huizinga, J.: Homo ludens: o jôgo como elemento da cultura. Coleção estudos. Editora da Universidade de S. Paulo, Editora Perspectiva (1971). https://books.google.com.br/books?id=BHQLAAAAYAAJ
7. Zichermann, G., Cunningham, C.: Gamification by Design: Implementing Game Mechanics in Web and Mobile Apps. O'Reilly Media (2011). https://books.google.com.br/books?id=zZcpuMRpAB8C. ISBN 9781449315399
8. Burke, B.: Gamificar: Como a gamificação motiva as pessoas a fazerem coisas extraordinárias. DVS Editora (2015). https://books.google.com.br/books?id=IIdbCwAAQBAJ. ISBN 9788582891070
9. da Silva, A., et al.: Gamificação na Educação. Pimenta Cultural (2014). https://books.google.com.br/books?id=r6TcBAAAQBAJ. ISBN 9788566832136
10. Bovermann, K., Bastiaens, T.J.: Towards a motivational design? Connecting gamification user types and online learning activities. Res. Pract. Technol. Enhanc. Learn. **15**(1), 1 (2020)
11. Andrade, F.R.H.: Gamificação personalizada baseada no perfil do jogador. Ph.D. thesis, Universidade de São Paulo, São Carlos, October 2018. http://www.teses.usp.br/teses/disponiveis/55/55134/tde-18102018-111511/
12. Deterding, S., Khaled, R., Nacke, L.E., Dixon, D., et al.: Gamification: toward a definition. In: CHI 2011 Gamification Workshop Proceedings, Vancouver BC, Canada, vol. 12 (2011)
13. Kapp, K.: The Gamification of Learning and Instruction: Game-based Methods and Strategies for Training and Education. Wiley (2012). https://books.google.com.br/books?id=GLr81qqtELcC. ISBN 9781118191989
14. Werbach, K., Hunter, D.: For the Win: How Game Thinking Can Revolutionize Your Business. Wharton School Press (2012). https://books.google.com.br/books?id=pGm9NVDK3WYC. ISBN 9781613630228
15. Aziz, A., Mushtaq, A., Anwar, M.: Usage of gamification in enterprise: a review. In: Proceedings of 2017 International Conference on Communication, Computing and Digital Systems, C-CODE 2017, November 2012, pp. 249–252 (2017). https://doi.org/10.1109/C-CODE.2017.7918937
16. Quiroz, V.B.A.R.d., et al.: Ucl go: um estudo qualitativo de uma plataforma gamificada adotada como prática em disciplinas do ensino superior (2019)
17. Belinazo, G., et al.: Uma abordagem gamificada para o gerenciamento de riscos (2019)
18. Thorpe, A.S., Roper, S.: The ethics of gamification in a marketing context. J. Bus. Ethics **155**(2), 597–609 (2019)
19. Neto, H.P.: Gamificação: engajando pessoas de maneira lúdica. FIAP, São Paulo (2015)
20. C. Dictionary: Electronic resource (2018). https://dictionary.cambridge.org/dictionary/english/slang
21. Basten, D.: Gamification. IEEE Softw. **5**, 76–81 (2017)
22. Chou, Y.: Actionable Gamification: Beyond Points, Badges, and Leaderboards. Packt Publishing Ltd., Birmingham (2019)
23. Mora, A., Riera, D., Gonzalez, C., Arnedo-Moreno, J.: A literature review of gamification design frameworks. In: 2015 7th International Conference on Games and Virtual Worlds for Serious Applications (VS-Games), pp. 1–8. IEEE (2015)

24. Reynolds, L.: Measuring Intrinsic Motivations. Handbook of Research on Electronic Surveys and Measurements, pp. 170–173 (2007)
25. Deci, E.L., Ryan, R.M.: Intrinsic motivation and self-determination in human behavior, vol. 36 (1985)
26. Deci, E.L., Ryan, R.M.: Intrinsic Motivation and Self-determination in Human Behavior. Springer, Boston (2013). https://doi.org/10.1007/978-1-4899-2271-7
27. Monteiro, V., Mata, L., Peixoto, F.: Intrinsic motivation inventory: psychometric properties in the context of first language and mathematics learning. Psicologia Reflexão e Crítica **28**, 434–443 (2015)

Designing a Workplace Violence Reporting Tool for Healthcare Workers in Hospital Settings

Meagan Foster[1](✉), Karthik Adapa[1], Amy Cole[1], Amro Khasawneh[1], Anna Soloway[2], Jeffrey Francki[2], Nancy Havill[2], and Lukasz Mazur[1]

[1] University of North Carolina at Chapel Hill, Chapel Hill, NC 27514, USA
meagan.foster@unc.edu
[2] UNC Health, Chapel Hill, NC 27514, USA

Abstract. Workplace violence (WPV) is severely underreported among healthcare workers (HCWs) who are four times more likely to experience WPV compared to professionals in other industries. The frequency in which HCWs are exposed to acts of WPV prompts urgency in the development of improved reporting tools to capture HCWs' experiences more adequately and provide evidence for strategic improvement initiatives. Techniques in human-centered design and product discovery techniques can enhance the redesign process and help teams identify what HCWs value and deliver value, faster.

Purpose: To explore how human-centered design and product discovery techniques can be used to inform feature recommendations for a minimum viable product (MVP) for an improved WPV reporting tool for HCWs.

Methods: We used a mixed-methods approach based on human-centered design and product discovery techniques.

Results: Two distinctive themes emerged, informing our design objectives: (1) To increase reporting of WPV and (2) to provide access to WPV support resources. The recommended set of product features was well-received among HCWs.

Conclusion: Human-centered design and product discovery techniques can be used in a complementary fashion to design WPV reporting tools that better align with the values of HCWs.

Keywords: Workplace violence · Hospital · Product discovery · Human-centered design · Cognitive ergonomics · Case study

1 Introduction

Hospitals are one of the most dangerous work environments in the country [1]. Among contributing factors is workplace violence (WPV), defined by the Occupational Safety and Health Administration (OSHA) as "any act or threat of physical violence, harassment, intimidation, or other threatening disruptive behavior that occurs at the worksite". This includes acts of verbal and physical abuse such as yelling, cursing, grabbing, kicking, and scratching [2]. Evidence shows healthcare workers (HCWs) in the United States experience intentional acts of physical violence that result in injury, at higher

F. Fui-Hoon Nah and K. Siau (Eds.): HCII 2022, LNCS 13327, pp. 368–387, 2022.
https://doi.org/10.1007/978-3-031-05544-7_28

rates compared to other industries combined [3]. Nurses and physicians are most likely to experience WPV among HCWs [4, 5]. It is estimated that 82% of nurses have or will experience WPV at least once during their career [2] and 71% of physicians have or will experience verbal assault [6]. The effect of WPV impacts the day-to-day roles of HCWs, their mental and physical wellbeing [7, 8], and the quality of care provided to patients [9]. HCWs who experience WPV perform on-the-job errors more frequently [10] and exhibit decreased performance and productivity [7]. In turn, health care organizations suffer from low staff morale, staff shortages, and increased healthcare expenses [6].

According to OSHA, each year a total of 18,000 WPV events are reported in the United States among nurses and other HCWs [11]. However, evidence shows that rates are severely underrepresented as the majority of WPV events go unreported [12]. Reporting behaviors are impacted by various complex socio-technical factors including the belief that such events are "part-of-the-job" [7, 13–15]. One study showed that 76% of nurses would not report a WPV event if the harm was unintentional [16]. Another study reported 77% of nurses agreed that assault is an expected consequence of their role [17]. Another commonly held belief among HCWs is that WPV reporting does not lead to improved worker protections [12], which is not unsubstantiated. For instance, a 2020 investigative report revealed that no corrective action was performed for 76% of violent events in California because there was "no continuing threat" [18]. These studies suggest a culture within healthcare that reinforces an unspoken expectation that HCWs demonstrate "resilience" by suppressing their well-being in order to fulfill the duties of their role. Underreporting is also associated with technical barriers such as the perception that WPV reporting tools are complicated, time-consuming, impractical, and onerous [12]. Each barrier negatively influences the choices, feelings, and fears HCWs have regarding WPV reporting. HCWs' aversion to WPV reporting, leads to surveillance data that is scarce and inadequate in providing comprehensive evidence to support the effective allocation of resources for strategic interventions.

As hospital systems strategize to collect more data from HCWs on WPV events, WPV reporting tools must be designed in a way that offers value to HCWs and considers user characteristics and tasks. Human-centered design techniques are commonly used to engage users and facilitate the learning experience throughout the product lifecycle. However, to manage delivery risks, validate assumptions, and systematically prioritize user value, product discovery techniques are used. The primary output of product discovery is. Otherwise, product teams face high costs to fix design problems once the project has been delivered. Therefore, it is critical to examine the contexts of the design problem from both a human perspective and user value perspective to avoid common design fallacies that result in unsatisfied users and unused products [19]. The combined use of human-centered design and product discovery techniques are well-suited to deliver an improved WPV reporting tool that provides the most value to HCWs, as early on as possible.

2 Purpose

The purpose of this article is to: (1) present in detail the activities conducted by our design team in the development of a product feature set for an improved WPV reporting tool and (2) explore how human-centered design and product discovery techniques can be used in combination to inform feature recommendations for a MVP. We used human-centered design and product discovery techniques to perform mixed-methods in three phases; (1) analysis, (2) synthesis, and (3) evaluation based on the John Chris Jones design philosophy [20].

3 Background

An employee survey conducted in 2019 revealed a 1.4% WPV reporting rate at an academic medical institution from August 2019 to October 2019. Representatives from the Workplace Safety Committee (WSC) raised concerns that technical barriers associated with the WPV reporting tool were negatively impacting WPV reporting behavior among HCWs. To address this problem, an interdisciplinary design team consisting of experts in workplace safety, human factors engineering, agile software development, and lean management practices was formed. This team operated in conjunction with an ongoing system-wide effort to mitigate WPV and duly represented various stakeholder perspectives as HCWs, supervisors, quality and safety personnel, and hospital leaders. The design team met every week except for periods of user testing. Other meetings were held with subsets of design team members on an as-needed basis.

We based our design process on the design exploration philosophy presented by John Chris Jones [20] in the 1970s publication, Design Methods. In this text, Jones, a pioneer in contextual design, outlines three universally agreed upon and essential stages of design: analysis, synthesis, and evaluation, more colloquially described as "breaking the problem into pieces", "putting the pieces together in a new way", and "testing to discover the consequences of putting the new arrangement into practice" [20].

The design team set goals for each phase of the design process. The goal of the analysis phase was to assess the current state of WPV reporting. In the synthesis phase, we aimed to confirm a shared understanding of user value and cultivate a product feature set for the MVP. Finally, in the evaluation phase, we validate the set of recommended product features by eliciting perspectives from HCWs on the priorities of the design. Figure 1 illustrates the design process.

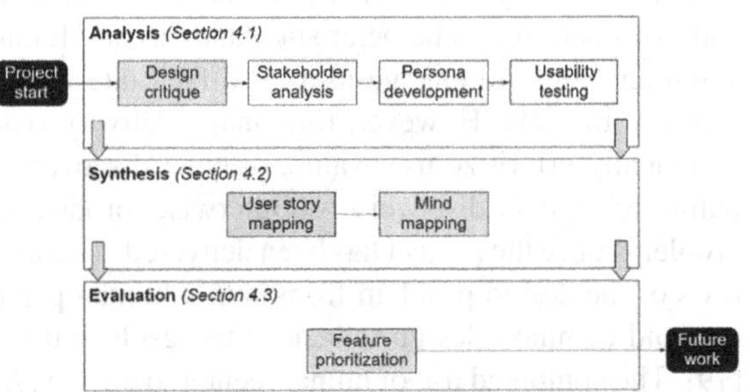

Fig. 1. Design process overview. Human-centered design methods are illustrated using white boxes and product discovery techniques are illustrated in grey.

4 Case Study

4.1 Analysis

Design Critique. To initiate the analysis phase, we performed a standalone design critique [21]. Design critiques are attended by business stakeholders and designers and are performed to elicit feedback on how well an existing design aligns with the design objectives [21]. Given that few members of the design team were familiar with the reporting tool, the design critique aimed to provide a basis for the context of the design problem and associated socio-technical barriers, before engaging primary users in more rigorous assessments.

The design critique was conducted over Zoom. The WSC representatives, who were also members of the design team, led a walk-through of the existing reporting tool using real-world examples to demonstrate use cases and technical barriers associated with WPV reporting. The design team discussed barriers associated with secondary usage of the WPV reports, in addition to, factors influencing poor reporting behavior. One member of the design team served as the note taker to document the qualitative insights. Barriers associated with WPV reporting were coded using the interrelated constructs contained in the DeLone and McLean Information System Success Model [22]: information quality, system quality, service quality, intention to use, usage, user satisfaction, and net system benefits. The result is shown in Table 1.

Table 1. Qualitative analysis of feedback captured the design critique categorized by the DeLone and McLean Information System Success Model.

Design critique summary
Information quality: Critical report elements include the type of event, type of person(s) affected, and where the assault occurred within the hospital The most pertinent information is captured as free text in the detailed description field and therefore must be manually reviewed The severity score assigned by the user is lower than the severity expressed in the description of the event
System quality: Difficult to access from the intranet HCWs mistakenly report WPV events in the patient safety reporting tool HCWs can bypass critical fields HCWs cannot select more than one event type, so they must submit a separate report for each type HCWs are unable to link related reports No event type for disruptive, intimidating, or hard to describe behaviors HCWs commonly choose "other location", particularly in cases where the event occurred outside of a unit, i.e., in the hallway or elevator Some dropdown option lists contain over 100 options Dropdown option lists cannot be searched Dropdown option lists are not comprehensive or well-organized

(continued)

Table 1. (*continued*)

Design critique summary
Service quality:
No user instructions are provided
No definitions for WPV or types of WPV events are provided
No information regarding WPV policies is provided
No information on how the report will be used is provided
No information on what to expect next is provided
Intention to use:
Many external barriers to reporting exist, including education
Most HCWs are highly empathetic to patients and will not report if they believe the patient will get in trouble
Most HCWs feel WPV is a part-of-the-job
Some HCWs feel WPV events occur too frequently to report them all
Managers commonly report on behalf of HCWs
Usage:
Reporting tools can be accessed from the intranet or the EHR
Reporting is voluntary
Reports must be submitted when filing for workers' compensation
User satisfaction:
Reporting takes too long
The WPV reporting tool is confusing
The WPV reporting tool offers little feedback on errors in completing the report
HCWs have privacy concerns regarding who receives the report
Net system benefits:
HCWs do not see the benefits of reporting
Reporting data is not useful for decision-making because it is incomplete or incorrect and does not adequately represent the prevalence of WPV

Stakeholder Analysis. A stakeholder analysis was performed to align the design team with the stakeholders associated with the design of an improved WPV reporting tool. We documented a description of the role of each stakeholder, their level of interest (low, moderate, high), their power or influence (low, moderate, high), and their position or level of support for the development of the tool (low support, moderate support, high support, neutral).

Stakeholders of the WPV reporting tool were identified based on their use of the existing WPV reporting tool. We categorized the stakeholders into five groups: hospital leadership, quality and safety personnel, managers, witnesses of WPV, and affected persons. These groups were further divided into two user groups: primary users and secondary users. Affected persons, managers, and witnesses of WPV are the intended users of the WPV reporting tool and therefore are primary users. Secondary users include hospital leadership and quality and safety personnel. Secondary users access the data from WPV reports to track and trend the prevalence of WPV across work environments and to inform organizational WPV mitigation initiatives. As secondary users of the reporting tool, adoption and fidelity of the reporting tool directly impact their ability to derive useful insights to achieve these objectives. Table 2 shows the relative interest and influence of each stakeholder regarding the design of an improved WPV reporting tool.

Secondary users of the system were omitted from further analysis in the design process and subsequent design activities focused on satisfying the primary users of the WPV reporting tool.

Table 2. Stakeholder analysis results.

Stakeholder	User type	Interest	Influence	Position/ Support level
Affected persons	Primary	High	Low	Moderate
Managers	Primary	High	Low	High
Witnesses of WPV	Primary	Moderate	Low	Moderate
Quality and safety personnel	Secondary	High	Moderate	High
Hospital leadership	Secondary	High	High	High

Persona Development. Based on literature and qualitative analysis performed in earlier steps of the design process, we documented the characteristics and goals of primary users of the WPV reporting tool using personas. The intention is to establish a shared understanding of the primary users so that design teams members can bring a user-value perspective to the synthesis phase of the design process. Personas included age, job role, work location, a background statement on the user, and a personal quote aimed to reflect a distinctive perspective of the user. We capture associated motivations or goals in addition to related frustrations.

A commonly shared and fundamental principle among our primary users is tactical time management as HCWs are often juggling multiple critical tasks within a fast-paced, clinical environment. Despite the frequent interruptions and being simultaneously pulled in many different directions, HCWs take pride in their resilience and ability to provide quality care. Unfortunately, WPV occurs often; so much so that WPV events have been normalized as a byproduct of their roles as HCWs. When reporting on WPV events, HCWs expect that the time to complete the report will be minimal as to not disrupt their workflow. Ultimately, by providing a report of the WPV, HCWs anticipate that some level of corrective or preventative action

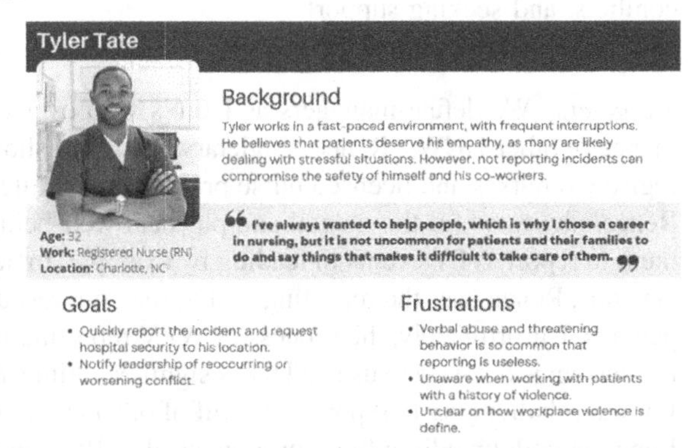

Fig. 2. User persona representing an affected person.

will follow. HCWs also expect that their privacy preferences will be respected, particularly when they prefer to remain anonymous or when their manager is the aggressor or assailant.

Affected Persons. Affected persons are HCWs who are impacted by WPV. Tyler Tate (persona shown in Fig. 2) has been impacted by WPV on many occasions. Tyler believes that patients deserve empathy as patients are often dealing with high levels of stress associated with the reason behind their hospital visits. When it comes to reporting, Tyler is motivated to report because he requires a real-time response to protect himself or his colleagues from assailants or to notify leadership of a WPV event that has worsened or is reoccurring. Tyler's reporting goals include the documentation of injuries sustained during the WPV event, notifying leadership of inappropriate or violent behavior, or reoccurring or worsening conflicts, and seeking support and resources when impacted by WPV.

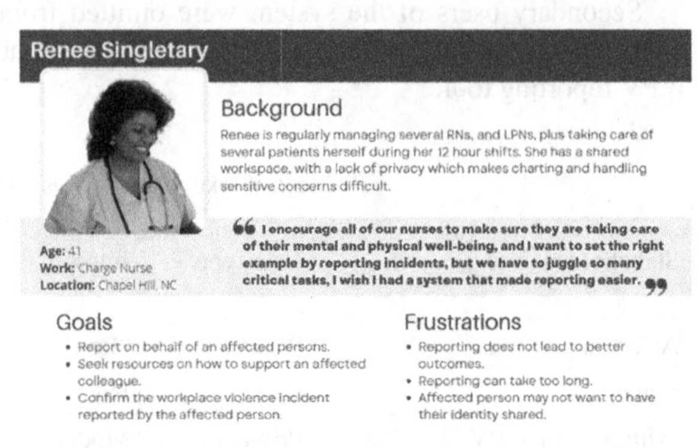

Fig. 3. User persona representing a manager.

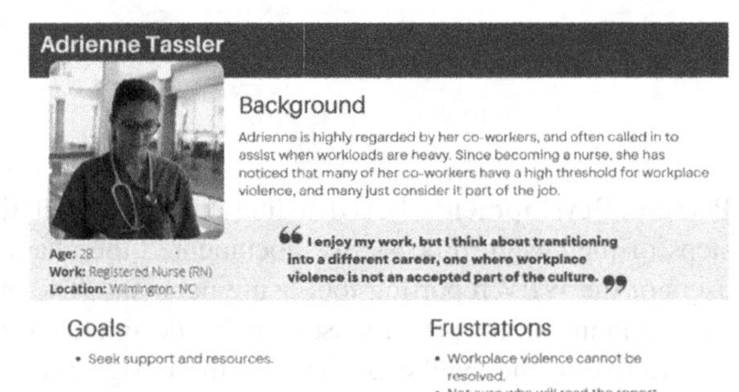

Fig. 4. User persona representing a witness of WPV.

Managers. We define managers as HCWs who oversee the job functions and performance of other HCWs. Renee Singletary (personas shown in Fig. 3) oversees a team of registered nurses and licensed nurse practitioners while also providing care to patients. Renee advocates for the mental and physical well-being of her nurses and encourages them to report WPV events in leading by example. In addition to reporting on behalf of her staff, Renee uses the reporting tool to seek out resources on how to support affected persons. Unfortunately, the process of WPV reporting is frustrating because the reporting system is not easy to use and takes significant time away from other important tasks. It is also challenging to report on behalf of others who would prefer to stay anonymous. Renee is undoubtedly a huge proponent of WPV reporting but is disappointed in the lack of improved outcomes for her and her staff.

Witnesses to WPV. We defined witnesses to WPV as HCWs who are not directly impacted by WPV but are a bystander or a witness to a WPV event. Adrienne Tassler (persona shown in Fig. 4) is often called in to assist in areas where the workloads are heavy. Adrienne has noticed a pattern among many coworkers that reflects a nonchalant or minimizing attitude toward WPV events. As it relates to WPV reporting, Adrienne, like Renee, is interested in seeking support and resources for herself and colleagues to mitigate WPV. While Adrienne enjoys many aspects of nursing, she considers careers, where WPV is less tolerated, appealing. Adrienne worries that reporting will increase the risk of retaliation or jeopardize relationships with coworkers based on who has access to the report. Adrienne is frustrated and feels WPV cannot be resolved.

Usability Testing. Primary users were invited to participate in a usability study examining the perceived usability, cognitive workload, and performance associated with completing WPV reporting tasks from the existing WPV reporting tool at an academic medical center employing ~ 7500 HCWs. Institutional review board (IRB) approval was required and all HCWs voluntarily consented to participate.

Twenty-one HCWs participated in the usability study online via Zoom or in-person in our Human Factors Laboratory. In the initial testing sessions, HCWs were asked to think aloud to provide feedback on issues related to functionality and ease of use. This technique aims to detect errors users encounter when performing a task. In subsequent sessions, participants were asked to submit a WPV report as they would in a real-world setting, using scenarios. Scenarios were used to elicit feedback and serve as a guideline for what elements of the tasks were carried out successfully or in error. Excerpts from the scenarios used in the usability study are shown in Table 3.

Table 3. Excerpts from workplace violence scenarios used in usability testing.

	Event descriptors	Scenario description
1	Patient-to-staff: sexual assault & harassment	"…You leave the bedside and return later to speak to the patient about their upcoming meeting with their doctor. The patient says they can't focus on the conversation because they are distracted by your chest"
2	Patient-to-staff & family-to-staff: intimidation, harassment & verbal abuse	"…While you are walking to the nurse's station, you see a member of the patient's family walking toward you. As they approach the nurse's station, they begin taking pictures of you and your computer screen. You ask them to stop but they continue. The charge nurse approaches the nursing station to help. They introduce themselves to the patient's family member who gets very close to the charge nurse's face and begins to yell 'You are messing with the wrong family. This nurse needs to be fired right now!'"
3	Visitor-to-staff: threat & intimidation	"…You start to explain the hospital rules and they say 'I hope you are not as dumb as you look but I will talk slowly: I'm not here for you. If you get in my way, you're going to be sorry'"

(continued)

Table 3. *(continued)*

	Event descriptors	Scenario description
4	Patient-to-staff: verbal assault & threats	"…You continue to try to explain that you will have to page the doctor to get the message to them, but she won't stop yelling. She calls you "an idiot" and says, 'I am recording this phone call and it's going to be all over the news.' You continue to try to ask the person to give you a moment to call the doctor for them. After five minutes of yelling, the person hangs up"
5	Staff-to-staff: hate speech, bullying & lateral violence	"…A surgeon you haven't met before comes in and loudly says 'This needs to be redone. This is an embarrassment.' Before you can respond the surgeon turns to another employee and says, 'they should be grateful for everything we have done for them.' You understand this comment to be a reference to your race"
6	Patient-to-staff: verbal assault & hate speech	"…This is your third day working with this patient, Blair Brooks, who has a diagnosis of dementia. During the three days the patient has: spit at you, grabbed your wrist, called you a "dumb b!*#$", told you that you had a "beautiful body", pushed an intravenous pole at you, thrown a plate onto the floor, and asked multiple times for a "person from this country" to take care of them"
7	Visitor-to-staff: verbal assault & physical assault	"…The person has a child in a stroller with them but visiting for children under 12 is restricted due to flu season. You stop and start to explain the visitation restrictions. The family member cuts you off and says, 'you're as bad as all of them!' and pushes the stroller so that it hits you in the shins"
8	Patient-to-staff: sexual assault	"…You board the elevator to take the patient to the ground floor. The elevator stops on the third floor and two people get on. You step closer to your patient to make room. Suddenly, you feel a hand squeeze your rear end. You realize it is the patient"

Validated instruments such as the Post-Study System Usability Questionnaire (PSSUQ) and the National Aeronautics and Space Administration Task Load Index (NASA-TLX) were used to quantify perceived usability and perceived cognitive workload, respectively. Performance was quantified by the time taken to complete the reporting task and the rate of correct responses based on expected responses for each scenario.

Results. Initial results of this usability study with 10 participants were published in the 2021 Med Info Conference proceedings [23]. The usability study concluded with a total of 21 participants. The initial 3 out of 21 participants were asked to think aloud while completing the report to identify usability barriers that were not captured during the design critique. All other participants completed the report as they would in a real-world setting and their perceived usability, CWL, and performance were assessed. We present the results of the usability study in the following sections.

Participant Characteristics. Participants included HCWs ranging from 30 to 60+ years old. 11 of the 21 participants were nurses, including one senior nurse manager. Two participants worked as administrative/clerical staff and two participants were quality and safety personnel. One participant was represented through the following roles: auxiliary/ancillary, physician, a profession allied to medicine (therapists/radiographers/assistants), chaplain, social work manager, and program manager.

Pre-screener Questionnaire Results. The results from the pre-screener questions are shown in Table 4. Additional actions taken by HCWs who experienced or witnessed a WPV event include calling the behavioral response team, notifying law enforcement, reporting the event to HR, and assisting with de-escalation.

Think aloud Results. Most feedback provided by HCWs echoed what had been captured during the design critique. HCWs expressed preferences for features that would make reporting easier, such as searchable and easier to browse dropdown option lists, and links for training or learning on WPV. (For more details, please refer to our previous work [23]).

Perceived Usability Results. The mean total PSSUQ score (standard deviation [SD]) and recommended human-computer interaction (HCI) standard [24] are shown in Table 5, along with mean subscale scores.

Table 4. Pre-screener results.

Question	Yes (n = 21)
Received incident report training	52%
Experienced or witnessed violence in the workplace	86%
Type of WPV experienced:	
Hostility or intimidation	81%
Verbal violence	81%
Threats	67%
Harassment	62%
Assault	48%
Battery	43%
Response to any experienced or witnessed WPV event:	
Told the person to stop	67%
Reported it to a senior staff member	62%
Completed incident/accident form	52%
Told a colleague	48%
Tried to defend myself physically	33%
Told friends/family	28%
Took no action	19%
Sought help from association	14%
Completed a compensation claim	10%
Tried to pretend it never happened	10%
Sought counseling	10%
Transferred to another position	9%
Sought help from the union	0%
Pursued prosecution	0%

Table 5. Results from the post-study system usability questionnaire (PSSUQ).

Measures	Mean (SD)	Recommended HCI Standard [24]
Total score	**3.05** (1.06)	<2.82
System usefulness	**2.80** (1.22)	<2.8
Information quality	**3.19** (1.10)	<3.02
Interface quality	**3.17** (1.13)	<2.49

Table 6. Results from the post-study system usability questionnaire (PSSUQ).

Measures	Mean (SD)	Recommended HFE standard [25]
Global NASA TLX score	**30.6** (17.7)	> 35 and <54
Weighted subscale scores:		
Mental demand	6.96 (6.55)	
Physical demand	0.22 (0.32)	
Temporal demand	4.88 (4.56)	
Performance	10.96 (7.91)	
Effort	4.22 (4.32)	
Frustration	3.35 (5.46)	

Perceived Cognitive Workload. The mean global NASA TLX score (standard deviation) and recommended human factors engineering (HFE) standard [25] are shown in Table 6, along with mean subscale scores.

Performance Results. Performance results are shown in Table 7 and Table 8.

Table 7. Performance is measured by time to completion where *n* represents the number of completed tasks.

	Overall (n = 31)	Received training (n = 17)	Did not receive training (n = 14)
Time to complete WPV report (SD) (in mins)	6.61 (2.40)	6.46 (2.44)	6.57 (2.35)

Discussion. The results of the usability study demonstrate that many participants had experienced WPV, although only half had received training on how to report a WPV event. Participants' responses to the WPV event offered insight into support or resources most utilized among participants, such as reporting the event to a senior staff member which was reported by over half of the participants. Counseling stood out to the design team as an underutilized resource. Quantitative analysis confirms perceived usability and

Table 8. Performance is measured by the rate of error where *n* represents the number of completed tasks.

	Overall (n = 31)	Received training (n = 17)	Did not receive training (n = 14)
Event type	13%	6%	6%
Type of person affected	19%	3%	16%
Event location	35%	23%	13%
Involved parties[1]	55%	35%	19%

[1] *Participants categorized involved parties, i.e., patients, visitors, or staff as a bystander, notified party, or assailant.*

perceived cognitive workload were higher than the recommended standards, indicating the need for improvements. Similarly, rates of error in performance indicate the need for an improved data collection strategy.

4.2 Synthesis

User Story Mapping. The user story mapping technique is a collaborative process used in agile development frameworks to sort user stories into groups based on the value a feature offers customers and business stakeholders. To use the user story mapping technique, we translated insights collected in the analysis phase into user stories, a lightweight requirement artifact uniquely designed to help teams capture the user's perspective on value. User stories are expressed using the canonical template: "As a [role], I want [feature] so that I can [benefit or value]".

Once the user stories were established, we assigned each user story to a task and activity. The tasks and activities reflected the critical data elements captured during the WPV reporting process. To identify user stories that should be considered in the initial iteration of product delivery, also referred to as the MVP, we group our user stories into two buckets: features for the MVP, and features for future releases. During the sorting process, similar user stories were merged, some were re-written for clarity, and others were discarded if redundant. The user story map finally contained 22 user stories. 12 of the 22 were recommended by the design team for the MVP, shown in Table 9.

Mind Mapping. To facilitate the process of generating design ideas for an improved reporting tool, the design team conducted a brainstorming session using the mind mapping technique. Mind maps are cognitive maps used to represent a central topic and its subtopics illustrated as branches that flow from the central topic. We defined the central topic of the mind map as "design objectives" and encouraged the design team to re-imagine an ideal WPV reporting tool using the human-centered perspectives captured in the previous steps of the design process. The results of this exercise offered two distinctive themes from which we derive our design objectives: (1) To increase reporting of WPV and (2) to provide access to WPV support resources. The set of proposed features for the MVP are shown in Table 10a and Table 10b.

Table 9. User stories selected for inclusion in the development of the MVP, organized by activity and task.

Activity	Task	User stories selected for the MVP
Access the reporting tool	Log in to reporting portal and view options on the landing page	As a primary user: • I want to understand how the information I share will be used so that I am aware of privacy implications • I need to know if I am in the reporting tool for workplace violence so that I can confidently submit my report • I want to understand what to expect after I report so that I can decide if reporting is worthwhile • I need to quickly connect with hospital security and support specialists in real-time, so I can request support when WPV occurs
Complete a new report	Indicate the type of violence that occurred	As a primary user: • I need easy-to-understand dropdown option lists so that I can accurately select the type of violence that occurred • I need to select more than one type of WPV even so that I can avoid submitting multiple reports
	Add names of person(s) involved and classify as an assailant, bystander, or affected person	As a primary user: • I need to indicate if I am reporting my own experience or on the behalf of another so that my involvement in the WPV event is accurately understood • I want the option to omit the name of the affected person(s) so that I can honor their privacy preferences
	Document when and where the event occurred	As a primary user: • I need to describe a location that does not appear in the dropdown option list so that I can indicate when a WPV event occurred outside of a particular unit, i.e., the hallway
Submit report	Elect to submit or save draft	As a primary user, • I would like the option to submit a report anonymously so that I can hide my identity • I need an option to "save and continue later" so that I can prioritize the duties of my role as needed • I want confirmation that I have completed the form correctly so that I can feel good about reporting

Table 10. a. Aim1 – proposed set for product features for MVP. b. Aim2 – proposed set for product features for MVP.

a

			Continuous analysis results				Discrete analysis results[1]						
ID	Feature Description	Rationale	Category	Dysfunctional (SD)	Functional (SD)	Importance (SD)	Category	M	P	A	I	R	Q
Aim 1. Increase reporting of workplace violence													
F1	Progressive web application (PWA) technology.	To demonstrate empathy for the workload of HCWs, HCWs will access the improved WPV reporting tool from a desktop shortcut or from their mobile device using progressive web application technology.	-	-	-	-	-	-	-	-	-	-	-
F2	Anonymous reporting.	To demonstrate empathy for HCWs privacy preferences, all reports will be submitted anonymously, by default. HCWs can provide more detailed information if they choose to.	Performance	2.25 (1.82)	2.92 (1.78)	5.33 (2.43)	Attractive	16.67%	25.00%	41.67%	16.67%	0.00%	0.00%
F3	Ease of reporting.	To demonstrate empathy for the workload of HCWs, we limit data collection to critical data elements including (1) the type of WPV that has occurred, (2) the type of people involved, (3) where the event occurred, and (4) when the event occurred. Each screen will contain discrete option choices. Once responses are provided for each critical data element, the report will be considered complete and can be submitted.	-	-	-	-	-	-	-	-	-	-	-
F4	Safety alerts.	To demonstrate commitment to the safety of HCWs, the improved WPV reporting tool will provide alerts to HCWs in the vicinity of an active threat or risk of violence. The safety alert will provide instructions on safety precautions that should be taken to protect themselves and those in their care.	-	-	-	-	-	-	-	-	-	-	-
F5	Reporting statistics dashboard.	To demonstrate organizational commitment to change, HCWs will have access to reporting statistics where they can monitor organization-level progress on WPV reporting initiatives.	Performance	2.42 (1.88)	2.83 (1.59)	6.17 (2.51)	Attractive	25.00%	25.00%	33.33%	16.67%	0.00%	0.00%
F6	Submit daily or weekly totals.	To demonstrate empathy for the workload of HCWs, the improved WPV reporting tool will allow HCWs to submit daily or weekly totals of WPV events as an alternative to individual, detailed reports.	Attractive	1.42 (1.62)	2.08 (2.11)	5.33 (2.56)	Indifferent	8.33%	8.33%	33.33%	41.67%	8.33%	0.00%

[1]Must-haves (M), performance (P), attractive (A), indifferent (I), reverse (R), and questionable (Q)

(*continued*)

Table 10. (*continued*)

b

ID	Feature Description	Rationale	Continuous analysis results: Category	Dysfunctional (std dev)	Functional (std dev)	Importanc e (std dev)	Discrete analysis results: Category	M	P	A	I	R	Q
Aim 2. Provide access to WPV support resources													
F7	Auto-generated recommendation for next-steps based on the report.	To demonstrate the value of WPV, HCWs will receive recommendations based on the information reported. For example, cases of verbal assault will be prompted to connect to counseling resources and those who experienced physical assault will be prompted to seek medical attention.	Performance	2.67 (1.78)	2.50 (2.07)	6.33 (2.05)	Performance	25.00%	33.33%	25.00%	16.67%	0.00%	0.00%
F8	24-hour support specialists.	To demonstrate support for safety of HCWs, options to connect with hospital police or access additional resources, such as counseling services or medical examiners, will be available from the home screen and top menu panel.	Performance	2.58 (1.93)	2.50 (1.93)	6.83 (2.37)	Performance	25.00%	33.33%	16.67%	16.67%	8.33%	0.00%
F9	Exclusion lists.	To demonstrate respect for the privacy concerns of HCWs, HCWs will be able to create exclusion lists with specific people, units, or departments they do not want their report shared with.	Attractive	1.08 (1.44)	2.83 (1.99)	5.00 (2.31)	Attractive	0.00%	8.33%	58.33%	25.00%	8.33%	0.00%
F10	In-app messaging system.	To demonstrate empathy for the privacy preferences of HCWs, the improved WPV reporting tool will include a private messaging center where HCWs can interact with support specialists from within the application to avoid receiving and sending sensitive communications from their work email account.	Attractive	1.17 (1.34)	2.33 (2.06)	3.33 (1.80)	Attractive	0.00%	8.33%	50.00%	41.67%	0.00%	0.00%
F11	Patient safety flags.	To demonstrate support for HCW safety, HCWs will have the option to flag patients so that other HCWs are notified when providing care to patients with a history of violence. Transparent governance and validation mechanisms will be required to protect patients from unfair biases or unfair evaluations based on characteristics such as race, ethnicity, age, or gender.	Performance	3.27 (1.98)	2.45 (1.68)	7.00 (2.64)	Performance	33.33%	41.67%	0.00%	16.67%	0.00%	8.33%
F12	Ticket numbers for report status tracking.	To demonstrate organizational commitment to follow-up, once a report has been submitted, HCWs are presented with a ticket number to track the report status and respond to requests for additional information, while remaining anonymous.	Performance	2.42 (1.88)	2.83 (1.34)	6.17 (1.91)	Must-have, attractive	33.33%	16.67%	33.33%	16.67%	0.00%	0.00%

[1]Must-haves (M), performance (P), attractive (A), indifferent (I), reverse (R), and questionable (Q)

4.3 Evaluation

Feature Prioritization. We distributed a survey among HCWs at an academic medical institution based on the Kano model. The Kano model was designed to help teams systematically prioritize product features by understanding how users feel about the availability of a given feature and how important it is that a feature (represented with a symbol F) is included in the development of the product. Thus, this survey study aimed to systematically assess HCWs agreement that the recommended set of product features are valued by HCWs. Institutional review board approval was required and all HCWs voluntarily consented to participate.

At the start of the survey, participants were asked to provide yes/no responses to screener questions:

(1) Have you experienced WPV while at work?
(2) Have you received training on WPV reporting?
(3) Have you previously reported a WPV event?

Based on the Kano model, participants were asked to respond to three questions for each feature:

(1) How do you feel if the reporting tool includes this feature?
(2) How do you feel if the reporting tool does not include this feature?
(3) How important is it to you that this feature is included in the reporting tool?

Participants were prompted to provide ratings for questions (1) and (2) on a 5-point Likert scale. Responses were assigned a weighted value of negative 2 for 'I dislike it', negative 1 for 'I can tolerate it', 0 for 'I am neutral', positive 2 for 'I expect it', and positive 4 for 'I like it'. Question (3) utilized a 9-point Likert scale where 1 indicated the feature was 'not at all important' and 9, 'extremely important'.

Average scores and standard deviation were calculated for each set of questions for each feature. The average score for questions (1) and (2) was mapped to a feature plot where the average score for question (1) is marked along the y-axis (functional) and the average score for question (2) is drawn along the x-axis (dysfunctional). The average score for question (3) is denoted by the relative size of the plotted dots.

Features were further categorized based on their placement on the feature plot. "Performance" features (P) indicate users' expectations to ensure adequate functionality and satisfaction. "Attractive" features (A) are features users like having but are not expected. "Indifferent" features (I) are features users were neutral on. "Must-be" features (M) are features users expect but will not make them more satisfied; although, if missing, users will be dissatisfied. Features, where average responses for question (1) were 'I like it' and for question (2) on average, were 'I dislike it' were categorized as reverse (R). 'Reverse' features indicate that users want the opposite of what has been proposed and therefore responses from question (1) and question (2) can be swapped to find the true categorization. 'Questionable' features (Q) are those were question (1) and (2) have conflicting responses, such as 'I like it' and 'I like it', which could be the result of a poorly phrased question.

Out of consideration for HCWs' limited time availability, our team decided to remove features F1, F3, and F4 from the survey. We rationalize the removal of these three features with the agreed assumption that these features are highly valuable to our primary users. This change reduced the estimated time needed to complete the survey to 10 min.

Results. 12 participants were included in this part of the study. 75% of participants had experienced WPV in a clinical setting, 91% of participants had received training on how to report WPV, and 42% had previous experience reporting WPV. The results of the continuous and discrete analysis are shown in Table 9a and Table 9b.

Our results indicated that the features contained in the proposed product feature set were mostly viewed as performance-enhancing or attractive features. Patient safety flags (F11) were rated as the most important feature, the top dysfunctional feature, one of the top two must-have features, and the top performance feature. A ticketing system (F12) was rated by participants as the second of the top two must-have features. Auto-generated recommendations (F7) and 24-h support specialists (F8) were rated second and third in dysfunctional features, importance, and performance, respectively. Exclusion lists (F9) and in-app messaging (F10) were the top two attractive features, respectively, followed by anonymous reporting (F2). Conversely, participants indicated indifference for F10 and daily report submissions (F6). Participants rated a reporting statistics dashboard (F5) and anonymous reporting (F2) as performance features (continuous analysis) and attractive features (discrete analysis). F11 was rated as questionable by at least one participant and F8, F6, and F9 had at least one participant indicate reverse results. See Fig. 5 for the resulting feature plot.

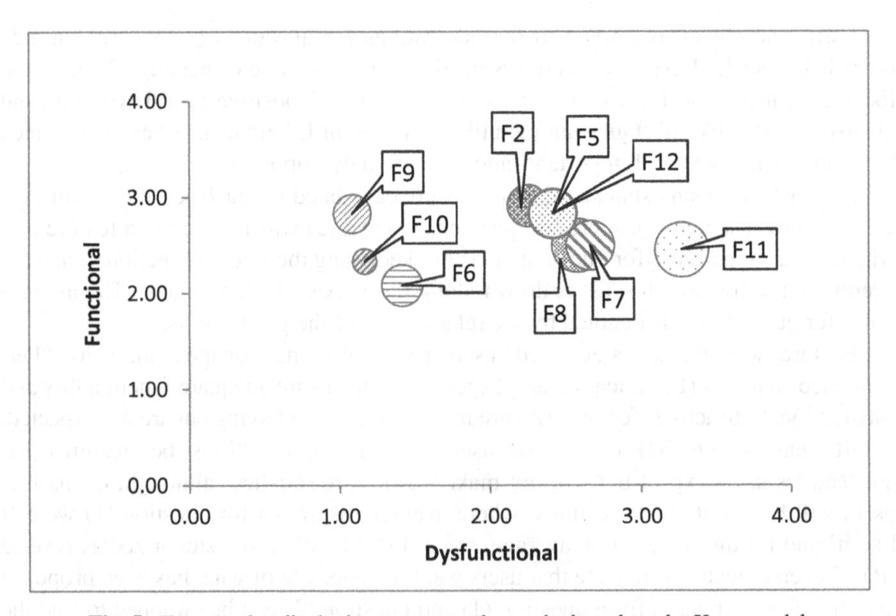

Fig. 5. Feature plot displaying survey study results based on the Kano model.

5 Discussion

The purpose of this case study was to (1) present in detail the activities conducted by the design team in the development of a product feature set for an improved WPV reporting tool and (2) explore how human-centered design and product discovery techniques can be used in combination to inform feature recommendations for a MVP. To achieve this, the design team evaluated the current state of WPV reporting using the existing technology, confirmed a shared understanding of user value to cultivate a product feature set for the MVP, and validated the set of recommended product features among HCWs. This study adds value by demonstrating how HCD and product discovery techniques can be used to explore, ideate, and validate design goals to deliver more impactful products. In our experience, we found that by using the two sets of techniques, the design team's recommendation for the MVP was well-received and supported by HCWs.

We recommend the following features be included in the development of the MVP for an improved WPV reporting tool: a ticketing system, patient safety flags, auto-generated recommendations, 24-h access to support specialists, exclusion lists, in-app messaging, and anonymous reporting in the development of a MVP for an improved WPV reporting tool. More information is needed to understand HCWs' indifference regarding the submission of daily or weekly totals.

We experienced several challenges during our efforts. In the analysis phase, it was difficult to form a generalized and unbiased perspective to represent the majority of the HCW population. Personas proved especially useful in acknowledging the bounds of our understanding of HCWs and mitigated over-generalization by developing archetypes for each type of primary user. In the synthesis phase, the design team needed to better understand the key drivers of reporting as seen by the HCWs. To achieve this, the design team needed to think beyond what HCWs specifically asked for and come up with innovative ideas to deliver what HCWs needed. While user story mapping provided a visual tool for organizing the raw list of requirements, the final prioritization relied heavily on the designs team's interpretation of value added by each feature. During the evaluation phase, due to the absence of wireframes or prototypes, the design team was challenged to describe the design features in a way that painted the picture of the improved reporting tool for HCWs. The Kano model provided structure to the feedback elicitation process, but how HCWs interpreted the description of the proposed features could vary given the lack of visual representations. Our future work with HCWs will include wireframes and functional prototypes.

5.1 Study Limitations

There were a few limitations to this study. The design team did not include developers who offer a technical feasibility lens to the solution space, design aims, and proposed features. Given HCWs are highly sought after as research participants and are prone to survey (research participation) fatigue, we shortened the survey used in the evaluation phase to ensure that HCWs could complete the survey with minimal interruption to the responsibilities of their roles. To do this, we made educated assumptions on the features that HCWs likely value as essential to the design of a reporting tool. These features, therefore, were not assessed by HCWs, and no feedback was collected to validate these assumptions. We also did not elicit feedback on the features that HCWs may have felt were missing from the list of proposed features. Finally, factors such as regulatory, financial, and organizational challenges, and missing interoperability standards must be overcome to increase HCWs adoption of official mobile devices in hospitals [26].

6 Conclusion

HCWs are experiencing WPV at an alarming rate. Efforts to mitigate underreporting must ensure that reporting tools do not further deter HCWs from reporting WPV events or seeking support. This case study demonstrates how the combined use of human-centered design and product discovery techniques can deliver WPV reporting tools that better align with what HCWs value.

References

1. Worker safety in hospitals | Occupational safety and health administration. https://www.osha.gov/hospitals. Accessed 16 Feb 2022
2. Speroni, K.G., Fitch, T., Dawson, E., Dugan, L., Atherton, M.: Incidence and cost of nurse workplace violence perpetrated by hospital patients or patient visitors. J. Emerg. Nurs. **40**(3), 218–228 (2014). quiz 295. https://doi.org/10.1016/j.jen.2013.05.014
3. United States Bureau of Labor Statistics: Workplace Violence in Healthcare (2018). Injuries, Illnesses, and Fatalities. https://www.bls.gov/iif/oshwc/cfoi/workplace-violence-healthcare-2018.htm. Accessed 11 Feb 2022
4. Liu, J., et al.: Prevalence of workplace violence against healthcare workers: a systematic review and meta-analysis. Occup. Environ. Med. **76**(12), 927–937 (2019). https://doi.org/10.1136/oemed-2019-105849
5. Dadfar, M., Lester, D.: Workplace violence (WPV) in healthcare systems. Nurs. Open **8**(2), 527–528 (2020). https://doi.org/10.1002/nop2.713
6. Watson, A., Jafari, M., Seifi, A.: The persistent pandemic of violence against health care workers. Am. J. Manag. Care **26**(12), e377–e379 (2020). https://doi.org/10.37765/ajmc.2020.88543
7. Sato, K., Wakabayashi, T., Kiyoshi-Teo, H., Fukahori, H.: Factors associated with nurses' reporting of patients' aggressive behavior: a cross-sectional survey. Int. J. Nurs. Stud. **50**(10), 1368–1376 (2013). https://doi.org/10.1016/j.ijnurstu.2012.12.011
8. Kahsay, W.G., Negarandeh, R., Dehghan Nayeri, N., Hasanpour, M.: Sexual harassment against female nurses: a systematic review. BMC Nurs. **19**(1), 58 (2020). https://doi.org/10.1186/s12912-020-00450-w
9. Roche, M., Diers, D., Duffield, C., Catling-Paull, C.: Violence toward nurses, the work environment, and patient outcomes. J. Nurs. Scholarsh. Off. Publ. Sigma Theta Tau Int. Honor Soc. Nurs. **42**(1), 13–22 (2010). https://doi.org/10.1111/j.1547-5069.2009.01321.x
10. Farrell, G.A., Bobrowski, C., Bobrowski, P.: Scoping workplace aggression in nursing: findings from an Australian study. J. Adv. Nurs. **55**(6), 778–787 (2006). https://doi.org/10.1111/j.1365-2648.2006.03956.x
11. Occupational Safety and Health Administration et al.: Guidelines for preventing workplace violence for healthcare and social service workers (OSHA, 3148-04R). Wash. DC OSHA (2015). Accessed 01 Mar 2021. https://www.osha.gov/sites/default/files/publications/osha3148.pdf

12. Arnetz, J.E., et al.: Underreporting of workplace violence: comparison of self-report and actual documentation of hospital incidents. Workplace Health Saf. **63**(5), 200–210 (2015). https://doi.org/10.1177/2165079915574684
13. Gates, D.M.: The epidemic of violence against healthcare workers. Occup. Environ. Med. **61**(8), 649–650 (2004). https://doi.org/10.1136/oem.2004.014548
14. Blando, J., Ridenour, M., Hartley, D., Casteel, C.: Barriers to effective implementation of programs for the prevention of workplace violence in hospitals. Online J. Issues Nurs. **20**(1), 5 (2015)
15. Marte, M., Cappellano, E., Sestili, C., Mannocci, A., La Torre, G.: Workplace violence towards healthcare workers: an observational study in the College of Physicians and Surgeons of Rome. Med. Lav. **110**(2), 130–141 (2019). https://doi.org/10.23749/mdl.v110i2.7807
16. National Advisory Council on Nurse Education and Practice: Violence against nurses. An assessment of the causes and impacts of violence in nursing education and practice (5th ed.) (2007). Accessed 01 Jun 2021. https://www.hrsa.gov/sites/default/files/hrsa/advisory-committees/nursing/reports/2007-fifthreport.pdf
17. Nachreiner, N.M., Gerberich, S.G., Ryan, A.D., McGovern, P.M.: Minnesota nurses' study: perceptions of violence and the work environment. Ind. Health **45**(5), 672–678 (2007). https://doi.org/10.2486/indhealth.45.672
18. Nguyen, C., Carroll, J., Myers, S., Rutanashoodech, A., Villarreal, M., Nious, K.: Violence against healthcare workers continues despite hospital safety law. NBC Bay Area. https://www.nbcbayarea.com/investigations/violence-against-healthcare-workers-continues-despite-hospital-safety-law/2238203/. Accessed 21 Jan 2022
19. Münch, J., Trieflinger, S., Heisler, B.: Product discovery – building the right things: insights from a grey literature review. In: 2020 IEEE International Conference on Engineering, Technology and Innovation (ICE/ITMC), pp. 1–8. June 2020. https://doi.org/10.1109/ICE/ITMC49519.2020.9198328
20. Jones, J.C.: Design Methods. David Fulton Publishers Ltd, London (1970)
21. W. L. in R.-B. U. Experience: Design critiques: encourage a positive culture to improve products. Nielsen Norman Group. https://www.nngroup.com/articles/design-critiques/. Accessed 20 Jan 2022
22. DeLone, W.H., McLean, E.R.: The DeLone and McLean model of information systems success: a ten-year update. J. Manag. Inf. Syst. **19**(4), 9–30, Springer, Cham (2003). https://doi.org/10.1080/07421222.2003.11045748
23. Foster, M., Adapa, K., Soloway, A., Francki, J., Mazur, L.: Evaluating usability of workplace violence tools, cognitive workload, and performance of healthcare professionals. Presented at the – Accepted as a paper at International Medical Informatics Association's Med Info 2021 virtual conference
24. Lewis, D.J.R.: Using the PSSUQ and CSUQ: in user experience research and practice. 1-7333392-0-5 (2019)
25. Young, M.S., Brookhuis, K.A., Wickens, C.D., Hancock, P.A.: State of science: mental workload in ergonomics. Ergonomics **58**(1), 1–17 (2015). https://doi.org/10.1080/00140139.2014.956151
26. Maassen, O., et al.: Future mobile device usage, requirements, and expectations of physicians in German university hospitals: web-based survey. J. Med. Internet Res. **22**(12), e23955 (2020). https://doi.org/10.2196/23955

Electronic Performance Monitoring: Review of Theories, Conceptual Framework, and Study Proposal

Thomas Kalischko[1] and René Riedl[1,2](✉)

[1] University of Applied Sciences Upper Austria, Steyr, Austria
{thomas.kalischko,rene.riedl}@fh-steyr.at
[2] Johannes Kepler University, Linz, Austria
rene.riedl@jku.at

Abstract. Use of electronic monitoring techniques in organizational settings has significantly gained in relevance in recent years, predominantly due to the increased availability of corresponding information and communication technologies. Moreover, the restricted social mobility due to COVID-19 and the resulting increase of home office has come along with an increased interest of employers to monitor the activities of their employees. Against the background of these developments, along with the general importance of theory-focused research, the goal of this review is to identify the theories used in the electronic performance monitoring (EPM) literature and to illustratively show how recently identified research gaps could be examined in future studies based on theory integration. A total of twelve theories were extracted from the literature and examined. Four of these theories were integrated in a conceptual framework. In addition, a study design for future research is outlined. The overall objective of the present contribution is to strengthen the theoretical foundation of the EPM research field.

Keywords: Electronic performance monitoring · Computer monitoring · Workplace surveillance · Theory · Review

1 Introduction

A report from the market research firm Gartner (2019) provided an outlook into the future of the employee monitoring market for 2020. They estimated that almost 80% of all companies will be using monitoring software to keep track of their organizational goals and their employees. This would have meant a significant increase, compared to the year 2018 where 50% of 239 surveyed companies used such techniques [1]. One year later a survey by Accenture supports the estimated increase, by reporting that 62% of the asked companies were using new technologies in order to gather data on their employees [2]. Those numbers were obtained before the corona pandemic and even though a market coverage of 80% seems high, recent newspaper reports such as the "BBC" [3] or "businessinsider" [4] report new ways of monitoring during home office hours. As part of an article, the "MIT Technology Review" conducted interviews with

F. Fui-Hoon Nah and K. Siau (Eds.): HCII 2022, LNCS 13327, pp. 388–406, 2022.
https://doi.org/10.1007/978-3-031-05544-7_29

large companies, which reported a 4-fold increase in inquiries on workplace-monitoring-technology [5]. Against this background, it is reasonable to conclude that the workplace-monitoring-market growth might even be underestimated, and that electronic monitoring of employees is a major phenomenon in today's economy and society.

Information and Communication Technologies (ICT) have a huge impact on individuals, organizations, and society. Major benefits are extensive communication, improvements in performance and productivity as well as increased access to information [6, 7]. These fast evolving and new technologies enable employers to conceive new ways to check the performance and the productivity of their employees. Physical monitoring of employees used to be the standard in many organizations [8]. However, recently a change was introduced by the digitalization of the workplace, big data, and other technologies such as artificial intelligence [9]. Historically, EPM started with electronic monitoring of call-center employees in the 1980s [10] and later became important in many other industries [11]. Importantly, during the corona pandemic many employees were forced to work from home, accompanied by increased use of monitoring software as employers at least want to have some "observation" of their employees. The interaction between EPM systems and employees represents a branch of Human Computer Interaction, which aims to improve this relationship in the long term. Since employee monitoring is a sensitive issue, this relationship between the monitoring systems and the employees is particularly important. Thus, the gap between the negative perceptions of the employee towards the system and the goal of the system to check the productivity of the employee should be reduced [12]. Improving this interaction is also in the interest of the company. As studies have already shown, the use of systems for the electronic monitoring of employees also has consequences [13]. For example, the use of such systems can have an effect on employee motivation [14, 15], which in turn is cited as a success factor for the effective implementation of omnichannel [16].

In order to better understand the complex nature of the EPM phenomenon, different theoretical foundations are useful and needed. Thus, a fundamental question arises: *Which theories have been used in the EPM literature and how can they be applied to address current research gaps?* EPM researchers have already used various theories to explain individual aspects associated with EPM. Knowing these theories, based on a systematic review of the literature, constitutes a valuable basis for future research. The need for a systematic identification and review of theories is also important if it is connected to existing research gaps. Based on the fact that such a systematic review of theories, to the best of the authors' knowledge, does not exist, we analyzed the scientific literature and identified twelve theories. Then, we integrated four theories and demonstrate how this integration contributes to a better understanding of current research issues which we identified in recent EPM research agenda and review papers. The overall objective of this article is to strengthen the theoretical foundation of the EPM research field and to form the basis for a better understanding so that decision makers who want to incorporate EPM in the company can establish a better human-computer interaction.

2 Theories Used in the EPM Literature

At the outset of this review, a literature search was conducted that included over 30 years of research since the term "Electronic Work Monitoring" was defined by the U.S.

Congress, Office of Technology Assessment, in 1987, and refers to the "computerized collection, storage, analysis, and reporting of information about employees' productive activities" p. 27 [17]. In the meantime, the term "Electronic Performance Monitoring" has become more established and is also used in this review. If EPM is used in a company, it can be used to monitor employees in different ways. On the one hand, individual employees and their work behavior, as well as private behavior [13] (via e-mail or social media monitoring along with attitudes and emotional states can be monitored [18, 19] (e.g., based on text analyses such as emails). This, in turn, makes possible the development of employee profiles [20]. On the other hand, monitoring can also take place at the team, department, or company level. Depending on the level, this also reduces the respective violation of employees' privacy [13].

The literature review was conducted on the basis of existing methodological recommendations [21, 22] in the period 10-1-2021–10-15-2021. The focus was on scientific sources. First, the search terms were defined, and these were used in an initial search in the Scopus and Web of Science databases. The literature search was conducted according to Webster & Watson (2002) [21] and vom Brocke et al. (2009) [22]. The search terms used to identify relevant scientific sources were derived from the following sources: [8, 13, 19, 23]. Specifically, the search terms "EPM", "electronic performance monitoring", "electronic monitoring", "workplace monitoring", "workplace surveillance", and "electronic monitoring" were identified as relevant. Following this initial search, a filtering strategy was used to identify the first relevant sources. Using these sources, both backward and forward snowballing was performed. Figure 1 summarizes the search process.

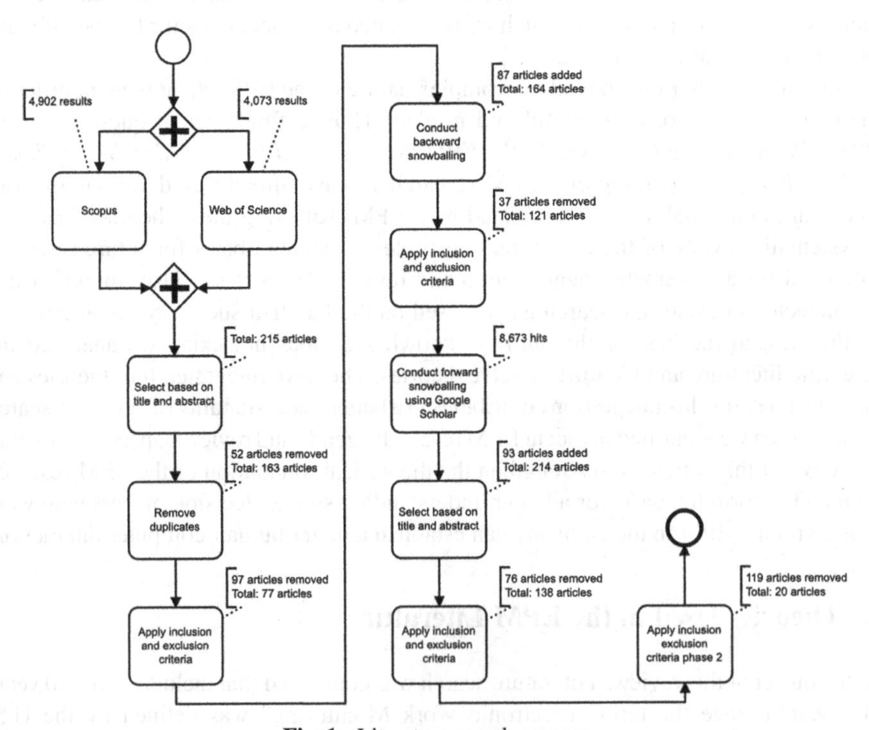

Fig. 1. Literature search process

The search yielded 8,975 hits (Scopus 4,902 and Web of Science 4,073); these were published between 1987 and 2021. In the first step, all unrelated papers were removed based on title and abstract, resulting in 215 articles. After duplicates were removed, the result was 163 unique articles. The remaining papers were then analyzed in detail, and the established filtering criteria were applied:

- Inclusion criterion: article focusing on EPM and/or examining factors that interact with EPM and/or EPM-related outcomes.
- Exclusion criterion: articles focused on other topics such as privacy, law, or ethics that did not clearly relate to the EPM literature were excluded.

Based on these criteria, 97 articles were identified that did not fit the subject matter studied. This left 77 articles for the backward snowballing process, through which 87 additional articles could be added. Applying the filtering criteria reduced the current state by 43 articles and thus 121 articles remained for the forward snowballing process. For this process, the results in Google Scholar were applied (8,673 hits) and the output results were reviewed. After a review of the titles and abstracts, 93 additional articles were added, resulting in a total of 214 articles. Finally, the established filtering criteria were applied again, and thus, after removing 76 articles, a total of 138 scientific publications were obtained. The search conducted up to this point is part of a large-scale study and thus also serves to update the literature used by the authors. By defining further inclusion and exclusion criteria, the focus of the study presented here, namely the focus on theories used within the EPM literature, is elaborated. The further procedure was carried out as follows.

- Inclusion criterion phase 2: the article and the methodology used in it draws upon a scientific theory.
- Exclusion criterion phase 2: The article does not build on a scientific theory in a substantial way and hence only mentions theory without provision of any details.

After applying our inclusion and exclusion criterions within phase 2 of our literature search process, we ended-up with 20 relevant articles. Within this literature base, 12 different theories could be identified, which are outlined in alphabetical order in the next chapters. If follows that theory has not played a major role in the extant EPM literature. Rather, most papers are empirical in nature without reference to existing theories. In the following presentation, each theory is divided into two paragraphs. The first paragraph of each theory describes the original theory. Then, the following paragraph discusses the respective work from the EPM literature and its findings in relation to the theory presented.

2.1 Agency Theory

Agency theory is based on Jensen and Meckling (1976) [24] and deals with the relationship between principal and agent. Accordingly, one or more persons (agents) are assigned to perform a task on behalf of the principal, including a certain decision-making authority. If one assumes in this theory that both the principal and the agent want to maximize their respective benefits, there is reason to believe that the agent will not always act in the interest of the principal [24]. As an example, consider a scenario where employees are in the home office, pursuing goals or engaging in activities unrelated to work and hence do not contribute to goal accomplishment [25]. In order to avoid detrimental actions by the agent, the principal could provide incentives to the agent to behave in her interest, but could also implement monitoring techniques to observe the agent's behavior [24]. Referring to the EPM context, this theory predominantly applies to the manager-employee relationship within an organization.

Based on agency theory, one study indicates that the more managers depend on employees and their performance, the more likely the manager is to increase monitoring measures towards his employees. Thus, the dependency of managers on employees correlates positively with the intensity of electronic monitoring. Specifically, the study found that within a virtual team, a team leader intensified electronic monitoring for two reasons. First, when she was dependent on their employees for critical information, and second, when she expected low work performance from employees based on earlier consistently low job performance perceptions over an extended period of time. In addition to these findings, it was found that monitoring was secretly increased for those employees on whom the team leader was more dependent or if they had a generally lower level of trust towards the employees [25].

2.2 Balance Theory

The balance theory by Smith and Sainfort (1989) [26] describes an approach that combines job design and stress in a holistic way. It integrates psychological and biological theories and puts them together in an ergonomic framework. It theorizes that working conditions and stress have psychological as well as physiological and behavioral effects on the employee. The model presented includes five elements (individual, technology, task, organization, environment) that need to be balanced in order to avoid stress [26].

There are no empirical studies in the EPM literature that rely on this theory. However, we found a call for future studies in combination with a brief reflection on the theory in the EPM literature. This includes a focus on specified constructs and measurements to examine the effects of technology, task, work environment, and organizational structure. It is argued that appropriate management and intervention regarding these factors should allow to control the individual stress of employees [27].

2.3 Communication Privacy Management Theory

The theory is based on the original idea of the communication boundary management theory by Petronio (1991) [28] and was further developed by the same author to the communication privacy management theory [29]. The theory can be used to explore a variety of personal communication problems. It includes principles and processes that explain how people communicate when they want to regulate or protect private information. These principles are conceptualized as follows: (1) Ownership: If information is considered one's own, it counts as private. (2) Distribution: If information is considered one's own, it is also considered one's own right to distribute it. (3) Personal privacy rules: The distribution of private information is often based on personally set rules. (4)

Collective privacy boundaries: Once private information is shared, a collective privacy boundary is formed. The rules are based on the original rules of the owner with consultation of the co-owners. (5) Boundary turbulence: This arises when the co-owner does not share the information flow with third parties in a compliant manner. However, this need for privacy clashes with the human need for social interaction and interpersonal relationships. This dichotomy thus acts as a driver of how decisions regarding one's privacy are made [29–31].

A study based on communication privacy management theory examined the effect of a supporting and controlling purpose for electronic monitoring and its impact on applicant responses to a fictitious job application. The distinction between supportive and controlling purpose of EPM is well documented in the literature [19]. The findings revealed that applicants rated an organization as less attractive when it engaged in monitoring based on controlling purpose. In addition, privacy concerns were rated as higher if compared to monitoring based on supportive purpose [32]. Another study found that a control-oriented corporate culture increased communication privacy turbulence in CPM. The instability in communication privacy in CPM had largely negative consequences on trust in employee monitoring policy, but not on trust in employee monitoring members. Employee commitment and compliance to employee monitoring were positively influenced by both trust in employee monitoring policy and trust in employee monitoring members [33].

2.4 Deterrence Theory

Deterrence theory has its origin in the 1968 work of Gary S. Becker entitled "Crime and Punishment: An Economic Approach" [34]. This theory "describes a process of offender decision making that consists of two linkages—one in which official sanctions and other information affect a would-be-offender's perceptions about the risks of criminal conduct, and another in which such perceptions influence the decision of whether or not to offend" [35] p. 1. According to deterrence theory, a distinction can be made between two types of deterrence: deterrence by punishment (the employee is made aware of incorrect behavior with the threat of consequences) and deterrence by denial (the employee is convinced that erroneous behavior will not lead to his or her goals) [36]. In summary, the theory states that criminal activities are committed when the expected success is higher than the expected costs. The use of optimally designed policies to prevent illegal behavior serve as the cornerstone for optimal cost allocation [34].

In the EPM literature, this theory has been applied in one source only, a conceptual study by Chen and Pfleuger (2008) [37]. It is argued that if the power of the employer is increased by implementing a monitoring system in the workplace, then implementing an information assurance policy will have a positive effect on employee behavior (because transparency dampens possible negative perceptions regarding monitoring) [37].

2.5 Ethical Theory

If one takes a closer look at the topic of EPM, ethical concerns immediately come to mind. The basis for many ethical aspects is provided by the ethical theory of Bauchamp and Bowie (1993) [38]. A major distinction made within this theoretical framework is

Utilitarianism versus Kantianism. Utilitarianism is based on the morality of activities, which can be judged ethically based on their consequences. Kantianism is based on inherent adjuncts of an action that make it right or wrong [38].

In the EPM literature there are numerous publications that advocate the importance of ethical aspects and present different approaches [13, 39, 40]. However, we found that no publication in the EPM literature presents empirical findings based on ethical theory. With regard to Utilitarianism, a benefit for organizations, employees, and society could be productivity increases of employees. Such an increase, in turn, could lead to higher employee salaries, and hence society could ultimately benefit from implementation of EPM [41]. However, these advantages must be viewed in light of the possible disadvantages of EPM, such as stress, reduced motivation, or distrust [13]. This balancing of advantages and disadvantages is at the core of Utilitarianism. Kantianism, in contrast, places a high importance on intrinsic values and holds companies accountable for preserving the dignity, autonomy, and right to privacy of employees. From this perspective, the use of EPM systems would be considered unethical [41].

2.6 Feedback Intervention Theory

This theory is based on a meta-analysis of feedback intervention by Kluger and DeNisi (1996) [42]. Feedback interventions can be defined as "actions taken by (an) external agent(s) to provide information regarding some aspect(s) of one's task performance" [42] p.255. The central assumption of the theory is that feedback interventions run the locus of attention along three general and hierarchical levels of control. These levels are tasks and meta-tasks (involving the self), motivation (involving the focal task), and task learning (involving the details of the focal task). It was found that the effectiveness of feedback interventions decreases as attention moves up the hierarchy and closer to the self and thereby away from the task [42]. Five main assumptions emerge from the theory of Kluger and DeNisi (1996). (1) Employee behavior is regulated when compared to standards and goals. (2) Standards and goals are hierarchically oriented according to task learning, task motivation, and meta-tasks. (3) Everyone's attention is limited and only feedback-standard gaps receive the level of attention to influence behavior. (4) Attention is usually directed to a moderate level of hierarchy. (5) Feedback interventions change behavior in that the locus of attention of each employee is changed [42, 43].

The original theory was applied and further developed within the EPM literature. Feedback intervention theory asserts that negative feedback increases employee performance, but only when employees focus on task motivation processes. This assertion was tested and advanced, among others, by suggesting that interpersonal fairness is a critical determinant of focus of attention. Moreover, it has been found that when feedback is intense, employees frequently become anxious and self-esteem is possibly lowered, which, in turn, guides the focus on meta-task processes [44]. In another study in the EPM literature, it was found that the relationship between the desire to improve and work performance was stronger when the focus was on the task-motivational level and the relationship was weaker when the focus was on the meta-task level [43].

2.7 Psychological Reactance Theory

Psychological reactance theory is a theory that originated in the work of Brehm in 1966 [45] and is based on the construct of freedom. It assumes that each person highly values his or her own freedom, choice, and autonomy. In essence, the theory states that when an external stimulus threatens individual freedom of choice, psychological reactance will follow. Reactance is triggered by a threat to one's freedom and manifests itself in anger and a negative attitude [45, 46]. Psychological Reactance Theory is based on four major constructs. (1) Freedom: The individual belief that everyone can act as she wishes. In Brehm's theory, freedom is conceptualized based on actions, emotions, and attitudes. (2) Threat to freedom: Includes everything that makes it more difficult for a person to live out her freedom. (3) Psychological reactance: A psychological state in which freedom is perceived as threatened, or non-existent. (4) Restoration of freedom: whenever individual freedom is threatened, or non-existent, motivation emerges to reestablish that freedom [45, 46].

Within the EPM literature, privacy invasion by EPM is assumed to be positively related to state reactance. This assumption was confirmed by an empirical study [47] and privacy, in general, is seen as an important freedom by many people. The study also found that people with high trait reactance tend to experience state reactance more often. Anger, which is a component of state reactance, was found to led to counterproductive work behavior (defined as "voluntary behavior that violates significant organizational norms and in so doing threatens the well-being of an organization, its members, or both" [48] p. 556) and reduced organizational citizenship behavior (defined as "behavior that is discretionary, not directly or explicitly recognized by the formal reward system, and in the aggregate promotes the efficient and effective functioning of the organization" [49] p. 8) [47]. Another study used the theory as conceptual basis and found that high levels of monitoring lead to negative work behavior, and that attitudes toward monitoring acted as a mediator in this relationship [50].

2.8 Reinforcement Theory

Reinforcement theory was formulated by Skinner in 1953 [51] and states, among other things, that positive reinforcement, based on rewards, leads to repetition of the desired behavior, while negative reinforcement, based on punishment, leads to termination of the undesired behavior. Behavior is divided into two categories: (1) desirable and (2) undesirable. (1) Desirable behavior should come along with positive reinforcement; thus, an employee will receive a reward if she demonstrates a desired behavior. (2) Undesirable behavior should come along with negative reinforcement; thus, an employee will receive a punishment if she demonstrates an undesired behavior [51].

In the EPM literature, this theory was used in one paper and the authors argue that individuals who are monitored have a higher level of motivation than individuals who are not, as they also do not fear consequences based on their own job performance. This hypothesis was confirmed based on experimental evidence [15].

2.9 Self-regulation Theory

Self-regulation theory has its origin in the work of Frederick Kanfer (1970) [52]. In addition to this theory, however, there are numerous researchers who have also contributed a great deal to this theory. Self-regulation is the ability to change or adapt one's own goal-directed activities under changing circumstances. The processes of self-regulation are set in motion when routine activities are hindered or when there are great challenges that deviate strongly from the usual. In this respect, self-regulation is divided into five interconnected and repetitive phases [53]: (1) goal selection, (2) goal cognition, (3) directional maintenance, (4) directional change and reprioritization, and (5) goal termination.

Within the EPM literature, one conceptual study draws upon this theory and indicates that the effect of self-regulation will occur in monitoring scenarios and that perception of monitoring will come along with employee behavior changes. Thus, the illusion of control over the monitoring system is assumed to have a positive effect on employee behavior toward monitoring technology in the workplace [37].

2.10 Social Facilitation Theory

The basic form of interindividual influences are represented by the paradigm around social facilitation and are based on social psychology studies such as in the Triplett's experiments on speed and competition from 1897 [54]. These ideas were further developed by Robert B. Zajonc and are now manifested in Social Facilitation Theory. in this domain, two paradigms are distinguished. The first paradigm deals with behavior observation in the presence of passive observers. The second paradigm deals with behavior of individuals with the presence of other individuals engaged in the same activity. In essence, the theory indicates that simple tasks can be performed more efficiently in the presence of passive observers. However, the opposite is true for cognitively more demanding tasks. Regarding co-action effects, it can also be deduced that if a task is well learned, the individual performance is better than if it is not well learned. Thus, in summary, the presence of others increases task performance on simple tasks and decreases task performance on complex tasks. Social facilitation theory focuses mainly on performance, but also points to side consequences such as stress [54].

Studies related to electronic performance monitoring show how social facilitation theory can be applied in the work context. Through EPM systems, personal on-site monitoring by the supervisor is replaced and can be done virtually at any time from any place. Aiello and Douthitt (2001) [55] developed a social facilitation integrative model. The model integrates multiple aspects and different perspectives and a variety of outcomes, such as presence factors, individual factors, performance factors, task factors, and situational factors. This integrative work can be complemented by another aspect, namely the factors that influence a manager to implement EPM in the company. It is argued on the basis of Social Facilitation Theory that managers who believe that their employees perform better through monitoring measures are also more inclined to implement such systems in the company [56].

One study in the EPM literature confirmed that task performance based on complex tasks was significantly worse under monitoring conditions, if compared to task performance based on simple tasks. By working in a group, this negative effect could be reduced by 40%. Moreover, people who had an external locus of control (the belief that instructions come from other people, for example) felt more anxiety than people with an internal locus of control during a monitoring setting [57]. However, another study based on social facilitation theory found no significant differences in performance when comparing three groups (alone, physical-monitoring, computer-monitoring) [58]. A social facilitation effect was demonstrated in another study [14].

2.11 Social Information Processing Theory

Social information processing theory assumes that an employee's immediate social environment has an influence on her attitude and behavior. It is thus based on the assumption that individuals adapt their own beliefs, attitude, or behavior to social conditions. According to Salancik and Pfeffer (1978) [59], attitude and behavior are influenced by the social environment by means of two mechanisms. (1) Individual attitude or behavior is influenced on the basis of information regarding on what is socially accepted. (2) Individuals' attention is given to that information which is necessary to understand relationships between behavior and its outcome [59].

We found the following application of this theory in the EPM literature: If a supervisor monitors one task, but not another one, then the employee assumes that the monitored task will be rewarded, whereas the unmonitored task will be not [60, 61]. One EPM study found that electronic monitoring serves as a social signal for the importance of performance. This applies to quantitative (e.g., task speed) and qualitative performance (e.g., accuracy). However, when qualitative performance is monitored, it is seen as more important [60].

2.12 Theory of Planned Behavior

The theory of planned behavior was coined by Ajzen (1986) [62] and is fundamentally based on the relationship between belief and the resulting behavior. Behavioral intentions are predicted based on three major constructs: (1) attitude toward the behavior (determined by a person's accessible beliefs about that behavior; belief refers to the subjective probability that the behavior will lead to a certain outcome), (2) subjective norms (perspective on a particular behavior that is influenced by the judgment of others), and (3) perceived behavioral control (individually perceived difficulty or ease of executing a specific behavior) [63].

The theory of planned behavior has been applied in the EPM literature in a conceptual paper [37]: The negative or positive feelings an employee has toward monitoring technology form her attitude. Moreover, subjective norms address the influence employees have on each other. If a new monitoring policy is perceived positively by most employees, then there is a high probability that the rest of the employees will also have a more positive attitude. Moreover, it was found that if employees have the opportunity to participate in the design of EPM systems, they are more likely to use accept monitoring technology. It is also reported that the higher the perceived control, the higher the likelihood that employees will be compliant with monitoring technology [37]. One study confirms this assumption that employees with higher levels of personal control have better attitudes toward monitoring technology [64].

2.13 Summary of Theories

In this chapter, the aforementioned theories and the corresponding EPM literature are summarized and categorized. Bariff and Ginzberg (1982) [65] developed a classification scheme that defines four levels of analysis in relation to behavioral IS research: individual, group, organization, and society. These levels are applied in several other seminal papers, such as Lai and Mahaparta (1997) [66] and Vessey et al. (2002) [67], to classify IS literature. Hence, we used this scheme to categorize the EPM theories regarding level of analysis. Table 1 summarizes our classification.

Table 1. Theory classification

Theory	EPM literature	IND	GRO	ORG	SOC
Agency theory	[25]		X		
Balance theory	[27]	X			
Communic. privacy management theory	[32, 33]		X		
Deterrence theory	[37]	X			
Ethical theory	[41]				X
Feedback intervention theory	[43, 44]	X			
Psychological reactance theory	[47, 50]	X			
Reinforcement theory	[15]	X			
Self-regulation theory	[37]	X			
Social facilitation theory	[14, 55–58, 70]		X		
Social information processing theory	[60]		X		
Theory of planned behavior	[37]	X			

Table 1 shows that we identified a total of 12 theories. Moreover, our study reveals that the analytical focus of a majority of theories is the individual level (7 out of 12; IND), followed by the group level (4 out of 12, GRO) and society (1 out of 12, SOC). The organizational level (ORG) is not the focus of any of the 12 theories. Note that we classified theories according to their basic description in the literature (summarized here in the first paragraph of each of the ten theory sub-sections) and we coded the major level of analysis in those cases in which a theory may refer to more than one level of analysis. As an example, ethical theory may refer to organization and society; however, because the main focus is society, we coded into this category.

Note that both the first and second author of this paper independently assigned each theory to one of the four levels of analysis. To calculate the inter-rater reliability we used Cohen's Kappa coefficient [68]. This value indicates the degree of consistency of coding between two persons, while the possibility of random match is already taken into account. According to Landis and Koch [69] values for the Cohen's Kappa coefficients are "substantial" between 0.61 and 0.80 and values above are "almost perfect". The value of Cohen's Kappa coefficient for the 12 classified theories was 0.89 [agreement: 11/12; possibility of random match: 1/4; see Cohen 1960, p. 40]. It follows that the result of the coding is highly reliable. The one case of disagreement was resolved through discussion.

Another finding of our analysis is that most theories are psychological theories. Thus, other potentially relevant reference disciplines such as organization theory or sociology have been applied less frequently in the EPM literature.

Because EPM is a rapidly developing phenomenon of global relevance, it is reasonable to conclude that this phenomenon concerns all four levels of analysis. It follows, then, that future EPM studies should: (1) expand its conceptual focus from the individual level to the group, organizational, and societal levels, (2) consider more intensively other reference disciplines in addition to the dominating discipline of psychology, such as organization theory or sociology, (3) integrate different analytical levels into one theoretical framework, referred to as multi-level theorizing (e.g., [71, 72]), and (4) integrate different theories in one conceptual framework.

In the following, we elaborate on point (4) and illustrate, based on integration of four different theories, how theories can be combined in the EPM context.

3 Theory Integration and Study Proposal

In order to show how theories can be integrated to advance EPM research, we first had to identify existing research gaps in the EPM field. To this end, we screened three recent research agenda and review papers [13, 19, 73] and identified the major research gaps which can be addressed using a combination of four of the presented theories. In order to be able to address the research gaps accordingly, theories were selected, some of which have already been extensively applied in the EPM literature (Communication Privacy Management Theory, Social Facilitation Theory), and some of which have not yet been sufficiently considered in this area, but which form a very promising basis (Expectancy Theory, Psychological Reactance Theory): attitude towards EPM and work behavior [13, 19], performance and stress (employee health) effects of EPM use in organizations [13], privacy issues in the context of EPM [13, 19], and connection of employee behavior and EPM outcomes (consequences) [19]. A more detailed explanation of the research gaps identified is provided within the theories.

The *psychology reactance theory* can be used to study the attitude towards EPM, perceived level of EPM, and work behavior of employees. Attitude towards EPM deals with positive, neutral, or negative inclination towards EPM and comes along with consequences, predominantly (1) organizational commitment, (2) organizational citizenship behavior (OCB), and (3) counterproductive work behavior (CWB). Work behavior deals with the basic processes of how a particular task, or a set of tasks, is accomplished by the employee [13].[1] Next, a bridge to outcomes (consequences) is needed. Such outcomes are conceptualized by other theories.

Social facilitation theory, for example, can be used to conceptualize outcomes such as job performance or job stress. Both outcomes are closely interrelated [15, 76, 77] and therefore they provide an interesting basis for further research. Monitoring employee performance is at the core of EPM systems [15, 57]. Due to rapid technological progress, there are many new ways to monitor this performance (e.g. via sentiment analysis or cyberloafing) [73]. That the use of EPM in the workplace can lead to stress has already been shown by several studies [14, 15, 78]. However, with the use of new technologies for monitoring purposes, there is again a risk of increased stress levels [73]. Moreover, based on *communication privacy management theory*, the foundation for research on privacy invasion can be laid. This is particularly relevant for future research due to the rapid advancement of technology and the increased use of EPM software (also as a consequence of increasing home office rates and employers' elevated interest to monitor their employees at home via software programs; for details, please see [3, 5]).

To build the bridge between behavior and outcomes, the use of *expectancy theory* is suggested. Here, the impact of expectation fulfillment on employee behavior and on resulting outcomes is examined. The basis of the theory is based on the assumption that behavior is motivated by expected outcomes. The theory is based on three pillars: (1) Valence – the individual value an employee places on an outcome. For example, more effort should lead to reaching certain performance goals. (2) Expectancy – each employee has different expectations and capabilities. It is based on the belief that a certain amount of effort will lead to a result and is usually based on past experience. (3) The employee assumes that if the performance is appropriate, then she will also achieve the desired outcome (e.g., promotion, higher salary). The combination of these three pillars leads to a certain motivation level in the employee, which promotes positive outcomes and prevents negative ones [79, 80].

To sum up, (1) psychological reactance theory can be used to conceptualize the study of employee behavior and attitude, (2) social facilitation theory can be used to conceptualize outcomes such as performance and stress, and (3) communication privacy management theory can be used for the conceptualization of phenomena related to privacy invasion. Finally, the bridge between behavior and outcomes can be explored by means of (4) expectancy theory. Figure 2 graphically summarizes the integrated use of these four theories in a conceptual framework.

To empirically test the presented research gaps, we propose an online survey. The use of survey as method is critical, as our framework in Fig. 2 includes several attitudinal and perceptual measures, which, by definition, need a self-report measure to be captured accurately. In the suggested pilot study, we also recommend measuring work behavior and performance as self-report to keep the complexity of data collection at a reasonable level. However, we suggest that future research captures these two constructs more directly. For example, in HCI user behavior, or work behavior, is frequently measured

[1] Organizational commitment can be defined as the behavior in which employees identify with the company and which positively exceeds the expected behavior [74]. OCB refers to positive behavior that is outside the employee's organizational duties, whereas CWB aims to harm the organization and its stakeholders [75].

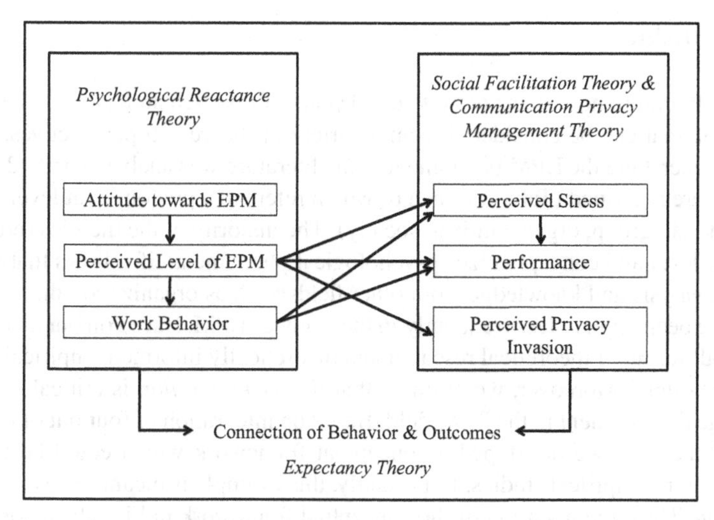

Fig. 2. Example of theory integration in the study of EPM

based on clickstream data or computer-based recording of human behavior in office settings. Moreover, performance can be measured objectively (e.g., based on number of executed tasks in a specific period).

To help potential participants understand what the study is about, we recommend an introductory text that explains EPM and the most common technologies behind it. This gives all participants a consistent picture and a common starting point. For a list of widely used technologies within an organization, we recommend the work of Ball (2021) for the European Commission [73] as well as a list of common ICT that is used at the workspace [81] (e.g. e-mail tools, word processing software).

According to Fig. 2, a total of six constructs must be measured. Next, we list measurement recommendations for these constructs based on citation of sources. (1) Attitude towards EPM: [82]. (2) Perceived level of EPM: [83] (e-mail monitoring is used here; we recommend extending the term to ICT here). (3) Work behavior: Since this term is very broad, the focus could be on measures that employees take as a counter-reaction to practiced monitoring measures. For example, the question of whether the presence status of collaboration tools is kept online during working hours or whether the company laptop is also used for private purposes. (4) Stress: [84, 85]. (5) Performance: [86]. (6) Perceived privacy invasion: [87].

Sample size is a critical element of any research design, because it affects statistical significance in hypotheses testing [88]. To be able to calculate a reasonable sample size for the proposed study, we recommend insights based on Bartlett et al. (2001) [89], who draw upon Cochran (1977) [90].

These recommendations should provide a good basis for empirically testing the research gaps identified.

4 Conclusion

The EPM literature now spans more than 30 years of research and addresses numerous phenomena that can be considered from a variety of theoretical perspectives. In order to better understand the EPM phenomenon, the literature was analyzed and 12 different theories were identified (Table 1), each of which refers to a predominant level of analysis (individual, group, organizational, society). The majority of the theories refers to the individual level and conceptualizes psychological phenomena. If follows that the other levels of analysis and knowledge from other fields such as organization theory or sociology have been applied less frequently in the study of EPM. Based on our analyses, we make a call for more theoretical research and theoretically informed empirical research in the EPM field. Moreover, we outlined that *theory integration* is critical to advance knowledge development in the EPM field. Based on integration of four out of the twelve identified theories, we developed a conceptual framework which could be used as a basis for future empirical studies. Importantly, this example is meant to be *one possible illustration.* Thus, the intention of the conceptual framework in Fig. 2 and corresponding descriptions is not so much to depict a fully testable theoretical model. Rather, the idea is to illustrate that it is possible to integrate several of the twelve outlined theories, along with theories that will potentially become relevant in the future. It is hoped that the present paper motivates more theoretical research, as well as more theoretically-informed empirical research in the EPM field. As a starting point, we proposed a survey design.

References

1. Gartner: Global IT spending to reach $3.8 trillion in 2019. https://www.gartner.com/en/newsroom/pressreleases/2019-01-28-gartner-says-global-it-spending-to-reach-3-8-trillion. Accessed 30 Oct 2021
2. Accenture: More responsible use of workforce data required to strengthen employee trust and unlock growth. https://newsroom.accenture.com/news/more-responsible-use-of-workforce-data-required-to-strengthen-employee-trust-and-unlock-growth-according-to-accenture-report.htm. Accessed 30 Oct 2021
3. BBC: How covid-19 led to a nationwide work-from-home experiment - BBC Worklife. https://www.bbc.com/worklife/article/20200309-coronavirus-covid-19-advice-chinas-work-at-home-experiment. Accessed 30 Oct 2021
4. Businessinsider: Companies are using webcams to monitor employees working from home. https://www.businessinsider.com/work-from-home-sneek-webcam-picture-5-minutes-monitor-video-2020-3?r=DE&IR=T. Accessed 30 Oct 2021
5. MIT Technology Review: This startup is using AI to give workers a "productivity score". https://www.technologyreview.com/2020/06/04/1002671/startup-ai-workers-productivity-score-bias-machine-learning-business-covid/. Accessed 30 Oct 2021
6. Melville, N., Kraemer, K., Gurbaxani, V.: Review: information technology and organizational performance: an integrative model of IT business value. MIS Q. **28**, 283–322 (2004)
7. Yunis, M., El-Kassar, A.N., Tarhini, A.: Impact of ICT-based innovations on organizational performance: the role of corporate entrepreneurship. J. Enterp. Inf. Manag. **30**, 122–141 (2017)
8. Ball, K.: Workplace surveillance: an overview. Labor Hist. **51**, 87–106 (2010)

9. Hamilton, R.H., Sodeman, W.A.: The questions we ask: opportunities and challenges for using big data analytics to strategically manage human capital resources. Bus. Horiz. **63**, 85–95 (2019)
10. Sprigg, C.A., Jackson, P.R.: Call centers as lean service environments: job-related strain and the mediating role of work design. J. Occup. Health Psychol. **11**, 197–212 (2006)
11. Ajunwa, I., Crawford, K., Schultz, J.: Limitless worker surveillance. Calif. Law Rev. **105**, 735–776 (2017)
12. Carneiro, D., Pimenta, A., Gonçalves, S., Neves, J., Novais, P.: Monitoring and improving performance in human-computer interaction: monitoring and improving performance in Human-Computer Interaction. Concurr. Comput. Pract. Exp. **28**, 1291–1309 (2016)
13. Kalischko, T., Riedl, R.: Electronic performance monitoring in the digital workplace: conceptualization, review of effects and moderators, and future research opportunities. Front. Psychol. **12**, 633031 (2021)
14. Aiello, J.R., Kolb, K.J.: Electronic performance monitoring and social context: impact on productivity and stress. J. Appl. Psychol. **80**, 339–353 (1995)
15. Bartels, L.K., Nordstrom, C.R.: Examining big brother's purpose for using electronic performance monitoring. Perform. Improv. Q. **25**, 65–77 (2012)
16. Grewal, D., Roggeveen, A.L., Runyan, R.C., Nordfält, J., Vazquez Lira, M.E.: Retailing in today's world: Multiple channels and other strategic decisions affecting firm performance. J. Retail. Consum. Serv. **34**, 261–263 (2017)
17. U.S. Congress, O. of T.A.: The electronic supervisor: new technology, new tensions (1987)
18. New York Times: Microchip implants for employees? One company says yes. https://www.nytimes.com/2017/07/25/technology/microchips-wisconsin-company-employees.html. Accessed 30 Oct 2021
19. Ravid, D.M., Tomczak, D.L., White, J.C., Behrend, T.S.: EPM 20/20: a review, framework, and research agenda for electronic performance monitoring. J. Manag. **46**, 100–126 (2020)
20. Montealegre, R., Cascio, W.F.: Technology-driven changes in work and employment. Commun. ACM **60**, 60–67 (2017)
21. Webster, J., Watson, R.T.: Analyzing the past to prepare for the future: writing a literature review. MIS Q. **26**, xiii–xxiii (2002)
22. Vom Brocke, J., Simons, A., Niehaves, B., Riemer, K.: Reconstructing the giant: on the importance of rigour in documenting the literature search process. In: AIS (ed.) Proceedings of ECIS 2009, p. Paper 161 (2009)
23. Backhaus, N.: Context sensitive technologies and electronic employee monitoring: a meta-analytic review. In: 2019 IEEE/SICE International Symposium on System Integration, pp. 548–553 (2019)
24. Jensen, M.C., Meckling, W.H.: Theory of the firm: managerial behavior, agency costs and ownership structure. J. Financ. Econ. **3**, 305–360 (1976)
25. Alge, B.J., Ballinger, G.A., Green, S.G.: Remote control: predictors of electronic monitoring intensity and secrecy. Pers. Psychol. **57**, 377–410 (2004)
26. Smith, M.J., Sainfort, P.C.: A balance theory of job design for stress reduction. Int. J. Ind. Ergon. **4**, 67–79 (1989)
27. Lund, J.: Electronic performance monitoring: a review of research issues. Appl. Ergon. **23**, 54–58 (1992)
28. Petronio, S.: Communication boundary management: a theoretical model of managing disclosure of private information between marital couples. Commun. Theory **1**, 311–335 (1991)
29. Petronio, S.: Communication Privacy Management Theory. Wiley, New York (2015)
30. Petronio, S., Durham, W.: Understanding and applying communication privacy management theory, pp. 309–322 (2008)

31. Margulis, S.T.: Three theories of privacy: an overview. In: Trepte, S., Reinecke, L. (eds.) Privacy Online. Springer, Heidelberg (2011). https://doi.org/10.1007/978-3-642-21521-6_2
32. Siegel, R., König, C.J., Porsch, L.: Does electronic monitoring pay off? J. Pers. Psychol., 1–11 (2021)
33. Chang, S.E., Liu, A.Y., Lin, S.: Exploring privacy and trust for employee monitoring. Ind. Manag. Data Syst. **115**, 88–106 (2014)
34. Becker, G.S.: Crime and punishment: an economic approach. The economic dimensions of crime, pp. 13–68 (1968)
35. Pogarsky, G., Piquero, A.R., Paternoster, R.: Modeling change in perceptions about sanction threats: the neglected linkage in deterrence theory. J. Quant. Criminol. **20**, 343–369 (2004)
36. Noll, J., Bojang, O., Rietjens, S.: Deterrence by punishment or denial? The eFP case. In: Osinga, F., Sweijs, T. (eds.) NL ARMS Netherlands Annual Review of Military Studies 2020. NA, pp. 109–128. T.M.C. Asser Press, The Hague (2021). https://doi.org/10.1007/978-94-6265-419-8_7
37. Chen, J.V., Pfleuger, P.: Employees' behaviour towards surveillance technology implementation as an information assurance measure in the workplace. Int. J. Manag. Enterp. Dev. **5**, 497–511 (2008)
38. Beauchamp, T.L., Bowie, N.E.: Ethical Theory and Business (1993)
39. Ball, K.S.: Situating workplace surveillance: ethics and computer based performance monitoring. Ethics Inf. Technol. **3**, 211–223 (2001)
40. Alder, G.S., Schminke, M., Noel, T.W., Kuenzi, M.: Employee reactions to internet monitoring: the moderating role of ethical orientation. J. Bus. Ethics **80**, 481–498 (2008)
41. Hawk, S.R.: The effects of computerized performance monitoring: an ethical perspective. J. Bus. Ethics **13**, 949–957 (1994)
42. Kluger, A.N., DeNisi, A.: The effects of feedback interventions on performance: a historical review, a meta-analysis, and a preliminary feedback intervention theory. Psychol. Bull. **119**, 254 (1996)
43. Alder, G.S.: Examining the relationship between feedback and performance in a monitored environment: a clarification and extension of feedback intervention theory. J. High Technol. Manag. Res. **17**, 157–174 (2007)
44. Alder, G.S., Ambrose, M.L.: Towards understanding fairness judgments associated with computer performance monitoring: an integration of the feedback, justice, and monitoring research. Hum. Resour. Manag. Rev. **15**, 43–67 (2005)
45. Brehm, J.W.: A theory of psychological reactance (1966)
46. Reynolds-Tylus, T.: Psychological reactance and persuasive health communication: a review of the literature. Front. Commun. **4** (2019)
47. Yost, A.B., Behrend, T.S., Howardson, G., Badger Darrow, J., Jensen, J.M.: Reactance to electronic surveillance: a test of antecedents and outcomes. J. Bus. Psychol. **34**, 71–86 (2019)
48. Robinson, S.L., Bennett, R.J.: Workplace deviance: its definition, its manifestations, and its causes (1997)
49. Organ, D.W., Podsakoff, P.M., MacKenzie, S.B.: Organizational Citizenship Behavior: Its Nature, Antecedents, and Consequences. Sage Publications, New York (2005)
50. Martin, A.J., Wellen, J.M., Grimmer, M.R.: An eye on your work: how empowerment affects the relationship between electronic surveillance and counterproductive work behaviours. Int. J. Hum. Resour. Manag. **27**, 2635–2651 (2016)
51. Skinner, B.F.: Science and Human Behavior. Simon and Schuster, New York (1953)
52. Kanfer, F.H.: Self-regulation: research, issues, and speculation. In: Michael (ed.) Behavior Modification in Clinical Psychology, pp. 178–220 (1970)
53. Karoly, P.: Mechanisms of self-regulation: a systems view. Annu. Rev. Psychol. **44**, 23–52 (1993)

54. Zajonc, R.B.: Social facilitation. Science **149**, 269–274 (1965)
55. Aiello, J.R., Douthitt, E.A.: Social facilitation from triplett to electronic performance monitoring. Group Dyn. **5**, 163–180 (2001)
56. Victor Chen, J., Ross, W.H.: The managerial decision to implement electronic surveillance at work: A research framework. Int. J. Organ. Anal. **13**, 244–268 (2005)
57. Aiello, J.R., Svec, C.M.: Computer monitoring of work performance: extending the social facilitation framework to electronic presence. J. Appl. Soc. Psychol. **23**, 537–548 (1993)
58. Griffith, T.L.: Monitoring and performance: a comparison of computer and supervisor monitoring. J. Appl. Soc. Psychol. **23**, 549–572 (1993)
59. Salancik, G.R., Pfeffer, J.: A social information processing approach to job attitudes and task design. Adm. Sci. Q. **23**, 224–253 (1978)
60. Stanton, J.M., Julian, A.L.: The impact of electronic monitoring on quality and quantity of performance. Comput. Hum. Behav. **18**, 85–101 (2002)
61. Larson, J.R., Callahan, C.: Performance monitoring: How it affects work productivity. J. Appl. Psychol. **75**, 530–538 (1990)
62. Ajzen, I.: From intentions to actions: a theory of planned behavior. In: Kuhl, J., Beckmann, J. (eds.) Action Control. SSSP Springer Series in Social Psychology, pp. 11–39. Springer, Heidelberg (1985). https://doi.org/10.1007/978-3-642-69746-3_2
63. Ajzen, I.: The theory of planned behavior. Organ. Behav. Hum. Decis. Process. **50**, 179–211 (1991)
64. Spitzmüller, C., Stanton, J.M.: Examining employee compliance with organizational surveillance and monitoring. J. Occup. Organ. Psychol. **79**, 245–272 (2006)
65. Bariff, M.L., Ginzberg, M.J.: MIS and the behavioral sciences. ACM SIGMIS Database **14**, 19 (1982)
66. Lai, V.S., Mahapatra, R.K.: Exploring the research in information technology implementation. Inf. Manag. **32**, 187–201 (1997)
67. Vessey, I., Ramesh, V., Glass, R.L.: Research in information systems: an empirical study of diversity in the discipline and its journals. J. Manag. Inf. Syst. **19**, 129–174 (2002)
68. Cohen, J.: A coefficient of agreement for nominal scales. Educ. Psychol. Meas. **20**, 37–46 (1960)
69. Landis, J.R., Koch, G.G.: The measurement of observer agreement for categorical data. Biometrics **33**, 159 (1977)
70. Stanton, J.M., Sarkar-Barney, S.T.M.M.: A detailed analysis of task performance with and without computer monitoring. Int. J. Hum. Comput. Interact. **16**, 345–366 (2003)
71. Zhang, M., Gable, G.G.: A systematic framework for multilevel theorizing in information systems research. Inf. Syst. Res. **28**, 203–224 (2017)
72. Zhang, M., Gable, G.G., Tate, M.: Overview of the multilevel research perspective: implications for theory building and empirical research. Commun. Assoc. Inf. Syst. **45**, 1 (2019)
73. Ball, K.: Electronic monitoring and surveillance in the workplace (2021)
74. Mowday, R.T., Steers, R.M., Porter, L.W.: The measurement of organizational commitment. J. Vocat. Behav. **14**, 224–247 (1979)
75. Jensen, J.M., Raver, J.L.: When self-management and surveillance collide: consequences for employees' organizational citizenship and counterproductive work behaviors. Group Org. Manag. **37**, 308–346 (2012)
76. AbuAlRub, R.F.: Job stress, job performance, and social support among hospital nurses. J. Nurs. Scholarsh. **36**, 73–78 (2004)
77. Bernardi, R.: The relationships among locus of control, perceptions of stress, and performance (1997)
78. Kolb, K.J., Aiello, J.R.: The effects of electronic performance monitoring on stress: locus of control as a moderator variable. Comput. Hum. Behav. **12**, 407–423 (1996)

79. Vroom, V.H.: Work and Motivation, Wiley, New York (1994)
80. Lawler, E.E., Suttle, J.L.: Expectancy theory and job behavior. Organ. Behav. Hum. Perform. **9**, 482–503 (1973)
81. Ayyagari, R., Grover, V., Purvis, R.: Technostress: technological antecedents and implications. MIS Q. **35**, 831–858 (2011)
82. Furnham, A., Swami, V.: An investigation of attitudes toward surveillance at work and its correlates. Psychology **06**, 1668–1675 (2015)
83. Snyder, J.L.: E-mail privacy in the workplace: a boundary regulation perspective. J. Bus. Commun. **47**, 266–294 (2010)
84. Siegrist, J., Wege, N., Pühlhofer, F., Wahrendorf, M.: A short generic measure of work stress in the era of globalization: effort-reward imbalance. Int. Arch. Occup. Environ. Health **82**, 1005–1013 (2009)
85. Motowidlo, S.J., Manning, M.R., Packard, J.S.: Occupational stress: its causes and consequences for job performance. J. Appl. Psychol. **71**, 618–629 (1986)
86. Koopmans, L., et al.: Development of an individual work performance questionnaire. Int. J. Prod. Perform. Manag. Emerald **62**, 6–28 (2013)
87. Fischer, T., Reuter, M., Riedl, R.: The digital stressors scale: development and validation of a new survey instrument to measure digital stress perceptions in the workplace context. Front. Psychol. **12**, 607598 (2021)
88. Peers, I.: Statistical Analysis for Education and Psychology Researchers: Tools for Researchers in Education and Psychology. Routledge (2006)
89. Bartlett II, J.E., Kotrlik, J.W., Higgins, C.C.: Determining appropriate sample size in survey research. Inf. Technol. Learn. Perform. J. **19**, 43–50 (2001)
90. Cochran, W.G.: Sampling Techniques, 3rd edn. Wiley, New York (1977)

Strategies for Working Remotely: Responding to Pandemic-Driven Change with Cross-Organizational Community Dialog

Elaine M. Raybourn(✉)

Sandia National Laboratories, Albuquerque, NM 87185, USA
emraybo@sandia.gov

Abstract. In response to the COVID-19 pandemic, the Exascale Computing Project's (ECP) Interoperable Design of Extreme-scale Application Software (IDEAS) productivity team launched the panel series *Strategies for Working Remotely* to facilitate informal, cross-organizational dialog in the absence of face-to-face meetings. In a time of pandemic, organizations increasingly need to reach across perceived boundaries to learn from each other, so that we can move beyond stand-alone silos to more connected multidisciplinary and multi-organizational configurations. The present paper argues that the unplanned transition to remote work, overuse of electronic communication, and need to unlearn habits associated with an overreliance on face-to-face, created unique opportunities to learn from the situation and accelerate cross-institutional cooperation and collaboration through online community dialog facilitated by informal panel discussions. Recommendations for facilitating online panel discussions to foster cross-organizational dialog are provided by applying the Simulation Experience Design Method.

Keywords: Remote work · COVID-19 · Community dialog · Computer mediated communication · Computer supported cooperative work

1 Introduction

The Department of Energy (DOE) Office of Science is the lead federal agency in the United States for fundamental research in energy. It has a core mission to develop new high performance computing (HPC) capabilities toward a capable *exascale* computing ecosystem. Exascale computing is a new class of high-performance computing systems capable of 1018 floating point operations per second [1] and whose power is measured in exaflops, or computing speed equal to one billion billion calculations per second, and one thousand times more powerful than today's petaflop machines. By 2023, it is posited that exascale computing will enable solutions to complex research problems, such as simulating the human brain, predicting space weather, cracking encryption codes, advancing medical research, analyzing millions of health records, and modeling the Earth's climate and the global economy [2, 3].

This mission has drawn over 1000 scientists to work together as members of the DOE and National Nuclear Administration's (NNSA) Exascale Computing Project (ECP), a

F. Fui-Hoon Nah and K. Siau (Eds.): HCII 2022, LNCS 13327, pp. 407–416, 2022.
https://doi.org/10.1007/978-3-031-05544-7_30

virtual organization with a portfolio of 24 specific, targeted applications, along with 70 software products being prepared for exascale as part of the nation's first exascale software stack (see exascaleproject.org). ECP, composed of over 82 high performance computing scientific software teams, has been characterized by the present author and others as a "team of teams," [4] or a network of coordinated, interdependent teams who share trust, transparent communication, and of a common purpose [5]. They interoperate through collaboration around an aligning narrative [6]. A more in-depth discussion on the application of team of teams to ECP is provided in [4].

ECP had mostly been communicating via face-to-face, augmented by electronic communication such as conference calls, documentation, and emails. However, shortly after a face-to-face annual meeting in February 2020 many ECP scientists, if not all, joined the global community in working remotely due to the COVID-19 pandemic. As a large virtual organization, the collective move to remote work did not adversely impact ECP's use of virtual collaboration tools, however, like many organizations ECP also relied heavily on travel to face-to-face meetings as a primary means to strengthen trusted relationships and build community. Therefore, the pandemic caused unforeseen disruptions to existing processes and practices, creating an opportunity to reimagine collaboration and community.

The present paper argues that the unplanned transition to remote work, overuse of electronic communication, and need to unlearn habits associated with an overreliance on face-to-face, created unique opportunities for ECP to learn from the situation, adapt, and accelerate cross-institutional cooperation and collaboration. One such opportunity was the creation of an informal panel series, *Strategies for Working Remotely*, which engenders online community dialog [7]. The panel series was launched by ECP's Interoperable Design of Extreme-scale Application Software (IDEAS) productivity team in response to the COVID-19 pandemic and the influx of previously co-located teams to working remotely. Based on lessons learned from having organized and moderated the *Strategies for Working Remotely* panel series, recommendations for facilitating online panel discussions to foster cross-organizational and community dialog are provided.

2 Why We Need Strategies for Working Remotely

Many scientific software teams have functioned as dispersed teams, or teams of teams [4] and are familiar with a variety of tools commonly used by software teams to stay connected, such as email, messaging, Slack, Gitter, GitHub, Skype, MS Teams, Zoom, etc. Additionally, if someone has traveled extensively, has colleagues at geographically distributed institutions, or has worked offsite, then they experienced virtual, or remote work at one time or another. While for some, working remotely had been an everyday occurrence, the global COVID-19 pandemic [8] thrust many of us into a new normal that involves extended remote work and/or reduced exposure to others while in co-located spaces and practicing social distancing. Several factors operating simultaneously have contributed to perceptions that working remotely now (during COVID-19) seems qualitatively different from working remotely before the pandemic. They include: unplanned and imposed transitions to working from home, degraded and overused electronic communication channels, and the need to learn new skills as well as unlearn old habits that over time may have shaped our views of what it means to be productive. Since it is likely that a combination of working from home and social distancing while at worksites is here to stay for a while in some form or another, there is an opportunity to share strategies for working remotely.

2.1 Unplanned and Imposed Transition

The transition from the office to working remotely at home for many of us was unplanned and imposed. We did not ask for it, we did not plan for it, and we certainly could not control it. Yes, we face challenges including caretaking and parenting while working remotely, transitioning our teams and operations to a fully virtual set-up, and virtually onboarding new team members. However, while the imposition of these unplanned changes may pose great stress for many of us, we should also nevertheless consider that there are many essential staff members who are required to report to work in co-located spaces while practicing physical distancing and in some cases putting themselves in harm's way for a greater good. The challenges faced by essential workers who are our colleagues and the need to move much of the workforce to working remotely, impact future policies and ways our organizations will respond in the near- and long-term. As our organizations shift back to what they consider "normal"—which was most likely already a hybrid arrangement of co-located and satellite team members who are remotely working, there will hopefully be increased awareness of the struggles faced by all. The trend of remote working will likely continue for some time especially since now we, as a community of computational scientists, have a more collective experience with working remotely. In our community, remote work arrangements are nothing new. What is new, however, is that we have *all been working remotely together*. This alone has reshaped our workplace to allow for increased awareness and, in many cases, to illuminate what it's like to experience work differently.

2.2 Degraded and Overused Communication Channel

Before the pandemic, we had a healthy combination of face-to-face and technology-mediated communication each day, both synchronous and asynchronous. We decided how much we wanted, and when. We had agency, or locus of control. The face-to-face communication channel is what communication scholars call stimulus rich—that is, we use our senses (auditory, visual, tactile, olfactory, and gustatory) to engage in verbal and nonverbal communication and are mostly unaware of it. Technology-mediated communication, on the other hand (except for some holographic and cross reality systems which may offer greater possibility for immersion), is still rather stimulus poor and therefore unappealing for many computational scientists.

Why? Perhaps because software development is deeply social. Research conducted at Microsoft found that developers share and maintain mental models largely through face-to-face communication and the code itself [9]. Many participants in this study reported avoiding email or formal design documents (including bug reports, specs, etc.) to generate or transfer knowledge among teams. Developers in this study also reported rarely using IM for code-related tasks, and instead used IM to connect socially with colleagues or family. If they needed to work out a problem, they were most likely to interrupt developers who were most knowledgeable about the issue. These one-on-one developer conversations often happen at the physical whiteboard [10].

Within the ECP scientific community the author observed that online collaboration was largely limited to email, messaging, Slack, Gitter, GitHub, Skype, MS Teams, Zoom, etc. Many scientists were explored online collaborative tools to support brainstorming or mind mapping—which have been a topic of decades of computer science and computer supported cooperative work (CSCW) research—with some of the earliest exploration in computational tools for sketch recognition and management dating back from the early 1960s [11]. Teams of designers and developers over several decades since have studied informal sketching practices at the whiteboard to inform the development of digital pens [12] collaborative tools [10, 13], models [14], large displays and groupware [15]. Decades of research notwithstanding, interactive groupware displays are still not widely used in practice [15]. The pandemic has illuminated opportunities to re-examine this research, especially in light of the overuse of electronic communication.

Since a greater number of us are working remotely, the overuse of electronic communication for extended periods of time often degraded by poor connectivity literally overwhelms the human brain as it attempts to process and share information via screens instead of through unmediated verbal and nonverbal channels. For example, "Zoom fatigue" characterized by our constant gaze into a video camera is exhausting [16]. Managing expectations when working in environments characterized by degraded and overused communication channels (mostly electronic) is the unfortunate current reality of working remotely. The good news, however, is that in the best cases productivity does not suffer [17], rather, it flourishes.

2.3 Need to Unlearn Old Habits

While most, if not all of us are open to learning new skills, we also should be open to *unlearning* old habits that no longer work for us in a new situation. Regarding working remotely and productivity, unlearning a habit in this case refers to a rethinking a familiar way of working that may no longer meet the collective expectations of productivity while we are working through a pandemic. It has been said that computational scientists may easily work from anywhere. However, this may only be true as long as certain conditions exist. Those conditions are different for each of us. The opportunity we have now with respect to working remotely is to determine for ourselves, and with our teams, what those conditions are. Even though some, if not most of us, may believe we are rather skilled at working remotely, we should be open to ways that newcomers to the remote working experience will impact perceptions of productivity, and the stability or comfort of the habits we once enjoyed. Very likely, we will find that our openness to new ways of working is rewarded with increased innovation, productivity, and satisfaction.

3 Facilitating Community Dialog with the ECP Panel Series

Many, if not all, of us crave spontaneous, synchronous group conversations – we miss humanity. A plethora of advice is available for organizing, moderating, and participating in virtual, or online conferences and webinars. For example, recent documentation was authored by members of ACM (Association of Computing Machinery) [18], IEEE Computing Society [19], and most relevant to this discussion, a checklist for producing a webinar series is provided by IDEAS-ECP [20, 21]. These resources provide information on the use of popular technical platforms and solutions for hosting conferences, webinars, and panel discussions; therefore, setting up the technical infrastructure is not discussed in the present paper.

To date, there have been twelve panel discussions (one conducted during a super computing conference), each ranging from 90–75 min [22] that have addressed advice from those who are experienced, parenting while working remotely, transitioning to virtual teams, onboarding and mentoring, how teams cope with disruption, hybrid organizational structures, creativity, innovation, and productivity. Most of the panels have focused on building individual resiliency skills to cope with working remotely during the COVID-19 pandemic. At the time of publication, over 1500 have participated in live discussions since April 2020. Survey response data indicate that many of the participants have attended 2 or more panel discussions (Fig. 1).

Fig. 1. Screen capture of cross-institutional panel discussion [7].

While the preparation checklist to producing an online panel series can be similar to that of a webinar [21], there are several nuances to hosting a realtime panel discussion that merit discussion. The major distinction between producing a panel series and a webinar is that a panel discussion is a live event focusing on spontaneous dialog and conversation, which introduces ambiguity into the production of the realtime event. A webinar is more structured than a panel discussion and follows a format more consistent with that concerned with the dissemination of information. Since the webinar format features an uninterrupted lecture, there is a significant amount of effort put toward preparation of webinar materials and artefacts. In producing a live event such as a panel discussion, on the other hand, the emphasis is on the facilitation of *dialog*, not the dissemination of information. Interruptions and redirections are often desired.

3.1 Application of the Simulation Experience Design Method

The present author applied the Simulation Experience Design Method [23] to create the setting and facilitate the opportunity for community dialog throughout the *Strategies for Working Remotely* panel series. The method, used by several to design serious games [24], training systems, and transmedia learning [25], treats a communication event, such as a panel discussion, as a system of experiences.

Using the Simulation Experience Design Method the panel organizer and/or moderator engenders community dialog by envisioning the desired perceptions, distributed cognition, and behavior among the audience through the application of 4 stages, 1) Interaction (modality), 2) Narrative (dialog), 3) Place (experience), and 4) Emergent Culture (audience participation and user generated content). At the core of the communication event is intercultural communication, and co-created meaning.

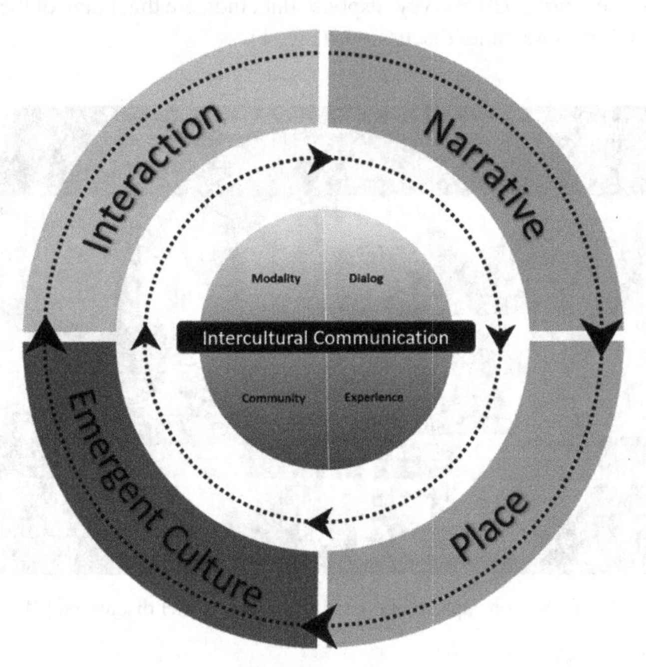

Fig. 2. Simulation experience design method & framework [23].

In the first stage, *Interaction*, a producer and moderator of a panel discussion aims to leverage the strengths of the communication channels in which interactions (utterances) will occur. For example, consider the use of a technology platform that the audience is using currently. Is it Zoom, Google Meet, MS Teams, or another? When choosing a technology think about how the audience will interact with the panellists—can they ask questions, vote for more popular questions, chat, turn video off or on, and participate with audio? How can you leverage the technology to incentivize participation? The goal at the Interaction stage is to leverage the strengths of the technology including when, where, and how the audience may access the panel, and interact with panellists and other audience members. At a minimum audio should exceed expectations, and videos of the live event should be easily accessed later.

In the second stage, *Narrative*, the moderator considers how to prepare panelists in advance with thought provoking questions and allow them to get to know other panelists prior to the event so that the dialog during the discussion is natural and authentic. Questions to consider at this stage are: What are the key take-aways of the panel for the audience? What stories should panelists be encouraged to share to meet the objectives of community dialog? While the panelists are the storytellers, the focus should be on the audience. If audience questions begin to shift, the moderator should be prepared to pivot. The moderator senses where the audience wants to go and leads the panelists in this direction.

The third stage, *Place*, encourages the moderator to account for the lowest common denominator and user experience—that is, the moderator imagines how the audience could be joining the panel discussion. A moderator of a virtual panel may assume or favor the visual medium. However, the panel should also be interesting for someone participating via audio only (e.g. phone, or only listening). Consider focusing on dialog and limit the use of materials such as PowerPoint slides. Think about the audio delivery—if the panel were a podcast would it be interesting? Coach panelists in their vocal delivery and encourage them to demonstrate vocal energy and passion throughout the discussion.

In the fourth stage, *Emergent Culture*, it is important to remember that in a live event neither the moderator nor the panelists "own" the outcome, instead the outcome is co-created by the audience, moderator, and panelists – *together*. In essence, the moderator of a panel discussion sets the dialog in motion and guides the realtime outcome by paying attention to the strengths of the panelists, rhythm of the dialog, audience interest, and energy level of all during the event.

Finally, to ensure effective intercultural communication, a moderator should be inclusive and produce content that is accessible for all. Provide captioning, or sign language interpretation when possible to augment online content. Read questions and chat messages aloud, and verbally describe any images that might appear during the discussion. If uploading videos to video sharing platforms such as YouTube, utilize automated captions. Online communication is persistent—and as such, panel discussions can serve as a relevant snapshot in time, or timeless contributions. Each have merit. Questions to consider are: Will the moderator and panelists be proud of their contribution over time? Did the moderator successfully serve the panel discussion to build community?

Producing a panel discussion is about responding to the audience – their concerns, topics of interest, and energy. The technology platform chosen should account for and make accessibility possible. As the communication is persistent, having high standards and allowing all involved to take pride in their contributions will go far in creating enduring, memorable content.

While survey feedback has been positive, we intend to further investigate the impact of an informal panel series in incentivizing cross-organizational community dialog within the context of a team of teams [4]. By the date of this manuscripts' publication, the panel series process for engendering community dialog has been replicated for conferences such as The International Conference for High Performance Computing, Networking, Storage, and Analysis (SC'21) and lessons learned from hosting the panel series has been featured in the Department of Energy Computational Science Graduate Fellowship (DOE CSGF) podcast *Science in Parallel,* hosted by the Krell Institute [26].

4 Conclusions

Unplanned and imposed remote work created a sea change that has altered the way we work now and will likely impact the way we work in the future. A panel series such as *Strategies for Working Remotely* [22] offers opportunities to scale and accelerate cross-institutional cooperation and collaboration in a community. Anticipating the pandemic's lasting impact, we have an opportunity to shape and influence the future of scientific work by collectively developing strategies for working remotely that build community. Our own habits and assumptions about productivity, togetherness, isolation, inclusivity, and communication have been challenged each day. The shift to remote work en masse during a pandemic and social change has given us an opportunity to be more inclusive and compassionate, opening doors for technological innovation to support how we work and communicate as teams of scientists, especially in those situations where we lack the most while absent from each other.

Acknowledgements. Sandia National Laboratories is a multimission laboratory managed and operated by National Technology and Engineering Solutions of Sandia, LLC, a wholly owned subsidiary of Honeywell International, Inc., for the U.S. Department of Energy's National Nuclear Security Administration under contract DE-NA-0003525. Image in Fig. 2 copyright by NTESS LLC and reproduced with permission. This work was supported by the U.S. Department of Energy Office of Science, Office of Advanced Scientific Computing Research (ASCR), and by the Exascale Computing Project (17-SC-20-SC), a collaborative effort of the U.S. Department of Energy Office of Science and the National Nuclear Security Administration.

References

1. Reed, D.A., Dongarra, J.: Exascale computing and big data. Commun. ACM **58**(7), 56–68 (2015). https://doi.org/10.1145/2699414
2. Simonite, T.: The US again has the world's most powerful supercomputer. Wired, 8 June 2018. https://www.wired.com/story/the-us-again-has-worlds-most-powerful-supercomputer/. Accessed 26 Mar 2021
3. Lohr, S.: Move over, China: U.S. is again home to world's speediest supercomputer. N. Y. Times (2018). https://www.nytimes.com/2018/06/08/technology/supercomputer-china-us.html. Accessed 4 Feb 2019
4. Raybourn, E.M., Moulton, J.D., Hungerford, A.: Scaling productivity and innovation on the path to exascale with a "team of teams" approach. In: Nah, F.-H., Siau, K. (eds.) HCII 2019. LNCS, vol. 11589, pp. 408–421. Springer, Cham (2019). https://doi.org/10.1007/978-3-030-22338-0_33
5. McCrystal, S., Collins, T., Silverman, D., Fussell, C.: Team of Teams: New Rules of Engagement for a Complex World. Penguin Random House LLC, New York (2015)
6. Fussell, C., Goodyear, C.W.: One Mission: How Leaders Build a Team of Teams. Penguin Random House LLC, New York (2017)

7. Raybourn, E.M.: Why we need strategies for working remotely: the exascale computing project (ECP) panel series. In: IEEE SC20 The International Conference for High Performance Computing, Networking, Storage, and Analysis, State of the Practice Talk, 17 November 2020
8. CDC Coronavirus (COVID-19). https://www.cdc.gov/coronavirus/2019-ncov/prevent-getting-sick/prevention.html. Accessed 24 Mar 2021
9. LaToza, T.D., Venolia, G., DeLine, R.: Maintaining mental models: a study of developer work habits. In: Proceedings of the 28th International Conference on Software Engineering (ICSE 2006), pp. 492–501. ACM, New York (2006). https://doi.org/10.1145/1134285.1134355
10. Cherubini, M., Venolia, G., DeLine, R., Ko, A.J.: Let's go to the whiteboard: how and why software developers use drawings. In: Proceedings of the SIGCHI Conference on Human Factors in Computing Systems (CHI 2007), pp. 557–566. ACM, New York (2007). https://doi.org/10.1145/1240624.1240714
11. Johnson, G., Gross, M.D., Hong, J., Yi-Luen Do, E.: Computational support for sketching in design: a review. Found. Trends Hum. Comput. Interact. **2**(1), 1–93 (2009). https://doi.org/10.1561/1100000013
12. Sra, M., Lee, A., Pao, S., Jiang, G., Ishii, H.: Point and share: from paper to whiteboard. In: Adjunct Proceedings of the 25th Annual ACM Symposium on User Interface Software and Technology (UIST Adjunct Proceedings 2012), pp. 23–24. ACM, New York (2012). https://doi.org/10.1145/2380296.2380309
13. Branham, S., Golovchinsky, G., Carter, S., Biehl, J.T.: Let's go from the whiteboard: supporting transitions in work through whiteboard capture and reuse. In: Proceedings of the SIGCHI Conference on Human Factors in Computing Systems (CHI 2010), pp. 75–84. ACM, New York (2010). https://doi.org/10.1145/1753326.1753338
14. Motta, A., Mangano, N., van der Hoek, A.: Lightweight analysis of software design models at the whiteboard. In: Proceedings of the 5th International Workshop on Modeling in Software Engineering (MiSE 2013), pp. 18–23. IEEE Press (2013)
15. Huang, E.M., Mynatt, E.D., Russell, D.M., Sue, A.E.: Secrets to success and fatal flaws: the design of large-display groupware. IEEE Comput. Graph. Appl. **26**(1), 10–17 (2006)
16. Fosslien, L., Duffy, M.: How to combat Zoom fatigue HBR. https://hbr.org/2020/04/how-to-combat-zoom-fatigue. Accessed 24 Mar 2021
17. Murph, D.: Five ways to build a more productive remote team. https://about.gitlab.com/blog/2019/12/10/how-to-build-a-more-productive-remote-team/. Accessed 24 Mar 2021
18. Matthews, J., et al.: ACM Virtual Conferences: A Guide to Best Practices. https://people.clarkson.edu/~jmatthew/acm/VirtualConferences_GuideToBestPractices_CURRENT.pdf. Accessed 16 Nov 2020
19. IEEE Virtual Conference Resource Guide and FAQs. https://www.computer.org/conferences/organize-a-conference/organizer-resources/hosting-a-virtual-event/cs-virtual-event-resource-guide. Accessed 16 Nov 2020
20. Marques, O.A., Bernholdt, D.E., Raybourn, E.M., Barker, A.D., Hartman-Baker, R.J.: The HPC best practices webinar series. J. Comput. Sci. Educ. **10**(1), 108–110 (2019). https://doi.org/10.22369/issn.2153-4136/10/1/19
21. Marques, O.A.: How to Produce a Webinar Series. https://github.com/betterscientificsoftware/Webinar-Process. Accessed 28 Oct 2020
22. Strategies for Working Remotely Panel Series. IDEAS-ECP. https://www.exascaleproject.org/strategies-for-working-remotely/. Accessed 24 Mar 2021
23. Raybourn, E.M.: Applying simulation experience design methods to creating serious game-based adaptive training systems. Interact. Comput. **19**, 207–214 (2007)
24. Raybourn, E.M.: Honing emotional intelligence with game-based crucible experiences. Int. J. Game Based Learn. **1**(1), 32–44 (2011)

25. Raybourn, E.M.: A new paradigm for serious games: transmedia learning for more effective training & education. J. Comput. Sci. 5(3), 471–481 (2014)
26. Science in Parallel: A Computational Science Podcast. https://www.krellinst.org/csgf/outreach/science-in-parallel. Accessed 11 Feb 2021

Designing a Worker Companion - Design Implications from On-Site and Remote Participatory Design in the Context of Industry 4.0

Jorge Ribeiro, Cristina Santos, Elsa Oliveira, and Ricardo Melo(✉)

Fraunhofer Portugal AICOS, Porto, Portugal
{jorge.ribeiro,cristina.santos,elsa.oliveira,ricardo.melo}@fraunhofer.pt
http://www.aicos.fraunhofer.pt

Abstract. The Industry 4.0 paradigm requires not only equipping the shop floor and the workforce with new digital tools, but also ensuring that digital technologies are well accepted and adopted by workers. This is achieved through the active involvement of those who will be affected by these new digital technologies. However, the involvement of end-users and workers in the design process is often confined to test fully developed solutions. Rarely are the workers fully involved in the design process, from preliminary research to the co-design of the proposed technologies. As such, and in order to involve workers on the design of a digital companion for shop floor operators, we applied a Participatory Design approach, with mixed-methods, consisting of fieldwork observations, interviews, and exploratory workshops with workers and other relevant stakeholders in order to understand their needs and desires in the context of Industry 4.0. These activities were conducted remotely and in situ, and aimed to identify issues concerning the worker, the workplace, and concerns regarding their physical and mental well-being. In this paper, we describe the process of designing the worker companion-the methods, main insights from the user research activities, and an initial prototype.

Keywords: Participatory Design · Industry 4.0 · Operator 4.0 · Human-centred design

1 Introduction

1.1 Background

The industrial work environment is experiencing a process of transformation propelled by advanced digitization within factories, the development of Internet technologies and the upsurge of "smart" objects [8]. This is often referred to as Industry 4.0 (i4.0).

The vision for i4.0 is one where networked, automated machines, through a layer of artificial intelligence programs, directly collaborate with human workers optimizing resources and increasing efficiency. This requires augmenting the

F. Fui-Hoon Nah and K. Siau (Eds.): HCII 2022, LNCS 13327, pp. 417–429, 2022.
https://doi.org/10.1007/978-3-031-05544-7_31

worker's physical, sensory, and cognitive capabilities through digital technologies, bringing forth the concept of Operator 4.0.

Embracing the i4.0 paradigm requires, therefore, not only equipping the workforce with new skills and tools but also ensuring digital technologies are accepted and adopted by the workers, being integrated into the industrial work environment in a socially sustainable way [14]. To this end, actively involving the workers in such a transformation is essential.

Endorsing this vision, the Augmanity project aims to address the challenges and opportunities brought on by industry 4.0 through the development of relevant technology with multiple objectives, such as the optimisation of work tasks and processes; the minimization of health issues from intensive work; the elimination of waste; re-skill existing workforce for the new digitization paradigm; measure and promote workforce engagement; integrate cyber-physical systems and high data rate connectivity; and to ensure efficiency reducing emissions. The Augmanity project is being implemented by a consortium with 22 partners, including universities, R&D centres, and companies from various areas, such as telecommunications, footwear manufacturing, furniture, water heating, among others.

In this paper we report on the process of conducting user research and participatory design with shop floor operators and on the design of a platform that caters to the needs of such workers called Worker Companion App (WCA). The WCA is being developed in context of the Augmanity project and intends to be a tool to support workers in i4.0 environments. Operators can use the WCA to receive and share information about the shop floor environment or other work-related information. This platform is being developed using a human-centric approach, with the direct involvement of operators and relevant stakeholders in its design, to ensure its adoption and acceptance by the workforce.

2 Literature Review

2.1 Industry 4.0 and Operator 4.0

Industry 4.0 (i4.0) describes a new industrial revolution, one that leverages digitisation and technological advances where machines and human workers are digitally connected among themselves, and operating under artificial intelligence programs to be more efficient, productive, and with increased quality. [6,10].

This vision is now a reality, with concepts of Industry 4.0 being implemented in today's factories [4,11]. This, according to Lorenz et al. [9], will be transformational. From an economic perspective, introducing collaborative machines on a shop floor is expected to signify a productivity increase and revenue growth, which will, in turn, create thousands of new job positions to perform tasks that require "flexible responses, problem-solving, and customization." However, a decrease in jobs requiring physical assistance or repetitive mechanical work is also expected. [9]

Authors argue that the role of the worker, more specifically of the operator, is also expected to change significantly, particularly as machines' roles in production become increasingly more embracing, with the operator transitioning from an execution responsibility to one of supervision [5–7]. In this new vision of the operator within i4.0, the human worker is assisted by technology, enhancing human abilities and surpassing limitations. As such, Romero et al., [14] have identified eight types of operators expected to integrate Industry 4.0: the super-strength operator, the augmented operator, the virtual operator, the healthy operator, the smarter operator, the collaborative operator, the social operator, and the analytical operator.

2.2 Participatory and Human-Centred Design in Industry 4.0

Participatory Design (PD) has, since its conception as part of the Scandinavian workplace democracy movement, been associated with transformations of the workplace [2,3,12]. Through the application of the theories, practices, and studies that characterise PD, the people affected by technologies are directly involved in their design [1]. Through ethnography-inspired methods such as observations, interviews, and co-creation/prototyping researchers and users collaborate to preserve the "tacit knowledge" of the users' workflows and work tools, rather than "doing away with them" [15].

Within the context of Industry 4.0, authors have identified the need for human-centeredness as an essential aspect of the design of technology for Industry 4.0, to ensure the creation of "flexible, inclusive, and safe workspaces, as well as better conditions, increased productivity, and improved quality" [16]. The authors also mention that human-centeredness will also signify "worker satisfaction and work wellbeing, more empowered and engaged workers". However, this human-centeredness is mainly concerned with adopting technology explicitly targeted for the worker on the shop-floor, in line with Romero's Operator 4.0. Villani and colleagues, for example, present a framework for "adaptive automation" that will assist the operator on work tasks, according to three different modules: measurement of human capabilities, the adaption of the interaction, and teaching and support. While this framework (called INCLUSIVE) tested the solution in real-world scenarios with representative users, there is no mention of the involvement of workers in the design of the framework itself. This is in line with Pfeiffer, Lee, and Held's findings that, while these technologies are aimed towards workers, design decisions about new automation solutions and digitization of manufacturing is "engineering departments often organized according to a strict division of labor (e.g., between mechanical engineering and IT), mainly following academic methods and thus integrating employees on the shop floor only in the last step" [13].

As these developed technologies can significantly impact the work and the lives of the workers, and in line with PD practice, those affected by technology should be directly involved in their design. For this reason, the design and development of the WCA is informed by Scandinavian Participatory Design practices and methods, directly involving the workers in different stages of the development of the WCA, from its conceptualization, into its design and validation. The methods utilized are described in the following section of this paper.

3 Methods

3.1 Study Design

To inform the design of the WCA and ensure that workers are active participants and decision-makers, we applied PD approaches to research and design, through employing a mixed-methods approach consisting of fieldwork observations, interviews, and exploratory workshops with relevant stakeholders, across 3 shop floor sites, on 3 different companies:

- *Industry 1* is a large producer of wood aggregate furniture
- *Industry 2* is a manufacturer of electrical household equipment
- *Industry 3* is a manufacturer of sanitary equipment

Table 1 provides an overview of all the activities carried out across the three locations. Between workers, management, and human resources we involved 36 stakeholders, across 34 sessions, in the different research activities. Participants were recruited through contacts on each company, and each participant signed informed consent before the start of the research activities.

Table 1. Overview of the user research activities

Activities	Industry 1	Industry 2	Industry 3	Total
Guided Tours	2	1	1	4
In situ interviews HR/HSE/Managers	0	0	0	0
Remote interviews HR/HSE/Managers	5	1	1	7
In situ interviews supervisors	1	1	0	2
Remote interviews supervisors	1	0	2	3
In situ interviews workers	3	1	0	4
Remote interviews workers	0	0	0	0
In situ workshos	7	5	0	6
Remote workshops	0	12	0	12

Guided Tours. The research team conducted guided tours on the three shop floors for a better understanding of the work context. On these tours, we were accompanied by a representative of the factory. Complementary material such as written notes, drawings, video, and photo capture was collected, in compliance with the security, privacy, and confidentiality of the workers and the company.

Semi-Structured Interviews. We conducted semi-structured interviews with HR/HSE/Managers, supervisors, and workers, both remotely and in situ. They were recorded, transcribed (nonverbatim), pseudo-anonymized, and analyzed. Participants were required to consent to the interviews and recording by signing a consent form or by verbally consenting in the case of remote interviews. The interviews' nonverbatim transcriptions were analyzed to identify common themes among participants. The following describes the topics explored in the interviews with each stakeholder:

- HR/HSE/managers. Interviews took place remotely and focused on: characterization of the production process; the number of workers and respective description in terms of sex, age, seniority, employment status, and professional activity; typology of working hours; evolution of the following indicators: accident rate, absenteeism rate, medical restrictions, occupational diseases, health complains; initiatives concerning worker engagement; opportunities for optimization and innovation.
- Supervisors (responsible of area, shift managers, others). Interviews took place both remotely and in situ and focused on: characterization of the production process; awareness, processes, and strategies related to workers' physical and mental well-being; metrics and productivity indicators; opportunities for optimization and innovation.
- Shop floor workers. Interviews took place in situ and focused on: characterization of work; relationship with work, supervisors, and colleagues; physical and mental implications of work; opportunities for optimization and innovation.

Co-Design Workshops. We conducted three sets of workshops with workers and supervisors, both remotely and in situ. The fist set of workshops aimed to establish workers' priorities (needs, preferences and expectations), and to identify their relationships with issues of data privacy and personal information disclosure. The second set intended to co-create solutions concerning data representation and visualisation, including iconography and color interpretation. In the third set of workshops, we focused on further exploring workers' needs, preferences and expectations, while testing the visual representation of data.

The qualitative data gathered from these methods were structured, analysed, and translated into insights concerning the role of workers and their insights. Participants were required to consent to the workshops and recording by signing a consent form, or by verbally consenting in the case of remote workshops.

The remote co-design workshops took place over Teams and the co-design activities were done on the Mural[1] (Fig. 1).

On the first set of workshops, which explored the relationship of the worker with the work, the workers' values and priorities, as well level of comfort (or discomfort) in disclosing personal information, the initial exercises of the first set asked the participant to drag specific concepts or words into three concentric circles, representing levels of agreement or disagreement to the participant (Fig. 2).

[1] www.mural.co - Web platform for remote collaboration.

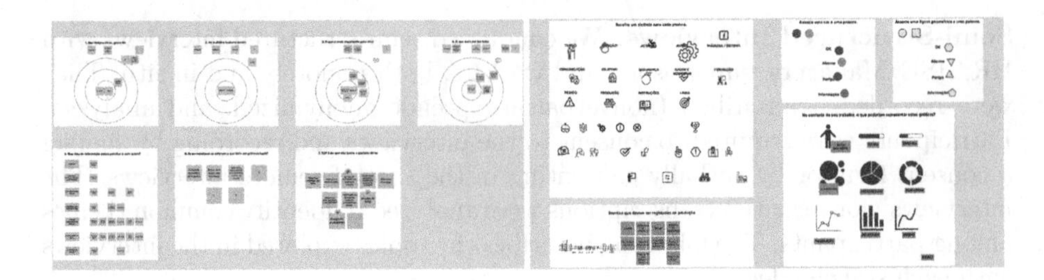

Fig. 1. Example of the results from the co-design sessions conducted remotely (First set of workshops on the left, second set on the right).

The first exercise served as an icebreaker, as well as a safe way for the participant to get familiar with the remote sessions platform. The following three exercises addressed 1) the issues of emotions felt at the workplace, 2) what is essential to the worker during work, and 3) what the worker is missing/needs the most. The fifth exercise dealt with the participants' comfort level regarding sharing personal information (physical and mental well-being, personal opinions, and professional information) with specific groups of people (team colleagues, HR, supervisors, health professionals). The sixth exercise asked the participants, "If I was in charge of the company for one day, what would I do first?". Some suggestions based on the results of the interviews were already pre-created, with blank cards where participants could fill in their answers. Lastly, the seventh exercise asked the participants to prioritize what they missed most at their work, from 1 to 3. Some options based on the interviews were given, allowing participants to add more if needed.

The second set of workshops, which aimed to identify and validate iconography that better-represented workers' mental models, consisted of five exercises, all dealing with the visual representation of concepts related to the worker and workplace. The first exercise consisted of several concepts and symbols, and the participants were asked to associate one symbol to each concept. The second exercise asked the participants to associate a color to the following words: "OK", "Alarm", "Danger", and "Information". The third exercise was similar to the second, with symbols instead of colors. The fourth exercise asked the participants to associate a different method of representation of data (such as a line graph, or a pie-chart) with a specific concept (such as Energy, Production, and Wellbeing). The fifth and final exercise presented a blueprint of a shop floor to the participants, asking them what information should be associated with it, and how they would represent it.

The third set of workshops, dealing both with the relationship between worker and work, as well as to test and validate different methods of visualisation and representation of work related data, consisted of eight exercises. The first, second, third, and fourth exercises were similar to those of the first set. Following these exercises, we asked the participants in what working situations they felt more fatigue, as well as when they felt less secure. The final three exercises

Fig. 2. Example of the results from the co-design sessions conducted in situ (First exercise on the left, third exercise on the right).

consisted of A/B tests. Two alternatives to representations of Fatigue, Security, and Posture were presented. The participants were asked what they interpreted of these different representations, and which they felt better represented these concepts.

4 Results

4.1 Insights from the User Research

Health and Well-Being. Workers report frequent health work-related problems and physical pain, in particular in the shoulders, back, arms, elbows, and wrists. Their work involves moving heavy weights and repeated movements for long periods of time, which leads to many of these problems. This also leads workers to self-medicate and work while in pain: "Today I'm ok [...] I took a pill [Voltaren 600 mg] before coming to work, or I would have no chance to work" (P09). Stretching or gymnastics as strategies to mitigate the negative effects of a demanding job was also brought up by workers. In the past, some companies tried to implement these types of activities, but without resolute success. While overall well regarded by workers, these activities took place during the existing breaks and by that time workers were already exhausted. Moreover, these activities took time from their *free time*: "It happened during the shift [...] during the break, but some people said that the break was to be freely enjoyed and should not be demanding [...]" (P08).

Work Practices and Conditions. The work conditions and practices was a frequent topic of discussion, with particular mentions of the environment at the factory, security, or non-regular working schedules. Much of the work was done in a noisy and hot environment, with bad air quality, and that requires workers to stand for many hours. Management and HSE can try to accommodate these problems (e.g. standing mats to reduce fatigue), but given the type of work, there is a limit to what can be done.

We have identified past projects that tried to explore how to use new technology to improve the working conditions; however, according to workers, without great success. For instance, an exoskeleton was tested in one of these companies to support workers lifting heavy loads. Although the exoskeleton would work well in this regard, it had limitations, namely its ergonomic dimensions, which resulted on low adoption by the workers:"it is uncomfortable" (P6). The device was uncomfortable because it restricted their movements and caused them to sweat: "although it helps carrying on weight it is uncomfortable to use it all day, even more with heat" (P06). Another experienced example was a smartwatch that would notify them when there was a problem with the machine or the task was done; however, because of frequent failures of the system, workers could not rely on the watch and still checked everything manually. As such, adoption of new technology will depend not only on the immediate benefits of the technology for the worker or company, but also on workers' perception of usefulness and reliability, ergonomics and ease of use.

There is a fine balance between security and production output. From shop floor workers to management, everyone considers security of most importance. However, during work, the security practices may be relaxed because it might be easier or faster to do some task without following the security guidelines or without wearing some kind of security equipment: "It also exists some laziness, even ours, and from other people:'this wont́ take long' - and changing that mindset is not easy." (P3).

Locus of Control. Machine faults, lack of material, the absence of a colleague, or even changes in pace are not uncommon in the shop floor. These types of disruptions, which are outside the worker's control, were brought up as a major cause of stress and frustrations in the workplace. It can result in a loss of production output, which in turn can result in overtime.

On the other hand, workers reported feeling satisfied when the job was well done, and they were contributing positively to its outcome or the team. For instance, by helping solve some problem in the production line: "I was glad because I watched the output, and I was able to call the quality team, and we figured out how to solve the problem." (P4).

Sharing Information. During our observations, it was possible to find on the shop floor, or near each workstation, company-provided instruments that workers could use to register work-related events, such as security, quality, or equipment issues. However, registering, updating and accessing information, e.g., processes, standards, changes, alerts, is time-consuming error-prone, and still relies heavily on paper.

While these instruments can create an official trail of information for management, workers still found the necessity to create their own communication channels for sharing production-related events with workers in the same shift, workers in the different shifts with similar roles, supervisors, or team leaders. Tools such as WhatsApp were being adopted by workers for work-related communication: "We now have WhatsApp groups [...] first came up the one from supervision [...] we realized that we had lots of information that could be lost [...] there is the paper sheet, but I may not be near the paper [...] even to share information timely, some events that could happen [...]." (P08).

Transitions between shifts require special care. Problems with the equipment or quality need to be reported to the worker in the next shift. While some worker may arrive a bit earlier to do this hand-off process in person, this does not always happen, which may result in lost information.

Open Communication. Open and direct communication between workers and supervision was deemed crucial for a proper job. It is important to be heard, to receive feedback, to share knowledge or problems on the job: "When we have an opened team, when we have an opened leadership, and we know our goals, what might possibly happen, the result of our work [...] it is very satisfactory." (P8). Supervisors have to be ready to listen to workers and manage any problem that may arise during the job, disagreement between workers, as well as other personal problems that workers might be going through. As it was noted by one of the supervisors, he had to take on the role of a "psychologist".

4.2 Requirements

The following requirements identify actions that the worker should be able to accomplish through the Worker Companion:

- Workers should be able to know their position, responsibilities, and rights within the company.
- Workers should be able to have at hand information concerning their job, workstation, or tasks.
- Workers should be able to easily share or report to co-workers or supervisors information about work, shifts, equipment, or the overall shop floor environment.
- Workers should be able to receive personal feedback or general information from supervisors or human resources
- Workers should be able to visualise personal information in an individual, private, and secure manner.
- Workers should be able to visualise and select who has access to their personal information.
- Workers should be able to interact with the companion, without it being too cumbersome and disruptive to their work/time.

4.3 System Design

With the insights and requirements collected during the user research activities, we have designed a system - Workers Companion App (WCA) - for a rich understanding of the worker within the context of an Industry 4.0 workplace. The WCA consists of two complementary applications. Through a web application workers will be able to access information that is relevant to them. At the same time, the WCA will offer a tool that HR managers and supervisors can leverage to broadcast important information or to collect user feedback.

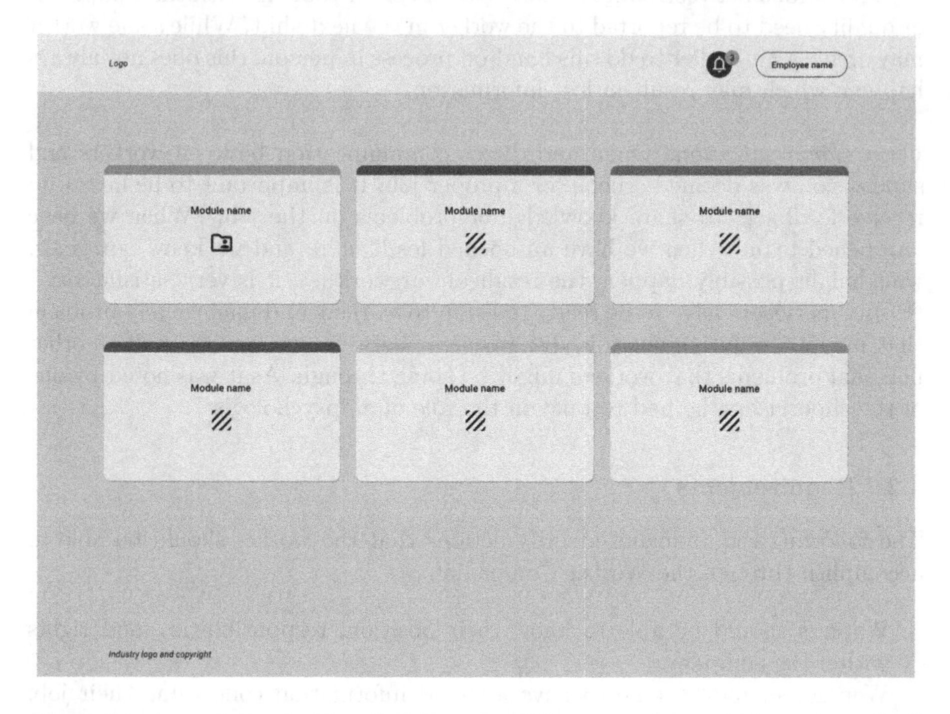

Fig. 3. Mockup of the dashboard listing the different modules of the system. Each company personalize this dashboard, enabling only the modules they need.

The WCA is built around a common structure that integrates with independent modules (Fig. 3), each one with a specific purpose and functionalities. WCA was designed so that companies could enable individual modules as necessary. This approach allows us to create personalized experiences based on the needs of the industry, company or worker. For instance, in the industries that are using sensors to measure the environment or workers, we can enable dedicated modules to visualize this information. Besides allowing us to provide a custom solution for each company, this also allows us to create personalized experiences for individual workers, since modules could be enabled on demand according to the requirements of individual workers or a group of workers. Another advantage of a modular approach is that it enable us to start with a smaller set of modules and include additional ones in the future as necessary. Based on the user needs identified in the research phase, an initial set of five modules were proposed:

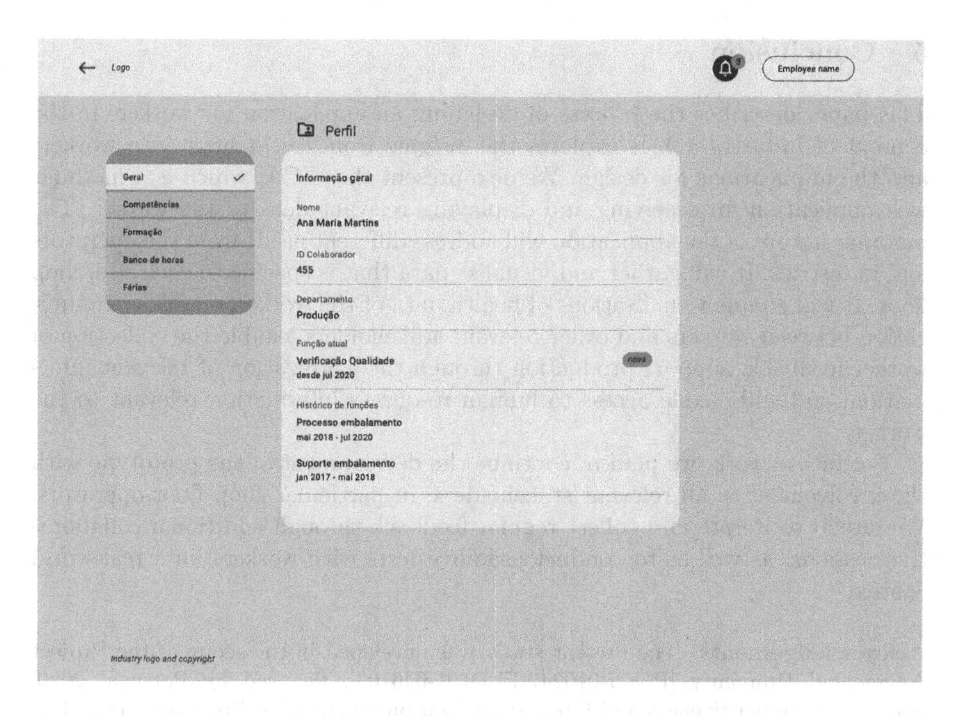

Fig. 4. Mock of the human resources module. Enables the worker to access its profile information.

- The **health module** will enable workers to monitor and understand their physical and mental well-being, and the ergonomics of their job. It will allow workers to visualize the data that is being collected by sensors (e.g., real-time ergonomic monitoring system that is being developed in PPS1) and receive reports or recommendations with actionable insights based on that same data.
- The **communication module** will offer a solution to facilitate the communication between workers, HR managers, and supervisors.
- The **workers feedback module** will be responsible for collecting feedback from the workers. This module will be responsible for implementing the interface for collecting feedback and for submitting the results back to the cloud server. This module intends to integrate with the HR platform to allow HR personnel to create and deploy on demand questionnaires on the workers' application.
- The **production module** aims to aggregate all the information relevant to the worker and the proper execution of their work, such the security information or other guidelines.
- The **human resources module** (Fig. 4) will offer information access to the worker's information within the company, such as salary, holidays, time banks, training acquired, and related.

5 Conclusion

This paper describes the process of designing an application for workers in the context of Industry 4.0. It explores the insights from researching with workers and the implications for design. We also present the WCA, which is a modular web application for receiving and displaying relevant data to the worker. The modular nature of the application will address different needs from the operators and industries. It will gather and visualise data that is specific to each site/shop floor. It will enable visualisations of health data of the worker, enable communication between workers and other relevant stakeholders, enable the collection of worker feedback, support production through the aggregation of relevant information, and will enable access to human resources information relevant to the worker.

For future work, we plan to continue the development of the prototype with the involvement of all relevant stakeholders, in particular shop floor operators. We intend to iterate and collect regular feedback through additional collaborative sessions, as well as to conduct usability tests with workers, in a real-world context.

Acknowledgements. The present study was developed in the scope of the Project Augmented Humanity [POCI-010247-FEDER-046103], financed by Portugal 2020, under the Competitiveness and Internationalization Operational Program, the Lisbon Regional Operational Program, and by the European Regional Development Fund.

References

1. Ehn, P.: The art and science of designing computer artifacts. Scandinavian J. Inf. Syst. **1**(9) (1991)
2. Ehn, P., Kyng, M.: The collective resource approach to systems design. In: Computers and Democracy, pp. 17–57 (1987)
3. Floyd, C., Mehl, W.M., Resin, F.M., Schmidt, G., Wolf, G.: Out of scandinavia: alternative approaches to software design and system development. Hum. Comput. Interact. **4**(4), 253–350 (1989)
4. Frank, A.G., Dalenogare, L.S., Ayala, N.F.: Industry 4.0 technologies: implementation patterns in manufacturing companies. Int. J. Prod. Econ. **210**, 15–26 (2019)
5. Gorecky, D., Schmitt, M., Loskyll, M., Zühlke, D.: Human-machine-interaction in the Industry 4.0 era. In: 2014 12th IEEE International Conference on Industrial Informatics, INDIN, pp. 289–294. IEEE (2014)
6. Kagermann, H., Helbig, J., Hellinger, A., Wahlster, W.: Recommendations for implementing the strategic initiative INDUSTRIE 4.0: securing the future of German manufacturing industry; final report of the Industrie 4.0 Working Group. Forschungsunion (2013)
7. Lall, M., Torvatn, H., Seim, E.A.: Towards Industry 4.0: increased need for situational awareness on the shop floor. In: Lödding, H., Riedel, R., Thoben, K.-D., von Cieminski, G., Kiritsis, D. (eds.) APMS 2017. IAICT, vol. 513, pp. 322–329. Springer, Cham (2017). https://doi.org/10.1007/978-3-319-66923-6_38
8. Lasi, H., Fettke, P., Kemper, H.-G., Feld, T., Hoffmann, M.: Industry 4.0. Bus. Inf. Syst. Eng. **6**(4), 239–242 (2014). https://doi.org/10.1007/s12599-014-0334-4

9. Loernz, M., Ruessmann, M., Strack, R., Lueth, K., Bolle, M.: How will technology transform the industrial workforce through 2025. Man and Machine in Industry 4
10. Neugebauer, R. (ed.): Digital Transformation. Springer, Heidelberg (2019). https://doi.org/10.1007/978-3-662-58134-6
11. Niewöhner, N., Asmar, L., Röltgen, D., Kühn, A., Dumitrescu, R.: The impact of the 4th industrial revolution on the design fields of innovation management. Proc. CIRP **91**, 43–48 (2020)
12. Nygaard, K., Bergo, O.T.: The trade unions-new users of research. Personnel Rev. **4**, 5–10 (1975)
13. Pfeiffer, S., Lee, H., Held, M.: Doing industry 4.0-participatory design on the shop floor in the view of engineering employees. Cuadernos de Relaciones Laborales **37**(1), 293–311 (2019)
14. Romero, D., et al.: Towards an Operator 4.0 Typology: A Human-Centric Perspective on the Fourth Industrial Revolution Technologies, October 2016
15. Spinuzzi, C.: The methodology of participatory design. Tech. Commun. **52**, 163–174 (2005)
16. Villani, V., et al.: The inclusive system: a general framework for adaptive industrial automation. IEEE Trans. Autom. Sci. Eng. **18**(4), 1969–1982 (2020)

Development and Evaluation of a Tangible Interaction Concept for Assembly Workstations

Swenja Sawilla(✉) and Thomas Schlegel

Institute of Ubiquitous Mobility Systems, University of Applied Sciences Karlsruhe, Moltkestr. 30, 76133 Karlsruhe, Germany
{Swenja.Sawilla,Thomas.Schlegel}@h-ka.de

Abstract. In the development and research of IT-supported systems, the user is increasingly in the center of attention. Therefore, we investigated the approach of how beneficial it would be to use tangibles (physical objects) as an interaction option for the user at an assemble workstation. For this purpose, we designed an augmented reality concept for the use of tangibles at a manual assembly workstation and evaluated it with an eye tracking study. Through the study, we found that the participants were very comfortable with the tangible interaction concept, learned very quickly, and found it very easy to use. Further, we present how it is possible to improve assistance systems, such as pick-by-light systems, with the use of augmented reality and tangibles. For this, we show that the use of tangibles offers many advantages like training new employees fast at an assembly workstation.

Keywords: Tangible user interfaces · Augmented reality · Eyetracking · Assembly workstation · Industry 4.0

1 Introduction

Natural user interfaces (NUI) have become increasingly important in the recent years.

NUIs adapt to the expectations and interests of the user. Tangible User Interfaces (TUI) are a variation of NUIs. Here the focus is on making digital content "tangible". A natural interaction with tangible user interfaces is based on ways of acting that people are familiar with from dealing with non-technological aspects of their real environment. In Industry 4.0, interactions between humans and machines or computers occur. A new conceivable solution approach for interaction with such systems would be the use of tangibles (physical objects) as interaction options for the user. We analyzed this form of interaction, development and use of such tangibles, especially in the context of assembly workstations in this work.

2 Literature Research

The term tangible can be translated as touchable, haptic, or dingable. Tangibles are also the physical embodiment of interactive control mechanisms [1]. The user continuously

F. Fui-Hoon Nah and K. Siau (Eds.): HCII 2022, LNCS 13327, pp. 430–442, 2022.
https://doi.org/10.1007/978-3-031-05544-7_32

receives haptic feedback and thus directly "touches" the interface to the digital world. The sense of touch when grasping a tangible plays an important role here. Unlike all other senses, the sense of touch is the only active human sense. For example, the user can feel the shape, the structure and the surface.

In 1997, Ischii and Ulmer defined the term Tangible User Interface (TUI) as user interfaces that extend the real physical world by connecting digital information with everyday physical objects and environments [2]. Later, Ishii (2008) describes this more specifically with a metaphor of an iceberg in the middle of a digital ocean. This "iceberg" is filled with information that users cannot grasp. This metaphorically described iceberg thus clearly stands out from the water and makes itself touchable and tangible [3].

2.1 Challenges for the Design of TUIs

TUIs promise faster learnability and easier usability compared to other interfaces. Hornecker (2008) says that developers of TUIs must ensure that human senses and abilities are addressed in parallel. TUI systems favor physical interaction with objects and environments [4]. Shaer et al. (2004) describe the following conceptual and technical criteria for designing such a user interface [5].

- **Virtual and real world are connected:** Virtual, as well as physical objects, fit well next to each other and exchange information. This poses the challenge that for each representation of the information, it must be weighed which of the objects is more suitable.
- **Multiple behavior:** TUI objects interact with other physical or digital objects. This could result in their own behavior changing as soon as another physical object is added.
- **Multiple actions:** In TUIs, physical three-dimensional objects are used. These provide considerably more action options than graphical objects. For example, tangibles can be shaken, pushed, pulled, pressed, deformed or thrown. This oversupply leads to the challenge of selecting the significant actions.
- **No standard input and output devices:** Different technologies are used for TUI systems and, consequently, different physical objects and implementations.
- **Distributed interaction:** A TUI allows multiple users to interact with different physical objects simultaneously. This implies that there is no single point of interaction. The input is parallel and not a serialized input stream.

2.2 Added Values When Using Tangibles

Klemmer (2006) describes a connection between motor actions and the human thought process. Tangibles can support the user in thinking, interacting and communicating. For example, the use of gestures as input supports the motor memory of a user. By having a person remember these interactions, the cognitive load during interaction can be lessened [6]. Shaer and Hornecker (2009) concluded that tangibles support epistemic actions. That means, when a user interacts with tangibles, the functions and possible uses are better understood. Moreover, these actions help to remember previous interactions and stimulate the memory [7].

Dourisch (2001) describes that tangibles have the advantage of being in the same world as the users. In particular, the physical tangibility of the object allows tangibles to be aligned, placed, and adapt to the needs of the user [8].

Ishii (2008) explains that a user can interact with the system ambidextrously. In addition, if multiple tangibles are provided as an input-abstract option, it is possible for multiple users to interact simultaneously [1]. Furthermore, the tangibles are visible to every user as well as viewers in the room and support coordination and collaboration in groups [7]. Another property of a tangible is that it can be handed down. This makes it possible to divide tasks and functions. Tangibles allow persistent mapping from the interface to the underlying data. The data is instantiated by the tangible and represented in a physical form. Thus, tangibles are both a representation and an input device [3].

A major advantage of interacting with tangibles is the ease of mapping. Ferscha et al. (2007) say that shapes, colors, material, and weight entice users to interact. They analyzed the influence of mappings and came to the conclusion that especially shape and size have a striking influence in the use of TUIs [9].

2.3 Limitation When Using Tangibles

Due to the physical fixed nature of tangibles, tangibles are limited in their interaction. Graphical user interfaces, on the other hand, provide the means to quickly create, duplicate, modify, or even delete objects. This limits a system's ability to modify the physical properties of tangibles [1].

Another limitation when interacting with tangibles can be the scalability of these objects. This is because tangibles occupy a physical space in a particular place or are tied to an installation. Thus, it is not possible to place and use an arbitrary number of tangibles. The complexity of tangibles in shape, size and their functions is another point in limiting the scalability. Especially the size and weight are very limited [9]. Kirk et al. (2009) concluded that mainly grasping, lifting, and moving tangibles is the primary form of interaction. This requires the user to perform different movements with more effort than is necessary with a mouse, for example [10]. Shaer and Hornecker (2009) therefore emphasize that when creating and developing tangibles, the load during use must also be considered. They noted that otherwise a user might quickly become fatigued while using tangibles. Not only the users, but also the material and texture of a tangible can "fatigue" and thus stop being used [7]. There is a risk of misplacing or even losing a tangible. If loss occurs, the functionality of a system may be limited. In addition, it may not be possible to regenerate a new tangible in a short period of time, which increases the limitation of the system [10].

3 Development of the Tangible Interaction Concept

To develop the concepts, we followed the norm DIN EN ISO 9241-210 that specifies a procedure for the design of human-centered activities. The procedure includes understanding and describing the context of use, specifying the requirements for use, designing the design solution as well as testing and evaluation the design [11]. In the following, we show how we applied this procedure in our concepts.

3.1 Context of Use and Pick-by-Light Realization

We developed an assembly workstation by following industry standards (see Fig. 1). The work surface (table) is 140 cm wide and 34 cm deep. On this work surface are three rows stacked on top of each other, each with seven small load carriers. These contain the individual pieces required to make the product "racer" (see Fig. 1). The worker has to complete eight work steps to produce the racer. The racer requires 48 picks. This stationary assembly workstation is designed for one worker.

Fig. 1. The assembly workstation and the product (racer) manufactured by a worker. (Color figure online)

In assembly, there are already many different assistance systems, such as pick-by-vision or pick-by-light, which support assembly workers in their work. Due to the positive resonance from the previous literature work about the use of AR in assembly [12–15], we decided for a pick-by-light implementation which is displayed by means of AR. For this pick-by-light implementation, we chose the following color pattern. If a small load carrier is lighted green by the beamer (see Fig. 1), the worker has to remove a piece from this small load carrier. If it is lighted red, no piece is required from the small load carrier. We used this color choice because these are complementary as well as signal colors and create the greatest focus of attention for the worker. If the worker has a red-green weakness, the projection is displayed in blue and orange.

3.2 Tangible Interaction Concept and Realization

We first established a structure to define the requirements for our tangible-interaction concept. Figure 2 describes the main tasks of an assembly worker at our workstation, which we later implemented in our interaction concept and developed suitable tangibles for it. The assembly worker receives an overview of the orders through the tangible interaction concept. The assembly worker can select an order and process it. Then, the assembly worker selects the guide to build the product racer. After selecting the racer guide, the assembly worker can start the guide and edit the guidance steps. If an instruction step is unclear to the assembly worker, the worker can activate a tip.

Based on this concept, we developed two tangibles (cube tangible and cuboid tangible) that enable a worker to interact with the system. With the cube tangible, the worker

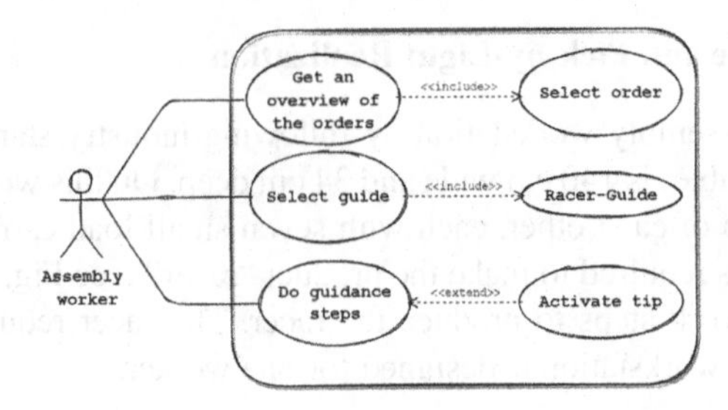

Fig. 2. Overview of the assembly worker task for our tangible-interaction concept.

selects the order and afterwards the guide (racer). It is also possible to go back and forth in the guidance steps with this tangible. For example, the worker has to put the cube tangible in one of the green fields on the surface projection (see Fig. 3 (a)). The cube tangible is 4 cm wide, 4 cm high, 4 cm deep and its material is wood. With the cuboid tangible, the worker activates the tips for the respective guidance step. This tangible is 8 cm wide, 6 cm deep and 2 cm high and its material is light plastic. We adjusted the size of the tangibles so that it is possible for the worker to easily operate with one hand. If a guidance step is unclear to the worker, the cuboid tangible has to be placed in the guidance that is unclear to the worker (see Fig. 3 (b)). The worker then receives more information about the guidance step with precise images and a short text (see Fig. 3 (c)). The cuboid tangible only operates in combination with the cube tangible. The tangibles can be rotated, pushed, lifted and placed to interact with the system.

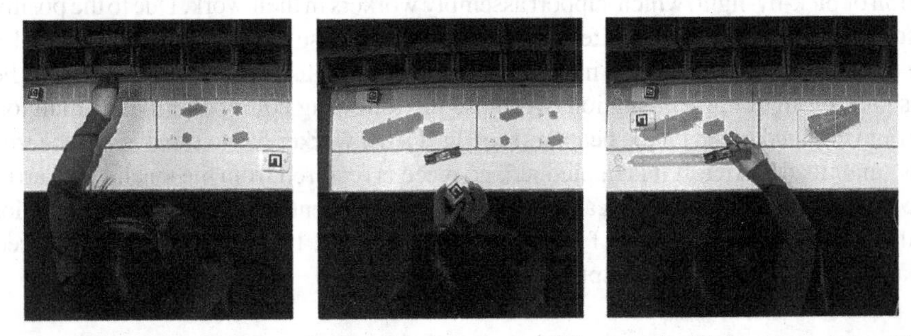

Fig. 3. **(a)** The participant is picking a piece from the small load carrier (lighted in green) and has placed the cube tangible on the first guidance step; **(b)** The participant wants to activate a tip with the cuboid tangible; **(c)** The participant has activated the tip and adjusts the workpiece. (Color figure online)

We used for the tangible detection ArUco markers. An ArUco marker is a square marker consisting of a wide black border and an inner binary matrix that determines its identifier (ID). In order for the tangibles to be recognized by the system or in this case by the Intel Real Sense (see Fig. 1), an ArUco marker must be placed on the front surface of the tangible.

4 Usability Study with the Use of Eyetracking Glasses

We evaluated the developed tangible interaction concept in a study to be able to analyze the user acceptance and the user behavior. With the use of eye-tracking glasses, we recorded the eye movements of the participants in order to obtain further conclusions about the user behavior. For this purpose, the participants were asked to complete the assembly process twice and to produce a racer after each run. We analyzed whether there were any complications during the processing of the racer. We also investigated whether there is a learning effect for the participants.

In the study, eleven participants attended. Here, seven were male and four were female. The participants had an average age of 26 years. The study consisted of two parts. In the first part of the study, the participants wore eye-tracking glasses and were asked to process and solve tasks at the assembly workstation. In the second part of the study, the participants filled out a questionnaire. For the questionnaire, the "Computer System Usability Questionnaire" (CSUQ) was used [16] and additional questions were added to get the feedback about the tangibles.

4.1 Results and Findings of the Questionare

Table 1 shows the participants' assessment of the usability of the system, the quality of the information and the user interface, the overall satisfaction and the usability of the tangibles. The participants were very satisfied with the usability of the system. All participants stated that they are satisfied with the implementation and that the system has all the functions and possibilities that the participants expect from the system. Additionally, all participants are overall satisfied with the tangible interaction concept. One initial finding is that all participants found it easy to interact with the tangibles. All participants stated that the interaction areas on the user interface were pleasant to reach. In addition, the participants stated that there was no place, where they found it difficult to interact with the tangibles. In summary, the participants rated that the user interface was understandable and they indicated that they enjoyed interacting with the system.

Table 1. Result of the questionnaire.

Questions	Strongly agree	Agree	Disagree	Strongly disagree
1) I find the system easy to use	63,6%	36,4%		
2) I was able to complete successfully the required tasks using this system	90,9%	9,1%		
3) I was able to complete the required tasks quickly using this system	45,5%	45,5%	9,1%	
4) I felt comfortable using this system	63,6%	36,4%		

(*continued*)

Table 1. (*continued*)

Questions	Strongly agree	Agree	Disagree	Strongly disagree
5) It was easy to learn to use this system	72,7%	27,3%		
6) I think I could become productive quickly with this system	72,7%	27,3%		
7) I found it easy to interact with the system	63,6%	36,4%		
8) If I made a mistake while using the system, I was able to fix it easily and quickly	63,6%	36,4%		
9) The information, such as tips, provided with this system was understandable to me	54,5%	45,5%		
10) It was easy to find the information I needed	45,5%	45,5%	9,1%	
11) The information provided by the system was easy to understand	63,3%	36,4		
12) The information helped me to solve the tasks	81,8%	18,2%		
13) The font on the user interface was easy for me to read	63,3%	9,1%	27,3%	
14) The user interface was understandable	54,5%	45,5%		
15) I enjoyed interacting with the user interface of this system	72,7%	27,3%		
16) This system has all the features and capabilities I expect from it	54,5%	45,5%		
17) Overall, I am satisfied using this system	63,3%	36,4%		
18) Personally, I find the use of AR at an assembly workstation helpful	81,8%	9,1%	9,1%	
19) I found it very easy to interact with the tangibles	72,7%	27,3%		
20) I could easily assign the tangibles by its shape on the user interface	72,7%	27,3%		
21) The interaction areas on the user interface were pleasantly accessible for me with the tangibles	90,9%	9,1%		

4.2 Evaluation of the Study Tasks

Since the participants produced the racer twice during the study, the study is divided into two runs. The total time of an instruction step and the corresponding tangible interaction was evaluated here. Figure 4 compares the two runs on average. In the first run, the participants needed an average of 13:38 min for the guidance steps and 02:58 min for the search times. To measure the search time, we evaluated the time how long it took a participant to remove the correct pieces from the storage boxes. In the second run, the participants needed 07:30 min for the guidance steps and 02:00 min for the search times. We observed that the participants were altogether 07:06 min faster in the second run.

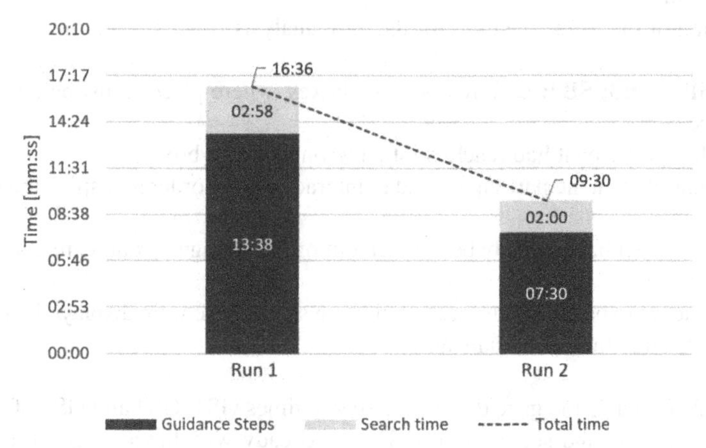

Fig. 4. Comparison of the two runs (on average).

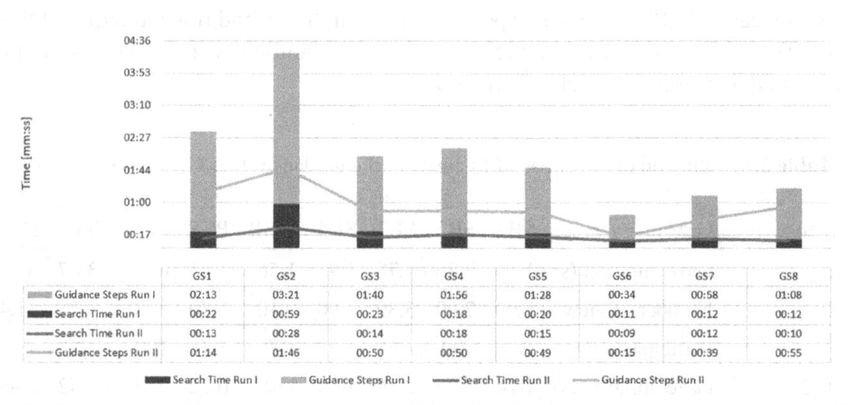

	GS1	GS2	GS3	GS4	GS5	GS6	GS7	GS8
Guidance Steps Run I	02:13	03:21	01:40	01:56	01:28	00:34	00:58	01:08
Search Time Run I	00:22	00:59	00:23	00:18	00:20	00:11	00:12	00:12
Search Time Run II	00:13	00:28	00:14	00:18	00:15	00:09	00:12	00:10
Guidance Steps Run II	01:14	01:46	00:50	00:50	00:49	00:15	00:39	00:55

Fig. 5. Comparison of the participants needed time for the guidance steps and the search time (both runs).

Figure 5 shows the time of the two runs that the participants needed for the search time of the eight guidance steps (GS). All participants needed the longest time for the search time of the second guidance step (GS2), because here the most pieces (seven pieces) had to be searched for and removed. All participants were faster in the second run. It is particularly visible that the participants for the search time in GS1, GS2, GS3 and GS5 were on average seven seconds faster in the second run. Why the participants were faster here will be analyzed with the help of the eye tracking data in the next section.

4.3 Evaluation of the Gaze Data

For the eye tracking analysis, we first created Areas of Interests (AOIs). Areas of Interests (AOIs) allow numerical and statistical analysis based on objects of particular inter-est of the stimulus.

Meaning of the AOIs, we used for the gaze analysis:

- **SB 1; SB2, SB3; SB4:** Different storage boxes, where pieces must be taken for the guidance step.
- **WSB:** If a participant had reached into a wrong storage box
- **Tan:** Tangible that the participant had to interact with in order to display the guidance step
- **Pcs:** Pieces that have already been taken out of the storage box and are placed on the work surface
- **AR-P:** Pieces to be taken out for the guidance step. These were displayed on the work surface through the AR visualization.

Table 2 compares the gaze data of the search times GS1, GS3 and GS5 of both runs (e.g. first run: GS1 and second run: GS1_2). Already with the average of all AOIs it becomes clear that the participants in the second run needed fewer views and looked at the AOIs for a shorter time. It is also interesting that the pieces (pcs) are looked at by fewer participants in the second run. It can also be seen that the participants looked at the AR pieces (AR-P) for a shorter period of time in the second run and required fewer views. This indicates that the participants are more familiar with the system and are more confident during the picking process.

Table 2. Evaluation of the Areas of Interests of the guidance steps GS1, GS3 and GS5.

Guidance step	Gaze analysis	SB1	SB2	SB3	SB4	WSB	PCS	AR-P	Tan	Ø
GS1	Gaze duration (s)	1,44	2,26	1,78	1,91	1,56	1,85	6,54	3,07	3,47
GS1	Number of views	5,00	7,00	5,30	4,90	9,10	6,80	27,0	4,20	11,44
GS1	Participants (%)	100	100	100	100	100	100	100	100	–
GS1_2	Gaze duration (s)	1,14	1,23	1,16	1,04	1,18	0,72	2,58	1,42	1,90
GS1_2	Number of views	2,60	4,10	3,80	3,00	6,50	2,78	11,2	2,10	6,25
GS1_2	Participants (%)	100	100	100	100	100	90	100	100	–

(continued)

Table 2. (*continued*)

Guidance step	Gaze analysis	SB1	SB2	SB3	SB4	WSB	PCS	AR-P	Tan	Ø
GS3	Gaze duration (s)	1,09	1,45	1,39	1,82	2,70	2,38	6,27	1,29	3,02
GS3	Number of views	3,80	4,60	3,67	4,11	12,1	7,44	23,5	2,00	9,68
GS3	Participants (%)	100	100	100	100	100	90	100	100	1,94
GS3_2	Gaze duration (s)	1,11	1,11	1,03	1,60	0,82	1,09	2,22	1,44	–
GS3_2	Number of views	3,10	3,10	3,50	2,90	4,60	4,40	9,60	2,00	6,12
GS3_2	Participants (%)	100	100	100	100	100	70	100	100	–
GS5	Gaze duration (s)	0,99	2,23	1,12	1,68	1,22	1,73	5,06	2,11	2,77
GS5	Number of views	3,90	6,40	3,00	4,50	6,80	6,44	20,40	3,30	6,8
GS5	Participants (%)	100	100	100	100	100	90	100	100	–
GS5_2	Gaze duration (s)	0,81	1,89	0,86	1,51	1,35	0,94	2,72	1,61	2,09
GS5_2	Number of views	3,60	4,00	3,00	4,10	6,90	3,80	12,2	2,50	6,90
GS5_2	Participants (%)	100	100	100	100	100	100	100	100	–

Figure 6 and Fig. 7 illustrate this finding in the form of a heat map of the first instruction step (GS1). In the heatmap of the first run, it is visible that the participantsmore often had their eyes on the pick pieces as well as on the AR pick pieces and that the participants control their pieces (see Fig. 6). The heat map of the second run shows that the control of the pick pieces with the AR visualization of the pick pieces by the participants became less (see Fig. 7).

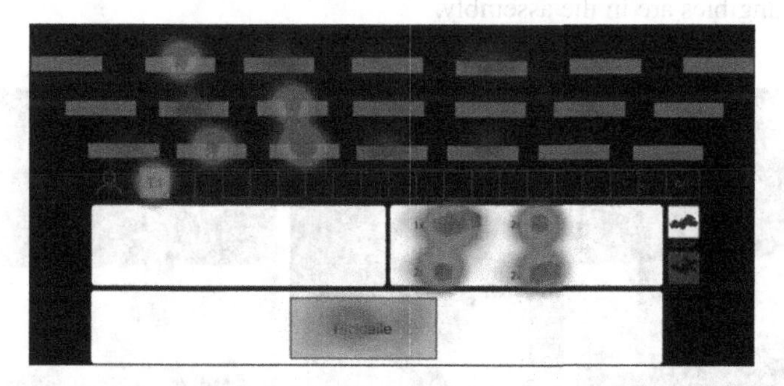

Fig. 6. Heatmap of the search time in the first run (GS1).

In summary, it could be determined that the participants in the second run already knew what they had to do and which tangible they needed to complete the task. In addition, the participants needed fewer glances and the duration of the AOIs' glances was shorter.

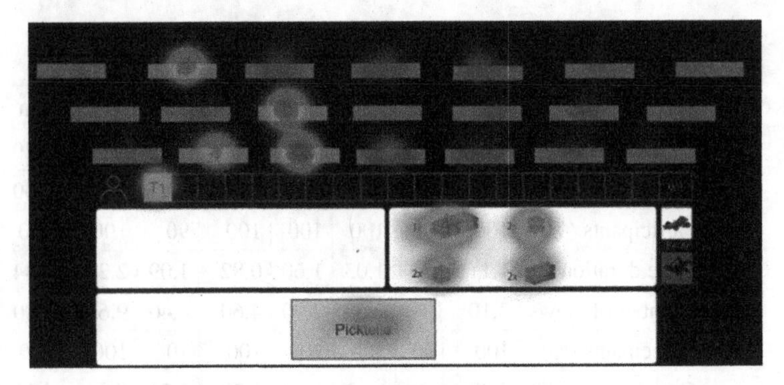

Fig. 7. Heatmap of the search time in the second run (GS2).

5 Analysis of the Potential of the Use of Tangibles in Assembly

During the use of the tangibles, there were several variations in how the tangibles were moved. One participant, for example, sometimes pushed the tangible with one finger and another participant, when he still had pieces in his hand, pushed the cube tangible further with, for example, three or two fingers (see Fig. 8). In the second run, we observed that the participants still held the pieces to be removed from the storage boxes in their hands and already pushed the cube tangible further and only then put the pieces down (see Fig. 8). In addition, we often saw that the participants used both hands at the same time to remove the pieces in the second run. Furthermore, it happened that the participants were almost finished with the guidance step or with the removal of the pieces from the storage boxes and were already pushing the cube tangible further (see Fig. 9). This is one reason why the participants were faster in the second run and it gives a view how beneficial tangibles are in the assembly.

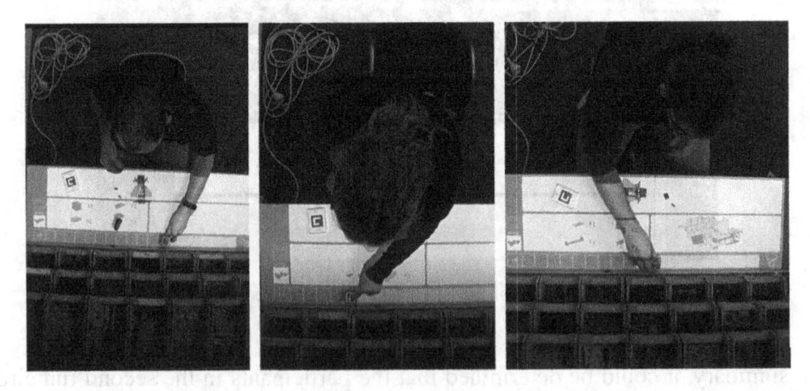

Fig. 8. Participants interacting with the tangible during the study.

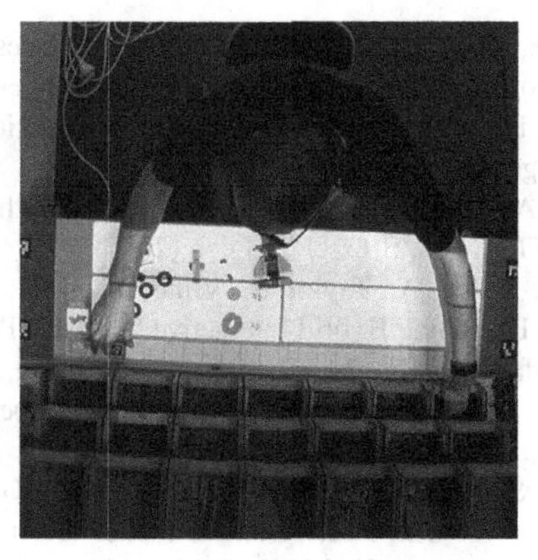

Fig. 9. Participants during a guidance step in the second run.

6 Conclusion and Outlook

The analysis of the eye tracking data showed that a learning effect could be seen especially in the second round and that all participants were significantly faster in the search times and guidance steps. Participants became faster by doing multiple steps at the same time, such as using both hands to remove pieces from two different storage boxes at the same time or advancing the tangible while working through a guidance step.

In general, the participants stated that they were satisfied with the tangible interaction concept and that they liked the interaction with the tangibles. In addition, participants found the use of AR in assembly very helpful. It is recognizable that the tangible interaction concept was very well received by the participants, and the participants learned to interact with the tangibles and the system very quickly, and a clear learning effect was visible. The results of the study showed that there are advantages to using tangibles in an assembly workplace and it gives a lot of potential for further improvement and development of assistant systems.

References

1. Ishii, H.: Proceedings of Tangible and Embedded Interaction (2008)
2. Ishii, H., Ullmer, B.: Tangible bits: towards seamless interfaces between people, bits and atoms. In: Proceedings of the ACM SIGCHI Conference on Human factors in computing systems, pp. 234–241 (1997)
3. Ishii, H.: Tangible bits: beyond pixels. In: Proceedings of the TEI – Tangible and Embedded Interaction, Bonn (2008)
4. Hornecker, E.: Die Rückkehr des Sensorischen: Tangible Interfaces and Tangible Interaction, Transcript Verlag (2008)
5. Shaer, O., Leland, N., Calvillo-Gamez, E., Jacob, R.: The TAC pradigma: specifying tangible user interfaces. Pers. Ubiquit. Comput. **8**, 359–369 (2004). https://doi.org/10.1007/s00779-004-0298-3
6. Klemmer, S., Hartmann, B., Takayama, L.: How bodies matter: five themes for interaction design. In: Proceedings of the 6th Conference on Designing Interactive Systems, New York, USA (2006)

7. Shaer, O., Hornecker, E.: Tangible user interfaces: past, present, and future directions. In: Foundations and Trends in Human Computer Interaction 3 (2009)
8. Dourish, P.: Where the Action Is - The Foundations of Embodied Interaction. MIT Press, Cambridge (2001)
9. Ferscha, A., Vogl, S., Emsenhuber, B., Wally, B.: Physical shortcuts for media remote controls. In: INTETAIN 08, Brüssel, Belgien (2007)
10. Kirk, D., Sellen, A., Taylor, S., Villar, N., Izadi, S.: Putting the Physical Into the Digital: Issues in Designing Hybrid Interactive Surfaces. British Computer Society, Swinton (2009)
11. Syberfeldt, A., Holm, M., Danielsson, O., Wang, L., Brewster, R.: Support systems on the industrial shop-floors of the future - operators perspective on augmented reality. Procedia CIRP **44**, 108–113 (2016)
12. Schlund, S., Mayrhofer, M.: Möglichkeiten der Gestaltung individualisierbarer Montagearbeitsplätze vor dem Hintergrund aktueller technologischer Entwicklungen. Zeitschrift für Arbeitswissenschaft **72**, 276–286 (2018). https://doi.org/10.1007/s41449-018-0128-5 (2018)
13. Tang, A., Owen, C., Biocca, F., Mou, W.: Comparative effectiveness of augmented reality in object assembly. In: Proceedings of the SIGGHI Conference on Human Factors in Computing Systems (2003)
14. Hou, L., Wang, X., Truijens, M.: Using augmented reality to facilitate piping assembly: an experiment-based evaluation. J. Comput. Civ. Eng. **29**(1), 05014007 (2015)
15. Funk, M., Kosch, T., Schmidt, A.: Interactive worker assistance: comparing the effects of in-situ projection, head-mounted displays, tablet, and paper instructions. In: Proceedings of the 2016 ACM International Joint Conference on Pervasive and Ubiquitous Computing, pp. 934–939 (2016)
16. Tullis, T., Albert, B.: Measuring the User Experience - Collecting, Analyzing, and Presenting Usability Metrics. Elsevier Inc., Amsterdam (2013)

Retail, Commerce, and Customer Engagement

Smart Fitting Rooms: Acceptance of Smart Retail Technologies in Omni-Channel Physical Stores

Larissa Brümmer and Silvia Zaharia(✉)

University of Applied Sciences Niederrhein, Krefeld, Germany
{larissa.bruemmer,silvia.zaharia}@hs-niederrhein.de

Abstract. In response to increasing growth rates in online retail and changing consumer behavior, many retailers are pursuing an omni-channel strategy. Smart retail technologies, such as smart fitting rooms, help to integrate online and offline channels and to create a strong, holistic customer experience.

This research investigates the drivers and barriers regarding the use of smart fitting rooms in German fashion retailing by extending the Unified Theory of Acceptance and Use of Technology 2 (UTAUT2) by the variables 'need for interaction' and 'willingness to provide personal information'. 'Age', 'gender' and 'experience' were examined as moderator variables. Data was collected using a quantitative online survey and analyzed by means of regression analysis.

The most significant and substantial factors influencing consumers' intention to use smart fitting rooms proved to be 'hedonic motivation', 'performance expectancy' and 'willingness to provide personal information'. The variables 'effort expectancy' and 'facilitating conditions' have a weak significant influence on the use intention. 'Social influence' and 'need for interaction' did not prove to be influential in this study.

The examination of moderator effects showed that 'age' only moderated the influence of 'willingness to provide personal information' while there were gender differences for 'performance expectancy' and 'hedonic motivation'. The results also show that, especially the predictor 'facilitating conditions' has a much larger effect for inexperienced users.

Keywords: Technology acceptance · Smart fitting rooms · UTAUT2 · Willingness to provide personal information · Digital technologies · Consumer acceptance

1 Introduction

E-Commerce is experiencing constant growth, leaving brick-and-mortar retailers in a precarious situation. These developments have perhaps been the most painful for the fashion industry. For the German fashion industry, online transactions accounted for 39.8% of the total market share in 2020 [11]. An increasing number of retailers are responding to the new buying behaviors of customers who, thanks to digitalization

F. Fui-Hoon Nah and K. Siau (Eds.): HCII 2022, LNCS 13327, pp. 445–462, 2022.
https://doi.org/10.1007/978-3-031-05544-7_33

and smart phones, are "always on". Their business models offer ever-present shopping options through a variety of distribution channels – so-called omnichannel retail. Although changes in buying behavior present major challenges for the clothing industry, physical stores will remain important for enabling a "unique, sensory shopping experience" [3] and offering high-quality service and immediate availability of the products [31]. Still, omnichannel retail will play a key role in fashion retail moving forward, facilitated by innovative technologies that must create a consistent and seamless shopping experiences across different channels [14]. Digitizing the point of sale (PoS), facilitated by innovative technologies, is therefore also of key importance [28].

Digital technologies can be integrated into existing retail settings and connect the physical and the digital world [32]. For consumers, digital technologies such as smart fitting rooms can expand service offerings through individualized interactions with the retailer, support their decision-making process, and improve their shopping experience through interactivity [45]. From a business perspective, digital technologies support an omnichannel retail strategy by enabling smooth switching between channels. Additionally, they provide information about the behavior and preferences of the consumers [31]. Even though the application of digital technologies at the PoS has enjoyed increasing popularity from a business perspective [21], questions still stand on the extent to which consumers are prepared to use them. Success is ultimately dependent on "acceptance" by consumers [19]. Innovations in retail are often met by resistance from consumers [20]. To overcome such resistance, it is necessary to understand the drivers of consumer acceptance [32]. A sizable number of studies on the acceptance of digital technologies at the PoS focus in particular on researching technologies in their totality without distinguishing between particular technologies [1, 2, 16, 32, 35, 42, 46]. Since digital technologies at the PoS differ with regard to their interactivity, presence and risk perception [44], many studies recommend future research focus on investigating single select technologies [2, 32]. For these reasons the present study focuses on a single technology, namely smart fitting rooms (SFR), to incorporate technology-specific factors into the research model.

SFR are (partially) automated, interactive self-service systems that consist of two components. The first component is a sensor technology - usually Radio-Frequency Identification (RFID) - which enables automatic identification and data collection of products with the help of a transponder [26]. The second component is a "magic mirror" - a mirror that is equipped with a touch function and also enables a virtual fitting with the help of Augmented Reality (AR) [10]. When a consumer enters the SFR with an item, the article of clothing is automatically identified based on RFID-technology and the SFR offer a range of functions:

- Additional product information and availability of alternate sizes and colors [12]
- Projecting the article of clothing into suitable environments [10]
- Connections to the online shop using customer profiles with the ability to place an "in-store-order" if customers agree to share their address and financial information [10].

Through the connection to the customer profile and the use of customer data (e.g. customer identity, purchase history), tried-on items can also be stored and additional personalized services offered [43].

The present study investigates the acceptance factors of smart fitting rooms from a consumer perspective within the framework of an acceptance model developed through theory and empirically validated. The following research question is derived for the study: *Which drivers and barriers exist regarding the use of SFR?*

2 Conceptual Framework and Hypothesis Development

2.1 Literature Review

SFR consist of a combination of several technical components, AR applications and RFID systems. Therefore, the state of research as far as the acceptance of SFR is concerned will be discussed as well as the acceptance of underlying technologies. AR applications have been integrated at the PoS for some time, which is why research on acceptance spans more than ten years [17, 18, 34]. Numerous studies concentrate on virtual fitting rooms [13] or glasses [27, 30]. Determinants of acceptance identified by such research are: the *quality of information* [13, 27], *aesthetics* [13, 27], *enjoyment* [27], *system quality* [13] as well as *response time* [27].

To date, the acceptance of RFID systems, which include SFR has been largely researched with respect to system perception and personal traits [26]. System characteristics suggested by the literature are the central determinants of the *Technology Acceptance Model (TAM)*: *perceived usefulness* and *perceived ease of use* as well as *data protection* [24]. Personal traits applied thus far are e.g. *technology anxiety* [26], *privacy risk harm* [23, 43], *trust* [43], *willingness to provide personal information* [43] as well as the UTAUT model's central determinants [26, 43].

Müller-Seitz et al. [24] investigated consumer acceptance of RFID systems in German electronic retailing. The results show that both *perceived ease of use* and *perceived usefulness* have a positive influence on the acceptance of RFID systems. The general *attitude towards data security* also has a positive effect on acceptance, while *security concerns* have a negative influence. No significant relationships were found between gender and acceptance of RFID technology, and age and acceptance of RFID technology.

Nysveen and Pedersen [26] investigated the acceptance of RFID technologies using an extended *UTAUT model*. Due to the novelty of RFID systems, consumers have little experience in dealing with such technologies. For this reason, the two additional determinants *technology anxiety* and *perceived privacy risk harm* were added. The results of the study show that technology anxiety has a significant negative influence on *attitudes towards the use* of RFID technologies. It should also be noted that *gender* moderates most of the relationships in the model: the *perceived benefit* is an important influencing factor for men, while the *facilitating conditions* play a greater role for women.

Weinhard et al. [43] researched the acceptance of SFR by applying an adapted model of the *Unified Theory of Acceptance and Use of Technology 2* (UTAUT2) by Venkatesh et al. [41] combined with Dinev and Hart's [8] *Privacy Calculus Model*. The results show that the *willingness to provide personal information* is a relevant driver of acceptance and should be accounted for in future research on the acceptance of SFR systems. If consumers are unwilling to divulge private data, they will not use SFR, regardless of their potential benefits. Since the sample only included young people (students), there are limitations to the generalizability of the findings. The moderating effects of age, gender and experience from UTAUT2 were also not studied.

2.2 Model and Hypotheses

The basis for developing this study's model is the *Unified Theory of Acceptance and Use of Technology 2* (UTAUT2) by Venkatesh et al. [41], which was appropriately modified for its application to this new field of research. The UTAUT2 pursues the goal of examining technology acceptance exclusively from the consumer perspective by investigating the intention to use (behavioral intention) and the adoption (use behavior) of a technology. Seven factors directly influence behavioral intention: performance expectancy, effort expectancy, social influence, facilitating conditions, hedonic motivation, price value and habit. Use behavior is influenced by behavioral intention, facilitating conditions and habit. The effect of performance expectancy on behavioral intention is moderated by the variables of age and gender. The effects of all other predictors (effort expectancy, social influence, facilitating conditions, hedonic motivation, price value and habit) on behavioral intention are moderated through the variables age, gender and experience [28].

SFR have only been available at the PoS for a few years and were quite unknown in Germany until recently [36]. For this reason, it cannot be guaranteed that respondents had gathered practical experience with SFR. In the context of this study, acceptance (dependent variable) was therefore measured according *behavioral intention (BI),* and the determinant *habit* was eliminated from the UTAUT2 model.

Additionally, to adapt the research model to the current context, the UTAUT2 model was expanded to include two further determinants. SFR are capable of increasing the quality of services by offering a variety of functions [31], such as additional product information and availability of alternate sizes and colors [12] or connections to the online shop using customer profiles [10]. A precondition, however, is that consumers are prepared to forgo interpersonal interactions [2]. According to studies, consumers' desire for personal interactions can lower their intention to use self-service technologies [2, 7]. For this reason, the determinant *need for interaction* was also adopted in the research model. Various studies on the acceptance of RFID technologies expanded the technology acceptance models to include determinants regarding data protection risk [4, 26, 43]. The present study therefore also included the construct *willingness to provide personal information (*WTPPI) from Dinev and Hart's [8] expanded Privacy Calculus Model. Because the use of SFR imposes no cost on consumers, the determinant *price value* was removed and replaced by the determinant *willingness to provide personal information.* Doing so takes into consideration that while consumers face no monetary costs, disclosing private data nevertheless poses a "burden" [43]. Additionally, in accordance with the UTAUT2 model, the study investigates whether the relationship between the dependent variable and independent variables are impacted by moderator variables such as age, gender and experience. The conceptualization of the determinants and the hypotheses are presented below.

Performance Expectancy (PE) is defined as the degree to which an individual thinks that using a certain technology will prove helpful, advantageous or valuable [41]. For example, SFR can personalize product recommendations, leading to more efficient decision making and other advantages [43]. Venkatesh et al. [40] posit that men are more strongly oriented toward tasks than women are. Additionally, *performance expectancy* has a stronger influence on *behavioral intention* for younger than for older users. These insights lead to the following hypotheses (**H1** marks the main hypothesis, **H1-Mx** marks the hypotheses on the moderating effects):

H1$_{\mathbf{(PE)}}$: PE positively influences BI.
H1-M1$_{\mathbf{(PE)}}$: The influence of PI on BI is moderated by the age of the user whereby there is a stronger effect for younger users.
H1-M2$_{\mathbf{(PE)}}$: The influence of PE on BI is moderated by the gender of the user whereby there is a stronger effect for men.

Effort expectancy (EE) is the measure of the ease of use of a technology [25]. Within the context of RFID services, Müller-Seitz et al. [24] determined that perceived user-friendliness, which is related to *effort expectancy*, holds positive sway over consumer acceptance. After considering the moderator variables *gender, age* and *experience* it was subsequently determined that expected effort is an especially significant factor for older women and inexperienced individuals [37]. Building upon this, the following hypotheses are postulated:

H2$_{\mathbf{(EE)}}$: EE positively influences BI.
H2-M1$_{\mathbf{(EE)}}$: The influence of EE on BI is moderated by the age of the user whereby there is a stronger effect for older users.
H2-M2$_{\mathbf{(EE)}}$: The influence of EE on BI is moderated by the gender of the user whereby there is a stronger effect for women.
H2-M3$_{\mathbf{(EE)}}$: The influence of EE on BI is moderated by the experience of the user whereby there is a stronger effect for inexperienced users.

The determinant *social influence* (SI) measures the extent to which a person perceives that other important people recommend the use of a new technology [25]. Blut et al. [2] research shows that SI exhibits a positive influence on use intention of self-service technologies. According to Venkatesh et al. [40] women and older users with less technology experience are more sensitive to social influence. As experience increases, this influence loses significance. This leads to the following hypotheses:

H3$_{\mathbf{(SI)}}$: SI positively influences BI.
H3-M1$_{\mathbf{(SI)}}$: The influence of SI on BI is moderated by the age of the user whereby there is a stronger effect for older users.
H3-M2$_{\mathbf{(SI)}}$: The influence of SI on BI is moderated by the gender of the user whereby there is a stronger effect for women.
H3-M3$_{\mathbf{(SI)}}$: The influence of SI on BI is moderated by the experience of the user whereby there is a stronger effect for inexperienced users.

The antecedent *facilitating conditions (FC)* measures the extent to which a person perceives whether the logistic or technical infrastructure is present to facilitate the use of the new technology [41]. Nysveen and Pedersen [26] determined that *facilitating conditions* have a significant positive influence on use indentions of RDIF capable services [26]. Additionally, they found that the effect of FC on BI is stronger for women and inexperienced users. This leads to the following hypotheses:

H4$_{\mathbf{(FC)}}$: FC positively influences BI.
H4-M1$_{\mathbf{(FC)}}$: The influence of FC on BI is moderated by the age of the user whereby there is a stronger effect for older users.
H4-M2$_{\mathbf{(FC)}}$: The influence of FC on BI is moderated by the gender of the user whereby there is a stronger effect for women.
H4-M3$_{\mathbf{(FC)}}$: The influence of FC on BI is moderated by the experience of the user whereby there is a stronger effect for inexperienced users.

Hedonic motivation (HM) is an affective factor describing the satisfaction and fun felt by someone while using a technological innovation [41]. Earlier empirical studies on acceptance of self-service technologies at the PoS show that the sense of enjoyment is especially relevant to a user's acceptance of a technology [2, 9, 35]. Research suggests that hedonic motivation dominates among women while utilitarian motivation prevails among men. Women seem to perceive shopping as a free time experience, while men perceive shopping as a task to be completed [29]. This leads to the following hypotheses:

H5$_{\mathbf{(HM)}}$: HM positively influences BI.
H5-M1$_{\mathbf{(HM)}}$: The influence of HM on BI is moderated by the age of the user whereby there is a stronger effect for younger users.
H5-M2$_{\mathbf{(HM)}}$: The influence of HM on BI is moderated by the gender of the user whereby there is a stronger effect for women.
H5-M3$_{\mathbf{(HM)}}$: The influence of HM on BI is moderated by the experience of the user whereby there is a stronger effect for inexperienced users.

Need for interaction (NFI) is defined as the "desire to retain personal contact with others (particularly frontline service employees) during a service encounter" [6]. Diverse studies confirm that consumers' desire for interpersonal interaction lowers the intention to use technology at the PoS [22]. Furthermore, the present study will explore whether moderators have an influence on NFI. This leads to the following hypotheses:

H6$_{\mathbf{(NFI)}}$: NFI positively influences BI.
H6-M1$_{\mathbf{(NFI)}}$: The influence of NFI on BI is moderated by the age of the user whereby there is a stronger effect for older users.
H6-M2$_{\mathbf{(NFI)}}$: The influence of NFI on BI is moderated by the gender of the user whereby there is a stronger effect for women.
H6-M3$_{\mathbf{(NFI)}}$: The influence of NFI on BI is moderated by the experience of the user whereby there is a stronger effect for inexperienced users.

Willingness to provide personal information (WTPPI): Providing personal information is necessary in order to utilize the different functions of interactive fitting rooms, e.g. customer ID, address, financial data, shopping history [8]. According to Weinhard et al. [43] the WTPPI construct has a significant positive influence on the use intention of SFR. The present research expands the model by Weinhard et al. [43] to study whether the moderators age, gender and experience influence this relationship. Since there are no existing empirical studies available, it is not possible to make a specific statement as to the nature of the moderating influence.

H7$_{(WTPPII)}$: WTPPI positively influences BI.
H7-M1$_{(WTPPI)}$: The influence of WTPPI on BI is moderated by the age of the user.
H7-M2$_{(WTPPI)}$: The influence of WTPPI on BI is moderated by the gender of the user.
H7-M3$_{(WTPPI)}$: The influence of WTPPI on BI is moderated by the experience of the user.

The developed research model, including relational pathways and interdependencies is pictured in Fig. 1. The hypotheses on the moderating effects, marked Hx-Mx were not drawn for reasons of clarity.

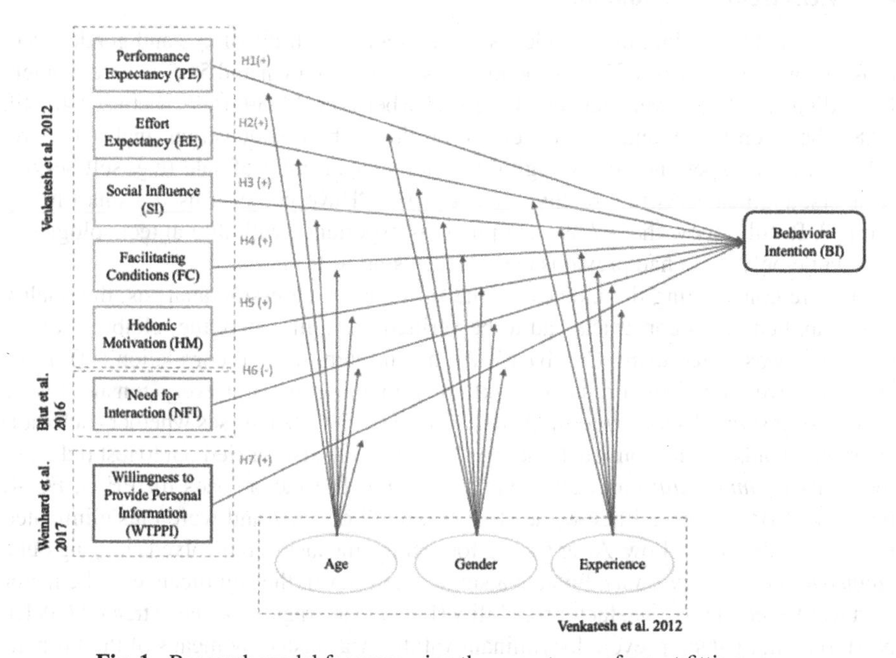

Fig. 1. Research model for measuring the acceptance of smart fitting rooms

3 Research Design

Data collection was implemented using a quantitative survey (standardized online survey). The selected population was German e-commerce users older than 18 years who purchased clothing offline at least once in the last 12 months and the participants were recruited via an online panel. Quotas were set in order to reach a sufficient number of respondents within each gender and age group. The sample is representative for German online shoppers with respect to age and gender.

The questionnaire begins with a presentation of the research object. The SFR and their functions were explained using pictures and the use cases for which necessary personal data is collected were clarified. Besides questions on respondents' use of self-service technologies and SFR at the PoS to date, the questionnaire largely focused on each construct.

To measure the latent constructs, existing applications from the literature were adapted and the research object adjusted accordingly. The constructs PE, EE, FC, HM, NF and WTPPI were measured across four indicators, while BI and SI were measured across three. Details of the application can be seen in Appendix 1. Each item was measured using a five-step Likert scale (1 = "strongly disagree" up to 5 = "strongly agree").

4 Results

4.1 Measurement Validation

Following data cleaning, the sample's sociodemographic traits of age and gender were broken down as follows: of 258 respondents, 48.5% were men and 51.5% were women. Regarding age, 20.9% were between 18–29, 14% between 30–39, 16.7% between 40–49, 12.8% between 60–69 and 16.3% were 70 or older. The first questions analyzed were: "Do you have experience using digital technologies at point of sale (e.g. self-service registers, augmented/virtual reality, scan & go)?"; "Have you ever used a smart fitting room?". Results show that 47.3% had previous experience with digital technologies at PoS while only 4.3% had previous experience using SFR.

Before constructing the indices for the subsequent regression analysis, the quality of the applied latent constructs had to be verified. As a first step, the reliability of the constructs was tested using Cronbach's Alpha and item to total correlation (ITC). All constructs were confirmed reliable. Next, the remaining items of every construct were tested using *single factor explorative factor analysis,* which assesses whether each indicator only loads on one construct (factor) [48]. This was also carried out to test indicator reliability (*commonality*) as well as *factor loading.* The factor loads for EE_3, FC_4, NFI_3 and BI_2 were all below the strict threshold of $\geq$0.7 and were thus eliminated from the analysis to follow. After testing for quality, the last step involved carrying out a *confirmatory factor analysis.* First, at a single factor level, the significance of the factor load was tested via t-value, factor reliability (FR) and average variance extracted (AVE). Next, on a multi-factor level, discriminant validity was tested by means of the Fornell-Larcker Criteria. All constructs fulfilled the criteria as deemed necessary in the literature [48]. The results are depicted in Table 1.

5 Hypotheses Tests

Direct Effects. Before conducting the regression analysis to test the proposed hypotheses, the necessary model conditions were verified. In examining the bivariate correlation matrix, it became apparent that the indexed regressors were partially highly significant inter-correlated, leading to distortions of the coefficients in the multiple regression. The factors *hedonic motivation* and *performance expectancy* ($r = 0.780$), *WTPPI* and *performance expectancy* ($r = 0.624$) as well as *facilitating conditions* and *effort expectancy* ($r = 0.793$) were especially highly correlated with each other according to Cohen [5]. This resulted in an inability to interpret the causal connections between every independent variable and the dependent variable within the multiple regression analysis [33]. For this reason, it was decided to consider the simple regression and single explained variations in order to obtain better insights into the influence of the predictors [47]. The detailed results can be seen in Table 2.

When simultaneously estimating all independent variables, the full model proved highly significant ($p < 0.001$), at a level of 82.6% ($adjR^2 = 0.826$), for explaining the use intention of SFR. Aided by the results of the simple regression and proportion of variance, the study was able to answer the research question, "*Which drivers and barriers exist regarding the use of SFR?*" (see Fig. 2).

The results show that *hedonic motivation* holds the greatest influence over *behavioral intention* ($\beta = 0.868$; $p < 0.001$) and explains a further 6.92% of additional variance above and beyond the other predictors. The second strongest factor was deemed to be *performance expectancy* ($\beta = 0.848$; $p < 0.001$). A more distant third place was the variable *WTPPI*, which had a significant impact on SFR use intention at ($\beta = 0.646$; $p < 0.001$). Furthermore, the variables *effort expectancy* ($\beta = 0.478$; $p < 0.001$) and *facilitating conditions* ($\beta = 0.479$; $p < 0.001$) also proved influential, but with weaker effects on use intention. The variables *social influence* and *need for interaction* did not prove to be influential in this study, causing hypotheses H3 and H5 to be discarded. All other main hypotheses were confirmed (see Table 3).

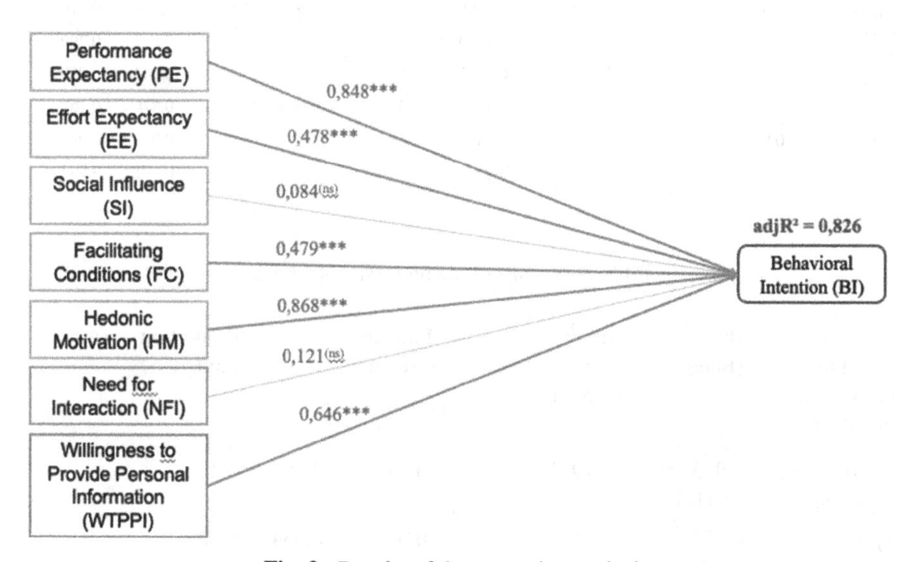

Fig. 2. Results of the regression analysis

Table 1. Confirmatory factor analysis results including quality criteria

Construct	Cronbach's α (≥0.6–0.7)	Average variance extr. (≥0.5)	Factor reliability (≥0.6)	Indicator	ITC (≥0.3–0.5)	α if Item is deleted	Sign. factor loading	Factor loading (≥ 0.7)	Commonalities (≥ 0.5)
Performance expectancy	0.920	0.808	0.944	PE_1	0.805	0,901	***	0,891	0,794
				PE_2	0.817	0,896	***	0,899	0,807
				PE_3	0.809	0,900	***	0,894	0,799
				PE_4	0.837	0,890	***	0,912	0,832
Effort expectancy	0.875	0.802	0.924	EE_1	0.766	0.822	***	0.897	0.805
				EE_2	0.735	0.849	***	0.880	0.775
				EE_4	0.789	0.789	***	0.909	0.827
Social influence	0.915	0.854	0.946	SI_1	0.830	0,875	***	0,925	0,856
				SI_2	0.835	0,871	***	0,928	0,861
				SI_3	0.819	0,884	***	0,919	0,845
Facilitating conditions	0.782	0.702	0.875	FC_1	0.650	0.672	***	0.862	0.743
				FC_2	0.695	0.629	***	0.883	0.780
				FC_3	0.526	0.819	***	0.764	0.583
Hedonic motivation	0.869	0.726	0.921	HM_1	0.805	0,796	***	0,907	0,823
				HM_2	0.773	0,812	***	0,884	0,782
				HM_3	0.560	0,817	***	0,719	0,719
				HM_4	0.761	0,893	***	0,876	0,767
Need for interaction	0.822	0.722	0.912	NFI_1	0.747	0.684	***	0.899	0.809
				NFI_2	0.690	0.742	***	0.870	0.758
				NFI_4	0.599	0.831	***	0.807	0.652
Willingness to provide Personal information	0.920	0.809	0.944	WTPPI_1	0.791	0,906	***	0,883	0,779
				WTPPI_2	0.808	0,899	***	0,894	0,799
				WTPPI_3	0.890	0,871	***	0,943	0,889
				WTPPI_4	0.779	0,908	***	0,876	0,767
Behavioral intention	0.938	0.943	0.971	BI_1	0.884	/	***	0.971	0.942
				BI_3	0.884	/	***	0.971	0.942

Table 2. Results of the regression analysis

Dependent Variable: behavioral intention	B[a] (Beta)	adjR^2 (ordinary linear regression)[a]	Multivariate correlation		Proportion of variance Expl.
			partial	Part	
Performance expectancy	**0,963*** (0,848)**	0,719***	0,345	0,151	2,28%
Effort expectancy	**0,657*** (0,478)**	0,225***	0,083	0,034	0,12%

(continued)

Table 2. (*continued*)

Dependent Variable: behavioral intention	B[a] (Beta)	adjR^2 (ordinary linear regression)[a]	Multivariate correlation		Proportion of variance Expl.
			partial	Part	
Social influence	0,084$^{(ns)}$ (0,077)	0,002^{ns}	−0,097	−0,040	0,16%
Facilitating conditions	**0,717*** (0,479)**	0,226***	0,003	0,001	0,01%
Hedonic motivation	**1,072*** (0,868)**	0,753***	0,538	0,263	6,92%
Need for interaction	0,140$^{(ns)}$ (0,121)	0,011^{ns}	0,050	0,021	0,04%
Willingness to provide Personal information	**0,672*** (0,646)**	0,414***	0,307	0,132	1,74%

(ns) = not significant/ *significance on 5%-level/ ** high significance on 1% level/ *** highly significance on 0.1% level.
[a]p-values for significance assessment of the regression coefficients, betas and corrected coefficients of determination are based on heteroskedastic-consistent standard error estimators (HC3).

Table 3. Test result of the main hypotheses

Hypothesis	Supported
H1$_{\mathbf{(PE)}}$ PE positively influences BI	Yes
H2$_{\mathbf{(EE)}}$ EE positively influences BI	Yes
H3$_{\mathbf{(SI)}}$ SI positively influences BI	No
H4$_{\mathbf{(FC)}}$ FC positively influences BI	Yes
H5$_{\mathbf{(HM)}}$ HM positively influences BI	Yes
H6$_{\mathbf{(NFI)}}$ NFI positively influences BI	No
H7$_{\mathbf{(WTPPI)}}$ WTPPI positively influences BI	Yes

Moderating Effects. The variables age, gender and experience were incorporated into the study model as moderating effects. Experience was measured based on the question: "How experienced do you consider yourself in using digital technologies at the point of sale?" To study the moderating effects, the sample was divided into two groups by means of the hypothesized moderators [39]. A multi-group analysis of partial samples followed, testing for significant difference between groups using the independent samples t-test [15].

For the moderator *age*, only hypothesis H7-M1 was confirmed. The results of the analysis show that there is a significant difference at a 5% level ($p = 0.019$) in the subgroups with WTPPI. This is in keeping with the hypothesis that the effect is moderated by age whereby a larger effect was seen in older users ($ß = 0.848$; $p < 0.001$). *Gender* confirmed the hypotheses for *performance expectancy* and *hedonic motivation.* It can be inferred from the results that the sense of utility is more significant for men than for women when it comes to engendering use intention. For women, however, *hedonic motivation* proved to be a stronger driver of acceptance. The degree of *experience* moderated the effect of *facilitating* conditions and *WTPPI* as hypothesized. While the variable *facilitating conditions* had no significant influence for experienced users, it proved highly significant for inexperienced users ($\beta = 0.612$; $p < 0.001$). The subgroups of the predictors WTPPI showed a stronger effect for inexperienced users at a level of 0.1% ($p < 0.001$). This means that the willingness to disclose personal information plays an especially important role for inexperienced users as opposed to experienced users considering their use intention. The remaining partial hypotheses were not confirmed (see Appendix 2).

6 Conclusion

6.1 Discussion and Business Implications

This study aimed to answer the following research question: What drivers and barriers exist regarding the use of SFR? The variable *hedonic motivation* has the largest influence on customers' behavior intention. As a result, retailers should focus on integrating entertainment value into SFR technology and prioritize communicating and positively influencing this aspect. Possible examples might be entertainment elements like projecting pictures of different environments or to reflect various occasions. Or, it could mean connecting to social media so that friends and family can participate in the buying process.

The second greatest influence belongs to the variable *performance expectancy*. The more a person expects to experience an advantage with SFR compared to using traditional fitting rooms, the more likely they are to use the technology. This finding is highly relevant in practice. Retailers utilizing SFR should give great attention to function that consumers perceive as useful, such as personalized product recommendations.

Additionally, results show that the predictor *willingness to provide personal information* offers an explanation beyond the UTAUT2 variables regarding the use intention of SFR. This study confirms once again that data protection risks are considered a decisive factor when integrating SFR at the POS. The results imply that consumers who aren't willing to provide personal information will not use SFR, in spite of their potential benefits. Retailers are advised to build a positive reputation to reinforce a foundation of trust and to be transparent about data privacy policies.

The variables *effort expectancy* and *facilitating conditions* have a highly significant positive influence on behavioral intention. Still, the effect for both predictors is weak. Since they constitute hygiene factors, retailers should thus place the focus for SFR on user friendliness as well as intuitive operability.

The variable *social influence* did not prove to be influential in this study. This means that the opinions of other people have little influence on a consumer's willingness to use SFR. An individual's rating of an SFR therefore appears to be largely based on personal factors, such as perceived enjoyment or perceptions of personal utility. Moreover, the postulated negative effect of the variable *need for interaction* could not be confirmed in the present study.

The analysis of the moderating effects presented in this study also reveal gender-specific differences. They indicate that men are more benefit-oriented and motivated by utility while women are led by hedonic motives. Employing gender-specific marketing strategies therefore shows considerable promise for successfully enticing both men and women's acceptance of SFR. It was also apparent that generally segmentation of consumers according to age is unnecessary. Regarding data protection, retailers who target older or inexperienced consumers are recommended to handle communication regarding personal data especially lightly. The results also show that the predictor *facilitating conditions* has a far greater effect for inexperienced users. This suggests that the provision of technical and logistical infrastructure (e.g. video tutorials that explain the functions of the interactive fitting room) is particularly relevant.

6.2 Limitations and Future Research

The present study is the first to research the factors influencing the use intention of SFR among a representative sample of the German population. By incorporating UTAUT2 variables and expanding the model to include other predictors, the research model proved capable of predicting 82.6% of the variance for use intention. Still, this means that 17.4% of variance remains unaccounted for. Furthermore, the results of the study show that the UTAUT2 variable *social influence* has no significant interdependencies with the research object. The fact that studies by Weinhard et al. [43] as well as Nysveen and Pedersen [26] also did not find a significant, or only very weak effect of social influence on the acceptance of service-based RFID technology indicates that acceptance is increasingly dictated by individual influencing factors. In future studies, it would therefore be interesting to explore further individual factors. That way factors such as *personal innovation level* or *fear of technology* could be adopted instead.

The study further demonstrated that concern over data protection is an important factor influencing the acceptance of SFR. Next to SFR, many other self-service technologies (e.g. self-service terminals or interactive window displays) use personal data. It is therefore recommended that the *willingness to provide personal information* predictor be taken up in future studies on the acceptance of various PoS technologies.

Beyond this, future studies to validate the research model should examine the extent to which the significance of influencing factors are related to respondents' previous experience with SFR. Experience proved to be a moderating effect in most path connections. This may only gain in importance considering the rapid expansion of SFR at the PoS, making it worthwhile to include the UTAUT2 variable *habit* as an independent variable in future studies. Additionally, as awareness and recognition of this technology grows, it would be advisable to study real use behavior as a dependent variable in several years.

In conclusion, a critical appraisal of the study's outlook shows that additional research continues to be necessary to best understand the acceptance of SFR and other self-service technologies at the PoS.

Appendix 1: Operationalization of the Constructs

Behavioral intention		Source
BI_1	If I had the chance in the future, I would use SFR	Nysveen/Pedersen (2014)
BI_2	I cannot imagine using SFR in the future	Venkatesh et al. (2012) Weinhard et al. (2017)
BI_3	If you had the chance, how likely would it be that you would use SFR?	Nysveen/Pedersen (2014)
Performance expectancy		
PE_1	I would find SFR useful for trying on clothing	Venkatesh et al. (2012)
PE_2	Using SFR would help me to try clothing on quicker	Venkatesh et al. (2012)
PE_3	Using SFR would help me make easier and more targeted decisions on articles of clothing	Venkatesh et al. (2012) Weinhard et al. (2017)
PE_4	The use of SFR would improve the experience of trying on clothes for me (e.g. through personalized product suggestions)	Nysveen/Pedersen (2014)
Effort expectancy		
EE_1	I would find it easy to use SFR	Venkatesh et al. (2012) Weinhard et al. (2017)
EE_2	I think using SFR is easy and straightforward	Venkatesh et al. (2012) Weinhard et al. (2017)
EE_3	I imagine the use of SFR is complicated	Venkatesh et al. (2012)
EE_4	I think that I could operate SFR without issue	Venkatesh et al. (2012)
Social influence		
SI_1	*Whether I use SFR in the future will be influenced by...* ... friends' or family members' recommendations	Venkatesh et al. (2012) Nysveen/Pedersen (2014)
SI_2	... friends' or family members' previous positive experiences	Venkatesh et al. (2012) Nysveen/Pedersen (2014)
SI_3	... whether friends or family members have used SFR in the past	Venkatesh et al. (2012) Nysveen/Pedersen (2014)

(*continued*)

(*continued*)

Facilitating conditions		
FC_1	With the help of a tutorial ("directions") that explains the functions for operating SFR, I think I would be capable of using one	Venkatesh et al. (2012)
FC_2	I think that my technical know-how is sufficient for using SFR	Venkatesh et al. (2012)
FC_3	With assistance from sales associates, I think I would be capable of using SFR	Venkatesh et al. (2012)
FC_4	I know how to find out more about operating SFR	Venkatesh et al. (2012)
Hedonic motivation		
HM_1	I imagine using SFR is entertaining	Venkatesh et al. (2012)
HM_2	I think it would be fun to use SFR	Venkatesh et al. (2012)
HM_3	I think it would be boring to use SFR	Venkatesh et al. (2012)
HM_4	It would be an interesting experience to use SFR	Tyrväinen et al. (2020)
Need for Interaction		
NFI_1	I like to receive personal recommendations when trying on clothes	Demoulin/Djelassi (2015)
NFI_2	Interacting with sales associates makes trying on clothes more fun for me	Demoulin/Djelassi (2015)
NFI_3	Receiving personal recommendations from sales associates when trying on clothes is not important to me	Demoulin/Djelassi (2015)
NFI_4	It bothers me when I do not get personal recommendations from sales associates when trying on clothes	Demoulin/Djelassi (2015)
Willingness to provide personal information		
WTPPI_1	I would provide my personal information (address and financial data) in order to place an online order through SFR	Dinev/Hart (2006)
WTPPI_2	I would register my customer account to use the SFR	Weinhard et al. (2017)
WTPPI_3	In order to take advantage of all the features of SFR, I would provide my personal information (e.g. customer ID, address & financial data)	Weinhard et al. (2017)
WTPPI_4	I would provide my personal information in order to access personalized content (e.g. personalized product recommendations)	Weinhard et al. (2017)

Appendix 2: Results of Hypotheses Tests for the Moderating Effects of Age, Gender and Experience

Age	Supported	*Gender*	Supported	*Experience*	Supported
H1-M1(PE)	No	H1-M2(PE)	Yes		
H2-M1(EE)	No	H2-M2(EE)	No	H2-M3(EE)	No
H3-M1(SI)	No	H3-M2(SI)	No	H3-M3(SI)	No
H4-M1(FC)	No	H4-M2(FC)	No	H4-M3(FC)	Yes
H5-M1(HM)	No	H5-M2(HM)	Yes	H5-M3(HM)	No
H6-M1(NFI)	No	H6-M2(NFI)	No	H6-M3(NFI)	No
H7-M1(WTPPI)	Yes	H7-M2(WTPPI)	No	H7-M3(WTPPI)	Yes

References

1. Balaji, M.S., Roy, S.K., Nguyen, B.: Consumer-computer interaction and in-store smart technology (IST) in the retail industry: the role of motivation, opportunity, and ability. J. Mark. Manag. **36**, 299–333 (2020)
2. Blut, M., Wang, C., Schoefer, K.: Factors influencing the acceptance of self-service technologies: a meta-analysis. J. Serv. Res. **19**, 396–416 (2016)
3. von Briel, F.: The future of omnichannel retail: a four-stage Delphi study. Technol. Forecast. Soc. Change **132**, 217–229 (2018)
4. Cazier, J.A., Jensen, A.S., Dave, D.S.: The impact of consumer perceptions of information privacy and security risks on the adoption of residual RFID technologies. Commun. Assoc. Inf. Syst. **23**(14), 235–256 (2008)
5. Cohen, J.: A power primer. Psychol. Bull. **112**, 155–159 (1992)
6. Dabholkar, P.A.: Consumer evaluations of new technology-based self-service options: an investigation of alternative models of service quality. Int. J. Res. Mark. **13**, 28–51 (1996)
7. Demoulin, N., Djelassi, S.: An integrated model of self-service technology (SST) usage in a retail context. Int. J. Retail Distrib. Manag. **44**(5), 540–599 (2016)
8. Dinev, T., Hart, P.: An extended privacy calculus model for e-commerce transactions. Inf. Syst. Res. **17**(1), 61–80 (2006)
9. Eyuboglu, K., Sevim, U.: Determinants of consumers adoption to shopping with QR-Code in Turkey. J. Int. Soc. Res. **9**(43), 1830–1839 (2016)

10. Hauser, M., Bardaki, C.: IoT-enabled customer experience in retail fashion stores: opportunities & challenges. In: 8th International Conference on Interoperability for Enterprise Systems and Applications (2016)
11. Handelsverband Deutschland – HDE e.V. (ed.): Online-Monitor 2021, Köln, 12 (2021)
12. Heinemann, G.: Die Neuerfindung des stationären Einzelhandels. Kundenzentralität und ultimative Usability für Stadt und Handel der Zukunft. Springer, Wiesbaden (2017). https://doi.org/10.1007/978-3-658-15862-0
13. Huang, T.-L., Liao, S.: A model of acceptance of augmented-reality interactive technology: the moderating role of cognitive innovativeness. Electron. Commer. Res. **15**(2), 269–295 (2015)
14. Huré, E., Picot-Coupey, K., Ackermann, C.L.: Understanding omni-channel shopping value: a mixed-method study. J. Retail. Consum. Serv. **39**, 314–330 (2017)
15. Janssen, J., Laatz, W.: Statistische Datenanalyse mit SPSS. 9, überarbeitete und erweiterte Ausgabe. Springer, Heidelberg (2017). https://doi.org/10.1007/978-3-662-53477-9
16. Kallweit, K., Spreer, P., Toporowski, W.: Why do customers use self-service information technologies in retail? The mediating effect of perceived service quality. J. Retail. Consum. Serv. **21**, 268–276 (2014)
17. Kim, J., Forsythe, S.: Adoption of virtual try-on technology for online apparel shopping. J. Interact. Mark. **22**(2), 45–59 (2008)
18. Lee, H.-H., Fiore, A.M., Kim, J.: The role of the technology acceptance model in explaining effects of image interactivity technology on consumer responses. Int. J. Retail Distrib. Manag. **34**(8), 621–644 (2006)
19. Maisch, B., Palacios Valdés, C.A.: Kundenzentrierte digitale Geschäftsmodelle. In: Fend, L., Hofmann, J. (eds.) Digitalisierung in Industrie-, Handels- und Dienstleistungsunternehmen, pp. 29–51. Springer, Wiesbaden (2018). https://doi.org/10.1007/978-3-658-21905-5_2
20. Mani, Z., Chouk, I.: Drivers of consumers' resistance to smart products. J. Mark. Manag. **33**, 1–21 (2017)
21. Metter, A.: Mit virtual promoter zum point of experience. In: Knoppe, M., Wild, M. (eds.) Digitalisierung im Handel, pp. 59–78. Springer, Heidelberg (2018). https://doi.org/10.1007/978-3-662-55257-5_4
22. Meuter, M.L., Bitner, M.J., Ostrom, A.L., Brown, S.: Choosing among alternative service delivery modes: an investigation of customer trial of self-service technologies. J. Mark. **69**(2), 61–83 (2005)
23. Mukherjee, A., Smith, R.J., Turri, A.M.: The smartness paradox: the moderating effect of brand quality reputation on consumers' reactions to RFID-based smart fitting rooms. J. Bus. Res. **92**, 290–299 (2018)
24. Müller-Seitz, G., Dautzenberg, K., Creusen, U., Stromereder, C.: Customer acceptance of RFID technology: evidence from the German electronic retail sector. J. Retail. Consum. Behav. **16**, 31–39 (2009)
25. Nistor, N., Wagner, M., Heymann, J.O.: Prädiktoren und Moderatoren der Akzeptanz von Bildungstechnologien. Die unified theory of acceptance and use of technology auf dem Prüfstand. Empirische Pädagogik **26**(3), 343–371 (2012)
26. Nysveen, H., Pedersen, P.E.: Consumer adoption of RFID-enabled services. Applying an extended UTAUT model. Inf. Syst. Front. **18**, 293–314 (2014)
27. Pantano, E., Rese, A., Baier, D.: Enhancing the online decision-making process by using augmented reality: a two country comparison of youth markets. J. Retail. Consum. Serv. **38**, 81–95 (2017)
28. Pantano, E., Timmermans, H.: What is smart for retailing? Procedia Environ. Sci. **22**, 101–107 (2014)
29. Pratibha, R.: Does consumer's motivation differ across gender? A study in shopping malls. Int. J. Manag. Stud. **2**(5), 23–28 (2018)

30. Rese, A., Baier, D., Geyer-Schulz, A., Schreiber, S.: How augmented reality apps are accepted by consumers: a comparative analysis using scales and opinions. Technol. Forecast. Soc. Change **124**, 306–319 (2017)
31. Rese, A., Schlee, T., Baier, D.: The need for services and technologies in physical fast fashion stores: generation Y's opinion. J. Mark. Manag. **35**, 1–23 (2019)
32. Roy, S.K., Balaji, M.S., Quazi, A., Quaddus, M.: Predictors of customer acceptance of and resistance to smart technologies in the retail sector. J. Retail. Consum. Serv. **42**, 147–160 (2018)
33. Schwarz, J.: Einfache und multiple Regressionsanalyse/Logistische Regressions- analyse. Skript im Modul Statistik im Masterstudiengang Pflege an der Berner Fachhochschule an der FHS St. Gallen (2012). https://www.yumpu.com/de/document/read/29990249/lehreinheit-5-regressionpdf-schwarz-partners-gmbh. Accessed 14 Jan 2021
34. Spreer, P., Kallweit, K.: Augmented reality in retail: assessing the acceptance and potential for multimedia product presentation at the PoS. Trans. Mark. Res. **1**(1), 20–35 (2014)
35. Srinivasan, R.: An investigation of Indian consumers' adoption of retail self-service technologies (SSTs): application of the cultural-self perspective and Technology Acceptance Model (TAM). Graduate Theses and Dissertations. Iowa State University (2014)
36. Szameitat, C., Blank, D.: Verbraucher nutzen digitale Tools am POS (2016). https://www.stores-shops.de/technology/verbraucher-nutzen-digitale-tools-am-pos/. Accessed 6 Sept 2020
37. Taiwo, A.A., Downe, A.G.: The theory of user acceptance and use of technology (UTAUT): a meta-analytic review of empirical findings. J. Theor. Appl. Inf. Technol. **49**(1), 48–58 (2013)
38. Tyrväinen, O., Karjaluoto, H., Saarijärvi, H.: Personalization and hedonic motivation in creating customer experiences and loyalty in omnichannel retail. J. Retail. Consum. Serv. **57**, 102233 (2020)
39. Urban, D., Mayerl, J.: Angewandte Regressionsanalyse: Theorie, Technik und Praxis. Studienskripten zur Soziologie. Springer, Wiesbaden (2018). https://doi.org/10.1007/978-3-658-01915-0
40. Venkatesh, V., Morris, M.G., Davis, G.B., Davis, F.D.: User acceptance of information technology: toward a unified view. MIS Q., **27**(3), 425–478 (2003)
41. Venkatesh, V., Thong, J.Y.L., Xu, X.: Consumer acceptance and use of information technology: extending the unified theory of acceptance and use of technology. MIS Q. **36**(1), 157–178 (2012)
42. Wang, C.: Consumer acceptance of self-service technologies: an ability–willingness model. In: Int. J. Mark. Res. **59**(6), 787–802 (2017)
43. Weinhard, A., Hauser, M., Thiesse, F.: Explaining adoption of pervasive retail systems with a model based on UTAUT2 and the extended privacy calculus. In: Twenty First Pacific Asia Conference on Information Systems (2017)
44. Wünderlich, N., et al.: "Futurizing" smart service: implications for service researchers and managers. J. Serv. Res. **29**, 442–447 (2015)
45. Wünderlich, N., Wangenheim, F.V., Bitner, M.J.: High tech and high touch: a framework for understanding user attitudes and behaviors related to smart interactive services. J. Serv. Res. **16**, 3–20 (2013)
46. Zielke, S., Toporowski, W., Kniza, B.: Customer acceptance of a new interactive information terminal in grocery retailing: antecedents and moderators. IGI Global (2011)
47. Zimmermann, V.: Der Konsument in der digital-kollaborativen Wirtschaft. Eine empirische Untersuchung der Anbieterseite auf C2C-Plattformen. Best Masters. Springer, Wiesbaden (2017). https://doi.org/10.1007/978-3-658-16652-6
48. Zinnbauer, M., Eberl, M.: Die Überprüfung von Spezifikation und Güte von Strukturgleichungsmodellen – Verfahren und Anwendung. In: Schriften zur Empirischen Forschung und Quantitativen Unternehmensplanung (EFOplan), 21 (2004). https://www.en.imm.bwl.uni-muenchen.de/04_research/schriftenefo/ap_efoplan_21.pdf. Accessed 8 Dec 2020

Unfolding the Practices of Live Streaming: A Dramaturgical Theory Perspective

Tsai-Hsin Chu[1], Yi-Ling Shen[1], and Yen-Hsien Lee[2](✉)

[1] Department of E-learning Design and Management, National Chiayi University, Taiwan, Republic of China
thchu@mail.ncyu.edu.tw

[2] Department of Information Management, National Chiayi University, Taiwan, Republic of China
yhlee@mail.ncyu.edu.tw

Abstract. Live steaming become an emerging significant phenomenon for creating huge market value in business. When it is reposed successful stories about the live streaming economy, researchers are interested in how a live streaming can be successful. Current studies explore this issue by investigating the motivations that can drive audiences watching a live streaming. However, it remains unclear about how and why the liver streamers perform particular practices to attract audiences and manage their live streaming. This study bridges this gap by applying the dramaturgical theory to explain the process where live streamers manage consumer's sensemaking process with image creation. The dramaturgical theory regards individual daily behaviors as 'performances in front of others.' Specifically, an individual is like an actor, who manage her image by performing certain behaviors in front of people. This theory provides an opportunity to analyzes the live streaming practices by revealing the live steamers' understandings that constitute and support their 'performances' in the live streaming platform. Our findings identify three types of performance, namely idols, master players, and market seekers. Our analysis discusses the three types of performance by describing the live stream understandings on the stage (i.e. the live streaming industry), on the audience they faced, and on the roles they played (i.e. as live streamers) to unfold the logic embedded in live streaming practices. Our findings can contribute to deepen current understanding on live streaming practices and provide insights to aim live streamers to leverage audience's sensemaking for image creation.

Keywords: Live streamer · Dramaturgical theory · Stage · Role played · Performance

1 Introduction

Live streaming is noticed by organizations with repaid growth and high potential on creating market value. Many successful stories are reported to present the power of live streaming in business. In the United States, for example, Ninja who is a well-known live streamer earns more than US$500,000 per month by live streaming the game 'Fortnite'

F. Fui-Hoon Nah and K. Siau (Eds.): HCII 2022, LNCS 13327, pp. 463–480, 2022.
https://doi.org/10.1007/978-3-031-05544-7_34

on Twitch [1]. Also, another well-known live streamer on Twitch, SypherPK, received about US$203,702 per month from subscriptions, advertising, and endorsements [2]. For another example, a Swedish live streamer, PewDiePie was estimated to earn US$4 million in 2013 and US$12 million in 2014. PewDiePie's channel on Youtube has more than 54 million the subscriptions in 2017 [3].

When business draws the eyes on the amount of value brought by live streaming, we know little about how to create a success live stream. Current literature studies the live streaming phenomenon by examining audiences' behaviors and motivations on watching live streaming[4–6]. However, it remains unclear about how liver streamers manage their image to attract people in live streaming. This gap might let us neglect the important mechanism that live streamers manipulate to develop strategies and practices on create a para-social relationship with the audiences.

Current studies suggest that audiences actively endow and use the cultural meaning of celebrity to build the para-social relationship [7, 8]. These studies suggest that celebrities are the source and carriers of cultural meanings (e.g., fashion) given by consumers, and endorsements refers to transform those cultural meanings into specific items (e.g., leather bags) into customers' self-concept through consumption (e.g., I am a stylish person) [7]. In this process, consumers use the meaning brought by celebrities as a means to establish self-identity, and establish a para-social relationship between themselves and celebrities [7].

In live stream context, however, the creating of cultural meaning is a challenge task because of the nature of live streaming. First, live streaming is highly competitive sine everyone can conduct live stream with a low cost [9]. There are up to 3.8 million streamers broadcasting on Twitch in 2020 [10]. Second, the audiences can easily switch among one live streaming to another with a low cost. Third, live streaming creates authenticity and is not edited as the live streamer and the audiences interact directly and immediately. Fourth, the attention of audiences to live streaming is rather short. According to a Chinese report, audience spent only 3 min in average to decide whether they switch to another live steaming channel. These data collectively show that live streamers compete the audiences' attention and retention in a very short time. If the live streamers cannot successfully make audiences create a desirable cultural meaning in minutes, the audience can easily switch to another channel and the live streamer lose the audience. Live streamers can management the audience meaning creation by providing appropriate cues. The meaning creation process, also phased as sensemaking process, is triggered by the external cues an individual perceive and observe [11]. In this sense, live streamer had to strategically plan the content and the way of interaction in order to facilitate image creating.

This study applies Goffman's (1956) dramaturgical theory to explores a research question: how live streamers manage audience's sensemaking process to build particular image in live streaming. The dramaturgical theory emphasizes that every individual is an actor in daily life as she or he performs particular behaviors "in front of people" to shape her or his image. This theory explains how an individual act in front of people to manage impressions [12, 13]. The dramaturgical theory provides an opportunity to analyze how the live streamers manage image, and to provide live streamers perspective to live streaming phenomenon to fill the theoretical gap. Here, we regard live streaming practices as a "performance", and explore the relationship between liver streamers' understanding on stage, audience, and the role they played as a streamer.

2 Literature Review

2.1 Live Streaming

Live streaming allows live streamers to deliver real-time video and audio to the audiences, as well as to have concurrent two-way interactions with them [14]. The live streamers can respond to the audiences, and the audiences can actively participate in the live streaming by posting messages in chatroom, making donations, and sending virtual gifts to the live streamers [14]. In business, live streaming creates a disruptive way to interact with customers as audiences. As live streaming can be accessed by mobile phone with a low budget, it attracts the broader audiences on Internet [9]. Also, live streaming can build direct and real connections between liver streamers and the audiences. Through the real-time live streaming, the liver streamer can interact with the audiences and get feedback immediately [9]. In addition, live streaming allows audiences to build a community where the audiences establish contacts with each other [9]. In this light, the community in live streaming share a particular atmosphere to create specific collective behaviors among the audiences. Moreover, live streaming builds a sense of authenticity and transparency to enhance audiences' trust since it is done as an improvisation without editing [9].

2.2 Research on Live Streaming

For live streamers who expect to leverage the benefits of live streaming, how to compete and retain audiences to win the market is an important issue. Many studies investigate the way to be a successful live streamer. One sort of research concerns IT affordances that live streaming technology can provide. For example, Sjöblom, Törhönen, Hamari and Macey [4] conclude the critical IT features shared across 100 successful live streamers on Twitch. Their findings indicate that the most commonly used features are a superimposed screen with the live streamer and the gaming screen with the live streamer's explanation on microphone. In this sense, the camera and microphones serve as the affordances that create a stage for live streamers to position themselves as celebrities. In addition, a pop-up window to nominate the latest subscribers and donators is also popular feature used. Pop-up window is the affordance to create a sense of honor to encourage audiences to donate and subscribe, thereby increasing the live broadcasters' income.

Another sort of research focuses on the audiences' motives and behaviors of watching live streaming. For example, Diwanji, Reed, Ferchaud, Seibert, Weinbrecht and Sellers [5] point out that reactive action is the most common type of information behavior found in the live streaming room. Therefore, live streamer can attract audiences' interactions by producing and responding messages through chat room to invite reactive actions. The study also suggests that the live streaming frequency, time of duration, and dialogue with the audience are important factors of retaining audience participation. In addition, it reports that the audience's cognition and emotion affect the time and the number of live streamers that they want to watch [15]. Research also shows that the audience's sense of identifications with the live streamer affect the audience's willingness on continuously watching the live streaming [6]. Stress release, social integration, and the audience's emotional factors are positively correlated to the time the audiences spent on watching a live streaming [16].

When current literatures increase our understanding on live streaming phenomenon by highlighting the importance of technological features and audience behaviors, few of them reveal the perception of live streamers. This ignorance might be regrettable because we might neglect the important dynamics on how live streamers create and manage the live streaming practices. To bridge this gap, this study explores the live streamer's perception on how they create their live streaming by unfolding the governance logic and the embedded rationality through Goffman's dramaturgical theory.

2.3 The Dramaturgical Theory

The dramaturgical theory, suggested by the American sociologist Erving Goffman, advocates that human society is a big stage and everyone in the society can be regarded as an actor. In her or his daily life, each individual in human society performs in front of others by a specific appearance, talk, and behaviors to shape the specific image [17]. In the dramaturgical theory, Goffman distinguishes the social situations into "frontstage" and "backstage," where the frontstage refers to the field that an individual interacts and communicates with others while the backstage is the occasion for the individual to be alone [13, 18]. Individuals (as actors) perform in front of people through leveraging environmental settings (such as backgrounds, props, decorations, etc.) and personal appearances (such as appearance and behavior) to present the image that she or he wants to be perceived by people [12, 17]. In other words, an individual create performance as way of impression management to shape and reinforce their self-image in front of others through showing appropriate appearance, behaviors, and background props that match the ideal character [12, 17]. Here, performance refers to the methods used by actors in the frontstage to make the audiences believe the roles they played are real and to convince the audiences that the actor in front of them is such a person [12]. A successful performance needs to include three elements: dramatization, idealization, and mystification [12]. Dramatization means that actors apply exaggerated movements, tone, or specific behaviors to highlight the personality or characteristics of a role. Idealization refers to performing in a way that meets the audiences' expectations toward the role. The actor will present behaviors that match the expectations of the role, conceal the behaviors that are inconsistent with the role expectations, to shape the role's positive impression [17]. Mystification means that the actors maintain proper social distance to create a sense of mystery, so that the audience can earn respect and awe [12, 18].

An effective performance is not fanciful but needs to be designed. In order to convince the audience of the authenticity of the role impression, actors perform based on their interpretations toward the role, the stage, the script, and audience expectations [13, 19]. Before the performance, the actor makes sense of the role with script and perform the role-played accordingly. According to their interpretations on the script, actors posit particular posture with specific tones to develop the expected impression and to deliver meanings to the audiences. When the script is ambiguous, the actors fill the gap by their own interpretation on the stage and role to adapt to the uncertainty [13, 19]. As actors may have different interpretation to a role due to her or his experience and training background, the individual's performance vary during the process of impression management [17]. Therefore, to better understand an actor's performance, we must unfold that person's interpretations of the role and script.

To our study, the dramaturgical theory provides a perspective to explore the underlining logic of how live streamers manage their live steaming show. Here, we regard the live streaming show as a "frontstage," and the live streamer as the "actor". Specifically, we analyze the liver streamers' (e.g. the actor) performance by revealing the live streamers' understanding of the live stream industry (e.g. stage), of a live streamer (e.g. the role) and the fans (e.g. the audiences).

3 Research Methods

This study conducted an interpretive case study to explore how live streamers manage their impression that can be perceived by the audiences through performance. We applied theoretical sampling for selecting cases. Theoretical sampling referred to a sampling method guided by theory; that is, the selected case phenomenon had the potential to explore specific research issues [20]. We selected the cases from Twitch platform as the target platform for the study. There are based on four reasons. First, Twitch is a platform dominated by live streaming, thus the observed phenomenon might exclude potential disturbance brought by other features created for social media platform. Second, Twitch was the largest live streaming platform and has a high rate of usage. In 2017 and 2018, Twitch became the 30th most viewed website in the world, with approximately 15 million users going online every day [21]. Third, live streaming on Twitch was critical to live streamers because the number of views, subscriptions, and donations were directly relevant to live streamer' revenue of live. And fourth, there are many senior and experienced live streamers on Twitch.

The research design was iterative, which that the research questions were emerged along with the field study. With the continuous dialogue between data and theories, we gradually converged and obtained research findings. This research design not only consist to the principle of theoretical sampling, but also allowed researchers to determine which groups to target, what data to collect, and where to obtain these data based on the results of the preliminary analysis[22]. It can make it more flexible to conduct an in-depth discussion on the concepts emerging from field data [20].

3.1 Data Collection

This research uses interviews and achieve data to collect our data. For interview, we sought live streamers' understanding on the live streaming industry, themselves, and audiences. For archive, we collect relevant articles to extend our understanding toward the live stream industry, as well as enhance the observation on live streamer's performances on live streaming.

This study applied the follow steps to recruit the potential interviewees. First, we search for the live streamers' information on Twitch based on the number of views. We distinguished three group of liver streamers with by accumulative views (i.e. above

50,000, 10,000 to 50,000, and less than 10,000). We created a pool of 20 live streamers for each group, and sent interview invitations to these live streamers' e-mail or Facebook fan page. In this round, a total of 65 interview invitations were sent but received only 2 replies. Then, we create another round of recruiting process and sent out 23 interview invitations, and had 5 more live streamers agreement on the interview. Second, for live streamers who agree to be interviewed, we schedule interviews with them individually. Due to the COVID-19 pandemic, all the interviews were performed through online communication.

The interviews were conducted in a semi-structured manner and open-ended questions. Before the interview, the researcher draws up an interview outline to guide the conduct of the interview, but the topics of the interview are not limited to those listed in the interview outline. The interview outline included the background of the live streamers, the live streamers' understanding of the live streaming and being a live streamer, how the live streamers view the relationship with live streaming management and fans, and the live streamers' online management. A total of 7 live streamers were interviewed in this study, and the average length of the interview was 80 min. The basic information of the interviewees is shown in Table 1.

Table 1. Basic information of interviewees

ID	Gender	Tenure	Number of followers	Average number of monthly streaming	Average hours of streaming	Average cumulative views per video
S1	Male	9 years	465,000	2.4	7 h 4 m	22,892
S2	Male	10 years	127,000	1.3	8 h 17 m	20,541
S3	Male	1 years	51,000	1.1	6 h 55 m	7,134
S4	Female	1–2 years	27,000	1.05	3 h 30 m	8,800
S5	Female	4–5 years	22,000	1.37	8 h 33 m	4,678
S6	Male	4–5 years	22,000	1.1	2 h 35 m	4,474
S7	Male	1 year	8,498	1.1	5 h 51 m	2,394

3.2 Data Analysis

In data analysis, we used two levels of data analysis methods [22, 23]. In the first-order analysis, open coding is the first step. In all the transcripts, the researcher reviewed and extracted important concepts sentence by sentence, and marked the label that could express the concept. Next, we group label with similar concept into categories. From the categories, we sorted out a preliminary outline to emerge the themes. The second-order analysis focused on the dialogue between theory and our field data. In this process, the researcher used dramaturgical theory as a theoretical lens to interpret the field data. With this, our analysis focused on live streamers' understandings on the stage (i.e.

the live streaming industry), on the audiences, on the role (i.e. as a live streamers), and the performance (i.e. live streaming practices). We figured out the linkage among these themes in order to generate our interpretation on how the live streamers manage impression in live streaming.

4 Research Findings

By analyzing the live streamers' interpretations, we find the three types of live streaming performances to explain the path of impression management. The three types of practices were named idols, master players, and market seekers. The three types of performances were explained by describing the live streamers' understandings on the stage (i.e. the live streaming industry), on the role played (i.e. the live streamer), on the audience they faced, and on their performance at frontstage (i.e. live streaming practices).

4.1 Idols

The stage: what is the nature of the live streaming industry

How an actor performed her role was relevant to her understanding toward the setting. In our research context, a live steamer's understanding on the nature of the live streaming industry might frame the way of position and strategies acted in the industry. How did the idols understand the stage? To balance work between interest was the noticed benefit that the idol players could leverage by live streaming. However, high degree of uncertainty and competition made it challenging of being a successful live streamer.

Live streaming industry is uncertain and unpredictable

The idols interpreted the live streaming industry as a high competitive market with full uncertainty. In this sense, live streamers might not have comparable return with the effort they spent. Therefore, a success could be a mystery and was attributed to timing. Two idols noted:

> "The live streaming industry, I think… It is like bubble... It is unpredictable… everything is hard to predict. You need to hold on the chance and seize the opportunity when the timing comes. It is important." ~ S2

You need stable traffic

Traffic referred to the number of visit on a website, and was one of the important indicators of the live streaming performance. To live streamers, how to create a stable traffic to their channel was critical to win the competition. For this, idols emphasized to create memory points to impress audiences. For example, an idol explained:

> "Live streamers need traffic for survive. It is a well-known fact. …You posited memory points to impress audiences for creating traffic." ~ S5

The audience: what is the nature of my audiences

Live streamers' interpretations toward the audiences might also frame the practices that they acted in live streaming. Idols noted their audiences were a group of young men who liked to watch gaming playing and with a high sense of Internet culture. The constitution of the group might be diverse: some were skillful game players, some were attracted by appearance of live streamer, and some were sugar daddies who sponsored generously.

The audiences liked to watch me playing games

Idols noted that most of audiences visited their channel for watching them played games, as well as interacted with them. One noted:

> "Most people come to see you playing games. They didn't expect to learn from your operation... Most of them just want to watch your live streaming. They also wanted to interact with you." ~S5

The role: what is my position as a live streamer

The live streamers would make sense to the role they played as a live streamer with their interpretations on the stage and the audiences. They sought role position to impress the audiences by creating particular para-social relationship between them and the fans. The idols attracted and retained audiences can by his (her) good appearance, nice personality, and particular characteristics, such as good game players.

I am a popular idol

The idols managed their image as an online celebrity who were famous for good looking and nice personality. For example, one streamers mentioned:

> "Fans liked me. The word "liked" was more likely as adoration. I was an idol." ~S4

The live streamers leveraged their advantages to become an idol. Some female observed the fans preferred to interact with girls and made good use of female identity to impress the audience. Instead of presenting as a distanced super star, these female live streamers develop their image as a girl-next-door. One live streamer said:

> "You had to find good personal characteristics to attract the audiences. The first one, your appearance should be attractive. For girls, the second one, you had lovely voice. Then, you paid attention to the way you speak. You needed to figure out the personality you set for the role. This is very important when you got start as a live streamer." ~S4

The male live streamer builds his gentle and sincere impression through calm and gentle voices when he replies to the audience. One live streamer said:

> "The personality I set was calm and gentle. I didn't make suddenly noises and my voice is stable and comfortable. Therefore, I could comfort the audiences when they participated in my live streaming." ~S2

The performance: how to do at live streaming

Sensitive to the uncertainty of the live streaming industry, idols concentrated on making personal characteristics salience to impress the audiences. Good appearance, sweet or gentle voice, and positive personality were the elements theses live streamers used to shape the image as an idol. Also, as an idol, the liver streamers spent more effort on maintaining relationship with the audiences. Showing intimacy to get close to fans as possible was the major way of their performance.

Leverage personal characteristics to create traffic

The idols presented her or his personal characteristics to positively manage the audiences' impressions. They performed as sweet person or gentlemen with high skill of playing games. One idol described her performance as an example:

> "What I could attract the audiences was that I am a girl who played games well. ... I'm pretty and a skillful game players. My record reached to Diamond in League of Legends." ~S5

Get close to the audiences to maintain 'likes'

The idols concerned to build good para-social relationship between them and the audiences. With the expectation to be liked and be adored, they usually performed intimacy to get close to the audiences. They observed and responded to chatroom messages along with game playing. They also created and joint the small talks among the fans. A live streamer said:

> "In my live streaming, most of time I would focus on the game I play. I would also create a subject of a talk with the game to invite discussion with the audience. .... I would ask them, for example, "how should I play the game now? Any comment?" ...By this, I could invite a bit more interactions with the audiences. I played the game seriously. And, at the same time, I would respond to the chat room discussion as much as possible." ~S5

The idols also perform intimacy by disclosing personal affairs. They often joked with the audience to show their closeness to the people. For example, a live streamer said:

> "The topics I shared in my live streaming channel were more relevant to daily life. ... It would make you closer to the audiences. For example, many of my audience knew that I'm twins. The audiences were surprised when I invited my sister performed together in my live streaming. After that, once I live streamed alone, the fans would say whether my sister substituted me today. They would make jokes to me." ~S5

However, the idols manage intimacy with appropriate distance to create a sense of mystery and avoid harassment. One live streamer mentioned the importance of keeping an appropriate distance from the audience:

> "if you got too close, some people might not treat you as a celebrity, and it might be troublesome. …I maintained a distance between live streamer and the fans. I knew they were the people who liked me, and I appreciated for their supports. …My way to return their favor was to provide a better channel for comforting them online." ~S2

4.2 Master Players

The stage: what is the nature of the live streaming industry

Most of the master players were professional e-sports players who won some online competition. They understood the live streaming as the substituted career path when they retired from the e-sport field. They sought the opportunity for leveraging their mastery on games to create income in live streaming industry.

Live streaming can increase visibility

As a professional e-sport player, the master players sought visibility in order to have more chances to be recruited by e-sport teams, as well as to get the opportunity of being sponsored. Observing the popularity of live streaming, the master players interpreted it as an effective channel to win visibility. By live streaming, the master players could easily present their mastery of games to community and could recruit a group of followers. One live streamer explained that:

> "The connection between e-sports games and live streaming was to win visibility. You might be a good e-sport player who won games. But only the people in e-sport industry knew who you were. If you went live streaming, you would go public and many people might become your fans by click-and-watch. …Business opportunities came, too. Computer equipment and game companies might ask you for endorsement." ~S7

Live streaming can balance interest and income

To the master players, they admired the benefits of live streaming with that it could balance their interests of playing games and making income. They leveraged their experience as professional players to demo advanced skills and answered the audience's questions to extend the sources of income. One said:

> "You could play games and interact with fans, it's pretty good. … I thought this work quite interesting … People would donate and subscribe to support your performance in live streaming. Live streaming brought extra income and create the opportunity of endorsement." ~S3

Also, the master players understood live streaming as a substitute career path. Because e-sport player usually retired in young age, they had to develop another career path for making income. Therefore, live streaming was regarded as the appropriate market for them to leverage past experiences to create a new future. For example, one live streamer mentioned the motive that he started live streaming, he said:

> "The career of an e-sport player would be terminated about 25 and 26 years old. Now, I am 23 years old. … I did live streaming for develop my career for the next stage. There was no age limitation for live streaming, I could switch to this industry." ~S7

The audience: what is the nature of my audiences

The master players understood their audiences as a group of people who came to see and learn advanced skills. Although most of the audiences were learners, the master players found that some audiences wanted throw down the gauntlet to them.

People who want to see advanced game operations.

Master players explained the reason why audiences came to their live streaming by their identity of professional e-sports players. They understood that audiences were curious about how a profession played a game with excellent skills. To master players, they interpreted that to watch an exciting game was the audiences' expectation. Master players also noted that the audiences were more satisfied if they could learn some skills from live demonstration or answer of questions. One live streamer said:

> "My live streaming attracted people who wanted to see my excellent operations. It mainly depended on how capable I showed. They can learn a lot from watching my live demonstration. ... Many people came to my channel to see an exciting game. They didn't like to watch I playing game in a weird way." ~S3

The role: what is my position as a live streamer

The master players interpreted their role of live streamers as masters with stable and excellent gaming skills. Although the task of playing game was serous, they didn't want to create too much distance with the audiences. Instead, they wanted the audiences perceived that they were interesting people in daily life.

I am a master

The master playerss were proud of their identity of professional players and liked to present the excellent game skills in live streaming. While demonstrating operations, the master players explained the strategies they applied to deal with the situations. One live streamers said:

> "The appealing point was that I played games pretty well. But, I wanted the audiences know why I could play games well. That was why I explain my practices in live streaming. I hope others could see why I play games well." ~S7

I am easygoing

Compared with other e-sports players, the master players believed that they had a chance to be successful live streamers because of the easygoing and open-minded personality. For example, two live streamers explained that:

> "They said that I'm an easygoing person. They liked to learn something in my live streaming...Different from many introverted professional players whose live were only games, I'm quite extroverted and talkative. ... Being a successful live streamer, you must be also good at chatting with the audiences. I could do it very well." ~S7

> "I could play game very well. It was my advantage to attract people to watch my live streaming...However, I'm introverted and a little bit quiet in a team. It seemed to be grim when I played games seriously. I kept turning over this image by saying something funny." ~S3

The performance: how to do at live streaming

To seek career developing in live stream industry, the master players managed their performance by showing their mastery. To prevent the atmosphere of live streaming went too serious, in addition, they made joke to the audiences to present intimacy.

Theses live streamers wanted to present the images of profession and funny person to their audiences.

Show an exciting game to present profession

The master players showed an exciting game by presenting excellent operations to perform their role as masters. They knew that many audiences came for their profession entertainment on playing games. These live streamers didn't hesitate to show their capabilities to attract and retain the audiences. They also told the audiences about the game strategy to show the level of mastery of the game. One live streamer explained what he presented the mastery:

> "I knew each character in this game very well. For example, what skill it mastered, and how fast it moved. I would inform these to the audiences. I explained my strategy on leverage this information on controlling situation, and the signal I used to let my team member moved at the best timing." ~S7

Perform a challenging goal to enhance the perception of mastery

The master players understood that the audience expect a wonderful and exciting game. In their performance, these live streamers chose to challenge the master teams on advanced server. For example, a live streamer chose to play the League of Legends on South Korea server as a regular content of live streaming. He explained that:

> "I usually chose to play game on South Korea server because it was challenging. The game would be excited since there were many good players on that server. When I play game in South Korea server [Note: League of Legends Korea Server], I would become very serious because it is my job." ~S7

Another live streamer also mentioned his performance by setting goal conquered. He noted:

> "I would set a goal in my live streaming today. For example, how many places we wanted to reach and clear. Or I set a goal to reach a particular ranking, such as Top 50. It attracted lots of audiences to follow and support me for completing the challenging goal." ~S3

Show funny to build intimacy

The master players performed intimacy to the audiences through funny ways of speaking. The live streamers leverage this way of speaking to make the audiences impression of humor. For example, a live streamer said:

> "I performed funny in live streaming. I was not good in speaking English, for example, I made fun by directly translating a Chinese term into English and pronounced that word with weird tone. It brought good feedback by the audiences. They said that I was a dork. …This way of interaction would be like daily chats among friends…The audience liked this and regarded you an interesting person." ~S7

4.3 Market Seekers

The stage: what is the nature of the live streaming industry

The market seeker perceived the live streaming industry as a competitive market, and seeking an appropriate segment to provide service was important for survival. They observed that the competition of live stream industry was cruel, and they must have enough traffic to win the opportunities of business endorsement invitation. To them, live streaming could be an effective instrument to make money.

Live streaming is a competitive and cruel market

Market seeker understood that live streaming was full of competition. It was also a cruel market because the platform only supported the famous live streamers with amount of traffic. Without enough traffic, the platform would not share revenue with the live streamer. Also, the platform set limitation to live streamers, such as shorter duration of keeping their live video. For them, the income would only rely on audiences' donation. Two interviewees talked about their observations and said:

> "The famous live streamers could sign a contract with the platform. The platform paid him a fixed salary every month with requiring certain hours of live streaming per month as a condition. …I'm not famous enough on the platform, so I didn't get this contract. I received no money unless the audiences donated me. …" ~S6

Traffic brings business opportunities.

The market seeker emphasized the importance of traffic to run business around live streaming. If their live streaming had a stable amount of traffic, it could attract companies to invite for advertising and endorsement. For example, a live streamer said that:

> "When your traffic became stable, you could receive more companies' invitation for advertising. You could endorse some products. They might also sponsor you with products for rewarding your audiences. You could get many benefits." ~S6

The audience: what is the nature of my audiences

The market seeker believed that their audiences were opened to the multiple contents in live streaming. Most of them were loyal fans who supported and donated the live streamers for a long time.

Audiences have various interests

Market seekers understood their audiences as people who opened to different types of content and their interests were diverse. By monitoring the number of views and participants among live streams, market seekers found that the audiences not only liked to watch game playing but also were interested in conversations about political event and life matters. One streamer mentioned his observation:

> "The audiences' favorites were uncertain. Sometimes political contents were also popular. You might have 7,000 to 10,000 audiences in your regular live streaming. But you would have 35,000 when you talk about hot events about politics and vote" ~S1

The audiences are passionate fans

The market seekers found a group of fans who followed them for years. These fans built a close connection with the live streamer. The fans would help the live streamers for solving problems, and they also support the live streamer at offline events. For example, one streamer said that:

"The fans were very enthusiastic. …Once we hosted an offline e-sports event, fans came from distanced cities. The event was open in the afternoon, but the fans gathered at 8 am in the morning." ~S1

The role: what is my position as a live streamer

The market seekers sought to create a positive impression as they were worth-to-trust celebrity. For them, to develop the audiences' sense of trust was important for endorsement.

I am a trustworthy person

Market seekers understood their positive personality traits attracting audiences and getting trust from them. A live streamer talked about the role expectation as a live streamer:

"As a celebrity, we would often promote something good. … We want to lead people to a positive way of thinking. It was my expectation for being a live streamer." ~S1

In addition, the market seekers wanted to be trustworthy persons in the audience's mind. It was because the success of endorsement depended on the trust of the audiences. The market seekers develop the sense of trust by showing their sincere. One live streamer mentioned the importance of building trust and said:

"If the audiences did not trust you, they would not believe you. They would not want to watch your live streaming or buy anything you endorsed. Thus, I thought that live streamers had to be serious about this. I managed the impression by presenting sincere to them." ~S6

I am a skillful game player

The market seekers understood their uniqueness of game playing to particular market. They sought to attract the particular audience with specific interests. For example, a live streamer said:

"When I was making a video of this game (i.e. Tom Clancy's Rainbow Six), I was just a normal game player without any title. And then, I became a master. I represented Taiwan in a world competition and won the highest ranking." ~S6

The performance: how to do at live streaming

The market seekers interpreted the live streaming industry full of cruel competitions. To won the competition, they had to select appropriate market segment and to grasp timing for gathering traffic in order to generate revenue. The market seekers acted on their performance in live streaming by establishing an identity of being a trustworthy master game player in the live streaming room.

Identify market opportunities for creating traffic

The market seekers acted on their performance for making traffic. These live streamers noticed that all the business invitations would be depended on the traffic of their live streaming channels. Therefore, market seekers paid much of attention on looking for opportunities that could attract audiences. For example, a live streamer said:

> "In 2017, I started making video to teach audiences how to play Tom Clancy's Rainbow Six [Note: a tactical shooting game]. That that time, I searched on the Internet and found that no videos in Chinese. I knew I could make those videos, and I found a business opportunity. An instructional video had high potential to be popular because people needed it when they played the game. d it." ~S6

Try new games to maintain traffic

The market seekers also demonstrated new games to extent customer bases for maintaining traffic. For example, a live streamer mentioned the strategy of successful live streaming, and he said:

> "Don't keep your contents on only one game. It would make your audiences less and less because the audience might get tired of the game. So, you must bring new ideas and always create new themes for retaining your audiences. You had to let them feel that your live streaming would bring new fun to them." ~S6

5 Discussion

This research applied dramaturgical theory to analyzed liver streaming practices by unfolding liver streamer's interpretations on the industry, audiences, and themselves as a live streamer to explain how they conduct live streaming. In this line, we regard live streaming practices as a performance that live streamers leverage to manage image in front of the audiences. Our research findings present three logics that live streamers hold to perform their live streaming, namely idol player, master players and market seeker. The liver streamers' interpretations on stage, audience and role, as well as their way of performance are listed in Table 2.

Table 2. Research findings

	Idols	Master players	Market seeker
Understanding of the stage: the live streaming industry is…	A high uncertainty and unpredictable industry It needs to create traffic	A way to increase visibility A work can balance interest and income	A high competitive and cruel market Traffic can bring business
Understanding of the audience: the people gathering here are…	Interested in watch me playing games Attracted by my appearance	Interested in watching an excited game Attracted by my advanced and excellent operation	Interested in diverse topic of contents Passionate fans
Understanding of the role: I am…	A lovely idol A sweet or gentleman who comfort people	A maser game player An easygoing person	A trustworthy person A skillful game player
The performance Practices	Leverage personal characteristics to create traffic Get close to the audiences to maintain 'likes' Keeping some distance for making mystery	Showed an exciting game to present profession Performed a challenging goal to enhance the perception of mastery Showed funny to build audience intimacy	Identify market opportunities for creating traffic Try new games to maintain traffic

This research contributes to academy by three theoretical implications. First, our findings unfold the process where liver streamers manage their live streaming practices with their understanding on the industry, audiences and themselves. The dramaturgical theory provides the opportunity to explain live streaming practices in some detail to deepen our knowledge on live streaming strategies. As we can find In Table 2, the three group of live steamers perform differently when they make different senses on the live steam industry, audience and their role as live streamers. Our findings provide an illustration on the relationship between live streamer's understandings and their practices on image management.

Second, our findings illustrate three types of performances to clarify the practical logic behind the live streaming strategies of the live streamers. The three types of performance focus on different goals and strategies on image management in live streaming. For example, the idol players manage their image as popular idol, and their performance put much emphasis on increasing 'likes' from the audiences. To them, the larger 'likes' creates the larger traffic to support the success in the live streaming industry. Another example, the market seekers manage their image for endorsement that can bring incomes. Their strategies on performance are to look for opportunity to enhance traffic. To them, the amount of traffic brings huge commercial benefits, such as sale product, advertising, endorsement, and sponsorship.

Third, our findings extend the dramaturgical theory from sociology to business discipline. Our analysis not only shows the potential of dramaturgical theory on investigating a new business phenomenon, but also illustrates the way of using it to study human behaviors under commercial context.

To practice, this study can enhance live streamers practices by reminding the logics embedded in live streaming performance. Instead of improvisation, live streamers need to design their performance to drive the audience' sensemaking process for image management. In other words, an effective live streaming is far from reproduce best practices proposed by successful live streamers. It is more likely an image management process. Live streamers, like actors, leverage appearance, posture, voice, and particular actions in their performance to make the image salience to the particular audiences.

6 Conclusion

This study investigates live streaming practices with the dramaturgical theory to unfold the relation between liver streamer's understandings and their live streaming practices. Apply the dramaturgical theory as lens of analysis, we present three kinds of performance in which the live streamers act differently with distinct understandings on the live streaming industry, the audience they faced, and their role as live streamers. This research can bridge the existing theoretical gap by providing live streamer perspective. It can also deepen our understanding on live streaming, and inspire live streamers on how to manage image in their live streaming performance.

References

1. Kim, T.: Tyler "Ninja" Blevins explains how he makes more than $500,000 a month playing video game 'Fortnite', vol. 2021. CNBC (2018)
2. Mavrix, J.: How Much Does SypherPK Make? Net Worth, Age & More!, vol. 2021. GET ON STREAM (2021)
3. Fägersten, K.: The role of swearing in creating an online persona: the case of YouTuber PewDiePie. Discourse Context Media **18**, 1–10 (2017)
4. Sjöblom, M., Törhönen, M., Hamari, J., Macey, J.: The ingredients of Twitch streaming: Affordances of game streams. Comput. Hum. Behav. **92**, 20–28 (2019)
5. Diwanji, V., Reed, A., Ferchaud, A., Seibert, J., Weinbrecht, V., Sellers, N.: Don't just watch, join in: Exploring information behavior and copresence on Twitch. Comput. Hum. Behav. **105**, 106221 (2020)
6. Hu, M., Zhang, M., Wang, Y.: Why do audiences choose to keep watching on live video streaming platforms? An explanation of dual identification framework. Comput. Hum. Behav. **75**, 594–606 (2017)
7. Banister, E.N., Cocker, H.L.: A cultural exploration of consumers' interactions and relationships with celebrities. J. Mark. Manag. **30**, 1–29 (2014)
8. Escalas, J.E., Bettman, J.R.: Connecting with celebrities: how consumers appropriate celebrity meanings for a sense of belonging. J. Advert. **46**, 297–308 (2017)
9. Hall, J.: Going Live: A Content Marketers Guide to Live Video Streaming. vol. 2021. FIVECHANNELS (2018)
10. Yanev, V.: 37+ Live Streaming Statistics Every Marketer Should Keep in Mind in 2021. vol. 2021. techjury (2021)

11. Weick, K.E.: Sensemaking in Organizations. CA: Sage, Thousand Oaks (1995)
12. Corrigan, L.T.: Budget making: the theatrical presentation of accounting discourse. Crit. Perspect. Acc. **55**, 12–32 (2018)
13. Perkiss, S., Bernardi, C., Dumay, J., Haslam, J.: A sticky chocolate problem: impression management and counter accounts in the shaping of corporate image. Crit. Perspect. Acc. **81**, 102229 (2020)
14. Hamilton, W., Garretson, O., Kerne, A.: Streaming on Twitch: fostering participatory communities of play within live mixed media. In: Human Factors in Computing Systems (2014)
15. Hilvert-Bruce, Z., Neill, J.T., Sjöblom, M., Hamari, J.: Social motivations of live-streaming viewer engagement on Twitch. Comput. Hum. Behav. **84**, 58–67 (2018)
16. Sjöblom, M., Hamari, J.: Why do people watch others play video games? An empirical study on the motivations of Twitch users. Comput. Hum. Behav. **75**, 985–996 (2017)
17. Solomon, J.F., Solomon, A., Joseph, N.L., Norton, S.D.: Impression management, myth creation and fabrication in private social and environmental reporting: insights from Erving Goffman. Acc. Organ. Soc. **38**, 195–213 (2013)
18. Goffman, E.: The presentation of self in everyday life. Anchor Books, New York (1956)
19. Brommelsiek, M., Kanter, S.L., Sutkin, G.: An ethnographic study examining attending surgeon persona in the operating room and influence on interprofessional team action. J. Interprofessional Educ. Pract. **20**, 100359 (2020)
20. Mason, J.: Qualitative Researching. Sage, London (2002)
21. Johnson, M.R., Woodcock, J.: "And today's top donator is": how live streamers on Twitch.Tv Monetize and Gamify their broadcasts. Soc. Media + Soc. **5** (2019)
22. Strauss, A., Corbin, J.: Grounded theory methodology: an overview. In: Denzin, N.K., Lincoln, Y.S. (eds.) Handbook of Qualitative Research, pp. 273–285. CA: Sage, Thousand Oaks (1994)
23. Dutton, J.E., Worline, M.C., Frost, P.J., Lilius, J.: Explaining compassion organizing. Adm. Sci. Q. **51**, 59–96 (2006)

Research on the Design of New Retail Service System Based on Service Design Concept – Setting Electronic Product Recycling Service System as an Example

Wei Ding and Qian Wu(✉)

East China University of Science and Technology, 130 Meilong Road, Xuhui District, Shanghai, China
1757746362@qq.com

Abstract. In the era of consumer upgrading, people's consumption psychology and consumption behavior are different from traditional ones. Traditional brick-and-mortar operations or traditional e-commerce sales methods can no longer meet consumer needs, and new retailing has emerged. New retailing is an innovative retail model that is still in the process of exploration and needs to be guided by more scientific and reasonable theories. Service design is an effective methodology and strategic tool across all industries, and its focus is consistent with the focus on user needs in new retail. Therefore, introducing service design in the new retail field to gain insight into user needs and reframe problems or opportunities can achieve comprehensive customer experience and employee experience improvement, and make the new retail industry more systematic and standardized. This paper explores and investigates a better new retail service system model from the perspective of service design. The necessity of combining service design factors with new retailing factors is explored, and the breakthrough point of combining the two is found. On this basis, new retail design principles and design methods based on service design concept are proposed. Taking electronic product recycling service system as an example, this paper focuses on the service design strategy and new retail model of electronic product recycling system. Through visiting observation method, literature search method, comparison analysis method and other research methods, this paper proposes corresponding design strategies for existing problems such as weak user awareness of recycling, high degree of regional demand differentiation and inefficient transaction mode, and carries out design practice.

Keywords: New retail · Service design · Recycling · Electronic product

1 Theoretical Research on New Retail and Service Design

1.1 Overview of New Retail Research

Retailing refers to: the total activity of selling consumer goods and related services to an individual or social group of final consumers for their final consumption. Three

F. Fui-Hoon Nah and K. Siau (Eds.): HCII 2022, LNCS 13327, pp. 481–492, 2022.
https://doi.org/10.1007/978-3-031-05544-7_35

elements must be included in retail activities: the field (scene), the goods (or services), and the people (or institutions). When the Internet became popular around the world, the development of the retail industry gradually changed and e-commerce emerged in the 1990s. During this period, many e-commerce shopping sites were born, such as Amazon, Alibaba, etc. The impact of these e-commerce sites on offline physical stores was very big. But with the development of e-commerce in China for more than ten years, the Internet online dividend has gradually disappeared, and retail enterprises are forced to look for new development models in the future. The new retail is a revolution triggered by new technology with the background of consumer upgrading, from the connection of online and offline, to big data, cloud computing, efficient logistics, and the innovation of the whole retail industry chain. The concept of new retail was first proposed at the Yunqi Conference in June 2017, where Jack Ma further explained the concept that new retail is not only about selling goods, but also about improving consumer experience and providing high-quality services to consumers [1].

The literature search and measurement and visualization service of China National Knowledge Infrastructure (CNKI) were used to analyze the research status. See Fig. 1 for the results of a precise subject search using "New Retail" as the keyword. From the content of the research themes, the main focus is on the basic theoretical aspects, such as the interpretation of the connotation and characteristics of the concept, the analysis of the reasons for the development, the development path and the development trend. In addition, domestic and foreign scholars have also discussed new retailing models, strategies, and new retailing influencing factors, and the rest involve research on related issues in the context of new retailing, such as logistics, consumers, enterprise competitiveness, and supply chain. From the keyword co-occurrence network, there are more co-occurrence of keywords such as retail model, new retail, online retail, traditional retail, brick-and-mortar store, supply chain, and e-commerce.

Y Feng analyzed the motives, ways and implementation methods of business model innovation in traditional retail enterprises and proposed strategies for integrating online and offline resources to achieve synergistic development of binary business models [2]. Sorescu A et al. argued that retail business model innovation is determined by several components, including the way retail activities are organized, the type of activities performed, and the level of customer involvement [3]. By increasing the scenarios of customer-company interaction, the ultimate experience of customer products and services is enhanced, so that customers can feel products and services beyond expectations and enhance satisfaction and loyalty to the brand. Now after a period of retail fever, it is easy to find that many retail businesses are starting to rethink the new retail industry development, profit model, and technology drive.

1.2 Research Overview of Service Design Theory

In the 1990s, the world economy entered a period of transition. With the economic transformation, service design entered people's vision and became a new term in the contemporary design field, and received more and more attention. The International Design Association in the 2008 Dictionary of Design gives a clear definition of service design: service design needs to start from the user's point of view, and then set the function and form of the service. The goal is to make the user feel the usefulness and effectiveness of the service. In addition to this rather formal definition, scholars from different fields have also dug deeper into the concept of service design, including exploring the content and objects of service design, emphasizing collaborative innovation among stakeholders, emphasizing interdisciplinarity, and summarizing the methods, tools and principles of service design.

Service design contains five basic elements: actor (the set of all stakeholders in the service), touch point, offering, need, and experience [4]. According to the gradual progression of emotional levels, service design is divided into three levels, which are ontology level (the original attributes of the service), behavior level (human-computer interaction, functional operation and usability), and value level (emotional value, social value and co-creation value) [5]. The highest level of service design is the process of co-creation by users [6]. The co-creation process requires multi-stakeholder collaboration [7]. Service design emphasizes innovative ways to coordinate the interaction between people, institutions, and technological systems. Creating a good service experience requires comprehensive consideration of the needs of all stakeholders, and the design allows all stakeholders to efficiently and joyfully complete the service process. In the context of systems and experiences, service design follows the stakeholder-centered approach and principles, and is designed for multi-role stakeholders, including service providers and recipients [8]. Service design integrates a variety of skills and methods in design, management, and engineering; service design includes practices in different fields such as administration and social innovation; service design is a cross-disciplinary research direction that requires collaborative efforts [9].

The research status was analyzed by relying on the literature search and measurement and visualization services of CNKI. The results of a precise thematic search using the keyword "service design" are shown in Fig. 2. From the content of the research topics, the main focus is on the basic theoretical aspects, such as the interpretation of the connotation and characteristics of the concept itself. In addition, domestic and foreign scholars have also discussed service design systems, models, strategies and micro-innovations, and the rest of the research involves related issues in the context of service design, such as new retail, stray cat rescue, university libraries, and the elderly.

In general, scholars have focused on different conceptual studies, but basically agree that service design needs to take a global perspective. Through the design of systems and processes, service experiences can be provided that meet different needs. Although service design integrates various fields and includes a series of activities and processes, the most important thing is that consumers can have a memorable experience in this series of activities and processes. In addition, theoretical development cannot be achieved without the boost and verification of practice, and it is imperative to develop innovative practices of localized service design [9]. At the same time, service design is also a new design thinking, and with the active promotion of universities and government departments at home and abroad, service design is rapidly becoming a research hotspot, and its application to new retail is the future trend, which can make the new retail industry more systematic and standardized.

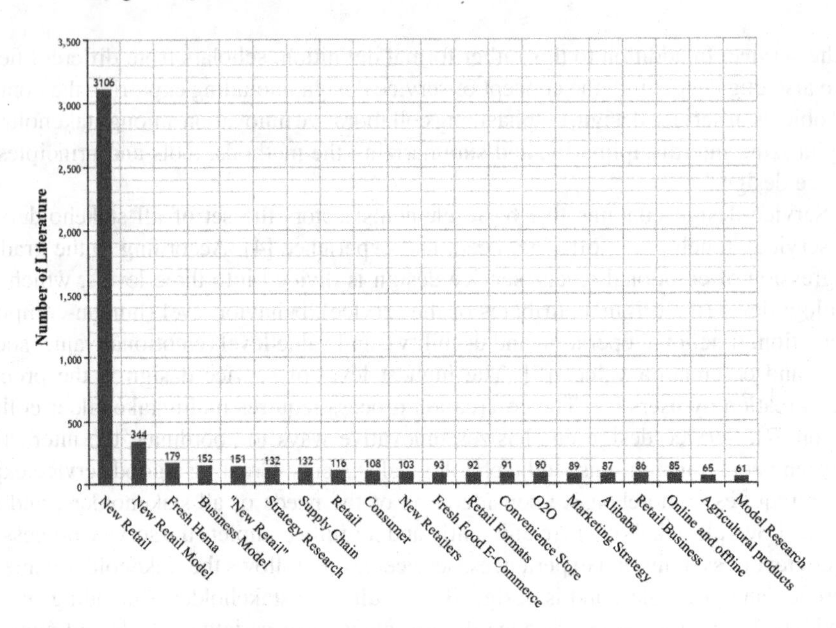

Fig. 1. Literature topics related to new retail

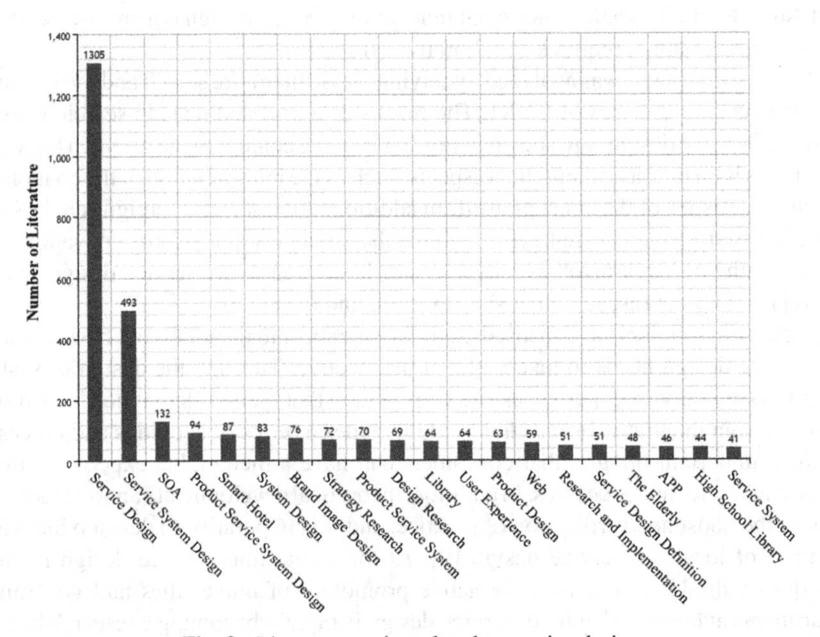

Fig. 2. Literature topics related to service design

2 Service Design Strategy in New Retail Industry

2.1 The Necessity of Combining Service Design Factors with New Retail Factors

Service design focuses on many factors, including technology, design, business, society, management and other horizontal aspects, and emphasizes a holistic and global perspective to identify and solve problems. It also focuses on the whole process of each level, and the user touch points in each process, and creates a better service experience through the analysis and optimization of the touch points. Service design is an effective methodology and strategic tool. By introducing service design in the new retail field, we can gain insight into customer needs and reconstruct problems or opportunities, so as to achieve comprehensive customer experience and employee experience improvement.

User-centered design for new retailing. In the design of new retail, whether it is the preliminary demand research, solution design, or the subsequent implementation and promotion of the solution, it should be closely focused on the users and put the users' needs in the first place, only in this way can we better ensure the accuracy and effectiveness of the solution. New retail advocates to provide users with a better experience, therefore, the collection and analysis of users' needs are extremely important. Use service design tools to enhance the user experience of new retail. Tools such as user portraits and user interviews in service design can collect and analyze user needs in a more scientific and reasonable way. The improvement of user experience needs to be made clear and concrete. By using user experience maps and service blueprints, we can visualize the various aspects that affect user experience, thus helping designers to better identify user pain points and user needs.

Create rich experience details by analyzing for key touchpoints. A complete service system is composed of multiple service stages and touchpoints together, following a certain service process. New retail seeks to create a new business model and shopping form, so as to improve users' shopping experience comprehensively. By using the key touch point analysis method in service design to describe and analyze the key touch points in the new retail chain, it can be more convenient for designers to grasp the user needs and improve the user experience.

Use the service design process to build the new retail design process. The service design process is concerned with all aspects from demand research to product verification, so the service design process can be applied to the design of new retail, thus improving the standardization of the new retail design process.

Validate and analyze new retail solutions more rationally through usability analysis. New retailing is a self-transformation and breakthrough in the retail industry, and there is no standardized model yet. Therefore, there will be many innovations of solutions, and it is very important to verify the value and evaluate the feasibility of the solutions. The usability analysis method in service design can be used to evaluate solutions more scientifically and rationally, so it can be well applied to new retail design.

2.2 New Retail Design Principles and Methods Based on Service Design Concept

User-Centered Design Principle. Grasp the user needs. user-centered design principles need to focus closely on users, analyze user needs, so as to maximize the satisfaction of user needs. A wide range of service design tools and methods can be used to try to grasp user needs more accurately. Inspire users to participate in service design. focusing on co-creation, stimulating user participation not only strengthens the connection between users and the product, and can obtain the most realistic feedback from users. simplify the operation process. Simplifying the user process as much as possible and removing redundant operations can be a great way to improve the user experience. Online retailing data has shown that meaningless operations can cause user churn, so the user process needs to be simplified as much as possible.

Scene-Guided Design Principle. Scenario is one of the important breakthrough points of new retail. Mobile e-commerce has caused a huge impact on offline retail by virtue of its convenience and efficiency, and the lack of scenario-based has been an important bottleneck for e-commerce, so offline stores need to shape more high-quality scenarios to attract users. The new retail advocates scenario-based services, which seek to disassemble and analyze needs around user scenarios. The service design emphasizes the analysis of user behavior in the user shopping process to better meet user needs. Therefore, in the design of new retail, it is necessary to build user scenarios, discover the user needs under the scenarios, and then carry out specific functional and formal design around the scenarios and needs.

Touchpoint-Based Design Principle. New retail focuses on all aspects of the user shopping experience. Using the theory related to service touchpoints in service design, the key contact points between users and the system are marked and analyzed to serve as key touchpoints for user experience. By dismantling the complex user experience process into the form of key touchpoints, the analysis of requirements and product design can be carried out more conveniently.

3 New Retail Model Research and System Construction

3.1 Analysis of Existing New Retail Models

Not only there is no unified standard for new retailing in terms of its conceptual boundary, but also its business models and business types are diverse and complex, and different scholars have given different classifications from different perspectives. Wang Jiabao and Huang Yijun categorize new retail into three types of models, including supplier-led, platform-led, and third-party-led [10]. Wang Yu and Li Rongjin believe that there are three main types of new retailing models: first, technology-driven, with the direction of laying out offline innovative technology experience to provide a good shopping experience with technology; second, optimized retailing, with traditional retail transformation as the main focus, optimizing their own retail ecology by optimizing the supply chain advantage offline and adopting SKU (Stock Keeping Unit) to enhance service experience; third, industry integration, aiming to innovate user experience and attract consumer consumption by creating a characteristic user experience environment [11].

In short, the current new retail business models are still in the exploration stage, and each company has developed different new retail business models according to its own advantages. They can be categorized into three models: first, the upgrading of traditional retail stores by relying on the mature resources of retail giants, such as data services, cloud computing, supply chain and logistics, to transform traditional retail stores into innovative smart stores; second, the provision of one-stop experience consumption services to consumers to realize scenario-based consumption; and third, unmanned retail self-service stores carried out by using big data, artificial intelligence and unmanned technology [12].

3.2 Construction of New Retail System

Among the many new retailing models, unmanned retailing is arguably the most novel retailing model. Based on the analysis in the previous section of this topic, the model of unmanned retailing is chosen to unfold in the new retailing model. In terms of service scenarios, the common activity scenarios of urban residents mainly include: communities, offices, shopping malls, subway stations and other open scenarios on the way. The community has a high population density and high comprehensive quality of residents, which has the value of more ideal business. Community-type scenes are relatively stable, have a certain radiation range, are closer to users, and have a higher frequency of contact with users. The mobility of office users is relatively low, the user base is relatively stable, and the office population has better consumption ability and consumption habits. On-the-way scenes mainly refer to scenes where users pass by on their way to and from work and scenes where users spend money and go shopping on weekends. These scenes are relatively broad and include a variety of user scenes and forms. The community scenario is highly replicable, the on-the-way scenario is more dependent on location and foot traffic, while the office scenario is more costly to communicate and promote, and less replicable. The market space of community-based scenes and on-the-way scenes is relatively high, and the market space of office scenes is relatively low. In terms of user needs, it mainly meets users' needs for more efficient, more convenient, more guaranteed quality and better experience.

4 New Retail Design Practice Under Service Design Concept

4.1 Background and Challenges

The consumer demand of the mobile Internet era and the continuous concern of society for environmental issues have directly brought about the launch of "Internet+ environmental protection" products. The secondary recycling of cell phones reduces the pollution of the environment and is a new experience for users. By tapping into the pain points that exist in the process of cell phone recycling, raising people's awareness of cell phone recycling and increasing the convenience of cell phone recycling, the waste of idle resources can be greatly reduced. There are currently three problems with the global used digital product trade. First, users' awareness of recycling used cell phones is weak, and regional demand is highly differentiated, with consumers in each region having different preferences for operating systems, brands and price points. Second, transaction prices are not open and transparent enough. Finally, the transaction model is relatively inefficient. Cell phone automatic recycling machine is not only a new product, but also a new business model of circular economy under the background of new retail.

4.2 Strategic Positioning and New Retail Model

User Requirement Analysis. The use habits and psychological changes of target users need to be focused when using electronic product recycling equipment. Through theoretical analysis and research of the target users, we can comprehensively grasp their knowledge of the electronic product recycling system, their functional needs and potential service needs of the whole service system. In this paper, various user research methods are used to build a user role model, and user needs are identified according to Maslow's hierarchy of needs theory, and the results are shown in Fig. 3.

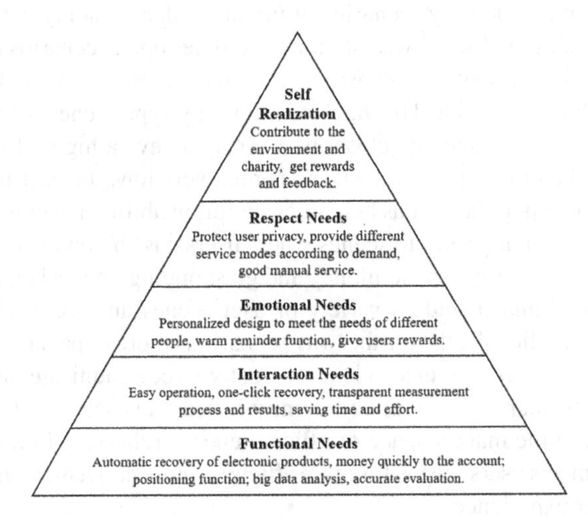

Fig. 3. Target user needs insight of electronic product recycling service system

Service Process and Contact Analysis. Electronics recycling is a complex system engineering, involving many stakeholders and the design of processes and experiences (see Fig. 4). Considering the attributes of the target users and combined with the service design theory, the principles to be followed in the design of the electronic product recycling service system include: the principle of recycling system integration design, the principle of design centered on the needs of the target users, and the principle of user participatory design.

Key Person Map. Key Person Map refers to a diagram that systematically shows the relationship between each key person. By studying the map in the electronics recycling system, it was determined that the main target users were office workers and students. Then, we studied and analyzed their needs and contact points, and presented each key person and character relationship in a visual form, as in Fig. 5.

Strategic Positioning. According to the specific needs of users for recycling time, information transfer and service process, the following innovative strategies were developed: matching strategy for recycling time and touch points, timely and interactive strategy for recycling service, and convenient and standardized strategy for recycling process.

Matching strategy for recycling time and touch points: Commuters and students have a strong sense of time, so the recycling process is as simple and convenient as possible, with a smooth operation process that conforms to users' behavioral habits and reduces the occurrence of errors. Real-time monitoring of the location of the device to reduce the waste of time caused by users running to the wrong place. By matching the contact point with the recycling time, it can improve the overall efficiency, reduce the time users wait in the process, relieve their anxiety, improve the user experience and thus increase the recycling rate.

Timely and interactive strategy for recycling service: Immediate delivery of information is required at all stages of the recycling process. Inform the user of the current recycling progress and status and allow for appropriate interaction between the user and the device.

Convenient and standardized strategy for recycling process: The existing electronics recycling industry lacks uniform standards, different platforms will develop different recycling standards and pricing models, and there are problems such as unprofessional and incorrect product testing, which will directly lead to user distrust and cause users to be reluctant to recycle.

New Retail Model. Based on the previous analysis, the model of unmanned retail is chosen to unfold. The laying of automatic recycling machines in the used electronics recycling industry is based on the following three considerations.

First is low cost: automatic recycling machines save both site rental costs and employee hiring costs, and can eliminate the need to invest resources in laying down offline stores.

Second is the convenience of the transaction: the emergence of self-service recycling machine will greatly shorten the physical distance between people and cell phone recycling channels, the traditional recycling industry "people" to "field" mode into "field" to "people". Cell phone recycling behavior can only be integrated into life through more scenes, so that more people can accept it.

Third is the improvement of scene power: automatic recycling machine through the coverage of multiple scenes, cultivate user awareness, improve user experience, and then realize efficient second-hand transactions. Recycling machines are mainly laid in supermarkets, convenience stores, commercial office buildings, schools, subways and community scenes. These scenes are all high traffic scenes, and entering such scenes can directly and effectively cultivate user habits and provide convenient and fast cell phone recycling. Moreover, each machine is equivalent to a billboard, which can achieve strong stickiness and strong connection in the entrance scene with high flow of people.

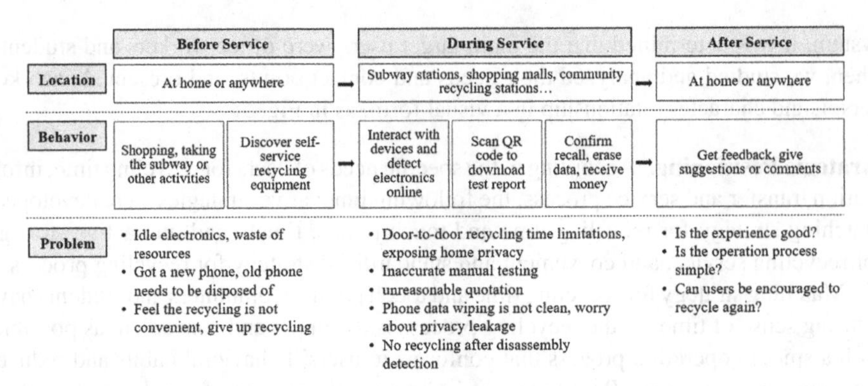

Fig. 4. Flow chart of electronic product recycling service

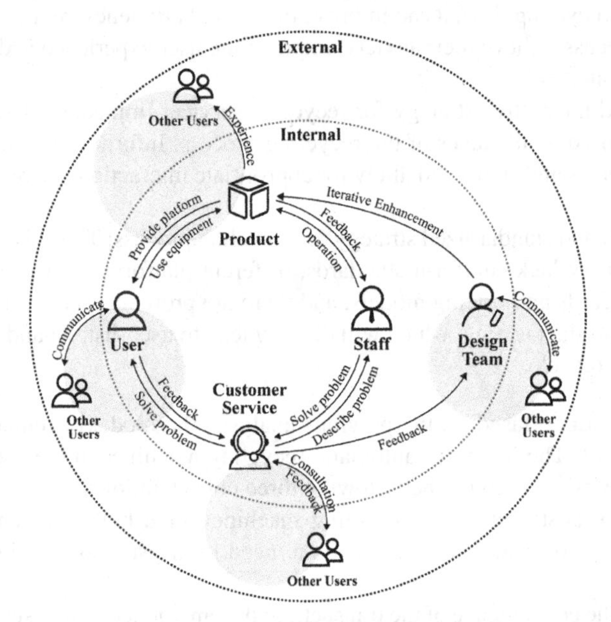

Fig. 5. Map of key people in the electronics recycling system

4.3 Functional Positioning and Service Blueprint

Through user research, combined with the service design methodology, the following functions are positioned for the electronics recycling service system.

Functions Positioning. For office workers and students, they are more concerned about whether the equipment is easy to find, whether recycling is convenient, whether the operation is simple, whether privacy is well protected, etc. the details are as follows.

1. Resell idle electronic products to protect environment and reuse resources.
2. Fast and accurate positioning of equipment, recommend the nearest one.

3. Operate the device smoothly and simply, with good operation guidance.
4. Detection reports can be transmitted between the device and the cell phone.
5. Erase data on-the-spot to protect user privacy.
6. Efficient and accurate detection of the whole machine with a uniform standard.
7. Immediate payment on the spot, so that users are more assured.
8. Provide information feedback and follow-up service after recycling.
9. Reasonable recovery price and various activities rewards.

Service Blueprint. Throughout the service blueprint, the service touchpoints that users come into contact with at each stage and the relationships with the touchpoints are clearly shown. The service blueprint allows designers to have a global overview of the entire service, helping the designers involved to better design and develop the service to achieve the desired outcome, as shown in Fig. 6.

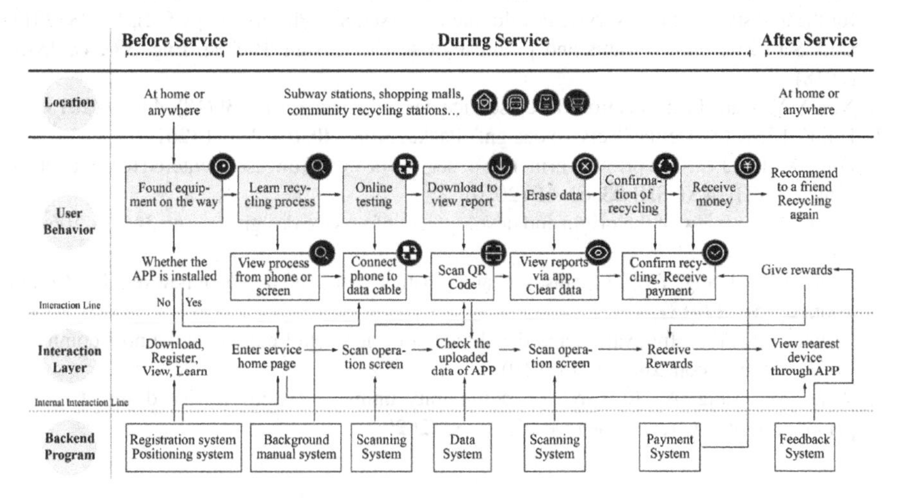

Fig. 6. Service blueprint for electronics recycling

5 Conclusion

This paper introduces service design in the field of new retailing, and explores and researches a better new retail service system model from the perspective of service design. The necessity of combining service design factors with new retail factors is explored, and new retail design principles and design methods based on service design concepts are proposed. The electronics recycling service system is used as an example to explore the construction of new retail model and system. Electronics recycling system's new retail layout combines online drainage with offline stores and recycling machines. Online, through background big data analysis, targeted and accurate user diversion can be achieved. Through online drainage to offline stores and automatic recycling machines, mobile phone recycling and down payment can be completed. Cell phone recycling self-service integrated operation design is a practical solution to the difficulties encountered in cell phone recycling without manual service. The reconfiguration of "people", "goods" and "field" facilitates the disposal and circulation of used electronic products.

References

1. Zhao, S.M., Xu, X.H.: The meaning, pattern and development path of "new retail." China Circul. Econ. **31**(5), 12–20 (2017)
2. Xia, Q., Feng, Y.: The conflict and coordination of traditional retailing enterprises about online and offline dual business model innovation——in the case of Suningyun. Econ. Manag. **30**(1), 64–70 (2016)
3. Sorescu, A., Frambach, R.T., Singh, J., et al.: Innovations in retail business models. J. Retail. **87**(7) (2011)
4. Clatworthy, S.: Service innovation through touch-points: development of an innovation toolkit for the first stages of new service development. J. Exer. Sci. Physiotherapy **5**(2), 15–28 (2011)
5. Luo, S.J., Zou, W.Y.: Status and progress of service design. Packag. Eng. **39**(24), 43–53 (2018)
6. Xin, X.Y., Cao, J.Z.: Location Based service design. Packag. Eng. **39**(18), 43–49 (2018)
7. Hu, F., Li, W.Q.: Define "service design." Packag. Eng. **40**(10), 37–51 (2019)
8. Ding, X., Du, J.L.: The primary principle of service design: from user-centered to stakeholder-centered. ZHUANGSHI **03**, 62–65 (2020)
9. Wang, P.: A review of the origin and development of service design. Design **34**(21), 106–109 (2021)
10. Wang, J.B., Huang, Y.J.: The causes, characteristics, types and trends of new retail. J. Comm. Econ. **23**, 5–7 (2018)
11. Wang, Y., Li, R.J.: Innovation and development of new retail models in the post-commerce era. J. Comm. Econ. **16**, 27–31 (2019)
12. Wang, W., Yang, A.: Research on community unmanned store service design from the perspective of new retail. Idea Des. **2**, 25–33 (2020)

Consumers' Trust Mechanism and Trust Boundary on Humanizing Customer Service Chatbots in E-commerce

Yimeng Qi(✉), Rong Du, and Ruiqian Yang

School of Economics and Management, Xidian University, Xi'an 710126, Shaanxi, China
yimengqi@stu.xidian.edu.cn

Abstract. Humanizing customer service chatbots have sparked significant interest for companies across industries. These years have witnessed some controversy on trust issues of such booming application. Previous researches have proposed some antecedents of customer service chatbots adoption (e.g., anthropomorphic features, algorithm aversion, emotional state). However, consumers' trust mechanism and trust boundary on humanizing customer service chatbots are not clear. Hence, we pay attention to personalization and contextualization grounded on above antecedents of customer service, incorporating personal habit, task creativity and social presence to investigate trust mechanism and trust boundary. We propose a research model, in which personal habit and task creativity are captured as independent variables, trust in humanizing customer service chatbots as dependent variable, and social presence as moderating variable. Hypotheses are developed and between-subjects scenario experiments are conducted to test hypotheses. Results of analysis of covariance (ANCOVA) and moderating effect test show that there exists positive effect between personal habit and trust in humanizing customer service chatbots, giving insights on complementary and substitutive influences on the interaction of independent variables and social presence for trust boundary. This paper provides practical and theoretical implications for e-commerce practitioners to improve the collaboration performance of intelligent customer service and human customer service.

Keywords: Trust mechanism · Trust boundary · Humanizing customer service chatbots · Personalization · Contextualization · e-commerce

1 Introduction

As the development of robot process automation (RPA) applications, customer service chatbots have emerged in communication, finance, retail and other fields, and they are revolutionizing the trust mechanism and trust boundary of the entire customer service industry. Since most of the customer service work is massive and repetitive, customer service chatbots, which adopt the applications of artificial intelligence (AI) in work processes, can help to reduce the number of manual receptions and eliminate repetitive work of human employees, while increasing response efficiency and decreasing labor

F. Fui-Hoon Nah and K. Siau (Eds.): HCII 2022, LNCS 13327, pp. 493–509, 2022.
https://doi.org/10.1007/978-3-031-05544-7_36

costs. However, in the electronic commerce (e-commerce) scenarios, the service is still mainly done by human, especially in the after-sales service stage, and the demand of "turning to human customer service" frequently appears in reality. Why are customer service chatbots not widely accepted by consumers in e-commerce? Since a large body of literature have demonstrated that anthropomorphism is found to widely adopted and positively relate to consumer adoption of chatbots [1] leading to more effective conversations [2], beneficial for transaction outcomes as well as contributing to significant increases in offer elasticity [3]. Hence, it is inevitable to examine trust mechanism and boundary of humanizing customer service chatbots from different angles.

In order to explore trust mechanism of consumers on humanizing customer service chatbots in e-commerce, firstly, trust is subjective and related to contexts. Customer traits and predispositions (e.g., computer anxiety), sociodemographic (e.g., gender), and robot design features (e.g., physical, nonphysical) have been identified as triggers of anthropomorphism [4]. In view of these, consumers' perception of trust may be relevant to their personalized contexts. Hence, it is critical to differentiate the subjective from objective factors that affect consumers' trust in humanizing customer service chatbots for predicting their subsequent trust. Additionally, positive interaction between social presence and trust in technology adoption area has been explored over years, short of other possible relationships between them [5]. Grounded on the current studies, we distinguished personal habit and task creativity as antecedents, aiming to examine how social presence moderate different consumers' trust in humanizing customer service chatbots other than direct effects of both independents. Within our work, trust mechanism was exploration of personalization and contextualization (i.e., personal habit and task creativity) for consumers' trust, while trust boundary was defined as imagined lines that mark the limits or edges of consumers' enhanced trust and diminished trust (i.e., interaction of antecedents and social presence). The two questions that we seek to address are:

RQ1: How do consumers' personal habit and task creativity affect their trust in humanizing customer service chatbots?

RQ2: How can social presence enhance consumers' trust under different consumers' personal habits and tasks creativity?

To answer the two research questions raised above, we designed a between-subjects scenario experiment to simulate the interaction situations between consumers and humanizing customer service chatbots after a preliminary study for task attributes test. We hired 141 subjects, who were from an IS experimental curriculum. According to personal habits, the subjects were divided into two categories: (1) subjects with interaction habit and (2) subjects without interaction habit, while high creativity and low creativity were set for task creativity respectively. The two-way ANCOVA was used to test the relationship between consumers' personal habit, task creativity and trust in humanizing customer service chatbots. Furthermore, the moderating effect of social presence was also validated. This paper explores consumers' trust mechanism and trust boundary on humanizing customer service chatbots, giving implications that help practitioners to make use of the findings to achieve the optimal human-machine coordination effect in e-commerce.

2 Literature Review and Hypotheses

2.1 Humanizing Customer Service Chatbots

Chatbots are autonomous software agents that support text-based exchanges with human users, drawing on tools and techniques from the domain of Natural Language Processing [6]. Customer service is a domain where chatbots have achieved strong and growing interest [7]. A central component in the area on the effective design of autonomous agents has been the role of anthropomorphism [8]. With this, some chatbots are designed to incorporate some attributes such as language style and name to enhance their human likeness [9]. Numerous scholars have provided evidence that humanizing customer service chatbots improve customer evaluation, as an illustration, social cues of anthropomorphism as humor, communication delays or social presence are beneficial for transaction outcomes in retailing setting [3]. Both anthropomorphism as well as the need to stay consistent significantly increase the likelihood that users comply with a chatbot's request for service feedback [10]. Nevertheless, humanizing customer service chatbots appears to only work contextually, e.g., when customers enter a chatbot-led service interaction in an angry emotional state, chatbot anthropomorphism has a negative effect on customer satisfaction, yet this is not the case for customers in nonangry emotional states [11]. Besides, consumers with high social phobia prefer anthropomorphic chatbots to less anthropomorphic chatbots [12].

Additionally, due to consumers not being able to identify their conversational partner when interacting via chats online, companies face the challenge of whether to disclosure the identity of chatbots. Prior research pertaining to the chatbot disclosure dilemma indicates that chatbot disclosure does not only have negative consequences, but can lead to positive outcomes as well depending on service context [13], thus setting expectations through cues of AI capability congruent to or less than the actual AI capability can increase engagement of customer service chatbots [14].

2.2 Antecedents of Customer Service Chatbots

Customer Traits. To clarify contextual circumstances in which anthropomorphism impacts customer intention to use chatbots, some studies have investigated relationships between humanizing customer service chatbots and their antecedents [4].

Since personalization is an aspect which is strongly emphasized in IS recently, and consumer personality can be predicted during contextual interactions, customer traits are becoming promising [15]. Specifically, need for interaction with the service employee will have a negative effect on the expected quality of the technology-based self-service option [16], so the anthropomorphism-adoption relationship is stronger on conditioning that a consumer's need for human interaction is higher [1]. Prior research has introduced two related yet distinct constructs about customer traits, namely experience and habit into technology adoption. Experience reflects an opportunity to use a target technology and is typically operationalized as the passage of time from the initial use of a technology by an individual [17]. Habit is defined as learned sequences of acts that become automatic responses to specific situations, which may be functional in obtaining certain goals or end states [18]. Not only the effect of facilitating conditions on behavioral intention to be moderated by age, gender, and experience, but also the adoption of habit have been clarified [19]. Moreover, experience is a necessary but not sufficient factor for habit formation, and the passage of chronological time can result in different levels of habit contextually. Thus, we adopt personal habit as an antecedent of trust in humanizing customer service chatbots for extending research of personalization to the emerging AI application.

H_1: Consumers' personal habits of interacting with customer service chatbots (vs. without such habit) enhance their trust in humanizing customer service chatbots.

Algorithm Aversion and Attributes of Task. Algorithm aversion is a phenomenon that consumers have a tendency to prefer humans over algorithms [20]. Specifically, when the error is committed by an algorithm, gut reactions are harsher (i.e., less acceptance and more negative feelings) and justice cognitions weaker (i.e., less blame, less forgiveness, and less accountability) [21]. Robots are invariably viewed as lacking human nature abilities (which are emotional) but not human uniqueness abilities (which are cognitive), thus algorithms are trusted and relied on less for tasks that seem subjective in nature [22]. While participants are considerably more likely to choose to use an imperfect algorithm when they can modify its forecasts [23]. Prior findings suggest that consumers' trust vary significantly depending on the type of tasks for which the algorithm is used. In terms of customer service chatbots, there has been little exploration or validation taking attributes of tasks as antecedents of technology adoption.

Attributes of task plays a significant role in consumers' trust, which have been proposed as task objectivity, awareness of algorithms performance and affective ability [22]. Besides, for high-creativity tasks, consumers exhibit lower willingness to adopt AI (vs. human) recommendation, in contrast, consumers exhibit higher willingness to adopt AI (vs. human) recommendation for low-creativity tasks [24]. In our work, we took task creativity into consideration for examining its effect on consumers' trust in humanizing customer service chatbots.

H_2: High task creativity (vs. low task creativity) reduces consumers' trust in humanizing customer service chatbots.

H_3: Consumers with personal habit of interacting with customer service chatbots (vs. without such habit) enhance trust in humanizing customer service chatbots more for high task creativity (vs. low task creativity); while consumers without the habit (vs. with such habit) reduce trust in humanizing customer service chatbots more for high task creativity (vs. low task creativity).

2.3 Social Presence and Consumers' Trust

Social presence theory describes the extent to which a medium allows a user to experience others as being psychologically present [25]. The degree of social presence in an interaction is posited to be determined by the communication medium: the fewer the channels or codes available within a medium, the less attention is paid by the user to the presence of other social participants. For example, transmitting more nonverbal visual or auditory codes such as facial expression, posture, dress, or vocalics would be more closely to Face-to-Face communication, thereby enhancing the degree of social presence [26]. A growing body of research in IS have investigated the influencing factors of social presence, the learning performance affected by social presence, the group decision-making caused by social presence and consumers' intentions attributed to social presence [27]. Among the areas raised above, multiple scholars have emphasized the important role of trust in forming consumers' intentions, so it's inevitable to analyze the relationship between social presence and trust.

Most current studies paid attention to the interaction between social presence and trust. Specifically, social presence-information richness (SPIR) affects consumers' trust and that trust subsequently has a stronger effect on purchase intentions than TAM beliefs [5]. Besides, by validating a four-dimensional scale of trust, the influence of social presence on these dimensions of trust and its ultimate contribution to online purchase intentions were clarified [28], and a set of three social presence variables (social presence of the web, perception of others, social presence of interaction with sellers) were proposed to have positive impacts on trusting beliefs [29]. Moreover, social presence was found to not only influence initial trust in the website, but also participants' enjoyment and perceived usefulness of the site [30], while social presence does enhance P2P customer trust via both utilitarian and hedonic engagement [31]. In addition, results gathered among Facebook users indicates that trust of a social networking site increases users' information seeking in informational channels, which elevates the sense of social presence [32]. There have been some studies exploring other possible relationships between social presence and trust recently, as how information support moderates the relationship between different social presence dimensions and trust in social commerce [33]. Some scholars also posited that social presence served to enhance/attenuate the influence of institutional trust building antecedents on that trust [34]. Therefore, we aim to validate whether the moderating effect of social presence on trust in humanizing customer service chatbots exists under different levels of personal habit and task creativity.

H_{4a}: For consumers without personal habit of interacting with customer service chatbots, social presence positively moderates the relationship with trust in humanizing customer service chatbots, while this effect does not manifest for consumers with personal habit of interacting with customer service chatbots.

H_{4b}: For tasks with high creativity, social presence positively moderates the relationship with trust in humanizing customer service chatbots, while this effect does not manifest for tasks with low creativity.

Figure 1 illustrates our research model.

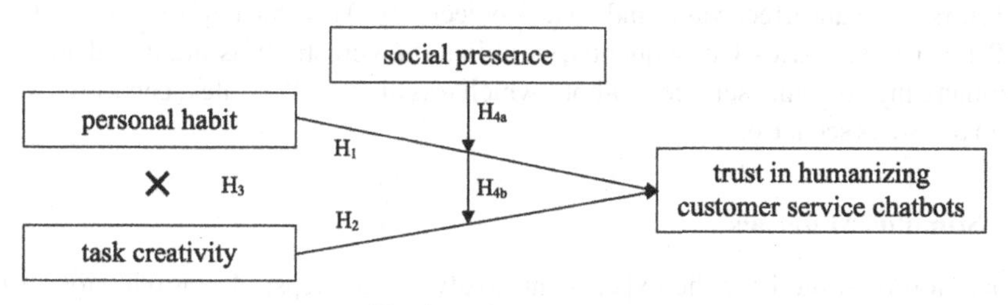

Fig. 1. Research model

3 Methodology

The goal of this experiment was to examine how personal habit and task creativity impact consumers' trust in humanizing customer service chatbots. To explore this, we conducted a 2 (with interaction habit vs. without interaction habit) $\times$ 2 (high task creativity vs. low task creativity) between-subjects scenario experiment. To implement realistic manipulation, real-life online chats on e-commerce platform were evaluated prior to designing experiment. We chose to use scenario experiments, to be able to control confounding influences and ensure high internal validity.

3.1 Preliminary Study

The goal of preliminary study was to distinguish the tasks creativity of two online chatting scenarios. We recruited 94 participants across the universities in Xi'an, Shaanxi (42 males) from November 5 to November 12, and the average age of them was 21. According to realistic scenarios, the pre-sales consulting and after-sales service scenarios were set for them respectively.

1. You found a down jacket on Taobao platform and want to know some details. So, you click on the "customer service chatbot" button and send the link of the product to the chatbot, hoping to ask for the size, material, delivery and coupon information.
2. You had bought this down jacket on Taobao platform. After receiving it, you found that the clothes smelled pungent and the price decreased after a few days. Hence you want to ask the customer service chatbot to explain your problem and propose that you want to refund the difference of price or return the clothes.

Based on the prior research, we selected three attributes of tasks appropriate for our work. Participants rated each of the tasks on how objective versus subjective it seemed, how much creativity they feel this task need, and how difficult versus easy it seemed, assessing constructs on 7-point-likert scales ranging from strongly disagree (1) to strongly agree (7). The tasks as well as the dimensions being rated were presented in random order. The demographics were collected at the end of the scale.

Given that two tasks were evaluated by all the participants, paired samples t-test was conducted to validate the attributes raised above. Results revealed that task creativity ($M_1 = 4.15, M_2 = 5.89, t = -9.857, p < 0.01$) and task difficulty ($M_1 = 2.54, M_2 = 2.94, t = -2.561, p < 0.05$) for the second scenario were significantly higher than the first scenario, while no significant effect was found for task objectivity ($M_1 = 5.61, M_2 = 5.52, t = 0.647, p > 0.1$) across scenarios. Consequently, we utilized task creativity as an antecedent of trust in humanizing customer service chatbots, which was high in after-sales scenario as well as low in pre-sales scenario.

3.2 Stimulus Materials

Preparation of stimuli for the experiment involved two steps: (1) introduction of real humanizing customer service chatbot that was trained based on the current applications, and (2) classification of participants' personal habit according to prior scales. In the first step, we targeted stores selling clothes on Taobao platform, the top online shopping website around China. It is universally acknowledged that there are hedonic [35] and utilitarian [36] elements comprising attitudes toward product categories, and research has revealed that the influence of social presence on trust and reuse intention with respect to utilitarian products is less than that with respect to hedonic products [37]. For instance, unlike headphones, increasing a firm's social presence for clothes through socially rich descriptions and pictures had a positive impact on attitudinal antecedents to purchase [38]. Therefore, we designed our experiment on a basis of the shopping for clothes. Besides, V5 customer service chatbot that had been widely adopted by e-commerce stores was introduced to our scenario. Apart from training it based on realistic reply of clothes stores chatbots, we configured it with numerous anthropomorphic characteristics (like name it Marry and talking style) and inserted it into WeChat Public Platform for further investigation.

In the second step, personal habit was measured ahead of the main study. Respondents should report their perceptions about habit of interacting with customer service chatbots. "Using customer service chatbots has become automatic to me", "Using customer service chatbots is natural to me" and "When faced with a particular task, using customer service chatbots is an obvious choice for me" [18]. We took the mean of statements on 7-point-likert scales, anchored by 1 = strongly disagree and 7 = strongly agree. The participants whose scores were over 4 were considered as ones with habit of using customer service chatbots, while those who scored under 4 (including 4) were defined as ones without habit of using customer service chatbots.

3.3 Procedure

For the main study, participants were 141 undergraduate students (65 male) from an IS experimental curriculum at Xidian university from December 25, 2021 to January 10, 2022. They received monetary compensation for their participation, and no one failed to complete the study. Demographic measures indicated that 90% were juniors or above, and the average age was 21; about 95% bought products online every month; almost all of them (more than 98%) had experiences interacting with customer service chatbots during online shopping.

As a cover story, participants were introduced to a clothes store on Taobao platform, providing a situation of buying a down jacket. The cover story explained that the participants should follow the notes attached to reply to customer service chatbot. The pre-sales and after-sales situation were given for everyone in different habit groups randomly. Then participants were instructed to ask for the questions they were concerned about (see Fig. 2). Pre-sales consulting was comprised of the size, coupons, material, delivery and so on, and after-sales service was constituted of return of goods, exchange of goods, refund of price difference, etc. Once their problems continued to be unsolved, they could choose to switch to human service.

After interacting with the humanizing customer service chatbot, participants reported their perceptions of (1) trust in humanizing customer service chatbots, (2) social presence, (3) trusting disposition, (4) familiarity with e-commerce platform. As control variables, in addition to controlling for data collection mode, we further gathered measures on demographics.

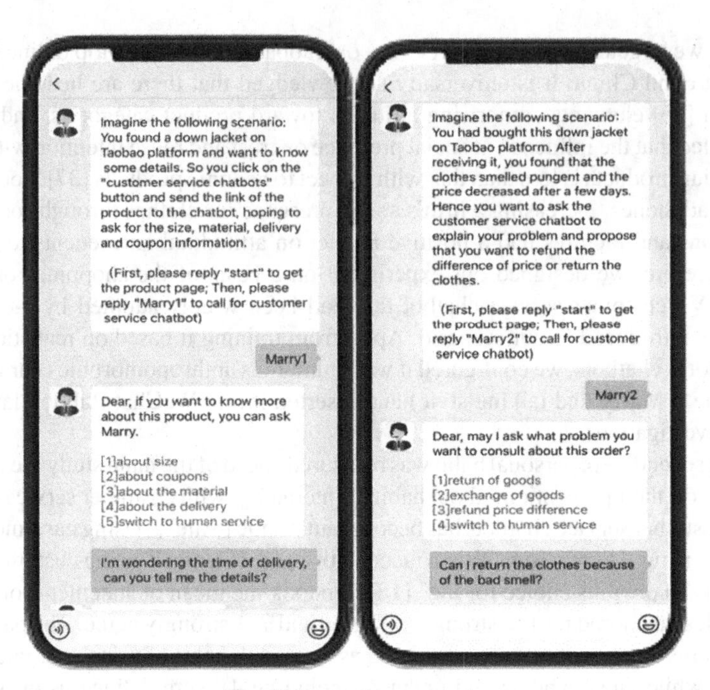

Fig. 2. Exemplary scenario

3.4 Measures

Trust in humanizing customer service chatbots was measured on a 7-point-likert scale, using items of integrity from Straub and Gefen [28], social presence was also measured on a 7-point-likert scale, using 5 items from Straub and Gefen [5]. Trusting disposition was measured on a 7-point-likert scale with 6 items from Straub and Gefen [28]. Familiarity with e-commerce platform was measured on a 7-point-likert scale with 4 items from Gefen [39].

We next examined the reliability and validity of major constructs in our study. Cronbach's alpha of all the constructs was 0.847, and Cronbach's alphas of social presence, trust in humanizing customer service chatbots, trusting disposition and familiarity with e-commerce platform were between 0.857 and 0.933, indicating high internal consistence reliability. According to KMO and Bartlett's test, the KMO measure of sampling adequacy was over 0.7, as well as the Bartlett's test of sphericity was significant, so it was well suited to make a factor analysis next. Then we conducted confirmatory factor analysis (CFA) to assess convergent and discriminant validity of all the constructs (see Table 1). Within standard estimate of factor loadings, the second construct of trusting disposition and the fourth construct of familiarity were below the cut-off value of 0.6, thus we eliminated them for retesting. The updated values were all beyond 0.6. Additionally, average variances extracted (AVE) of social presence, trust in humanizing customer service chatbots, trusting disposition and familiarity with e-commerce platform were 0.74, 0.741, 0.613 and 0.847, respectively, exceeding the cut-off value of 0.5. Besides, composite reliability (CR) for all the constructs were over 0.8, demonstrating high convergent validity. Finally, the square roots of AVE for all the constructs were greater than the correlations between them, indicating high discriminant validity (see Table 2).

Table 1. Composite Reliability and Convergent Validity

Construct	Measure	Factor loading	CR	AVE
Social presence	There is a sense of human contact interacting with customer service chatbots	0.86	0.934	0.740
	There is a sense of personalness interacting with customer service chatbots	0.861		
	There is a sense of sociability interacting with customer service chatbots	0.898		
	There is a sense of human warmth interacting with customer service chatbots	0.883		
	There is a sense of human sensitivity interacting with customer service chatbots	0.786		
Trust in humanizing customer service chatbots	Promises made by customer service chatbots are likely to be reliable	0.896	0.896	0.741
	I do not doubt the honesty of customer service chatbots	0.801		
	I expect that customer service chatbots will keep promises they make	0.888		
Trusting disposition	I generally trust other people	0.624	0.886	0.613
	I generally have faith in humanity	0.651		
	I feel that people are generally well meaning	0.874		
	I feel that people are generally trustworthy	0.932		
	I feel that people are generally reliable	0.831		
Familiarity with e-commerce platform	I am familiar with searching for products on the Internet	0.907	0.943	0.847
	I am familiar with buying products on the Internet	0.954		
	I am familiar with Taobao platform	0.899		

Table 2. Discriminant validity

	Social presence	Trust	Trusting disposition	Familiarity
Social presence	0.74			
Trust	0.374***	0.741		
Trusting disposition	0.29***	0.028***	0.613	
Familiarity	−0.151***	0.052***	0.282***	0.847
Square roots of AVE	0.86	0.861	0.783	0.92

Note: *** Significant at the 1% level; ** Significant at the 5% level; * Significant at the 10% level

4 Data Analysis and Results

4.1 Manipulation Test

To validate the main effect of our manipulation, we used analysis of covariance (ANCOVA). Personal habit and task creativity were taken as independent variables, and trusting disposition, familiarity with e-commerce platform, gender, experience of interacting with realistic chatbots as well as expenses of online shopping every month were used as covariates, and trust in humanizing customer service chatbots was defined as dependent variable. The results were shown in Table 3.

Table 3. Analysis of covariance

Source	MS	F	Sig
Trusting disposition	1.606	0.966	0.327
Familiarity	0.756	0.455	0.501
Gender	1.12	0.674	0.413
Experience	8.909	5.36	0.022**
Frequency of online shopping	1.498	0.902	0.344
Expenses of online shopping	3.113	1.873	0.173
Task creativity	0.863	0.519	0.472
Personal habit	16.8	10.108	0.002***
Task creativity * personal habit	12.524	7.535	0.007***
Adjusted R Square	0.129		

Note: *** Significant at the 1% level; ** Significant at the 5% level; * Significant at the 10% level

The results revealed a significant positive main effect of personal habit on trust in humanizing customer service chatbots ($M_{\text{without habit}} = 4.3276$, $M_{\text{with habit}} = 5.0964$, $F = 10.108$, $p < 0.01$). The above result provided support for H_1, which stated that consumers' personal habit of interacting with customer service chatbots enhances their trust in them. However, the main effect of task creativity on trust in humanizing customer service chatbots was not significant ($M_{\text{low creativity}} = 4.8378$, $M_{\text{high creativity}} = 4.7164$, $F = 0.519$, $p > 0.1$). Furthermore, the interaction effect of task creativity and personal habit on trust in humanizing customer service chatbots was significant ($F = 7.535$, $p < 0.01$). After simple effect analysis, consumers without habit of communicating with customer service chatbots were more likely to trust them under low creativity task ($M_{low\ creativity} = 4.7284$, $M_{high\ creativity} = 3.9785$, $F = 5.17$, $p < 0.05$), while the trust levels of consumers with habit of communicating with customer service chatbots were not reliably different from both types of tasks ($M_{low\ creativity} = 4.9007$, $M_{high\ creativity} = 5.3519$, $F = 1.49$, $p > 0.1$). From another perspective, consumers with habit of interacting with customer service chatbots were considered to trust them more under high creativity task ($M_{without\ habit} = 3.9785$, $M_{with\ habit} = 5.3519$, $F = 18.17$, $p < 0.01$), whereas no significant effect was found under low creativity task ($M_{without\ habit} = 4.7284$, $M_{with\ habit} = 4.9007$, $F = 0.47$, $p > 0.1$). (see Fig. 3), supporting H_3.

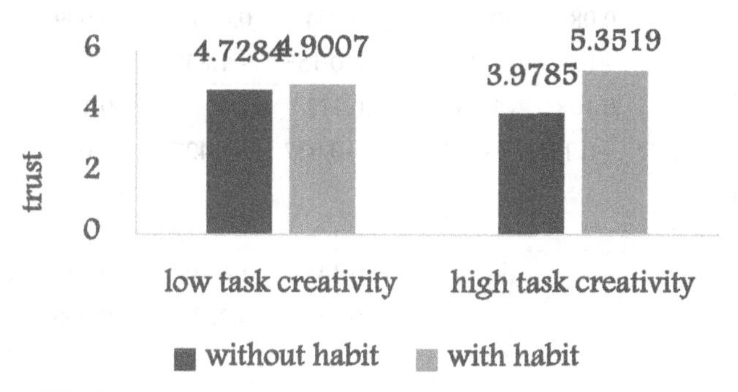

Fig. 3. Interaction effect of task creativity and personal habit

4.2 Moderating Effect Test

To better understand whether the different effects of personal habit and task creativity on trust in humanizing customer service chatbots were moderated by social presence, we conducted moderating effect test next. In line with the main analysis, personal habit and task creativity were confirmed as independent variables, trusting disposition, familiarity with e-commerce platform, gender, experience of interacting with realistic chatbots as well as expenses of online shopping monthly were regarded as control variables, and trust in humanizing customer service chatbots was defined as dependent variable. Finally, social presence was deemed as a moderating variable in this section. We mean-centered the values except personal habit and task creativity (discrete variable) prior to creating the interaction terms to reduce collinearity.

Table 4 presents the results of moderating effect test of social presence on personal habit and trust in humanizing customer service chatbots. When personal habit and social presence were introduced to the model, we found significantly positive effect of the both on trust in humanizing customer service chatbots ($\beta = 0.145$, $p < 0.1$; $\beta = 0.351$, $p < 0.01$). Then we entered the interaction term (personal habit*social presence) in the model, the result showed that social presence moderated the relationship between personal habit and trust in humanizing customer service chatbots ($\beta = -0.375$, $p < 0.01$). We also plotted the interaction effect of personal habit and social presence (see Fig. 4), a positive moderating effect of social presence on personal habit and trust in humanizing customer service chatbots was demonstrated. Specifically, a simple slope test was conducted. Results found that the moderating effect is significant only when social presence is low.

Table 4. Moderating effect test on personal habit and trust

	Model 1		Model 2		Model 3	
	Beta	t	Beta	T	Beta	t
Control variables						
Trusting disposition	−0.023	−0.26	−0.173	−1.984**	−0.206	−2.428**
Familiarity	0.066	0.737	0.15	1.775*	0.173	2.117**
Gender	0.084	0.937	0.034	0.413	0.009	0.111
Experience	−0.2	−2.243**	−0.153	−1.85*	−0.13	−1.613
Frequency	0.108	1.166	0.112	1.3	0.122	1.474
Expense	−0.126	−1.393	−0.037	−0.427	−0.01	−0.123
Independent variables						
Personal habit			0.145	1.678*	0.121	1.446
Social presence			0.351	3.727***	0.666	5.04***
Interaction term						
Personal habit*social presence					−0.375	−3.284***
F	1.737		4.639		5.627	
R square	0.072		0.219		0.279	

Note: *** Significant at the 1% level; ** Significant at the 5% level; * Significant at the 10% level

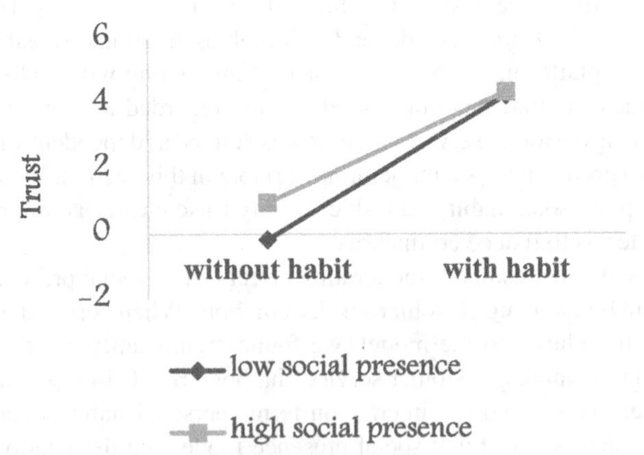

Fig. 4. Interaction plot of personal habit and social presence

Another test for moderating effect of social presence on task creativity and trust in humanizing customer service chatbots was performed (see Table 5). Consistent with the test above, social presence was validated significantly positive for trust in humanizing customer service chatbots ($\beta = 0.419$, $p < 0.01$). After the addition of the interaction term (task creativity*social presence) in the model, we found a significant moderating effect of social presence on task creativity and trust in humanizing customer service chatbots again ($\beta = 0.272$, $p < 0.05$). From the interaction plot (see Fig. 5), social presence remains positively moderating the relationship between task creativity and trust in humanizing customer service chatbots. In accordance with the simple slope test, the moderating effect of social presence is merely significant when social presence is high.

Table 5. Moderating effect test on task creativity and trust

	Model 1		Model 2		Model 3	
	Beta	t	Beta	t	Beta	t
Control variables						
Trusting disposition	−0.023	−0.260	−0.167	−1.875*	−0.145	−1.648
Familiarity	0.066	0.737	0.150	1.748*	0.124	1.453
Gender	0.084	0.937	0.037	0.436	0.044	0.535
Experience	−0.200	−2.243**	−0.161	−1.929*	−0.193	−2.312**
Frequency	0.108	1.166	0.120	1.382	0.120	1.403
Expense	−0.126	−1.393	−0.017	−0.196	0.001	0.007
Independent variables						
Task creativity			0.032	0.393	0.028	0.348
Social presence			0.419	4.611***	0.206	1.607
Interaction terms						
Task creativity*social presence					0.272	2.320**
F	1.737		4.221		4.475	
R square	0.072		0.204		0.235	

Note: *** Significant at the 1% level; ** Significant at the 5% level; * Significant at the 10% level

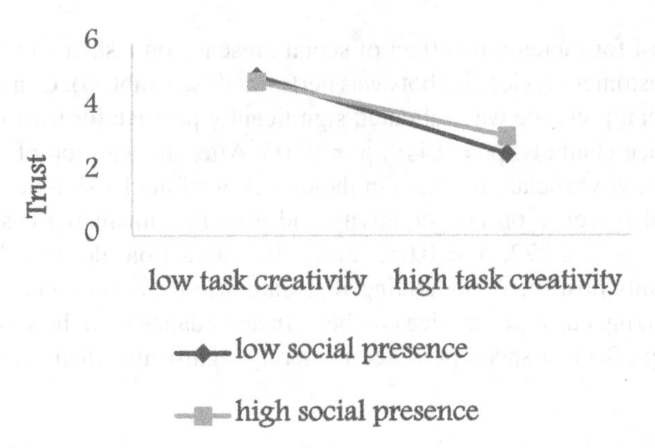

Fig. 5. Interaction Plot of Task Creativity and Social Presence

4.3 Results

By integrating personal habit, task creativity and social presence into conceptual model, this study aims to explore the mechanism and boundary of consumers' trust in humanizing customer service chatbots. The results of ANCOVA suggest that personal habit plays an important role for consumers' trust in humanizing customer service chatbots, whereas no significant effect was found from the perspective of task creativity. Consumers with personal habit of interacting with customer service chatbots seem to trust in humanizing customer service chatbots more for high task creativity, while consumers without habit reduces trust in them more for high task creativity. To further investigate the trust boundary, the moderating effect shows that social presence enhances relationship between consumers without personal habit and their trust, only significant for low social presence; while social presence positively moderates relationship between high task creativity and trust, only significant for high social presence.

5 Discussion and Conclusion

From completely new perspectives, this study advances the researches of trust in human-machine coordination in e-commerce scenarios to a new and promising area.

On a theoretical level, first, by leveraging guidelines for personalization and contextualization proposed by recent research, we bring a fresh perspective on trust model of emerging humanizing customer service chatbots. Grounding our arguments in specific antecedents of trust (i.e., personal habit and task creativity), we explain the mechanisms through which consumers' trust can also be fostered by manipulating the personalized and contextual factors. Second, the studies pertaining to humanizing customer service chatbots are enriched. We break through the great body of current research about whether to accept anthropomorphic chatbots, such as attributes, adoption, working context and disclosure. The hypotheses are established directly on humanizing customer service chatbots, broadening horizons for the research in this topic. Third, we offer insights on other possible relationships between social presence and trust. Although previous literature has extensively examined the relationship between social presence and trust, this study is one of the first to explore the moderating effect of social presence with trust.

From organizational perspective, there are several managerial implications. First, our study highlights the importance of increasing popularity of customer service chatbots for e-commerce platform, building up consumers experience, thereby facilitating them to form relevant habit. With this, the goal of enhancing trust in humanizing customer service chatbots can be achieved over time. Second, it is critical to incorporate functions distinguishing consumers without (vs. with) habit of interacting with customer service chatbots by monitoring their operating proficiency or history for service provider. Consumers without habit are more keen on human customer service assisted under high creative task, in conjunction with certain visual or auditory stimulus for social presence. Moreover, especially for high creative task, 3D product display, voice communication together with virtual reality can be synthesized to maintain high social presence for all consumers. Finally, managers ought to not only emphasize on the anthropomorphic features and response rate of customer service chatbots, but also dedicate more time to training their employees to coordinate with chatbots appropriately.

Our study has some limitations should be noted. First, despite we designed and simulated scenarios, the data were self-reported and may be subject to respondents' cognitive bias or individual differences. Hence, realistic data can complement and extend our findings next. Second, although we have augmented interaction of habit and task creativity together with social presence to identify trust boundary, there is still much space to explore trust boundary in conjunction with qualitative study or experimental study. Third, this study validated hypotheses using cross-sectional data, however, trust is subjective and changing over time. Incorporating time series would be a more beneficial avenue for the future.

Acknowledgements. This research is supported by the National Natural Science Foundation of China through grant 72171187, and partially supported by the Key Project of Shaanxi International Science and Technology Cooperation through grant 2018KWZ-04.

References

1. Sheehan, B., Jin, H.S., Gottlieb, U.: Customer service chatbots: Anthropomorphism and adoption. J. Bus. Res. **115**, 14–24 (2020)
2. Roy, R., Naidoo, V.: Enhancing chatbot effectiveness: the role of anthropomorphic conversational styles and time orientation. J. Bus. Res. **126**, 23–34 (2021)
3. Schanke, S., Burtch, G., Ray, G.: Estimating the impact of "humanizing" customer service chatbots. Inf. Syst. Res. **32**, 736-751 (2021)
4. Blut, M., Wang, C., Wünderlich, N.V., Brock, C.: Understanding anthropomorphism in service provision: a meta-analysis of physical robots, chatbots, and other AI. J. Acad. Mark. Sci. **49**(4), 632–658 (2021). https://doi.org/10.1007/s11747-020-00762-y
5. Gefen, D., Straub, D.: Managing user trust in B2C e-services. e-Service 2(2), 7–24 (2003)
6. Shawar, B.A., Atwell, E.:Chatbots: are they really useful? Ldv Forum 22(1), 29–49 (2007)
7. Van Doorn, J., et al.: Domo arigato Mr. Roboto: emergence of automated social presence in organizational frontlines and customers' service experiences. J. Serv. Res. **20**(1), 43–58 (2017)
8. Touré-Tillery, M., McGill, A.L.: Who or what to believe: trust and the differential persuasiveness of human and anthropomorphized messengers. J. Mark. **79**(4), 94–110 (2015)

9. Araujo, T.: Living up to the chatbot hype: the influence of anthropomorphic design cues and communicative agency framing on conversational agent and company perceptions. Comput. Hum. Behav. **85**, 183–189 (2018)
10. Adam, M., Wessel, M., Benlian, A.: AI-based chatbots in customer service and their effects on user compliance. Electron. Mark. **31**(2), 427–445 (2020). https://doi.org/10.1007/s12525-020-00414-7
11. Cammy, C., et al.: Blame the bot: anthropomorphism and anger in customer–chatbot interactions. J. Mark. **86**(1), 132–148 (2022)
12. Jin, S.V., Youn, S.: Why do consumers with social phobia prefer anthropomorphic customer service chatbots? Evolutionary explanations of the moderating roles of social phobia. Telemat. Inform. **62**, 101644 (2021)
13. Mozafari, N., Weiger, W.H., Hammerschmidt, M.:The chatbot disclosure dilemma: desirable and undesirable effects of disclosing the non-human identity of chatbots. In: ICIS (2020)
14. Grimes, G.M., Schuetzler, R.M., Giboney, J.S.: Mental models and expectation violations in conversational AI interactions. Decis. Supp. Syst. **144**, 113515 (2021)
15. Shumanov, M., Johnson, L.: Making conversations with chatbots more personalized. Comput. Hum. Behav. **117**, 106627 (2021)
16. Dabholkar, P.A.: Consumer evaluations of new technology-based self-service options: an investigation of alternative models of service quality. Int. J. Res. Mark. **13**(1), 29–51 (1996)
17. Kim, S.S., Malhotra, N.K.: A longitudinal model of continued IS use: an integrative view of four mechanisms underlying postadoption phenomena. Manag. Sci. **51**(5), 741–755 (2005)
18. Limayem, M., Hirt, S.G., Cheung, C.M.K.: How habit limits the predictive power of intention: the case of information systems continuance. MIS Q. **31**, 705–737 (2007)
19. Venkatesh, V., Thong, J.Y., Xu, X.: Consumer acceptance and use of information technology: extending the unified theory of acceptance and use of technology. MIS Q. 36, 157–178 (2012)
20. Dietvorst, B.J., Simmons, J.P., Massey, C.: Algorithm aversion: people erroneously avoid algorithms after seeing them err. J. Exp. Psychol. Gen. **144**(1), 114 (2015)
21. Renier, L.A., Mast, M.S., Bekbergenova, A.: To err is human, not algorithmic–Robust reactions to erring algorithms. Comput. Hum. Behav. **124** 106879 (2021)
22. Castelo, N., Bos, M.W., Lehmann, D.R.: Task-dependent algorithm aversion. J. Mark. Res. **56**(5), 809–825 (2019)
23. Dietvorst, B.J., Simmons, J.P., Massey, C.: Overcoming algorithm aversion: People will use imperfect algorithms if they can (even slightly) modify them. Manag. Sci. **64**(3), 1155–1170 (2018)
24. Jifei, W., et al.: Impact of artificial intelligence recommendation on consumers' willingness to adopt. J. Manag. Sci. **33**(5), 29–43 (2021)
25. Fulk, J., et al.: A social information processing model of media use in organizations. Commun. Res. **14**(5), 529–552 (1987)
26. Walther, J.B.: Relational aspects of computer-mediated communication: experimental observations over time. Organ. Sci. **6**(2), 186–203 (1995)
27. Mao, C., Yuan, Q.: Social presence theory and its application and prospect in the field of information system. J. Intell. **37**(8), 186–194 (2018)
28. Gefen, D., Straub, D.W.: Consumer trust in B2C e-Commerce and the importance of social presence: experiments in e-Products and e-Services. Omega **32**(6), 407–424 (2004)
29. Lu, B., Fan, W., Zhou, M.: Social presence, trust, and social commerce purchase intention: empirical research. Comput. Hum. Behav. **56**, 225–237 (2016)
30. Ogonowski, A., et al.: Should new online stores invest in social presence elements? The effect of social presence on initial trust formation. J. Retail. Consum. Serv. **21**(4), 482–491 (2014)
31. Ye, S., et al.: Enhancing customer trust in peer-to-peer accommodation: A "soft" strategy via social presence. Int. J. Hospitality Manag. **79**, 1–10 (2019)

32. Hajli, N., et al.: A social commerce investigation of the role of trust in a social networking site on purchase intentions. J. Bus. Res. **71**, 133–141 (2017)
33. Jiang, C., Rashid, R.M., Wang, J.: Investigating the role of social presence dimensions and information support on consumers' trust and shopping intentions. J. Retail. Consum. Serv. **51**, 263–270 (2019)
34. Srivastava, S.C., Chandra, S.: Social presence in virtual world collaboration: an uncertainty reduction perspective using a mixed methods approach. MIS Q. **42**(3), 779–804 (2018)
35. Hirschman, E.C., Holbrook, M.B.: Hedonic consumption: emerging concepts, methods and propositions. J. Mark. **46**(3), 92–101 (1982)
36. Strahilevitz, M., Myers, J.G.: Donations to charity as purchase incentives: how well they work may depend on what you are trying to sell. J. Consum. Res. **24**(4), 434–446 (1998)
37. Choi, J., Lee, H.J., Kim, Y.C.: The influence of social presence on customer intention to reuse online recommender systems: the roles of personalization and product type. Int. J. Electron. Comm. **16**(1), 129–154 (2011)
38. Hassanein, K., Head, M.: The impact of infusing social presence in the web interface: an investigation across product types. Int. J. Electron. Commer. **10**(2), 31–55 (2005)
39. Gefen, D.: E-commerce: the role of familiarity and trust. Omega **28**(6), 725–737 (2000)

Online Shopping During COVID-19: A Comparison of USA and Canada

Norman Shaw[1(✉)], Brenda Eschenbrenner[2], and Ksenia Sergueeva[3]

[1] Ryerson University, Toronto, Canada
norman.shaw@ryerson.ca

[2] University of Nebraska at Kearney, Kearney, NE, USA
eschenbrenbl@unk.edu

[3] Drexel University, Philadelphia, USA
ks3788@drexel.edu

Abstract. During the COVID-19 pandemic, many governments restricted economic activity by imposing lockdowns or requiring capacity constraints, thereby impacting brick-and-mortar businesses. Consumers responded by staying at home and turning to online shopping. Some consumers were already familiar with online shopping, whereas for others it was a new experience. As restrictions are removed or reduced, consumers may permanently change their shopping habits and continue to buy online with greater frequency than prior to the pandemic. With empirical data from a cross section of Canadian and American consumers, this study investigates the factors that influence the continuation of online shopping. The results show that there is little difference between Canadians and Americans, with perceptions of convenience significantly influencing perceived usefulness, and efficiency being a significant factor as well but only for Americans. Perceived usefulness is important for continuance intentions, with hedonic motivation having a moderating effect. Our results provide guidance to practitioners who are interested in consumers' online shopping intentions after the pandemic and factors that can foster such activities.

Keywords: Online shopping · E-S-QUAL · Hedonic motivation · PLS · COVID-19

1 Introduction

The World Health Organization (WHO) declared the COVID-19 outbreak a pandemic on 11 March 2020 [1]. In the weeks that followed, various jurisdictions around the world declared states of emergency, locking down their economies and limiting in-store shopping to essential services only. The result was an increase in online shopping with global retail e-commerce increasing 25.7% to US$4.2 trillion [2].

Canada was no exception. Since the initial declaration of the COVID-19 pandemic in 2020, various Canadian provinces imposed curfews and strict self-quarantine policies which restrained in-store shopping [3]. These restrictions altered the shopping habits of Canadian consumers, leading to increased e-commerce: in 2020, Canada's online sales

F. Fui-Hoon Nah and K. Siau (Eds.): HCII 2022, LNCS 13327, pp. 510–525, 2022.
https://doi.org/10.1007/978-3-031-05544-7_37

grew by 75.0%, making it the second-fastest-growing e-commerce market worldwide [4]. E-commerce increased from $3.25 billion in 2019 (13.52% of total retail sales) to $4.21 billion in 2020 (17.83%) in 2019 [5]. Similarly in the USA, nonessential businesses were closed and retail business slowed, causing increases in e-commerce activity [6]. More specifically, the USA experienced growth in e-commerce retail trade sales to an annual volume in 2020 of US$762.68 billion, representing an increase of 31.8% [4]. According to the US Census Bureau, the electronic shopping and mail-order houses industry experienced a 35.2% increase in sales from 2019 to 2020, the most significant of any industry [7].

By the end of 2021, restrictions have started to relax. Stores are opening and consumers are returning to in-store shopping. However, because shoppers have experienced e-commerce, some for the first time and others to a greater extent, they may elect to continue shopping online. What is currently unknown is the factors that will influence consumers' intentions to continue shopping online, which is the objective of this research. Hence, the research question posed is: What factors influence consumers' intentions to continue shopping online as in-store shopping becomes available again?

Although prior studies have evaluated online shopping [8–10], little research has evaluated those factors that will influence consumers to continue their online shopping once the COVID-19 pandemic recedes. When consumers can resume in-store shopping similar to before the pandemic, it will be important for practitioners to understand the factors that will motivate consumers to continue the levels of e-commerce activity experienced during the pandemic. This study addresses this gap by studying online shopping continuance, comparing online activities prior to the pandemic with anticipated behavior after the pandemic. The theoretical foundation is E-S-QUAL [11], extended with convenience, security and the addition of hedonic motivation as a moderator.

This paper is organized as follows. The next section is the literature review which develops the hypotheses and concludes with the research model. The third section explains the methodology followed by the results. Section five details the results which are then discussed in section six.

2 Literature Review, Theoretical Foundation and Hypotheses Development

2.1 Service Quality

Many retail organizations are selling the same or similar products, whose quality characteristics can be measured by their durability and adherence to product specifications [12]. In order to attract consumers, these organizations need to differentiate themselves via the services that they offer [13]. The quality of this service can be defined as "the extent to which a service meets customers' needs or expectations" [14] In order to measure customers' perception of service, Parasuraman et al. [15] introduced SERVQUAL, comprising the dimensions of tangibility, reliability, responsiveness, assurance and empathy. This construct has been used to study the impact of service quality in various sectors of the economy, such as the airline industry [16], the auto aftersales market [17], healthcare [18], banking [19], hospitality [20] and retail [21].

Just as physical stores need to differentiate themselves through the services that they offer [22], websites must do the same [23]. In the early days of e-commerce, simply having a website with low prices was considered sufficient to drive sales, but as more retailers started to offer similar products online, service quality became a competitive factor [24]. Customers expect websites to facilitate their shopping, from the time they begin to search for products until their products are paid for and delivered [25]. Building on the research that lead to SERVQUAL, Parasuraman et al. [11] developed E-S-QUAL specifically to measure the service quality of e-commerce websites. With the aid of a random sample of Internet users of Walmart and Amazon, E-S-QUAL was measured across four dimensions: efficiency, fulfilment, system availability and privacy [11].

Because of its relevance to online shopping, E-S-QUAL forms the theoretical foundation for our research model [26], further extended with convenience, security and hedonic motivation. The following paragraphs in this section introduce our hypotheses.

2.2 Efficiency

Physical stores have their merchandise on display and, with a glance around the store, customers can quickly decide which departments to visit. The full characteristics of the product can be physically evaluated. The presentation of products within defined departments makes the consumer's visit efficient. Websites have to mimic the efficiency of the real world by focusing on the site design, ensuring it is easy to navigate and that product images are clearly and attractively displayed [23, 27]. Product information should be available with a simple click [28]. If the website is poorly organized, customers will lose patience and shop elsewhere, which they can do with the click of the mouse [29]. Hence, we hypothesize:

H1: Efficiency positively influences perceived usefulness of online shopping.

2.3 Fulfilment

Fulfilment refers to the receipt of the merchandise. In a physical store, the customer pays and walks out with the merchandise: there is no ambiguity with respect to the item and delivery is immediate. In contrast, when ordering online, the customer has to wait for delivery. Furthermore, the shipping costs may be high, the product delivered may not be what was ordered and its delivery may be later than anticipated [30]. Consumers may be discouraged to purchase because of the risk of delays and the hassle of returns [31]. Grewal et al. [32] suggested that "retailers must give more attention to fulfilment." Hence, our second hypothesis is:

H2: Fulfilment positively influences perceived usefulness of online shopping.

2.4 Security

Security refers to the perception that financial information shared through the website is safe. Consumers are familiar with using a plastic credit or debit card when paying in-store and they will have heard about security breaches via various news organizations [33]. However, when paying online, there are even more concerns because of the extra organizations involved: the Internet provider, the retailer's website and the financial institutions. Transaction details may be falsified or the credit card details could be stolen and used by an unauthorized third party [34]. Online shoppers will need assurances of security to find online shopping useful. Hence, we hypothesize.

H3: Perceived security positively influences perceived usefulness of online shopping.

2.5 Privacy

When shopping in-store, the consumer can remain anonymous. There is no need to provide any identifying information and, if cash is paid, there is no record of who purchased the merchandise. Also, shopping experiences such as the number of items viewed or amount of time spent in store can not be readily connected with the shopper. However, websites can track the number of items viewed or the amount of time spent on a webpage. Online websites require personal information: the consumer must provide name and address for delivery and order history is maintained to assist with future purchases. An email allows the retailer to send an order acknowledgement with tracking information for the customer to follow up. Consumers may be fearful that their personal information could be shared with other organizations without their permission or exposed in a manner that they do not wish [35]. They want to be confident that their personal data is kept private [23]. Therefore, we propose:

H4: Perceived privacy perceptions positively influence perceived usefulness of online shopping.

2.6 Convenience

One advantage of online shopping is the freedom to transact at any time and any place [36]. There is no need to visit a retail store or wait until the shops are open. Furthermore, prices can be readily compared via browsing different websites or even finding a website that consolidates products and prices from various sources. Larger objects can be purchased more conveniently because online delivery is included as part of the checkout process [37]. With government regulations restricting occupancy, online shopping is more convenient than driving to a store and then having to wait outside before the easing of capacity constraints permit entry. Due to regulations, many shoppers have had to buy some goods online and, as a result, they have found e-commerce to be convenient [38]. We hypothesize:

H5: Convenience positively influences perceived usefulness of online shopping.

2.7 Perceived Usefulness

Perceived usefulness has been established as a key variable influencing adoption [39]. Perceived usefulness has also been found to significantly influence continuance intentions in contexts such as massive open online courses (Daneji, Ayub, & Khambari, 2019). By browsing websites from the comfort of one's home, many online stores can be visited without having to travel. Products can be purchased from a variety of categories, and with delivery there is no need to carry items from store to home [40]. During the pandemic, online shopping offers an alternative especially for those who are more concerned about entering a physical store. A previous study of older adults online shopping found support for perceived usefulness influence on continuance intentions (Wu & Song, 2021). Therefore, we propose:

H6: Perceived usefulness influences intention to continue online shopping.

2.8 Hedonic Motivation

With improved communication speeds and website design, online shopping can be a pleasant experience. Consumers who find online shopping enjoyable may be more inclined to continue with e-commerce once the pandemic recedes [41] Arnold and Reynolds [42] identified six dimensions comprising hedonic motivation: the adventure, socializing, gratification, idea shopping, role shopping, and hunting for value. The prime motivation of shoppers may well be the utilitarian outcomes, but some of the shoppers will find the activity enjoyable [43]. We propose that hedonic motivation has a moderating influence on the relationship between perceived usefulness and continuance intention. Those consumers who find online shopping more enjoyable will also rate their utilitarian experience more highly and they will be more inclined to increase their online shopping. Hence, we propose hedonic motivation to have a moderating effect and hypothesize the following.

H7: Hedonic motivation moderates the influence of perceived usefulness on intention to continue with online shopping.

2.9 The Dependent Variable – Continuance Intention

Continuance intention, in an IS context, has been previously defined as "Users' intention to continue using OBD [online banking division]." (Bhattacherjee, 2001, p. 359). As in-store shopping becomes a possibility as restrictions are eased, consumers may choose to continue shopping online. The pandemic has provided motivation for shopping online due to the health concerns, restrictions (e.g., capacity), and alternate means to shop when existing options became unavailable (e.g., store closings). Even reluctant online shoppers will visit websites and place orders. Factors such as attitudes and experience may determine if they will continue to shop online when there is a semblance of normality that allows more in-store visits. Based on the context of this study, we are focused on intentions to continue shopping online and, thus, utilize continuance intentions as our dependent variable.

2.10 Research Model

The research model is presented in Fig. 1.

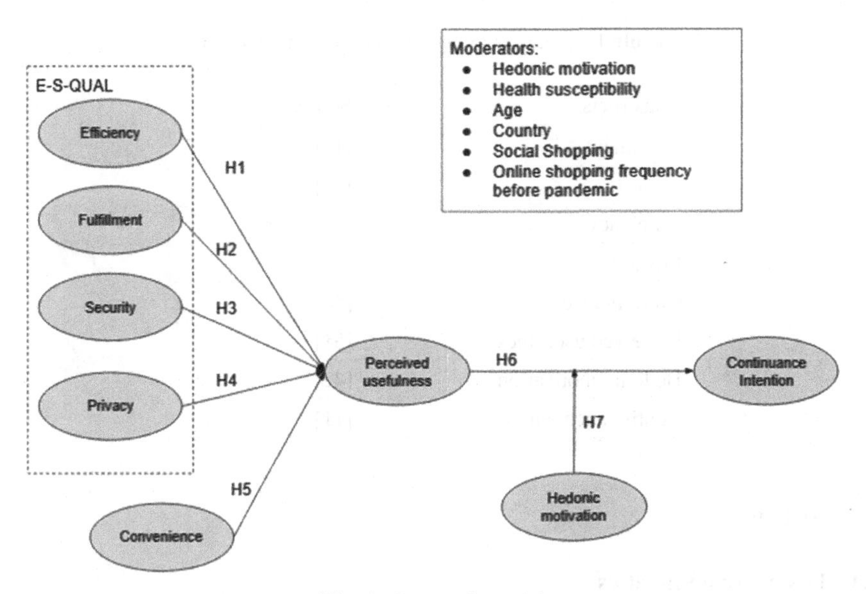

Fig. 1. Research model

3 Methodology

A survey research study was conducted to assess our research model. Construct measurement items were adapted from extant literature (see Table 1), and we utilized a 5-point Likert scale. Data was collected utilizing the services of Qualtrics [44]. Qualtrics provided panels with the distribution balanced across gender and four age groups. Only individuals 19 years of age or older were allowed to complete the survey.

The data were analyzed with SmartPLS [45]. The loadings of each indicator were calculated to ensure they were greater than 0.708, validating their convergence [46]. Discriminant validity was tested with the heterotrait-monotrait (HTMT) [47] and the Fornell-Larcker criterion [48]. Only indicators that satisfied the conditions of these tests (i.e., ratios were below .85 and the square root of the average variance extracted for each variable is greater than other correlation values, respectively) were included in further analysis.

The PLS algorithm was run for each country with a maximum of 300 iterations and the path weighting schemes to determine the path coefficients and the coefficient of determination, R^2. The next step was bootstrapping with 10,000 subsamples. Multigroup analysis was conducted to compare countries, gender, age groups and online shopping habits. Hedonic motivation was included in the model as a moderator.

Given that participants were asked to complete a single survey, there is a risk of common method bias [52]. To address this, we introduced a common marker variable [53]. Results showed that this variable was independent of the other variables.

Table 1. Source of construct measurement items.

Constructs	Source
Fulfilment	[11]
Security	[49]
Efficiency	[11]
Privacy	[11]
Convenience	[50]
Perceived usefulness	[51]
Hedonic motivation	[42]
Continuance intention	[11]

4 Results

4.1 Descriptive Statistics

The sample sizes were 535 from Canada and 509 from the USA (see Table 2). Participants were asked their age and were then divided into four groups: Gen Z from 19 to 26, Millennials or Gen Y from 27 to 44, Gen X from 45 to 56 and Baby Boomers from 57 plus.

Table 2. Sample gender and age by country

		Age group				
		19–26	27–44	45–56	57+	Total
Canada	Male	44	63	77	81	265
	Female	77	70	59	58	264
	Not specified	4	1	1		6
	Total	*125*	*134*	*137*	*139*	*535*
USA	Male	91	35	42	89	257
	Female	31	89	88	40	248
	Not specified	1	2	1		4
	Total	*123*	*126*	*131*	*129*	*509*

Participants' familiarity with online shopping prior to the pandemic was gauged by asking how frequently they had purchased goods online (see Table 3). Online shopping frequency was analyzed as a moderator.

Table 3. Online shopping frequency prior to pandemic

Online shopping frequency before the pandemic	Canada		USA	
	Count	%	Count	%
Never	28	5.2%	32	6.3%
Less than 1 time per month	177	33.1%	104	20.4%
1 time per month	106	19.8%	73	14.3%
2 times per month	90	16.8%	109	21.4%
3 times per month	49	9.2%	72	14.2%
4 times per month	28	5.2%	45	8.8%
5 times per month	23	4.4%	24	4.7%
6 times per month	6	1.1%	11	2.2%
7 + times per month	28	5.2%	39	7.7%

4.2 The Measurement Model

The PLS algorithm within SmartPLS calculated the outer loadings. One indicator for perceived usefulness was dropped because its loading was below 0.7. Subsequently, all indicators had values greater than 0.7 [46]. The reliability of the model was confirmed by showing that Cronbach's Alpha was greater than 0.6, that Composite Reliability was greater than 0.7 and Average Variance Extracted was greater than 0.5 [54]. See Table 4 for results.

Table 4. Reliability results by country

Construct	Canada			USA		
	Cr's alpha	CR	AVE	Cr's alpha	CR	AVE
Fulfilment	0.869	0.91	0.717	0.869	0.91	0.717
Convenience	0.883	0.915	0.682	0.883	0.915	0.682
Privacy	0.888	0.929	0.814	0.888	0.929	0.814
Security	0.9	0.93	0.769	0.9	0.93	0.769
Perceived usefulness	0.797	0.868	0.623	0.797	0.868	0.623
Continuance intention	0.897	0.936	0.83	0.897	0.936	0.83
Efficiency	0.92	0.939	0.756	0.92	0.939	0.756

Two tests were conducted for discriminant validity: the heterotrait-monotrait (HTMT) criterion [55] and the Fornell-Larcker criterion [48], both of which met the requisite criteria demonstrating that the constructs were discriminant.

4.3 The Inner Model

The PLS algorithm also calculated the path coefficients. The coefficient of determination (R^2) was calculated by the PLS algorithm for Perceived Usefulness and Continuance Intention. Table 5 shows the results by country.

Table 5. R-squared

	Canada	USA
Perceived usefulness	0.290	0.327
Continuance intention	0.524	0.520

Path significance was found by bootstrapping with 10,000 replacements. Each country was run separately. Path significance for each hypothesis is shown in Table 6.

Table 6. Path coefficients and significance

	Canada		USA	
Fulfilment to PU	−0.059		−0.02	
Convenience to PU	0.419	***	0.425	***
Privacy to PU	−0.03		−0.056	
Security to PU	0.231		0.137	
PU to Continuance Intention	0.345	***	0.459	***
HM to Continuance Intention	0.404	***	0.297	***
Efficiency to PU	0.095		0.198	***
Moderating effect of HM on CI	−0.12	**	−0.131	***

*** $p < 0.001$ ** $p < 0.01$

4.4 Moderating Role of Hedonic Motivation

As seen in Table 6, hedonic motivation moderated the effect of perceived usefulness on continuance intention. The path was significant for both Canada and USA. By plotting the slopes of the moderation effect for each country, the graphs showed that hedonic motivation had a stronger effect for US consumers. See Figs. 2 and 3.

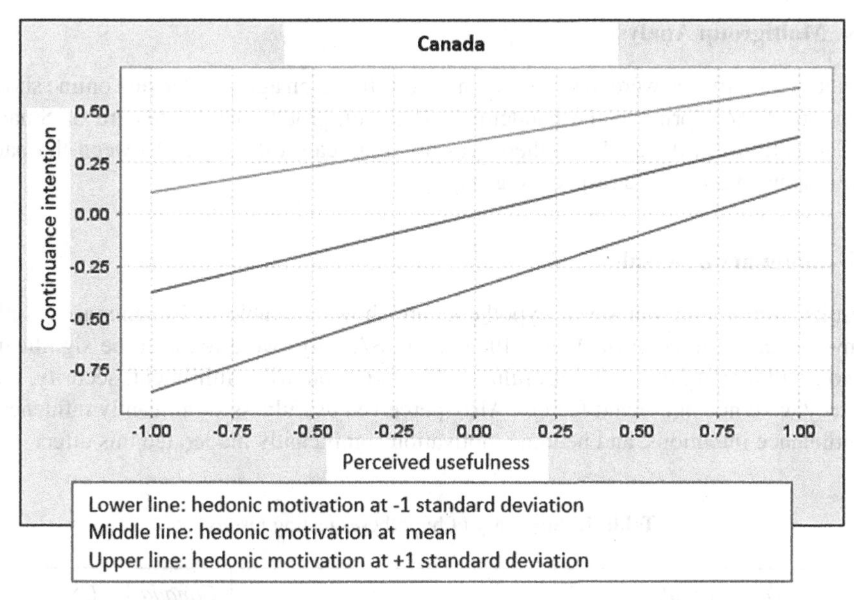

Fig. 2. Moderating effect of hedonic motivation for Canada

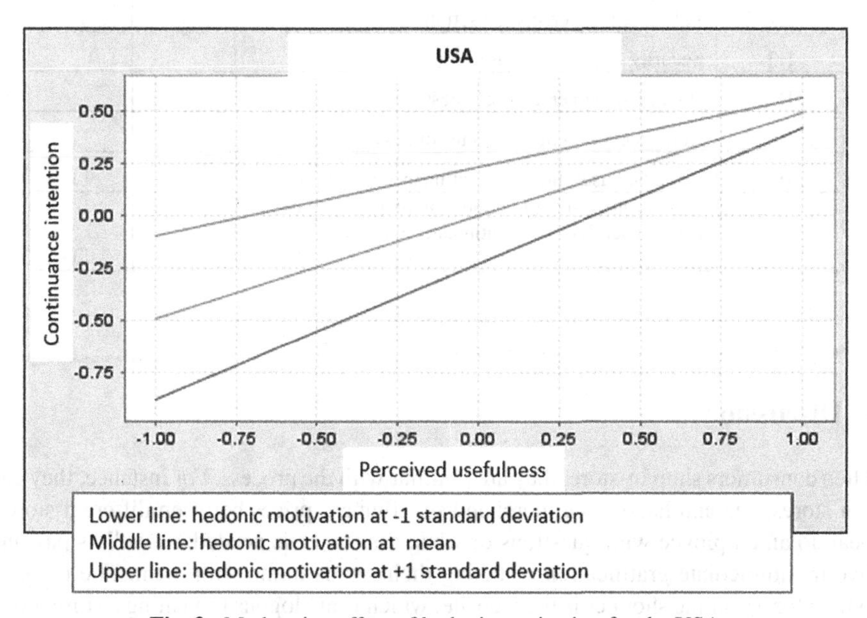

Fig. 3. Moderating effect of hedonic motivation for the USA

4.5 Multigroup Analysis

Various comparisons were made between groups based on age, gender, and online shopping experience (prior to the pandemic). The multigroup analysis feature of SmartPLS was used. In the analysis, there was no significant differences between the path coefficients of any of the respective groups.

4.6 Summary of Results

Results from testing our seven hypotheses are shown in Table 7. To summarize, only convenience (both countries) and efficiency (USA only) were found to be significant factors influencing perceived usefulness. For both countries, fulfilment, security, and privacy were not significant factors. Also, perceived usefulness significantly influenced continuance intentions, and hedonic motivation significantly moderated this effect.

Table 7. Summary of hypotheses testing results

Hyp.	*Path*	*Canada*	*USA*
H1	Efficiency -> perceived usefulness		☐
H2	Fulfilment -> perceived usefulness		
H3	Security -> perceived usefulness		
H4	Privacy -> perceived usefulness		
H5	Convenience -> perceived usefulness	☐	☐
H6	Perceived usefulness -> continuance intention	☐	☐
H7	Moderating effect of hedonic motivation on perceived usefulness -> continuance intention	☐	☐

5 Discussion

When consumers shop in-store, they are familiar with the process. For instance, they can visit stores, see and handle the merchandise, compare prices between different stores, speak to an employee with questions or when assistance is needed, as well as pay and have the immediate gratification of taking their goods home with them. The point of contact for an online shopper is the website, which is analogous to visiting a store. For a satisfying online experience, the website must be efficient, meaning that it is available, easy to navigate and responsive.

In our study, efficiency was significant for American shoppers but not for Canadians. In previous studies, Marimon [40] found that efficiency was not significant for shoppers of a Spanish supermarket, while Lee [23] found that website design and responsiveness were significant for undergraduate students in Taiwan. Efficiency may not be considered significant because the majority of websites today are well-organized and responsive. However, for Americans, the influence of efficiency on perceived usefulness was significant, which may be explained by the sample having more familiarity with online shopping. In our sample, 59% of Americans had shopped online two or more times per month prior to the pandemic, compared to 41% of Canadians.

Our research results suggest that online shoppers value convenience. This confirms the findings of Duarte et al. [50] who found Portuguese consumers appreciated that shopping from home saved them time and effort of going to stores. Convenience also solves the problem of time-starved people as they are able to shop 24/7. Also, our research results suggest that consumers value the usefulness of online shopping. They may have had little choice during the pandemic, but many of them stated they would increase their e-commerce activity once the pandemic was over. In a comparison of pre- and post-pandemic shopping, 78% of Americans and 63% of Canadians said they would shop more than once per month online, compared to 59% and 41% prior to the pandemic, respectively. Our results confirm that perceived usefulness is a key factor influencing adoption [39, 56].

However, neither security nor privacy concerns were significant. This is in contrast to past studies [57, 58]. One argument for these variables being nonsignificant is that Internet shopping is more common and those consumers that engage in online shopping have already overcome their misgivings [59].

A major difference between in-store and online shopping is the fulfilment of the order which refers to the final delivery of merchandise and the risk that the actual merchandise will not live up to the descriptions on the website. Pink and Djohan [60] found successful fulfilment had a positive effect on customer satisfaction. In our study, there was no significant influence. This may be due to the maturity of online retailers who have refined their processes such that deliveries are faster and more accurate. In addition, due to competition, some retailers have made returns easier, such as accepting returns without questions and often at no charge.

Hedonic motivation was found to be a significant moderating factor. Those consumers who found online shopping enjoyable engaged more frequently during the month. Kim et al. [61] found similar results when investigating wearable technology. The moderating effect for Americans was greater than that for Canadians. This could be because of the more extensive experience of US consumers with online shopping.

5.1 Theoretical Contribution

We offer a novel contribution by extending E-S-QUAL with convenience. In addition, our results suggest that the dimensions of E-S-QUAL influence perceived usefulness which then influences continuance intention, with hedonic motivation being a moderating factor. The research model is, therefore, novel and applied empirically in the context of online shopping during the time when health concerns and government regulations around COVID-19 are forcing consumers to reevaluate their attitudes towards purchases via the Internet.

5.2 Implication for Practitioners

The COVID-19 pandemic has brought about rapid changes in shopping habits. Practitioners can observe what has worked under such circumstances and then apply these lessons once the pandemic recedes and life returns to 'normal'. Consumers will continue to use online shopping so long as they judge it to be useful. Websites must be well-designed, easy to navigate and responsive. They must be available 24/7 with the means to obtain product information and price comparisons. Once all items are in the cart, checkout must be fast and straightforward. Websites must be designed with a 'fun' element so that consumers are engaged hedonically while shopping. This is important for both Canadian and American consumers, with an extra emphasis for the latter.

5.3 Limitations and Future Research

The challenge when collecting survey data is the sample selection, which in this study was via panels. Consequently, the participants are limited to those members who have listed their names on these panels and therefore do not represent a cross section of the population. Although various strategies were employed to ensure that questionnaires were answered forthrightly (e.g. attention filters and straight-line analysis), there is still the possibility that some participants answered the questions without much consideration other than to receive the reward. Further, when asking questions about future behavior (e.g. use of online shopping after the pandemic), participants may have false expectations.

Future research can adopt the same model and compare other countries. The model can be expanded to include expectation-continuation theory. Longitudinal studies will provide more accurate information about consumer habits and could be planned once the COVID-19 pandemic is nearing its end.

6 Conclusion

The COVID-19 pandemic has encouraged many consumers to engage in online shopping, some of them for the first time and others with increased frequency. When life returns to 'normal' once the pandemic is over, online shoppers may well continue to shop with greater frequency than before. For this to happen, their experience must be positive. This study extends an established model of website service quality and evaluates online shopping before, during and after the pandemic. The results can guide practitioners to ensure that their websites are responsive, well organized, easy to navigate and enjoyable to use to ensure continued success in their online presence. Customers will then be more likely to continue engaging in online shopping.

References

1. Cucinotta, D., Vanelli, M.: WHO declares COVID-19 a pandemic. Acta Bio Medica: Atenei Parmensis **91**, 157 (2020)
2. von Abrams, K.: Global Commerce Forecast 2021 (2021). https://www.emarketer.com/content/global-ecommerce-forecast-2021

3. The Canadian Press: Grim anniversary: A timeline of one year of COVID-19 (2021), https://www.ctvnews.ca/health/coronavirus/grim-anniversary-a-timeline-of-one-year-of-covid-19-1.5280617
4. Statista: eCommerce report 2021 (2021). www.statista.com
5. eMaketer: Canada's retail ecommerce sales (2021)
6. Ward, B., Sipior, J.C., Lombardi, D.R.: COVID-19: state sales and use tax implications. Inf. Syst. Manag. **37**, 343–347 (2020)
7. United States Census Bureau. E-commerce (2021). <https://www.census.gov/library/publications/time-series/e-commerce.All.List_267926749.html
8. Liu, F., Lim, E.T., Li, H., Tan, C.-W., Cyr, D.: Disentangling utilitarian and hedonic consumption behavior in online shopping: an expectation disconfirmation perspective. Inf. Manag. **57**, 103199 (2020)
9. Dayal, S., Palsapure, D.: A study on the individual's online shopping continuance intention on Amazon. in for consumer electronics. Int. J. Bus. Global. **24**, 240–255 (2020)
10. Sethuraman, P., Thanigan, J.: An empirical study on consumer attitude and intention towards online shopping. Int. J. Bus. Innov. Res. **18**, 145–166 (2019)
11. Parasuraman, A., Zeithaml, V.A., Malhotra, A.: ES-QUAL: a multiple-item scale for assessing electronic service quality. J. Serv. Res. **7**, 213–233 (2005)
12. Garvin, D.A., Quality, W.D.P.: Really mean. Sloan Manag. Rev. **25**, 25–43 (1984)
13. Thompson, P., DeSouza, G., Gale, B.T.: The strategic management of service quality. Qual. Prog. **18**, 20–25 (1985)
14. Arora, P., Narula, S.: Linkages between service quality, customer satisfaction and customer loyalty: a literature review. IUP J. Mark. Manag. **17**, 30 (2018)
15. Parasuraman, A., Zeithaml, V.A., Berry, L.L.: Servqual: a multiple-item scale for measuring consumer perc. J. Retail. **64**, 12 (1988)
16. Hapsari, R., Clemes, M.D., Dean, D.: The impact of service quality, customer engagement and selected marketing constructs on airline passenger loyalty. Int. J. Qual. Serv. Sci. **9**, 21–40 (2017)
17. Gencer, Y.G., Akkucuk, U.: Measuring quality in automobile aftersales: AutoSERVQUAL scale. Amfiteatru Econ. **19**, 110 (2017)
18. Turan, A., Bozaykut-Bük, T.: Analyzing perceived healthcare service quality on patient related outcomes. Int. J. Qual. Serv. Sci. **8**, 478–497 (2016)
19. Han, S.-L. Baek, S. Antecedents and consequences of service quality in online banking: an application of the SERVQUAL instrument. ACR North American Advances (2004)
20. Melinda, A.W., Arifudin, R., Alamsyah, A.: Implementation of the servqual method as a service support decision support system in hotels. J. Adv. Inf. Syst. Technol. **1**, 91–97 (2019)
21. Haming, M., Murdifin, I., Syaiful, A.Z., Putra, A.H.P.K.: The application of SERVQUAL distribution in measuring customer satisfaction of retails company. J. Distrib. Sci. **17**, 25–34 (2019)
22. Gaur, S.S., Agrawal, R.: Service quality measurement in retail store context: a review of advances made using SERVQUAL and RSQS. Mark. Rev. **6**, 317–330 (2006)
23. Lee, G.G., Lin, H.F.: Customer perceptions of e-service quality in online shopping. Int. J. Retail Distrib. Manag. **33**, 161–176 (2005)
24. Zeithaml, V.A., Parasuraman, A., Malhotra, A.: Service quality delivery through web sites: a critical review of extant knowledge. J. Acad. Mark. Sci. **30**, 362–375 (2002)
25. Yee, B.Y., Faziharudean, T.: Factors affecting customer loyalty of using Internet banking in Malaysia. J. Electron. Bank. Syst. **21** 1–22 (2010)
26. Ladhari, R.: Developing e-service quality scales: a literature review. J. Retail. Consum. Serv. **17**, 464–477 (2010)
27. Pousttchi, K., Hufenbach, Y.: Engineering the value network of the customer interface and marketing in the data-rich retail environment. Int. J. Electron. Commer. **18**, 17–42 (2014)

28. Parker, C.J., Wang, H.: Examining hedonic and utilitarian motivations for m-commerce fashion retail app engagement. J. Fash. Mark. Manag.: Int. J. **20**, 487 (2016)
29. Warden, C.A., Wu, W.-Y., Tsai, D.: Online shopping interface components: relative importance as peripheral and central cues. Cyberpsychol. Behav. **9**, 285–296 (2006)
30. Rao, S., Griffis, S.E., Goldsby, T.J.: Failure to deliver? Linking online order fulfillment glitches with future purchase behavior. J. Oper. Manag. **29**, 692–703 (2011)
31. Titiyal, R., Bhattacharya, S., Thakkar, J.J.: E-fulfillment across product type: a review of literature (2000–2020). Manag. Res. Rev. (2022, ahead-of-print). https://doi-org.ezproxy.lib.ryerson.ca/10.1108/MRR-04-2021-0254
32. Grewal, D., Iyer, G.R., Levy, M.: Internet retailing: enablers, limiters and market consequences. J. Bus. Res. **57**, 703–713 (2004)
33. Green, D., Hanbury, M., Cain, A.: If you bought anything from these 19 companies recently, your data may have been stolen (2019). https://www.businessinsider.com/data-breaches-retailers-consumer-companies-2019-1
34. Ranganathan, C., Ganapathy, S.: Key dimensions of business-to-consumer web sites. Inf. Manag. **39**, 457–465 (2002)
35. Kuo, T., Tsai, G.Y., Lu, I., Chang, J.-S.: Proceeding, the 17th Asia Pacific Industrial Engineering and Management System Conference, pp. 7–10 (2016)
36. Shaw, N., Sergueeva, K.: 26th Annual DIGIT Workshop (2021)
37. Bhatnagar, A., Misra, S., Rao, H.R.: On risk, convenience, and Internet shopping behavior. Commun. ACM **43**, 98–105 (2000)
38. Jensen, K.L., Yenerall, J., Chen, X., Yu, T.E.: US consumers' online shopping behaviors and intentions during and after the COVID-19 pandemic. J. Agric. Appl. Econ. **53**, 416–434 (2021)
39. Davis, F.D.: Perceived usefulness, perceived ease of use, and user acceptance of information technology. MIS Q. **13**, 319–340 (1989)
40. Marimon, F., Vidgen, R., Barnes, S., Cristóbal, E.: Purchasing behaviour in an online supermarket. Int. J. Market Res. **52**, 111–129 (2010). https://doi.org/10.2501/s147078531020 1089
41. Olsson, T., Lagerstam, E., Kärkkäinen, T., Väänänen-Vainio-Mattila, K.: Expected user experience of mobile augmented reality services: a user study in the context of shopping centres. Pers. Ubiquit. Comput. **17**, 287–304 (2013)
42. Arnold, M.J., Reynolds, K.E.: Hedonic shopping motivations. J. Retail. **79**, 77–95 (2003). https://doi.org/10.1016/s0022-4359(03)00007-1
43. Childers, T.L., Carr, C.L., Peck, J., Carson, S.: Hedonic and utilitarian motivations for online retail shopping behavior. J. Retail. **77**, 511–535 (2001)
44. Qualtrics: Qualtrics: What is Qualtrics? (2020). https://csulb.libguides.com/qualtrics
45. Ringle, C.M., Wende, S., Becker, J.-M.: SmartPLS3 (2015). http://www.smartpls.com
46. Hair, J.F., Ringle, C.M., Sarstedt, M.: PLS-SEM: indeed a silver bullet. J. Mark. Theory Pract. **19**, 139–152 (2011)
47. Henseler, J., Hubona, G., Ray, P.A.: Using PLS path modeling in new technology research: updated guidelines (2015)
48. Fornell, C. Larcker, D.F.: Evaluating structural equation models with unobservable variables and measurement error. J. Mark. Res. **18**, 39–50 (1981)
49. Alkhowaiter, W.A.: Digital payment and banking adoption research in Gulf countries: a systematic literature review. Int. J. Inf. Manag. **53**, 102102 (2020). https://doi.org/10.1016/j.ijinfomgt.2020.102102
50. Duarte, P., e Silva, S.C., Ferreira, M.B.: How convenient is it? Delivering online shopping convenience to enhance customer satisfaction and encourage e-WOM. J. Retail. Consum. Serv. **44**, 161–169 (2018)

51. Daragmeh, A., Sági, J., Zéman, Z.: Continuous intention to use E-Wallet in the context of the COVID-19 pandemic: integrating the health belief model (HBM) and Technology Continuous Theory (TCT). J. Open Innov.: Technol. Market Complex. **7**, 132 (2021)
52. Podsakoff, P.M., MacKenzie, S.B., Lee, J.Y., Podsakoff, N.P.: Common method biases in behavioral research: a critical review of the literature and recommended remedies. J. Appl. Psychol. **88**, 879–903 (2003). https://doi.org/10.1037/0021-9101.88.5.879
53. Simmering, M.J., Fuller, C.M., Richardson, H.A., Ocal, Y., Atinc, G.M.: Marker variable choice, reporting, and interpretation in the detection of common method variance: a review and demonstration. Organ. Res. Methods **18**, 473–511 (2015)
54. Hair, J.F., et al.: Partial Least Squares Structural Equation Modeling (PLS-SEM) Using, R, pp. 115–138. Springer, Heidelberg (2021). https://doi.org/10.1007/978-3-030-80519-7
55. Henseler, J., Ringle, C.M., Sarstedt, M.: A new criterion for assessing discriminant validity in variance-based structural equation modeling. J. Acad. Mark. Sci. **43**(1), 115–135 (2014). https://doi.org/10.1007/s11747-014-0403-8
56. Legris, P., Ingham, J., Collerette, P.: Why do people use information technology? A critical review of the technology acceptance model. Inf. Manag. **40**, 191–204 (2003)
57. Keisidou, E., Sarigiannidis, L., Maditinos, D.: Consumer characteristics and their effect on accepting online shopping, in the context of different product types. Int. J. Bus. Sci. Appl. Manag. (IJBSAM) **6**, 31–51 (2011)
58. Fihartini, Y., Helmi, R.A., Hassan, M., Oesman, Y.M.: Perceived health risk, online retail ethics, and consumer behavior within online shopping during the COVID-19 pandemic. Innov. Mark. **17**, 17–29 (2021)
59. Miyazaki, A.D., Fernandez, A.: Consumer perceptions of privacy and security risks for online shopping. J. Consum. Aff. **35**, 27–44 (2001)
60. Pink, M., Djohan, N.: Effect of ecommerce post-purchase activities on customer retention in Shopee Indonesia. Enrich.: J. Manag. **12**, 519–526 (2021)
61. Kim, J., Forsythe, S.: Adoption of virtual try-on technology for online apparel shopping. J. Interact. Mark. **22**, 45–59 (2008)

A Better Shopping Experience Through Intelligent Lists: Mobile Application and Service Design to Improve the Financial Lives of Young Adults

Jung Joo Sohn(✉) and Abhay Sunil

Purdue University, West Lafayette, IN 47906, USA
{jjsohn,sunil2}@purdue.edu

Abstract. Nowadays, many young adults who become independent lack the necessary financial knowledge and experience to succeed. This paper introduces a service called ListSmart which is aimed at helping young people create and manage shopping lists, save more on their purchases, and give them a seamless shopping experience in physical stores. Due to their lack of experience, they spend and save money inefficiently. We seek to design a smart shopping list application that can influence millennials and Gen Z and encourage them to build healthier financial lives. Through a user-centered design approach, we researched young adults and their financial behavior by reviewing existing literature, sending two rounds of surveys and user evaluations. From our research, we have created four main features: intelligent shopping suggestions, budgeting capabilities, the ability to link with local stores, and a list-sharing platform. With these features, we seek to explore how smart shopping lists budgeting tools can improve the financial lives of young people in the modern age.

Keywords: Smart shopping list · Financial management · Shareable platform

1 Introduction

Mobile applications have reached new heights of accessibility and use across the globe [1]. This growth in mobile application use can be seen impacting various aspects of modern life. One such aspect is with people's shopping experience. There is a rising trend in online purchases being made with the advent of mobile software solutions [2]. Mobile applications have not only changed how people shop, but they have also changed people's process of shopping, most notably with shopping lists. This phenomenon has inspired further insights into shopping behavior. Several studies have indicated that people who use shopping lists, digital or physical, have spent less and bought fewer items than those who do not [3, 4]. However, based on an initial survey we conducted with various residents in major metropolitan areas (ages 18–30), we saw that many only occasionally use shopping lists and they usually use their device's default notes application. Some reasons for the participants' answers stem from the convenience of

F. Fui-Hoon Nah and K. Siau (Eds.): HCII 2022, LNCS 13327, pp. 526–541, 2022.
https://doi.org/10.1007/978-3-031-05544-7_38

a notes application and that many find it too complicated to plan out their purchases. "Smart shopping list" is a term that adequately describes the synthesis of a mobile-based solution with the shopping list experience [5]. This term is a common theme we explore through various surveys, studies, and user tests.

2 Related Works

2.1 Problems in Young Adults' Spending Habits

When screening for literature, we decided to focus on two aspects of young adults' finances. We looked into the behaviors of young adults when shopping and how the platforms they use to shop affect their spending.

Firstly, millennials and Gen Z have a higher tendency to impulse shop than other generations. Since shopping has become easier than ever in history, companies actively benefit from this behavior. Studies indicate that 41% of Gen Z are impulse buyers with millennials following closely at 34% [6]. This is because young adults desire speed and efficiency when shopping. In addition, the sheer amount of content offered to young adults gives them more opportunities to buy items that they do not need when they go out to shop.

Secondly, shopping through online platforms increases the chances of young adults making impulsive purchases. Both Gen Z and millennials prefer online shopping to traditional brick-and-mortar stores. The convenience of browsing through products on mobile devices has been the main factor that has enticed young adults to shop online [7]. However, this greater convenience has brought an increase in people's susceptibility to spontaneously buy products. When looking through large catalogs, young adults tend to make quicker decisions and are more comfortable with buying items that elicit some sort of emotional response [2].

By understanding the factors that influence target demographic spending habits, we were able to begin looking into actions that can impact the impulsive tendencies of young adults. Planning purchases is a vital part of a responsible financial life [3]. If more young adults create lists for the items they need, they can begin the necessary steps to saving money in the long term.

2.2 Existing Solutions on the Market

We selected a few personal finance and shopping list applications to get an understanding of what features are on the market and how users feel about them. This initial research was foundational for realizing the potential of our solution (Fig. 1).

2.3 Goal-Setting as the Main Method

Goal-setting theory refers to the theory of motivation that posits that setting clear and measurable goals allows people to effectively set habits and gradually attain higher achievements [8, 9]. Applying goal-setting theory into applications has the potential for impacting the behaviors of certain target groups. Popular programs that utilize goal-setting are fitness applications. Fitness is an area where many struggle to meet their

Existing Applications	Listonic	Anylist	Listease	Grocery	Our Groceries	Default Notes app (iOS)
Collaborative Lists	•	•		•		
Cloud backup				•		•
Budgeting features						
Categorization	•	•	•		•	
Meal Planning		•		•		
Receipt Record			•			
Smart Reccomendations						
Account System	•	•		•	•	•
Store Integration						
Barcode Scanning			•		•	
Desktop Application	•				•	•

Fig. 1. Existing application chart on the Apple App Store and Google Play Store (1/11/2021)

goals. This has caused a great deal of research on the effectiveness of goal-setting in the fitness applications on the market. A study by the found that people who set difficult weight-loss goals were likely to give them up after the first week. The study found that the likelihood of a person achieving their goals could be predicted by looking at their first week's progress [10]. A feature that would help alleviate this issue is to notify users when they are unlikely to reach their goal. However, it was shown that if users set attainable goals and keep track of their trajectory, goal-setting proved to be effective.

We decided to implement goal-setting into this application as a means to help young adults shift their behavior with shopping towards more careful decision-making. Our adoption of goal-setting focused on attaining the core mechanisms of good goal-setting. We shifted our attention towards allowing users to set smart, attainable, realistic, and time-bound goals [11]. By giving young people a tool to conveniently plan their purchases and track their spending, a culture of responsible financial habits can be further promoted.

2.4 Design Goals

We realized two key goals for this project. The first is to help young adults save more on their purchases. This goal will be achieved through the implementation of budgeting features within the smart shopping list service. These budgeting features will allow users to see the total costs of their lists, set monthly categorical budgets, and set saving goals to further encourage the user to buy only what is necessary. We see great value for the presence of these features in a shopping list application as the current market for this type of application either has very limited budgeting capabilities or none at all.

Our second goal is to make the shopping list experience easier and more effective for users. This will be achieved by giving users suggestions for the best value options for the items they are buying making list creation much faster. This aspect works in tandem with our first goal as it helps the user save more money while also giving them a more efficient application experience. Another method of achieving our second goal is through the addition of a shareable platform for shopping lists. Users would be able to create lists and upload them for other users to see for a more convenient user experience. When uploading a list, users can decide to share it publicly or privately with friends. The implementation of this feature allows people to browse and download other people's shopping lists for their specific needs like recipes. In addition to this shareable list platform, users can collaborate with others by creating lists with their friends. This gives users a better shopping experience when buying items with a group.

Through these two goals, we seek to design a shopping list application and service called ListSmart with five main feature categories; intelligent shopping suggestions, a list-sharing platform, budgeting capabilities, the ability to link with local stores, and a reward system. We hope these features will help users' shopping behaviors which would, in turn, improve the financial lives of young adults.

3 Methods

To understand how we can design an application that best serves the shopping and financial needs of young adults, we need to understand what the current market is and what users are doing right now. We conducted two phases of research to give us this understanding. We began by conducting primary research by sending out two rounds of surveys to participants to see what their behaviors are with shopping and budgeting. Then we conducted secondary research by observing the current applications available regarding list creation and personal finances.

3.1 First-Round Survey

Our first-round survey aimed at understanding essential information about the habits of our target demographic with a total pool of 58 participants. The survey was split into four sections: general information, spending habits, finance tracking, and budgeting habits.

Some key findings we gathered from our first survey were that 87.9% of participants said that they would likely save more if they were to set saving goals (Fig. 2). This is important as it shows us that users would respond well to a saving goal feature in our ListSmart concept. When asked to rank the amount of stress caused by tracking their finances, 31.7% of participants ranked their stress a 4 out of 5 (Fig. 3). 81% said that they use a mobile or computer application to keep track of their finances (Fig. 4). We then saw that 50% make shopping lists when shopping. 50% of them track their finances 1–2 times a week and 61% use a mobile application. From that, we can gather that many use their phones for financial purposes and that ListSmart could have great accessibility to these users.

Our main purpose for the first survey was to get a sense of where our sample pool is in terms of their financial status. With this data, we were able to narrow down our questions and begin creating more clear insights.

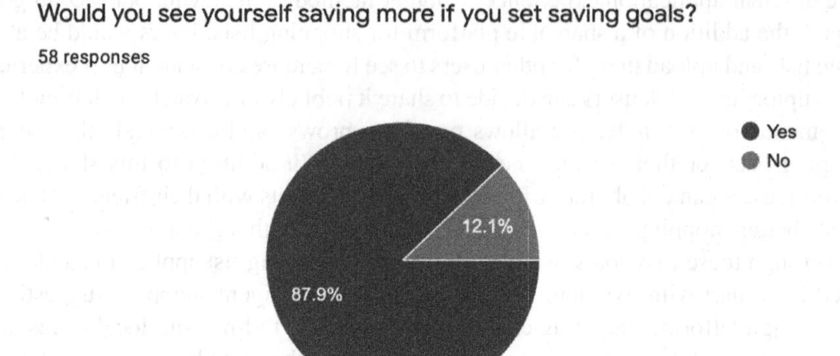

Fig. 2. Pie chart showing if users would see themselves saving more with saving goals

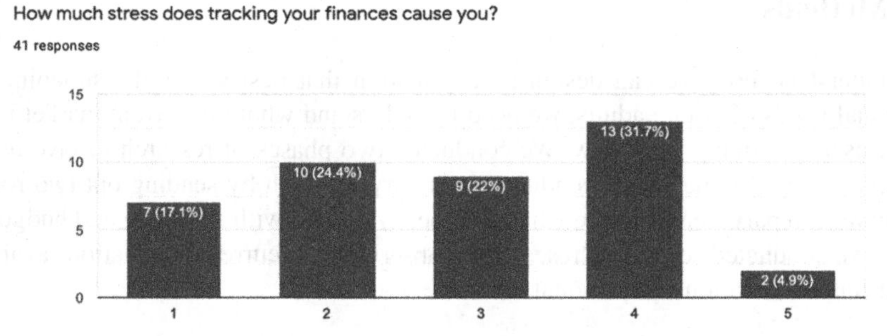

Fig. 3. Bar graph showing how much stress tracking finances causes users (only users who track their finances were presented with this question)

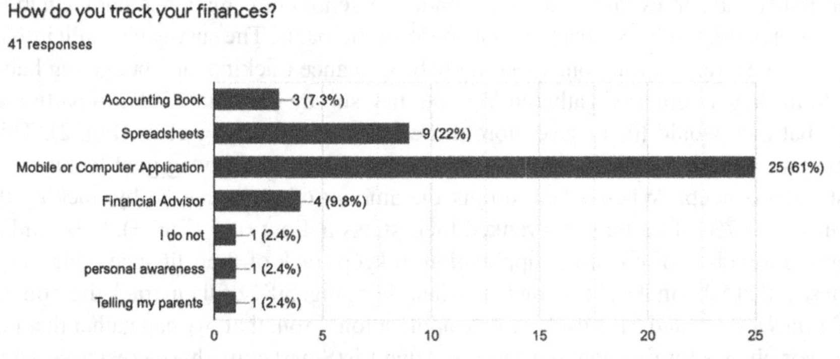

Fig. 4. Bar graph showing what methods participants use to track their finances (only users who track their finances were presented with this question)

3.2 Second-Survey

The second-round survey focused on understanding how our users use shopping lists and their experiences with similar applications. The second survey had a total of 31 participants. We shortlisted our participant pool so that we could get more valuable data from people who had more enthusiasm and insights into our concept. The three sections of the second survey are general information, spending habits, and shopping expenses.

While our second survey was given to a smaller sample size, the questions asked were designed to extrapolate more valuable information. From this, we were able to determine some notable findings. 87.1% of participants said that they make shopping lists for numerous purchases (Fig. 5). 73.3% said that they use their default notes application to make lists and 76.7% of participants make shopping lists for groceries (Fig. 6, and 7). 32.3% stated that they would rank themselves as a 3 out of 5 in how prone they are to impulse shopping. Of that 32.3%, 90% of them are between the ages of 18–23. 87.1% of participants stated that they use shopping lists when making large purchases and 73.3% of them said they use the default notes application on their phones. 48.4% said that price is the most important part when they are searching for products with customer reviews coming second at 32.3%. This sheds light on the potential benefit users could gain from using an application like ListSmart as smart recommendations are a key part of the process. Connecting with the previous statistic, we saw that 45.2% of participants answered that looking through product reviews is the most time-consuming part of the shopping experience. The final important finding is that 74.2% reported that they would use an application with a calculator and planning feature in it.

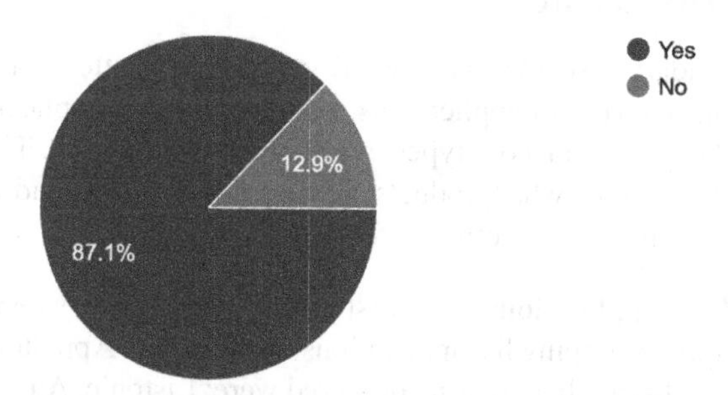

Fig. 5. Pie chart showing people making shopping lists for numerous items

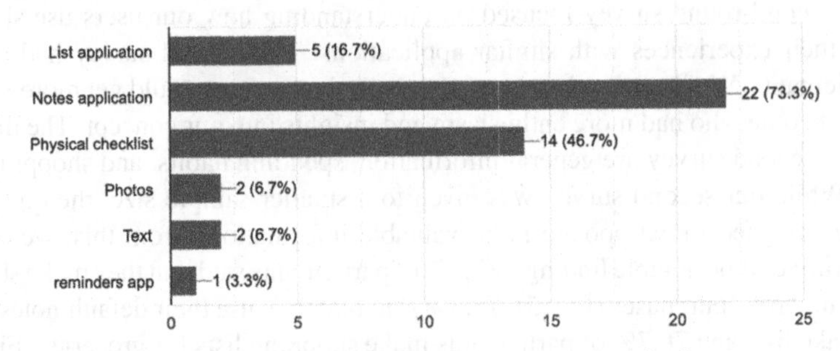

Fig. 6. Bar graph showing what platform users use to make lists

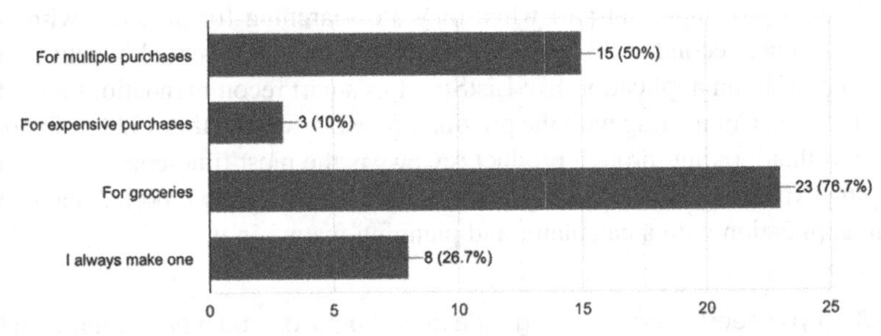

Fig. 7. Bar graph showing when users make shopping lists

3.3 Secondary Research

For our secondary research, we studied various applications on the market currently. We looked at two types of applications, five shopping list applications and five personal finance applications. For both types, we studied user reviews. This phase is critical for us as it allows us to see what products are currently available and what the users of those applications want to see in terms of features.

Shopping List Applications. The first part of our secondary research was looking into the most popular shopping list applications on the Apple App Store and the Google Play Store. The five list applications we observed were: Listonic, AnyList, Listease, Grocery, and OurGroceries. We began by reading and selecting the most helpful reviews for each application. Then from those, we were able to extrapolate the primary needs from the users for these applications. Our findings can be seen in (Fig. 8).

The two applications that stood out to us were Listonic and Anylist. For Listonic, users seemed to want the ability to create sub lists within larger lists and to have more specific categorization features. The reviews for Anylist pointed out that they would like more intelligent features that would allow the app to connect with grocery store apps. This stood out to us, and we sought to implement this feature through ListSmart's intelligent item recommendations.

From App Store Reviews

Application	Finding	Insights	Source	Ideas
Listonic	Users seem to want to have the **ability to create sublists within larger lists**. The primary context for this request is for recipes.	A hypothesis for this finding is that users seem to want to have more complex features for a dedicated shopping list applications. This could differ from people who only use their phone's default list application.	"Hi! First of all great app! The only thing that is missing from being a five star app from me is the fact that there are no "mini lists" that can be implemented inside a bigger list. Lets say you have a recipe for a bacon with eggs. I want to have that recipe easily added to my grocery list instead of adding the ingredients myself every time. But beside that the app is perfect! Thank you!" - **Sonadrin 5 star review (Google Play Store)**	For features it should be important to implement a sublist feature for recipes and other uses.
AnyList	Users seem to be content with the application's features but would like to **see some sort of integration with grocery store apps to add said items to their shopping cart.**	One key feature of default list applications on iOS and Android is that they implement each systems' respective cloud platforms. This could explain why users like seeing this feature in third party applications.	"My wife and I love this app. It stores recipes, it's easy to create meal plans and grocery lists, and best of all, it syncs between our phones. More than once one of us was able to add items to the list while the other was in the store, saving an extra trip. If it could integrate with my grocery store's app to put the items into a shopping cart and check out, that'd be next level for sure". - **Brian Lewis 5 stars**	An idea derived from this review is to design a cloud based shopping list appiciation. Integrating with say google drive would give the most device compatibility.
Listease	Mixed feelings. User appreciated the pantry feature but did not use it very often. User also stated that item check off animations are clunky and unessecary. App is also unable to allow users to input amounts less than 1 for products in a list.	The primary complaints from this user are based on the interface and features. These include clunky animations and unessecary features.	"I used the Groceries app for years and when it was no longer available I switched to this one. I like the detail offered for each line item, I like the idea of the Pantry section although I haven't found it useful yet. The reason I only gave it 3 stars is for these reasons. 1. The aisle labels are easily rearranged to match the layout of my grocery store, but they don't stay that way. They spontaneously rearrange themselves, so I'm constantly putting them back where I want them. 2. The checkbox avatar is cute but irritating when I have to wait to check off the next item until the animation is done. 3. It doesn't let me list less than one. For instance, I often buy half a pound of something at the deli counter but I can't use less than one in the item details. 4. While I like it telling me how much my whole grocery list is worth I don't like that it includes things that are checked off. I leave things I buy regularly in the checked off section as a reminder to check if I need it and to make it easier to add to my list as needed. That feature worked really well in my previous app, but needs work in this one. 5. Related to number 4, my previous app learned which items I used regularly and would suggest them without prompting in the search window. I could just go down the list and click on what I wanted. 6. I think the coupon section should suggest coupons based on my shopping list." - **k-em 3 stars (Apple App Store)**	An idea to derive from this is to implement a pantry feature of sorts which is integrated better. This would go alongside a budgeting feature.
Grocery	Utilizes iCloud and Apple Reminders for data back up. **Allows the user to organize lists by certain stores**. Color organizing is an an appreciated feature	A hypothesis for this finding is users who regularly go to stores and by the same things regularly would find great value in having default lists for certain stores in an application.	"This app is fantastic at what it does and I especially appreciate that it uses Reminders as the back-end data store. This makes sharing easy. Many of the reviews here which are complaining about items, lists, and stores are from not learning how to use the app. For me, what works is I have a Grocery list and a Hardware Store list. The Grocery list has two stores (the two nearest locations of H-E-B are what I use). When I shop, I select the store in the app that corresponds to the store location I'm at and voila! The grocery items are sorted in the order I typically go through the store. Obviously, this works best when you tend to go through a store location the same way each time. Likewise, my hardware store list has a Blue store and and Orange store each corresponding to the hardware store chains bearing the respective titular colors. Great app!" - **jubesz 5 stars (Apple App Store)**	The idea of organizing lists by certain stores could be an interesting idea to pursue.
OurGroceries	User complains that they **are unable to turn off the "suggested categories" feature.**	An insight gleaned from this review is that the ability to use the phone's camera has created an easier and quicker user experience with this list application.	"Edit: There is no obvious way to turn off "suggested categories". This has been annoying. Now, back to the original review... Love it! Bought the paid version and have no regrets. I'm sharing it with my family which is so convenient (but sometimes ice cream ends up on the list and nobody will admit to adding it). The picture capability on the paid app makes finding specific items so easy, too. I never hesitate to recommend this app when the topic of shopping comes up in conversation." **- Jason Peterson 5 stars (Google Play Store)**	An idea derived from this review is to implement camera and barcode capabilities into our design.

Fig. 8. Our findings from analyzing user reviews for shopping list applications

Personal Finance Application Research. An important aspect of ListSmart is its budgeting capabilities. If we want to understand how to best implement budgeting features in ListSmart, we deemed it useful to study applications specially made for personal finances. Much like the previous secondary research study, we focused on the three most popular personal finance applications which are: Intuit Mint Finance, Clarity Money, and Goodbudget. Our studies on these applications can be seen in (Fig. 9).

From App Store Reviews

Application	Finding	Insights	Source	What if
Mint Finance	There is a common complaint that the app is unable to show the difference between overspending and underspending.	An insight gleaned from this finding is that a visual and/or interactive display of overspending is very important to users.	iOS App store Review for Mint Finances by user stimartins. Rating at 4 stars.	What are your current methods of tracking if you are overspending or underspending in certain budgets?
Clarity Money	Users report that there should be more rewards for paying off debts.	This could lend itself to gamification. Users seem to be open to certain aspects of gamification.	iOS App store Review for Clarity Money by user Crazy Uncle Steve. Rating at 5 stars.	How do you currently track your debts? Would you be more encouraged to save if you had an active debt tracking method?
Mint Finance	User reports that the implementation of goals into the app itself would create a better experience.	Another aspect that supports the gamification insights. This finding shows that users are open to having saving goals and that they would activley benefit from it.	"I love this App. It has definitely helped me save more money over the past 8 years that I have been using it. I have to give them 4 (would be 4.5) stars simply because they dont have a few of the features from the website in the Android app. I would love to have the goals on the app so I can know where I stand on them without signing into the browser website. Also, the site doesn't store passwords as well as it could. I keep having to input passwords every few months. Other than that, great app." - **Ashley Vaughn [4.0 Stars] Google Play Store Review**	Ties into "Would setting saving goals encourage you to save more?"
Goodbudget	Users say that one feature they would like to see is the addition of a search function for their transactions.	An insight from this finding is that users want to have more control over the process of tracking their finances.	iOS App store Review for Goodbudget by user Tricksterinator. Rating at 5 stars.	Would you say that you often lose track of your past transactions?
Mint Finance	Users sight the addition of the ability to set monthly categorical budgets as very useful.	Users want to see more traditional aspects of tracking finances to translate over to their mobile applications.	"Biggest flaws: When an account is replaced with a new card the account shows as if it's a separate account and for some reason certain transactions are showing on that account while most transactions were able to transfer to the new account. In addition, it shows that the credit limit for the new account is "N/A" or "$0". There's an issue also where I accidentally input inactive on one of the replacement card and can't change it back to active. No way to migrate the accounts" - **J Ri [2.0 Stars] Google Play Store Review**	Do you set your monthly budgets categorically?

Fig. 9. Our findings from analyzing user reviews for personal finance applications

3.4 Key Insights

From our findings, we shortlisted three key insights out of 43 total insights. The main insights gained from our research are shown below:

Stress from Tracking Personal Finances. A majority of participants (37.1%) ranked the amount of stress they feel when tracking their finances a 4 out of 5 (Fig. 3). When asked what issues they experience in managing their finances, users stated that they experience stress and confusion when seeing their spending summaries from their bank's application. Oftentimes numbers and data do not translate well to people and can create a disconnect from their spending behaviors. Implementing an appealing but clear method of displaying a user's financial status could lead to beneficial results for our users.

A willingness to Follow Saving Goals. From our first survey, we found that 87.9% of participants said that they see themselves saving more if they set specific saving goals (Fig. 2). This was further expanded on in the second survey when a majority of participants (38.7%) said that they would rank themselves at a 4 out of 5 in how likely they would be to follow saving goals set by an application. These findings help tie into our intention of adding goal-setting to our project. Our users have shown that they would be receptive to this feature. This inspired further ideation regarding how this would be implemented.

Desire for Convenient List Creation and Management. When asked how often users make shopping lists, it was found that 87.1% of participants say that they make a shopping list when they shop for numerous items (Fig. 5). They were then asked what method of list creation they use. 73.3% stated that they use their mobile device's default notes application when making lists with 46.7% saying that they use physical checklists. What we see from this data is that many of our users are unaware of existing shopping list applications and opt for more quick and unorganized methods of list keeping. Giving our users a convenient and organized method of creating shopping lists could prove to have significant potential.

4 Design Process

After the research and analysis phase, we started to ideate and figure out key features based on our insights. Through several rounds of brainstorming, we developed a dual-platform service. The two platforms are a mobile application (Fig. 10) and kiosks in retail stores (Fig. 11). These platforms are closely connected, all the data is synchronized from the ListSmart data server. Users can create and manage all their lists, budgets, friends list, and profile on the mobile application. The kiosk makes it easy for users to find items from the lists by showing them a map of the store with the locations of all their items. Also, users can send the map from the kiosk to the mobile application for convenience.

Fig. 10. Mobile application key screens

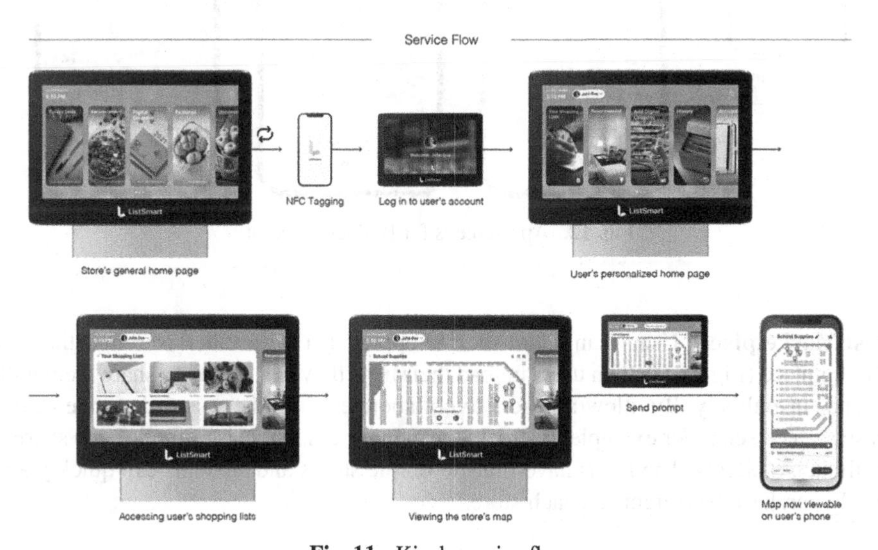

Fig. 11. Kiosk service flow

4.1 Main Features

From our three key insights, we created five main features for our mobile application and service that differentiate it from other products on the market. These include features include advanced budgeting capabilities, a shared platform, a cloud database, a reward system for user engagement, and smart recommendations which are aimed at helping the user to buy the best deals for various items.

Budget Forecast & Budgets Page. When users link their bank account to ListSmart, they can find their account balance and profile history on the application. It allows them to check the status of their situation and makes it easy to manage their finances on the app. The app visualizes the user's data and the progress of a user's budgets through different weather conditions. This feature is called the Budget Forecast (Fig. 12). If the user has a sunny forecast, that means that they are doing well with their budgets and are actively saving money. If the user is barely spending under their budget or a little over, they will have an overcast forecast. And if the user is consistently spending over their goals, they will be presented with a stormy forecast. Also, users can check on their spending and set new budgets on the Budgets Page (Fig. 12).

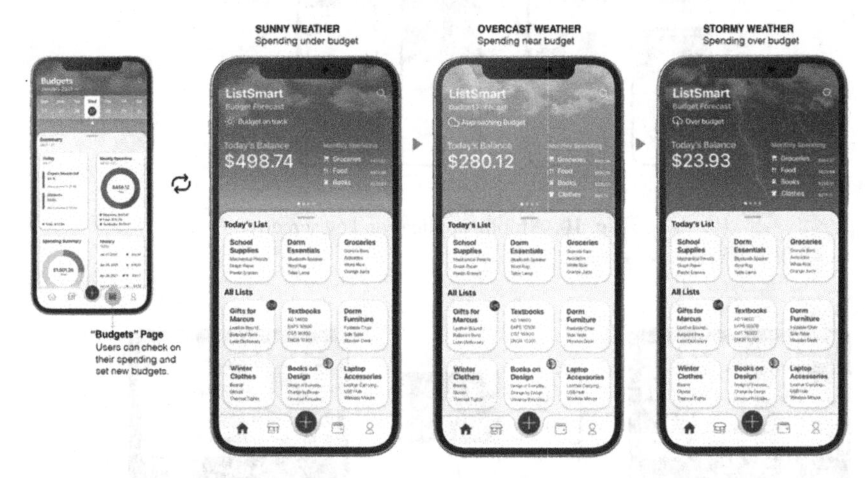

Fig. 12. App screens for Budget Forecast

List Marketplace. Users can share their shopping list with others through the List Marketplace (Fig. 13). When users upload their lists, they can decide to share them with friends or publicly. By allowing users to collaborate with others, a wide range of use cases can be seen. For example, instructors can make a list of class supplies and share it with their students. The items have links to online stores and students can quickly and easily purchase their items on each store.

Intelligent Recommendations. There are two types of intelligent recommendation features (Fig. 14). Type 1 is a For You recommendation and it is a smart suggestion based on the user's lists, purchase history, and other contextual information. Type 2 is when users are adding items to a list, ListSmart will give suggestions of where the user can get the best deal for their listed item.

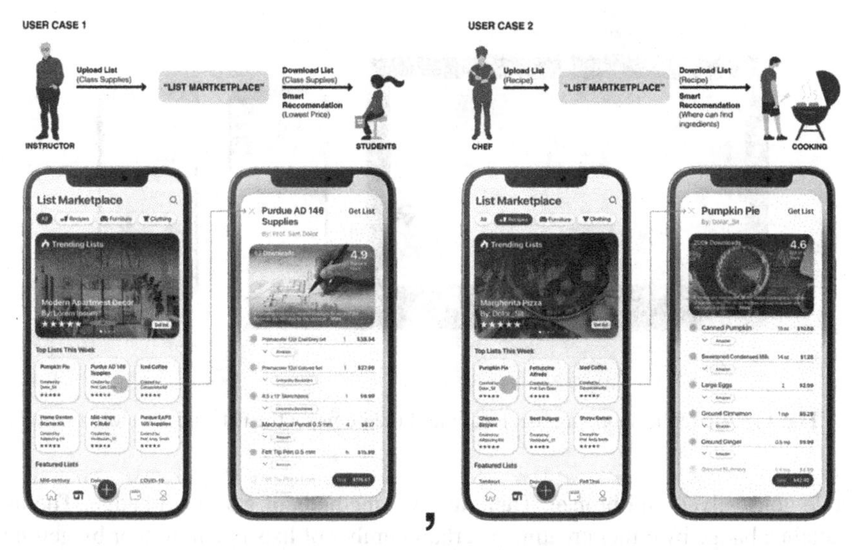

Fig. 13. App screens for List Marketplace

Fig. 14. 'For You' and 'Smart Suggestion' screens

Local Store Maps. By tapping their phone to the kiosk via NFC, users can access their lists on the kiosk. From the kiosk, users can see current deals at the given store and can access a store map with the location of their items marked. Users can send this map directly to their phones to create a seamless and efficient shopping experience (Fig. 15). By reducing the amount of time users take to find their items, they are less likely to get side-tracked and which ultimately reduces the likelihood of impulse shopping.

Reward System. We decided to add a reward system to the service to encourage users to continue using our service. Based on how the user sticks to their goals, they can attain various badges as a reward for their progress (Fig. 16). The three different badges

Fig. 15. Share the shopping list, location of the items on the list, and the store map

include gold, silver, and bronze. There are two methods of gaining a badge. The user can attain a badge by either creating a certain number of lists in a month or by getting a certain number of sunny forecasts in a month. For the list method, the requirements are 15 or more lists for gold, 10–14 for silver, and 5–9 for bronze. The criteria for the budget method are 20 or more sunny forecasts for gold, 10–19 for silver, and 5–9 for bronze.

Fig. 16. Badges in Profile page

5 Usability Test

After we finished designing, we conducted a usability test to see what users think of our application. We recruited 11 young adult users to test our prototype during one-on-one interviews. For the interviews, we showed our testers a three-minute video detailing the basics of our application and its core concept. Afterward, we asked screening questions to gain basic demographic data from our participants. The testers were then given a demo of ListSmart and were asked two questions. These questions were: 1) What features do you find most valuable and why? 2) Are the main features of the application easy to find?. After the demo was finished, participants were asked four questions. They include: 1) What are your thoughts on the look of the UI? 2) How would you describe your overall experience with the prototype 3) If there is one thing you could change about the product, what would it be and why? 4) Can you think of any real-world situation in which you would use the List Marketplace?

After gathering data from our usability tests, we organized it into two charts to help give us a clear view of what the general thoughts are from all of our tested users. The first chart organizes the demo into seven topics and ranks people's opinions into a 1.0–3.0 ranking scale where 1.0 is dissatisfied and 3.0 is very satisfied. These rankings were based on user responses (Fig. 17). We then created a table detailing what features our users preferred (Fig. 18). From our data, we can see that the most favored feature is the List Marketplace with the Budget Forecast/Page following behind it.

Described Problem	1.0	2.0	3.0
UI/GUI			
Navigation	0	1	10
Weather Metaphor	2	0	9
Color Theme	0	1	10
Problem Solving			
Convenience	0	2	9
Intuitive Budgeting	1	3	7
Creating/Managing Lists	0	2	9
In-Store Features	0	1	10

Fig. 17. Table showing the quantified results of the user interviews

Feature List	Preferred Feature
Budget Forecast/Page	4
List Marketplace	7
Intelligent Recommendations	0
Local Store Maps	1
Reward System	1

Fig. 18. Table showing users' preferred features

6 Discussion and Conclusion

To encourage young adults to manage their finances efficiently, we designed a service for managing shopping lists and budgets. We began with primary research and secondary research to learn about our target user. Through the user study, we gathered 43 insights and decided on three key insights for developing the features of our service. In this stage, the key insights were selected with these criteria; managing efficiency, intuitive visual information, a sharable platform with other users, and motivational methods. From our key insights, we designed a seamless service that has a connected mobile application and kiosk service in a retail store for a better shopping experience with five key features.

After developing the service, we did a usability test with users to understand what we need to improve. Three topics stood out to us as good points of discussion. Firstly, we focused on how to create, manage, and share a shopping list. However, users want to purchase the items on the lists directly and not from links to various online stores. The other topic is the prioritization of lists. Our users have stated that they would like to be able to add "urgency" to certain lists to help better organize their lists. And lastly, some users were confused as to what the Budget Forecast was trying to convey. Some thought that it was displaying actual weather. A suggestion we received was to have a tutorial when a person first uses the application which details what the Budget Forecast is and how it works. Our future work may consider addressing these points of discussion to help create the best solution for young adults to better their financial lives through our service.

References

1. MindSea Team: 28 mobile app Usage & Revenue Statistics to know in 2021. https://mindsea.com/app-stats/

2. Lissitsa, S., Kol, O.: Generation X vs. generation Y – a decade of online shopping. J. Retail. Consum. Serv. **31**, 304–312 (2016)
3. Davydenko, M., Peetz, J.: Shopping less with shopping lists: planning individual expenses ahead of time affects purchasing behavior when online grocery shopping. J. Consum. Behav. **19**, 240–251 (2020)
4. Arnaud, A., Kollman, A., Berndt, A.: The use of shopping lists by generation Y consumers in grocery shopping. http://hj.diva-portal.org/smash/record.jsf?pid=diva2%3A794801&dswid=1330
5. Harsha Jayawilal, W.A., Premeratne, S.: The smart shopping list: an effective mobile solution for grocery list-creation process. In: 2017 IEEE 13th Malaysia International Conference on Communications (MICC), pp. 124–149 (2017)
6. Djafarova, E., Bowes, T.: 'Instagram made me buy it': generation Z impulse purchases in fashion industry. J. Retail. Consum. Serv. **59**, 102345 (2021)
7. Dabija, D.C., Lung, L.: Millennials versus gen Z: online shopping behaviour in an emerging market. In: Văduva, S., Fotea, I., Văduva, L., Wilt, R. (eds.) GSMAC 2018. SPBE, pp. 1–18. Springer, Cham (2019). https://doi.org/10.1007/978-3-030-17215-2_1
8. Locke, E.A.: Goal setting theory: the current state. In: Latham, G.P. (ed.) New Developments in Goal Setting and Task Performance, pp. 623–630. Routledge/Taylor & Francis Group (2013)
9. Consolvo, S., McDonald, D.W., Landay, J.A.: Theory-driven design strategies for technologies that support behavior change in everyday life. In: Proceedings of the SIGCHI Conference on Human Factors in Computing Systems, pp. 405–414 (2009)
10. Gordon, M., Althoff, T., Leskovec, J.: Goal-setting and achievement in activity tracking apps: a case study of myfitnesspal. In: The World Wide Web Conference, pp. 571–582 (2019)
11. Tondello, G.F., Premsukh, H., Nacke, L.E.: A theory of gamification principles through goal-setting theory. In: Proceedings of the 51st Hawaii International Conference on System Sciences (2018)

Design of Engagement Platforms for Customer Involvement

Fang-Wu Tung(✉) and Yu-Wei Chen

National Taiwan University of Science and Technology, 43, Keelung Rd, Taipei 106, Taiwan
fwtung@ntust.edu.tw

Abstract. The concept of customer cocreation has emerged as a key advantage for companies wishing to maintain or gain competitiveness. In the Web 2.0 context, bidirectional communication flow creates a conversational environment in which people can share their ideas and opinions with companies and others. This study explored the use of visual interfaces and gamified elements in the development of online platforms for engaging users in providing their opinions. A total of 134 participants were recruited. The findings indicated that the participants' engagement and experience with a platform influenced their intention to participate in additional cocreation activities in the future and that gamified elements enhanced the participants' experience with and motivation to engage in cocreation activities. A visual interface in a platform can effectively guide and support users in cocreation activities. Furthermore, gamification can enhance the hedonic value of the cocreation experience and strengthen intention to participate.

Keywords: Customer cocreation · Engagement platform · Gamification · User experience · Visual-based interface

1 Introduction

Consumer behavior and the business environment are changing rapidly. Companies must maintain accurate information regarding consumer requirements in the ever-changing environment. Many companies have developed customer-centric strategies to obtain consumer insights and create product and service offerings that meet the needs of their target market [1]. Many companies have involved customers in new product development since the 1990s. Collecting customer opinions enables a company to effectively obtain ideas and insights for successful product adaptation. Prahalad and Ramaswamy [2] proposed the term "customer cocreation" to describe collaboration between customers and companies and suggested that customers should be recognized as partners who cocreate value for product and service offerings. In the increasingly complex competitive business environment, customer engagement in the internal product development process helps companies deliver offerings that can adapt to uncertain changes in market demand. Therefore, companies must learn to leverage customers' knowledge and experience to create value and drive innovation.

The concept of customer cocreation, defined as a set of methods for establishing collaborative processes with customers in the development of new products, has emerged

F. Fui-Hoon Nah and K. Siau (Eds.): HCII 2022, LNCS 13327, pp. 542–551, 2022.
https://doi.org/10.1007/978-3-031-05544-7_39

as a key advantage for companies wishing to maintain and gain competitiveness [3, 4]. The popularization of Internet and other information and communication technologies has led to the increasingly widespread application of online platforms for communication between companies and customers. Furthermore, users have become more active in expressing their opinions and posting comments on the Internet, which has inspired the development of various cocreation platforms designed to harness the knowledge and resources of users. Because Internet-based platforms are convenient and suitable for direct interaction with consumers, they facilitate value cocreation. Several enterprises have exploited the vast potential of the Internet by establishing platforms designed to facilitate customer involvement. Customers can contribute to value cocreation as "experts of their experiences" [1]. Given the increasing interest in tapping into user experiences through the Internet, companies must design online cocreation platforms that support and encourage customer involvement. Platforms with a well-designed interactive interface can guide customers to integrate their experience and knowledge into cocreation activities.

Virtual cocreation environments have been extensively researched and commonly applied, and numerous studies have explored how consumers can be motivated to participate in online cocreation. This study takes YouBike, which is an automated point-to-point bicycle rental service in Taiwan, as an example. The study aims to explore the use of online platforms with a visual-based interface and gamification concepts for engaging users in providing experience-based opinions and comments on YouBike. Two workable platforms were developed for YouBike users to participate in the feedback-giving activity.

More specifically, this study was conducted to complete the following three objectives:

1. explore the development of an online platform for engaging customers in feedback-giving activities;
2. investigate the effects of the developed cocreation platforms on participants' engagement, experience, and intention to use the platform in the future (henceforth referred to as "future intention"); and
3. investigate the effects of participants' engagement and experience on their future intention.

2 Literature Review

2.1 Value Cocreation with Customers

Customer involvement in the process of new product development enables companies to obtain external knowledge and thus align new products with market demands as well as increase new product success rates [5]. Companies have traditionally accomplished this by hosting activities such as lead-user workshops, focus groups, and idea competitions [6]. These activities enable companies to collect information regarding customers' experiences and perceptions by gathering customers' opinions, observations, concerns, and suggestions to improve products and services. Kristensson, et al. [7] reported that the involvement of ordinary customers in new product development results in the generation of original and valuable ideas.

Engagement platforms are essential for enabling active, creative, and social collaboration between producers and customers in cocreation processes [8]. An engagement platform can be described as a platform on which different parties can interact and exchange resources to create mutual value [9]. In the Web 2.0 context, bidirectional communication flow creates a conversational environment in which customers can share their ideas and opinions with companies and other customers. Internet-based platforms eliminate barriers to customers' interaction with producers and other customers [10]. Driven by advances in information communication technology, customers have become accustomed to sharing their ideas and opinions with others. The shared content provides valuable information for companies aiming to improve existing products or develop new products to meet unmet demands in the market. The platforms used to engage consumers in cocreation have expanded from events hosted in the physical world to virtual platforms on the Internet [11]. The Internet, which is widely accessible by individuals and businesses, has enabled the development of various virtual cocreation platforms and, in turn, enabled companies to integrate external knowledge in improving product and service offerings and customer satisfaction.

2.2 Virtual Cocreation Platforms

Virtual cocreation platforms enable companies to communicate and interact directly with customers, providing a relatively flexible and practical means of harnessing customer knowledge [12, 13]. Customers generally integrate their knowledge into virtual cocreation platforms by sharing ideas, design concepts, and prototype evaluations. Several enterprises have exploited the vast potential of the Internet by establishing online platforms designed to enable customer cocreation. For example, a component of the IKEA website invites consumers to participate in cocreation activities by offering their input on IKEA's latest projects [14]. Similarly, Starbucks Corporation implemented a website titled "My Starbucks Idea," which invited consumers to submit their ideas and suggestions in either text or image format and provided information on the company's current products, services, and social responsibilities [15]. In the same regard, the Lego Group invites consumers to submit their ideas through the "Lego Ideas" website [16].

The aforementioned cases demonstrate how Internet-based platforms have expanded the possibilities for value cocreation with customers. To augment the value of their product and service offerings, companies must focus on designing virtual cocreation platforms that can support and encourage customers to engage in the cocreation process. Bartl [17] developed an interactive platform on which users could create a future infotainment system for a vehicle by selecting preferred electronic devices—such as phones, displays, antennas, interfaces, and radio, sound, navigation, and control devices—to be visually displayed on the platform. The empirical study revealed that when a platform visually presenting realistic usage scenarios minimized users' participation efforts. The study's participants reported that the visual-based interface assisted them in designing an ideal infotainment system and made the task enjoyable.

In addition to visual presentation, interactive design can be used to enhance users' experiences with a virtual cocreation platform. Incorporating interesting elements into cocreation activities could enhance the cocreation experience, which, in turn, could increase participants' willingness to participate. Interesting elements can be developed on the basis of gamification, which refers to the application of game-like concepts or elements in nongame contexts. Gamification mechanisms—such as points, levels, badges, and leaderboards—have been applied on the Samsung Nation website to motivate customers to evaluate products and share opinions regarding their user experiences. This customer feedback can inform product revisions and new product development. An online ethnography study by Harwood and Garry [18] observed customers' engagement with Samsung Nation and reported that the gamification mechanisms promoted consumer engagement with the virtual cocreation platform.

3 Methodology

This study explored the design of two virtual platforms to facilitate customer participation in cocreation activities. Specifically, we investigated the development of online platforms with a visual-based interface and gamification concepts for users to provide their opinions based on their experiences. This study also investigated the effect of the platforms on the participants' engagement in and experience with the cocreation activity and their future intention as well as the effects of engagement and perceived experience on future intention.

3.1 Experimental Prototype

We took the YouBike system as an example and developed two online platforms for inviting YouBike users to share their opinions of the system. The rental system comprises bicycle post controllers and bicycles (Fig. 1). Users can use the controller to rent and return a bicycle and pay the rental fee by card.

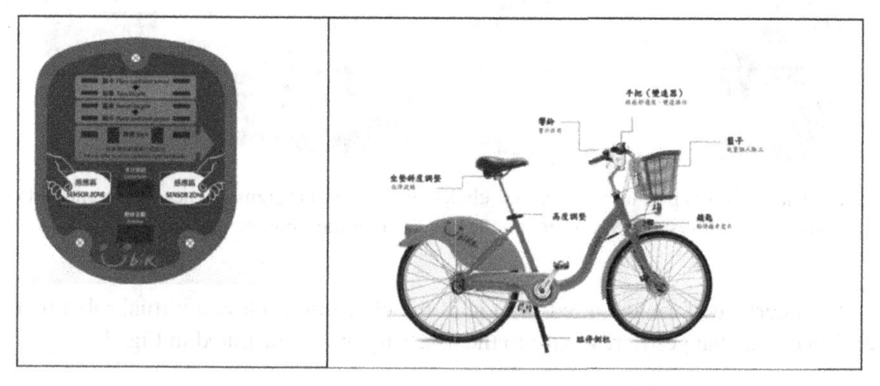

Fig. 1. YouBike post controller and bicycle*1

Customer feedback can be collected using surveys, service reports, or feedback forms. We aimed to develop virtual platforms that incorporated methods distinct from the traditional means of collecting user feedback. On the basis of the literature review, we incorporated a visual-based interface and gamification concepts into our platforms.

Two visual-based platform prototypes, one gamified and one nongamified, were developed. The visual interface platform is illustrated in Fig. 2. The platform provides close-up and mid-range views of the bicycle to enable users to provide feedback regarding the rental system, bicycle, and user interface. A user can provide their input through a dialog box by clicking the plus sign (+) near the different parts of the bicycle and the rental system. For example, as illustrated in Fig. 3, when the user clicks the plus sign near the brake lever, a dialog box appears and prompts them to input their opinions and suggestions regarding the bicycle's brake issues.

Fig. 2. Visual-based interface.

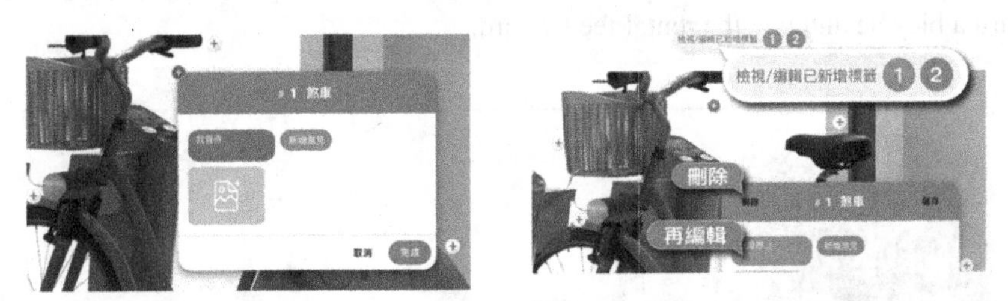

Fig. 3. Dialog boxes pop up when a user clicks one of the plus signs (+) near different parts of the bicycle or rental system. The user can then add or edit their comments.

The interface of the gamified version of the platform displays a virtual robot that is generated and changes in response to the user's input, as illustrated in Fig. 4.

Fig. 4. Virtual robot is generated and changes in response to the user input. The various virtual robots developed for the gamified platform are displayed on the right.

3.2 Participants and Procedure

A total of 134 participants were recruited for this study. Each of the participants was randomly assigned to one of the two platforms; 66 and 68 were assigned to the nongamified and gamified platforms, respectively. The participants were invited to provide their feedback regarding the YouBike system through their assigned platform. The participants could exit the platform at any time with a straightforward click. Thereafter, the participants were asked to complete an online questionnaire regarding their engagement, experience, and future intention.

3.3 Measurement Tools

The questionnaire contained four subscales: (1) engagement in the cocreation activity, (2) hedonic experience, (3) utilitarian experience, and (4) future intention. The participants' engagement in the cocreation activity was measured using the following items: "I concentrated on giving my opinion on YouBike," "I could express my opinions through this platform," "I enjoyed giving opinions on YouBike through this platform," and "I was involved in the process of giving my opinion." The hedonic experience subscale consisted of items assessing how interesting, pleasant, relaxing, and novel the participants found the platforms to be. The utilitarian experience subscale, which was related to perceived ease of use and usefulness, consisted of the following items: "The platform helped me express my feedback on YouBike," "The platform inspired me to provide feedback on YouBike," and "The platform is easy to use." The future intention subscale consisted of the following items: "I will use this platform to provide feedback in the future," "I will share this platform on social media with others," and "I would recommend this platform to others." Each item on the questionnaire was scored on a 5-point Likert scale ranging from 1 (very strongly disagree) to 5 (very strongly agree).

4 Results

The questionnaire data and feedback collected from the participants who used the two platforms were analyzed. Internal consistency was calculated to assess the reliability of the scales. The Cronbach's α values for the engagement, hedonic experience, utilitarian experience, and future intention subscales were all higher than 0.7. According to Nunnally and Bernstein [19], a Cronbach's α of 0.7 indicates adequate internal consistency and reliability; therefore, the measurement tool used in this study exhibited adequate reliability.

Multiple linear regression analysis revealed that the model was significant [analysis of variance F (3,130) = 50.02 and $p < 0.001$] with an adjusted R^2 of 0.53, indicating that engagement, hedonic experience, and utilitarian experience explained 53% of the variation in future intention. The remaining 47% of the variance resulted from variables not examined in this study. The adjusted R^2 was higher than the benchmark of 0.5 proposed by Hair Jr, et al. [20], indicating that the variance in the regression model was sufficiently explainable.

Table 1 presents the results of the multiple regression analysis, which reflect the effects of the independent variables on the dependent variable. As indicated in Table 1, each of the independent variables (engagement, hedonic experience, and utilitarian experience) in this model had a Sig. t value smaller than α (0.05). This indicated that each independent variable exerted a significant partial effect on future intention to participate in the cocreation activity.

Table 1. Results of multiple linear regression analysis and t values

Model	Unstandardized coefficients		Standardized coefficients	t	Sig.
	B	Std. Error	Beta		
(Constant)	−0.699	0.405		−1.728	0.086
Hedonic experience	0.393	0.093	0.322	4.230	0.000
Utilitarian experience	0.317	0.136	0.236	2.331	0.021
Engagement	0.400	0.148	0.277	2.702	0.008

An independent t test was performed to determine whether the two cocreation platforms affected the participants' engagement, experience, and future intention. Table 2 presents the results of the t test analysis. The mean scores for engagement, hedonic experience, utilitarian experience, and future intention assigned by the gamified platform group were all higher than those assigned by the nongamified platform group. This finding suggests that incorporating gamified elements into platforms can significantly enhance participants' hedonic experience of cocreation activities and their future intention. The two platform groups did not differ significantly in the subscales of engagement and utilitarian experience.

The participants' comments on the YouBike system received from the two platforms were calculated. The average numbers of comments per participant post on the nongamified and gamified platforms were 3.7 (SD = 3.02) and 4.9 (SD = 3.3), respectively. The results of the independent t test indicated a significant difference ($p < 0.05$) between the two platforms, suggesting that the gamified platform could motivate participants to

Table 2. Experimental results

Independent variable	Condition	N	Mean (SD)	t
Engagement	Non-gamified	66	4.42 (0.64)	−0.59
	Gamified	68	4.48 (0.50)	
Hedonic experience	Non-gamified	66	4.09 (0.72)	−2.50*
	Gamified	68	4.38 (0.61)	
Utilitarian experience	Non-gamified	66	4.38 (0.66)	−1.28
	Gamified	68	4.51 (0.57)	
Future Intention	Non-gamified	66	4.21 (0.71)	−2.48*
	Gamified	68	4.50 (0.63)	

*$p < 0.05$ (two-tailed)

offer more feedback. The average numbers of words per comment on the two platforms were not significantly different (18.9 words and 18.8 words, respectively; $p > 0.05$).

5 Discussion and Conclusion

This study explored the development of a virtual cocreation platform designed to engage customers in providing feedback on a product–service system. We developed two platforms as experimental tools for this study. Both of the platforms visually simulated the YouBike service system, but only one incorporated gamified elements. The effects of the platforms on the participants' engagement, experience, and future intention were investigated. The engagement and utilitarian experience subscale scores in the two platform groups did not differ significantly. In the follow-up interview, some participants mentioned that the designs of the platforms were intuitive and made it easy to input opinions and suggestions. The visual-based interfaces enabled them to recall their experiences of using YouBike, and the plus signs effectively guided them to input comments regarding specific parts of the YouBike system. One participant reported that "...the visual presentation of the YouBike scene engaged me in the realistic environment. It helped me provide feedback..." Another participant said, "...the plus signs around the bike and controller guided me to see the YouBike system in greater detail. When I clicked a plus sign, it would zoom into the corresponding part, and I would think about what could be improved in that part." Furthermore, the interactive design of the platforms motivated the participants to provide feedback. As mentioned by another participant, "...there were many plus signs on the platform. A feedback box would pop up when I clicked a plus sign, and the interactive design motivated me to click and input feedback..." According to the participants, the visual-based interfaces cultivated an immersive experience that guided them to recall their user experience, helping them provide feedback and devise new ideas.

The hedonic experience and future intention subscale scores were significantly different in the two groups. Pacauskas, et al. [21] identified hedonic experience as a motivating factor that guides participants to contribute to cocreation activities. A similar trend was observed in the present study. The hedonic experience subscale scores assigned by the gamified platform group were significantly higher than those assigned by the other group. The participants who used the gamified platform reported greater intention to use the platform and recommend the platform to others in the future than did those who used the nongamified platform. Also, the average number of comments and words per participant in the gamified platform group were higher than those in the nongamified platform group. This finding suggests that game-like elements can significantly enhance the hedonic value of cocreation activities and motivate participants to provide more feedback, thereby contributing to their future intention.

When a participant was using the gamified platform, a robot would be generated and evolve as the participant input more comments. According to the follow-up interviews, gradually assembling a robot enabled an immediate response and gave a sense of continuity. An interviewee stated, "I think it was quite interesting, seeing a robot taking shape slowly. It was low-pressure, and I was not affected by other individuals…". Another interviewee noted, "I posted four comments about the YouBike because I wanted to create a whole robot. I wanted to see what kind of robot I would assemble." Echoing the studies discussed in our literature review, we report that the use of gamified elements in cocreation activities can enhance the user experience and motivate users to participate in similar activities in the future.

A visual-based interface can create an immersive environment to engage and support users participating in cocreation activities. The platforms discussed herein visually display the YouBike system, thereby promoting immersion and motivating participants to provide more feedback. Compared with traditional feedback collection methods, this innovative means of collecting customer feedback allowed for a pleasant and engaging user experience. Furthermore, gamification can be used to improve the user experience and motivate future intention. The incorporation of a visual-based interface and gamified elements has the potential to frequently re-engage users in the cocreation activity, thereby making the customer feedback collection process more efficient.

Note 1. The two images were retrieved from https://taipei.youbike.com.tw/use/equipment?_id=5cc296e0083e7b58a8594e02.

Acknowledgments. This material is based upon work supported by the Ministry of Science and technology of the Republic of China under grant MOST 102–2218-E-011–018- and MOST 108–2410-H-011 -007 -MY2.

References

1. Breschi, R., Freundt, T., Orebäck, M., Vollhardt, K.: The expanding role of design in creating an end-to-end customer experience (2017)
2. Prahalad, C.K., Ramaswamy, V.: Co-creating unique value with customers. Strategy Leadersh. (2004)
3. Roser, T., DeFillippi, R., Samson, A.: Managing your co-creation mix: co-creation ventures in distinctive contexts. Eur. Bus. Rev. (2013)
4. Piller, F.T., Vossen, A., Ihl, C.: From social media to social product development: the impact of social media on co-creation of innovation. Die Unternehmung **65**(1) (2012_

5. Bolton, R., Saxena-Iyer, S.: Interactive services: a framework, synthesis and research directions. J. Interact. Mark. **23**(1), 91–104 (2009)
6. Cooper, R.G., Dreher, A.: Voice-of-customer methods. Market. Manage. **19**(4), 38–43 (2010)
7. Kristensson, P., Gustafsson, A., Archer, T.: Harnessing the creative potential among users. J. Prod. Innov. Manag. **21**(1), 4–14 (2004)
8. Ramaswamy, V., Gouillart, F.J.: The Power of Co-creation: Build it with them to Boost Growth, Productivity, and Profits. Simon and Schuster, New York (2010)
9. Ramaswamy, V., Ozcan, K.: The Co-creation Paradigm. Stanford University Press, Redwood City (2014)
10. Rayna, T., Striukova, L.: Open innovation 2.0: is co-creation the ultimate challenge? Int. J. Technol. Manage. **69**(1), 38–53 (2015)
11. Frow, P., Nenonen, S., Payne, A., Storbacka, K.: Managing co-creation design: a strategic approach to innovation. Br. J. Manag. **26**(3), 463–483 (2015). https://doi.org/10.1111/1467-8551.12087
12. Dahan, E., Hauser, J.R.: The virtual customer. J. Product Innovat. Manage. Int. Publicat. Product Dev. Manage. Associat. **19**(5), 332–353 (2002)
13. Poetz, M.K., Schreier, M.: The value of crowdsourcing: can users really compete with professionals in generating new product ideas? J. Prod. Innov. Manag. **29**(2), 245–256 (2012)
14. Koniorczyk, G.: Customer knowledge in (co) creation of product. a case study of IKEA. J. Econ. Manage. **22**, 107–120 (2015)
15. Hossain, M., Islam, K.Z.: Generating ideas on online platforms: a case study of "My Starbucks Idea." Arab Econ. Bus. J. **10**(2), 102–111 (2015)
16. Andersen, P., Ross, J.W.: Transforming the LEGO group for the digital economy (2016)
17. Bartl, M.: Co-creation in the automobile industry–The Audi virtual lab. Michaelbartl.Com, Munich (2009)
18. Harwood, T., Garry, T.: An investigation into gamification as a customer engagement experience environment. J. Serv. Market. (2015)
19. Nunnally, J.C., Bernstein, I.: Psychometric Theory. McGraw-Hill, New York. The Role of University in the Development of Entrepreneurial Vocations: a Spanish Study, pp. 387–405 (1978)
20. Babin, B.J., Anderson, R.E.: A Global Perspectivie. Kennesaw State University, Kennesaw (2010)
21. Pacauskas, D., Rajala, R., Westerlund, M., Mäntymäki, M.: Harnessing user innovation for social media marketing: case study of a crowdsourced hamburger. Int. J. Inf. Manage. **43**, 319–327 (2018)

Author Index

Printed in the United States
by Baker & Taylor Publisher Services